Cereal Biotechnology

Cereal Biotechnology

Edited by Callen Stokes

SYRAWOOD
PUBLISHING HOUSE

New York

Published by Syrawood Publishing House,
750 Third Avenue, 9th Floor,
New York, NY 10017, USA
www.syrawoodpublishinghouse.com

Cereal Biotechnology
Edited by Callen Stokes

© 2019 Syrawood Publishing House

International Standard Book Number: 978-1-68286-785-3 (Hardback)

Cataloging-in-Publication Data

Cereal biotechnology / edited by Callen Stokes.
    p. cm.
Includes bibliographical references and index.
ISBN 978-1-68286-785-3
1. Grain--Biotechnology. 2. Agricultural biotechnology. I. Stokes, Callen.
SB189 .C47 2019
633.1--dc23

TABLE OF CONTENTS

Permissions

List of Contributors

Index

PREFACE

The world is advancing at a fast pace like never before. Therefore, the need is to keep up with the latest developments. This book was an idea that came to fruition when the specialists in the area realized the need to coordinate together and document essential themes in the subject. That's when I was requested to be the editor. Editing this book has been an honour as it brings together diverse authors researching on different streams of the field. The book collates essential materials contributed by veterans in the area which can be utilized by students and researchers alike.

Cereals are the edible components of grains that are composed of the germ, the endosperm and the bran. Some examples of cereal crops are rice, maize, wheat, oats and barley. They are an important source of carbohydrates, proteins, fats, vitamins and minerals. Cereals are genetically modified using biotechnological methods in order to develop desirable characteristics. Rice can be genetically modified to increase vitamin A levels, enhance tolerance to herbicides and resistance to pests, accelerate photosynthesis, produce human proteins, etc. Specific strains of maize are also genetically engineered to increase nutritional value and drought resistance. Wheat can be genetically modified by transgenic, biolistic methods and Agrobacterium methods. This book elucidates the principles and techniques of cereal biotechnology in a multidisciplinary manner. The topics included herein are of utmost significance and bound to provide incredible insights to readers. With state-of-the-art inputs by acclaimed experts of this field, this book targets students and professionals.

Each chapter is a sole-standing publication that reflects each author´s interpretation. Thus, the book displays a multi-facetted picture of our current understanding of application, resources and aspects of the field. I would like to thank the contributors of this book and my family for their endless support.

Editor

Enhanced salt stress tolerance of rice plants expressing a vacuolar H+-ATPase subunit c1 (*SaVHAc1*) gene from the halophyte grass *Spartina alterniflora* Löisel

Niranjan Baisakh[1,*], Mangu V. RamanaRao[1], Kanniah Rajasekaran[2], Prasanta Subudhi[1], Jaroslav Janda[3], David Galbraith[3], Cheryl Vanier[4] and Andy Pereira[5]

[1]School of Plant, Environmental, and Soil Sciences, Louisiana State University Agricultural Center, Baton Rouge, LA, USA

[2]Southern Regional Research Center, Agricultural Research Service, United States Department of Agriculture, New Orleans, LA, USA

[3]Department of Plant Sciences, University of Arizona, Tucson, AZ, USA

[4]University of Nevada, Las Vegas, NV, USA

[5]Department of Crop, Soil, and Environmental Sciences, University of Arkansas, Fayetteville, AK, USA

*Correspondence
email nbaisakh@agcenter.lsu.edu

Keywords: gene expression, halophyte, rice, salt tolerance, vacuolar ATPase.

Summary

The physiological role of a vacuolar ATPase subunit c1 (*SaVHAc1*) from a halophyte grass *Spartina alterniflora* was studied through its expression in rice. The *SaVHAc1*-expressing plants showed enhanced tolerance to salt stress than the wild-type plants, mainly through adjustments in early stage and preparatory physiological responses. In addition to the increased accumulation of its own transcript, *SaVHAc1* expression led to increased accumulation of messages of other native genes in rice, especially those involved in cation transport and ABA signalling. The *SaVHAc1*-expressing plants maintained higher relative water content under salt stress through early stage closure of the leaf stoma and reduced stomata density. The increased K+/Na+ ratio and other cations established an ion homoeostasis in *SaVHAc1*-expressing plants to protect the cytosol from toxic Na+ and thereby maintained higher chlorophyll retention than the WT plants under salt stress. Besides, the role of *SaVHAc1* in cell wall expansion and maintenance of net photosynthesis was implicated by comparatively higher root and leaf growth and yield of rice expressing *SaVHAc1* over WT under salt stress. The study indicated that the genes contributing toward natural variation in grass halophytes could be effectively manipulated for improving salt tolerance of field crops within related taxa.

Introduction

Soil salinity is a major threat to agricultural productivity worldwide. At high concentrations of salts in the soil, plants experience a physiological drought because of the inability of roots to extract water, and high concentrations of salts within the plant can be toxic (Munns and Tester, 2008). Because salt tolerance is a quantitative trait governed by multiple genes, success with classical breeding has been low in developing salt-tolerant crops (Winicov, 1998). Genetic manipulation provides an alternative yet sustainable approach to engineering rice plants for salinity tolerance. Recently, however, four markers from candidate genes including SKC1 within the 'Saltol' QTL have been identified in rice and are being used in marker-assisted selective breeding of salt-tolerant rice (Thomson et al., 2010).

Halophyte models have been used as a source of genes for engineering salt tolerance in heterologous systems. Considering the differences in the anatomical features between dicots and monocots and also in their adaptation regulation networking, it is necessary to explore halophytic models among grass species as a source for superior alleles/regulation machinery through genetic engineering onto food crops, which are mostly grasses, such as rice (Tester and Bacic, 2005). Earlier we reported that *Spartina alterniflora* (smooth cordgrass), a monocot halophyte with reportedly all possible mechanisms of salt tolerance, shares 80–90% similarity with rice with regard to their DNA and protein sequences (Baisakh et al., 2008). Although there are a handful of reports documenting transgenic overexpressers showing some degree of salt tolerance, actual production of transgenic plants with demonstrably improved salt stress tolerance is few and slow (Flowers, 2004).

Among various mechanisms, control of ion movement across tonoplast (and plasma membrane) to maintain low Na+ concentration in the cytoplasm is the key cellular factor in salinity tolerance. Salt-tolerant plants differ from salt-sensitive ones in having a low rate of Na+ and Cl− transport to leaves and the ability to compartmentalize these ions in vacuoles to avoid salt toxicity by preventing their build-up in cytoplasm or cell walls. In halophytes such as common ice plant (*Mesembryathemum crystallinum*), vacuolar sodium sequestration under salinity is mediated by an active Na+/H+ antiporter energized by the proton motive force, which is generated and maintained by plant V-ATPase (EC3.61.34), which is one of the several members of the ATP-dependent protein pumps, through primary active H+ transport at the tonoplast (Barkla et al., 1995). The most abundant subunit of the V_0 complex of vacuolar ATPase is subunit c, which is encoded by the largest multigene family. It is present in six copies that form the part of the proton-conducting pore responsible for proton translocation (Sze et al., 1999), although the biological significance of this is unknown. Subunit c was the first multigene family reported in eukaryote V-ATPases (Sze et al., 1992). H+-ATPase acts as a primary transporter that pumps protons out of the cytoplasm, thus creating a pH and electric potential gradient across the vacuole that activates many secondary transporters involved in ion and metabolite uptake (reviewed in Serrano, 1989; Sussman, 1994; Michelet and Boutry, 1995; Palmgren, 1998). Subunit c is a highly hydrophobic protein in the V_0 domain. It is essential for the production of an active V-ATPase holoenzyme and is likely to be directly involved in H⁺-transport. Transcriptional changes of subunits of the vacuolar ATPase in response to salinity stress have been reported from a number of plants. For example, salt-induced transcriptional activation of V-ATPase subunit c has been observed in common ice plant (Dietz and Arbinger, 1996; Löw et al., 1996; Tsiantis et al., 1996), which included a considerable and fast increase in vacuolar-type H⁺-ATPase activity in tonoplast vesicles, irrigated with high NaCl concentrations (Ratajczak et al., 1994). This demonstrates the prime importance of V-ATPase in the adaptation of common ice plant to high sodium concentrations. In the halotolerant sugar beet, transcripts of the V-ATPase subunits A and c were found in root and leaf tissue, and NaCl treatment caused an increase in the transcript levels in leaves, but not in the roots (Kirsch et al., 1996; Lehr et al., 1999). In Porteresia coarctata, roots showed immediate pronounced (two-fold) increase in V-ATPase c transcript, whereas in the leaves, there was no significant increase until or after 5 h and it achieved twofold and threefold accumulation only after 24 and 48 h following salt stress (Senthilkumar et al., 2005). The subunit c transcript level declined in both leaf and root after 10 h of salt withdrawal. Recently, we demonstrated that an expressed sequence tag (EST) showing similarity to vacuolar H⁺-ATPase subunit c1 (SaVHAc1) was highly induced in both leaf and root tissues of smooth cordgrass (Spartina alterniflora) when plants were grown under hypersaline (500 mM NaCl) condition (Baisakh et al., 2008). These studies indicate that coordinated enhanced steady-state transcript levels of specific V-ATPase subunits in root and/or shoot are a characteristic for halotolerant plants, whereas salinity-induced up-regulation of V-ATPase subunits has also been demonstrated in non-halophytes (Tyagi et al., 2005) including resurrection plants (Chen et al., 2002).

The phenotypes caused by ectopic expression of a vacuolar pyrophosphatase (AVP1) in Arabidopsis suggested that manipulation of vacuolar proton pumps in economically important crops holds promise for the reclamation of farmlands lost to salinization and lack of rainfall (Gaxiola et al., 2002). Little, however, is known about the effect and biological role of orthologous expression of genes encoding V-ATPase subunits on the plants' ability to cope with salt stress, although its role in salt stress response has been indirectly shown in RNAi mutants (Padmanaban et al., 2004) or, as described earlier, by its increased transcript accumulation under salt stress. The present work was undertaken with an objective to understand the role that SaVHAc1 plays in plant's physiological response to salinity through its functional expression in transgenic rice.

Results

Sequence analysis of SaVHAc1

A cDNA clone (897 bp) containing the entire coding sequence of SaVHAc1 was isolated from the cDNA library of Spartina alterniflora, which contained a 498-bp-long open reading form (ORF). The SaVHAc1 was predicted to be a membrane protein of 165 (16.62 kDa) deduced amino acid residues under

transport and binding function GO category with subcellular localization in the inner tonoplast (Figure S1). The SaVHAc1 protein consisted of two hydrophilic and two hydrophobic alpha-helices with an average hydrophobicity of 0.979. The protein consisted of three transmembrane helices (each of 23 aa length) and a primary signal peptide (30 aa).

Comparison of the genomic sequences, isolated through standard genome walking procedure using primer specific to 5'- and 3'- UTR and ACP primer (Seegene Inc.), and cDNA-sequences of SaVHAc1 revealed five exons ranging from 28 to 289 nucleotides in length separated by four introns ranging from 14 to 1034 nucleotides in length with GT-AG conformation expected for eukaryotic nuclear genome (Figure 1a). Comparative analysis of the V-ATPase subunit c sequences from other plants showed that SaVHAc1 formed an independent group within the cluster that comprised of Eleusine glauca, Arabidopsis thaliana, Gossypium hirsutum and Mesembryathemum crystallinum (Figure 1b). It shared 87% and 99% identity with rice at the DNA and protein level, respectively. However, at the protein level, it showed 100% identity with the V-type proton ATPase (Asvat-P1) from Avena sativa. Restriction mapping (with

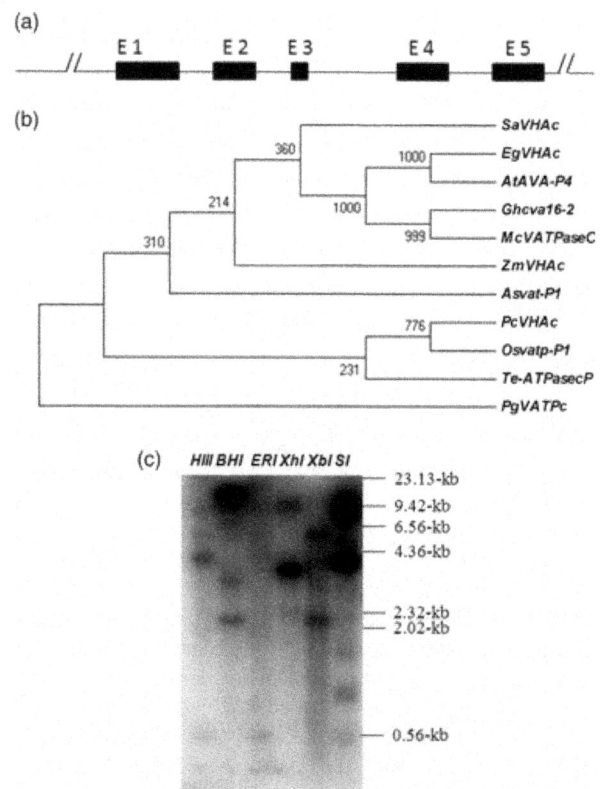

Figure 1 Characterization of SaVHAc1 from Spartina alterniflora. Genomic organization (a) showing 5 exons (E1…E5). DNA sequence similarity of SaVHAc1 with V-ATPase gene from other plants (b). Copy number analysis of SaVHAc1 in Spartina alterniflora (c). Size fragments of λ/Hind III marker in kb are shown next to the horizontal bars. Sa, Spartina alterniflora; Eg, Eleucine glauca; At, Arabidopsis thaliana; Gh, Gossypium herbaceum; Mc, Mesembryathemum crystallinum; Zm, Zea mays; As, Avena sativa; Pc, Porteresia coarctata; Os, Oryza sativa; Te, Triticum aestivum; Pg, Pennisetum galucum; HIII, HindIII; BHI, BamHI; EI, EcoRI; Xh, XhoI; Xb, XbaI; SI, SacI.

six different restriction endonucleases viz., HindIII, BamHI, EcoRI, XhoI, XbaI, and SacI) showed the presence of more than one copy of the SaVHAc1 in the genome of S. alterniflora (Figure 1c). The complexity of the restriction patterns detected by full length SaVHAc1 suggested that a multigene family encodes the 16-kD proteolipid SaVHAc1 in S. alterniflora, which may account for more than the fact that S. alterniflora in itself is a allohexaploid ($2n = 6x = 62$).

Stable integration and inheritance of SaVHAc1 in transgenic rice

The integration of SaVHAc1 expression cassette (Figure 2a) in the rice genome was initially identified by positive PCR signals using gene-specific primers and subsequently confirmed by Southern blot analysis. The copy number and integration was further validated by T-DNA-rice genomic DNA flanking sequence analysis. In CLA9 and CLA20, SaVHAc1 was integrated as a single copy (Figure 2b), whereas in CLA19, four copies were integrated. SaVHAc1 was mapped in the long arm of chromosome 7 of CLA20 (Figure 2c). The segregation of SaVHAc1 in the first selfing generation (T_1) progenies of CLA9 and CLA20 followed a single gene (3:1) Mendelian inheritance (Figure 2d). The SaVHAc1 locus was fixed to homozygosity in T_2 generation (Figure 2e).

SaVHAc1 expression conferred salt tolerance in rice plants

Upon imposition of salt stress at 100 and 200 mM NaCl, the SaVHAc1 plants showed greater tolerance than the WT plants with respect to leaf chlorophyll bleaching in leaf-floating assay, leaf rolling, leaf withering, tip burning and other physiological and agronomic traits (Figure 3). The SaVHAc1-rice plants exposed to seedling stage salt stress grew normally after recovery and set seeds in the greenhouse, whereas the WT plants had very stunted growth, abnormal/incomplete panicle exsertion and were very highly sterile (Figure 3c). Tobacco transgenics expressing SaVHAc1 also showed enhanced tolerance to salinity as compared to the WT plants (Figure S2). No apparent difference was observed between WT and SaVHAc1-rice plants with regard to their growth and development under normal (unstressed) conditions (data not shown).

After 72 h of salt stress, the SaVHAc1-rice plants maintained shoot growth as was evident by 22% reduction of shoot length compared to approximately 43% reduction of the WT plants over their respective unstressed control (Figure 4a). Similarly, the SaVHAc1 plants had a better root growth with a root length reduction of 9% under salinity than the WT (22%; Figure 4a). The reduction in the shoot dry weight of SaVHAc1-rice associated with stress was substantially less (16%) than the

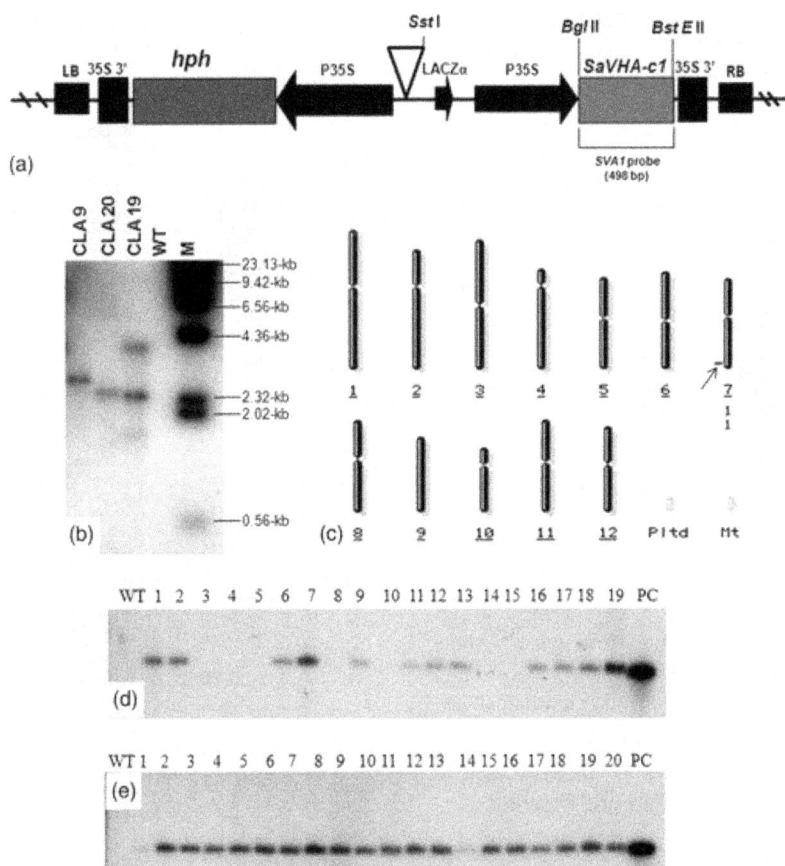

Figure 2 Generation and characterization of SaVHAc1 plants. Partial linear map of the plant transformation binary vector p35S:SaVHAc1 (a), Restriction analysis showing copy number of three independent SaVHAc1-rice (CLA9, CLA20 and CLA19), WT, wild type; M, λ/Hind III DNA size marker (b). Genome mapping of SaVHAc1 showing its integration (arrow marked) in Chromosome seven of rice (c). Single gene Mendelian segregation of SaVHAc1 in T_1 generation (d) and homozyosity of SaVHAc1-rice in T_2 generation (e). PC, plasmid positive control.

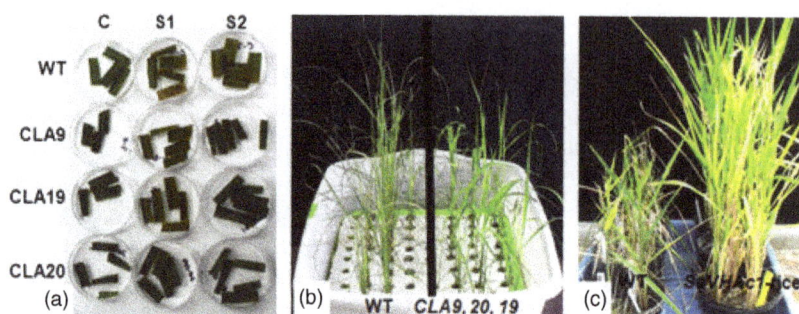

Figure 3 Salt tolerance of *SaVHAc1*-rice vis-à-vis WT rice under salt stress (S1 = 100 mM NaCl, S2 = 200 mM NaCl) in leaf disc assay (a), whole plant assay under hydroponics at 150 mM NaCl (b). The *SaVHAc1*-rice plants showed better growth and yielded more compared to WT rice under salt stress at reproductive stage (c).

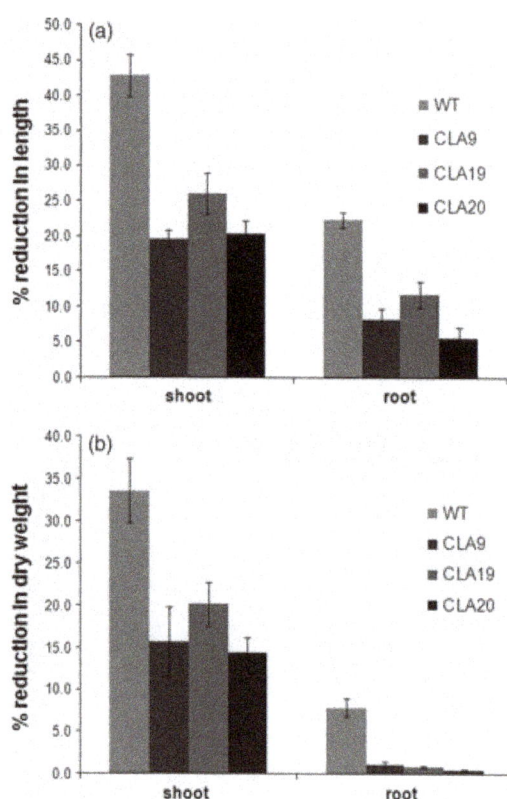

Figure 4 Reduction in shoot and root length (a) and dry weight (b) of *SaVHAc1*-rice vis-à-vis WT one week after salt stress (150 mM NaCl). Error bars represent standard error of means.

WT rice (34%; Figure 4b). The *SaVHAc1* plants did not show any significant reduction (0.8%) in root dry weight relative to the unstressed control plants, which was much lower than the 8% reduction of WT plants under stress (Figure 4b).

SaVHAc1-rice retained high chlorophyll, relative water content, and maintained higher yield under salinity

Rice plants expressing *SaVHAc1* maintained higher chlorophyll concentration over WT under saline conditions (Figure 5a). The loss of total chlorophyll because of salt stress (150 mM NaCl) was less (32%) in *SaVHAc1*-rice compared to WT plants (48%) when averaged over three time points (Figure 5a). The *SaVHAc1*-rice plants showed much less reduction (13%) in chloro-

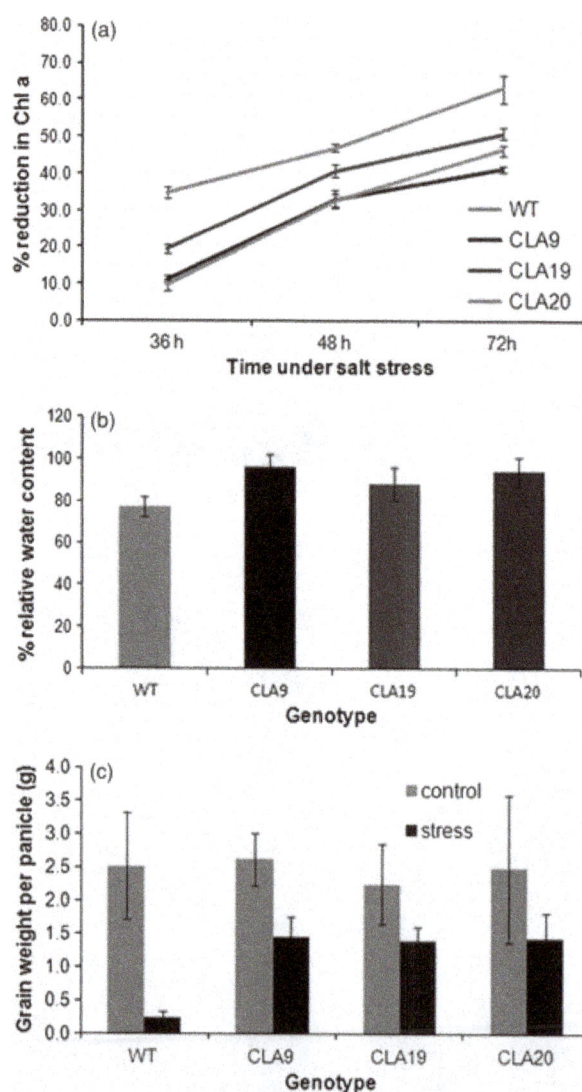

Figure 5 Chlorophyll a (a), relative water content (b) and grain yield per panicle of *SAVHAC1*- rice vis-à-vis WT plants under salt stress (150 mM NaCl) (c). Note that their yield was comparable under no-stress control condition. Error bars represent standard errors of means.

phyll a after 36 h of salt stress compared to the WT plants (35%). The loss of the chlorophyll was directly proportional to the duration of salt stress (Figure 5a). Chlorophyll concentration

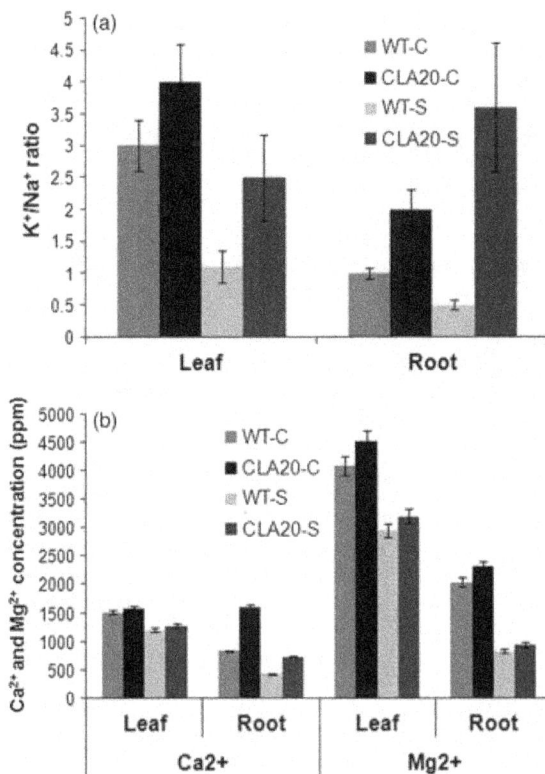

Figure 6 Inductive coupled plasma (ICP) analysis showing higher $K^+:Na^+$ (a), and Ca^{2+} and Mg^{2+} concentration (b) in leaf and root tissues of SaVHAc1-rice (CLA20) in comparison with WT rice under control (C) and salinity (S). LC, leaf control; RC, root control; LS, leaf stress; RS, root stress. Error bars represent standard error of means.

derived from the SPAD reading and estimated by acetone extraction method was comparable (data not shown).

Relative water content (RWC) is considered an appropriate measure of plant water status as well as osmotic adjustment under stress. RWC, which also reflects, in part, the transpirational loss of water, was estimated from the leaves of the SaVHAc1- and WT rice plants subject to salt stress. After a week under stress, the SaVHAc1-rice plants maintained higher RWC (93%) as compared with the WT (77%; Figure 5b).

Both SaVHAc1- and WT rice plants showed a reduction in grain yield per panicle upon salt stress during 2 weeks of the critical reproductive stage (i.e. panicle initiation). However, the SaVHAc1-rice produced nearly six times higher grain yield (1.42 g) per panicle than WT plants (0.24 g) under reproductive stage salt stress (Figure 5c).

SaVHAc1 expression led to accumulation of higher K^+ under salt stress

Only one SaVHAc1-rice line (CLA20) along with WT was included in the elemental analysis. The root Na^+ concentration of the SaVHAc1-rice was less than the WT plants, while the leaf Na^+ was high in both genotypes. In contrast, leaf K^+ concentration in SaVHAc1-rice was much higher than in WT, which was clear from the high K^+/Na^+ values in the transgenic lines (Figure 6a). Concomitantly, the SaVHAc1-rice lines also accumulated higher levels of Ca^{2+} and Mg^{2+} in their leaf and root tissues with or without stress (Figure 6b).

SaVHAc1 expression altered the expression of native rice genes upon salt stress

From the microarray experiment, many of the 43 311 probes were significantly different: 4287 (9.9%) between genotypes (WT and SaVHAc1-rice), 6761 (15.6%) between environments (saline or non-saline conditions) and 705 (1.6%) responded differently to the environment depending upon the genotype (Table S1). Although a large number of genes showed up-regulation in SaVHAc1-rice relative to WT (Table S1; Gene Expression Omnibus Accession no. GSE34724), eighteen genes that showed more than twofold increase or decrease in transcription in SaVHAc1-rice were considered significantly up- or down-regulated, respectively. Fourteen genes encoding proteins with either ion transport or metal binding function were significantly up-regulated (Table 1). The four genes that showed significant down-regulation in SaVHAc1-rice did not have any functional annotation (hypothetical or expressed proteins). Interestingly, none of the other subunits of V-ATPase were observed as being affected by SaVHAc1.

Semiquantitative expression analysis of genes (Figure 7) showed that SaVHAc1 transcript was much higher in the SaVHAc1-rice than the WT without or with salt stress imposition at all time points. Quantitatively, SaVHAc1 transcript accumulated

Table 1 Genes showing up-regulation (≥2-fold) by SAVHAC1-expression in rice under salt stress (analysed by microarray)

Probe ID	Gene	GO Slim ID	Term	Ontology	Fold increase
TR038922	Pyrophosphate-energized vacuolar membrane proton pump	GO:0005773	Membrane; vacuole, hydrolase activity	Cellular component	2.5
TR055074	Glutamine synthetase expressed	GO:0003824	Catalytic activity	Molecular function	2.7
TR066667	Metallothionein-like protein 1	GO:0046872	Metal binding	Molecular function	2.7
TR066578	Pathogenesis-related protein 10	GO:0035556	Intracellular signal transduction	Biological process	2.6
TR057433	Hypothetical protein				2.6
TR051207	Germin-like protein subfamily 1 member 7 precursor	GO:0005618	Cell wall	Cellular component	2.9
TR037990	Cortical cell delineating protein precursor	GO:0006810	Transport	Biological process	2.9
TR068459	Plastocyanin-like domain containing protein	GO:0005488	Binding, membrane	Molecular function	3.1
TR066580	Pathogenesis-related protein Bet v I family protein	GO:0035556	Intracellular signal transduction	Biological process	4.0
TR066664	Metallothionein-like protein 1	GO:0046872	Metal binding	Molecular function	3.8
TR067039	Cysteine synthase	GO:0003824	Catalytic activity	Molecular function	3.6
TR037504	Expressed protein				5.0
TR066668	Metallothionein-like protein 1	GO:0046872	Metal binding	Molecular function	5.0
TR047684	SCP-like extracellular protein	GO:0005576	Extracellular region	Cellular component	6.1

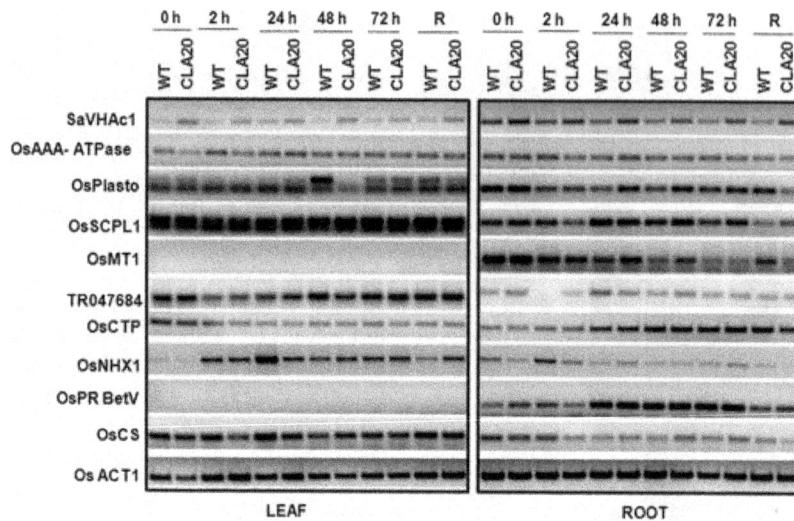

Figure 7 Semiquantitative RT-PCR analysis of rice native genes of rice in *SaVHAc1*-rice (CLA20) vis-à-vis WT under control (0 h) and at different time points under salt stress (150 mM NaCl) and 4 days after recovery (R) in both leaf and root tissues.

up to 25-fold higher in the *SaVHAc1*-rice under salt stress than the *SaVHAc1*-rice without salt stress and approximately 40-fold higher over the WT rice plants (Figure 8a). The overexpression of *SaVHAc1* in the transgenics also led to higher constitutive transcript accumulation of rice plasma membrane AAA-ATPase in the leaf tissues under no stress. Although there was no apparent change in its transcript in leaf tissues under salt stress, the root tissues showed comparable increase in its expression in both WT and *SaVHAc1*-rice under stress.

Other genes that were tested for their transcript accumulation showed up-regulation in *SaVHAc1*-rice as compared to WT plants under salt stress in a tissue and time-dependent manner (Figures 7 and 8b). One important function of the V-ATPase is to provide the driving force for H^+-coupled Na^+ antiporters, such as *NHX1*, which sequesters sodium into the vacuole (Apse *et al.*, 1999). Cation transporter genes such as *OsNHX1* and *OsCTP* showed little difference among *SaVHAc1*-rice and WT rice in leaf tissues, but their transcript was highly accumulated in the roots of *SaVHAc1*-rice compared to WT plants. *OsSCPL1* showed a slight decline in its transcript accumulation until 24 h of salt stress in their root, but overaccumulated after 48 h through the recovery stage. Cysteine synthase (*OsCS*) behaved similarly. Genes such as metallothionein (*OsMT1*) and PR protein (*BetV*) showed only root-specific expression with subtle difference between *SaVHAc1*-rice and WT plants.

Quantitative expression analysis showed up-regulation of *OsPLDα1* and *OsGPA1*, the two positive regulators of abscisic acid (ABA) signalling genes (Nilson and Assmann, 2010) in the *SaVHAc1*-rice over WT (Figure 8c) without and with stress. The leaves and roots of *SaVHAc1*-rice accumulated higher transcript of these two genes even without stress. Similarly, the expression of *OsSORK* (rice homologue of Arabidopsis *GORK*) was 2.8- and 1.9-fold higher in *SaVHAc1*-rice than WT with and without salt stress, respectively.

SaVHAc1 expression affected the density and opening of stomata

The *SaVHAc1*-rice had 39% fewer stomata per unit leaf area than the WT rice (Figure 9a,c). Further, the stomata were mostly closed in the *SaVHAc1* lines under salt stress as

compared with the WT (Figure 9b). However, there was no significant difference with respect to the stomata size as measured by their dimension. Cross-sections of leaf and root tissue did not show any apparent difference between the *SaVHAc1*- and WT rice (data not shown).

Discussion

The present results identified a major role for *SaVHAc1*, a gene similar to vacuolar H^+-ATPase, from a halophyte grass in several physiological processes particularly those related to salt stress. The vacuolar H^+-ATPase is the major proton pump that establishes and maintains an electrochemical proton gradient across the tonoplast, thus providing the driving force for the secondary active transport of ions and metabolites against their concentration gradient (Gaxiola *et al.*, 2007). The functional significance of V-ATPase in plant's ability to adapt to unfavourable stress was provided earlier (Dietz *et al.*, 2001). Transcriptional activation of V-ATPase at an early stage of salt stress has been shown in several instances (Baisakh *et al.*, 2008; Golldack and Deitz, 2000). The ability of a plant to change its gene expression profile to respond to salt stress, as we observed, might be a key mechanism for salt tolerance.

SaVHAc1 and ion homoeostasis

The capacity of plants to maintain a high cytosolic K^+/Na^+ ratio is likely to be one of the key determinants of salt tolerance. Halophytes utilize ion homoeostasis in the cytosol as one of the most important strategies for their adaptation to salinity. The *SaVHAc1* plants accumulated high level of Na^+ in roots and leaves; growth, however, was little affected by the toxic Na^+, which could be due to the sequestration of Na^+ at the tonoplast by a secondary Na^+/H^+ transporter that was energized by a proton motive force created by the overexpression of *SaVHAc1* (Apse *et al.*, 1999). Further, the increase in the concentration of other cations (K^+, Ca^{2+}, and Mg^{2+}) in the leaf and root tissues in the *SaVHAc1*-rice plants (Figure 6) could be due to the electrochemical gradient generated by the *SaVHAc1* expression, which is used by proton-coupled antiporters to accumulate more of Ca^{2+} and Mg^{2+} inside the lumen (Hirschi *et al.*, 1996)

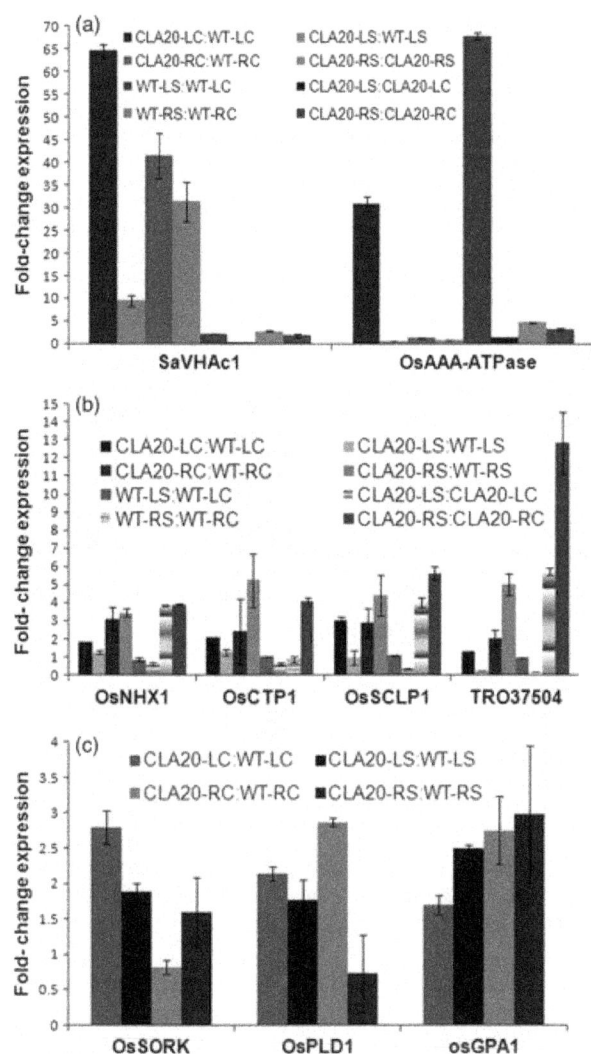

Figure 8 Quantitative RT-PCR of *SaVHAc1* and rice plasmamembrane ATPase (a), genes involved in ion transport/exchange (NHX and CTP), extracellular (*OsSCLP*) and expressed protein (TR037504) (b), and genes involved in ABA signalling (c). CLA20 = *SaVHAc1*-rice, WT, wild type; LC, leaf control; RC, root control; LS, leaf stress; RS, root stress. Error bars represent standard error of means.

and exclusion of toxic Na^+ from the cytosol to protect the transgenics. This led to a reestablished ion homoeostasis which was of critical importance for the adaptation of plants to salt stress (Niu *et al.*, 1995). It is widely established that Ca^{2+} ameliorates Na^+ toxicity in a variety of plant species (Hasegawa *et al.*, 2000). Elevated Ca^{2+} levels also stimulate net uptake of K^+, possibly by enhancing membrane integrity (Maathuis *et al.*, 2003), and thus may have modulated toward higher K^+/Na^+ ratio in the *SaVHAc1*-rice. Because, Mg^{2+} is a central molecule in chlorophyll, the increased accumulation of Mg^{2+} may also have contributed to higher photosynthesis in the *SaVHAc1* plants.

SaVHAc1 and vegetative growth and yield of plants under salinity

The *SaVHAc1* plants showed better root and shoot growth and improved dry weight and grain yield compared to the WT plants under saline conditions. *SaVHAc1*-rice possessed higher

leaf N concentration relative to WT rice as predicted from their higher SPAD values, which is known to be linearly correlated with leaf N concentration. Physiologically it is well established that primary proton pumps are crucial for plant growth and survival, and *VHAc* genes are expressed to support growth in actively growing cells and to supply increased demand for V-ATPase in cells with active exocytosis (Padmanaban *et al.*, 2004). In addition to maintaining ion homoeostasis, the *SaVHAc1* could be involved in protein sorting and membrane fusion events that are needed to promote growth, exocytosis of wall materials in certain cell types and tolerance to salt stress. Genetic evidence that V-ATPase influenced plant development and signalling came from the first V-ATPase mutant, *det3*, which had reduced subunit C transcript and V-ATPase activity (Schumacher *et al.*, 1999). The mutant was deetiolated when germinated under dark conditions and the mature plant was dwarf compared to wild-type plants. From our study and several earlier studies, it is clear that *VHAc1* has implications in cell expansion. Padmanaban *et al.* (2004) observed that high *VHAc1* promoter activity was consistently seen in the elongating zone of roots, expanding cotyledons or elongating hypocotyls, but not in non-expanding organs such as cotyledons of dark-grown seedlings or hypocotyls of light-grown seedlings. Further, they demonstrated that dsRNA-mediated RNAi mutant plants of *VHAc1* were more sensitive and showed reduced root and hypocotyl length relative to wild type after 4-days under salt stress (50 mM).

The expression of *SaVHAc1* played a role in the protection of plant's photosynthetic machinery from higher Na^+, which was evident by higher chlorophyll retention of the *SaVHAc1* plants in comparison with the WT plants under salinity (Figure 8). In principle, enhanced expression of vacuolar proton pumps can energize solute transport by increasing the availability of protons and thus increases vacuolar solute accumulation. The net increase in the cell solutes concentration of *SaVHAc1*-rice must have led to an increase in the uptake of water as supported by their higher leaf RWC such that these transgenic plants could maintain turgor under low soil water conditions.

SaVHAc1 and expression of transporter genes and other rice native genes under salinity

The expression of *SaVHAc1* was higher in both leaf and root tissues of *SaVHAc1*-rice with or without salt stress, which indicated that the *SaVHAc1*-rice already had the proton pump activated and prepared to adapt to the salt stress.

A number of rice native genes were affected by the *SaVHAc1* expression that was evident from the microarray experiment, although the number of genes that were significantly up- and/or down-regulated was low (Table S1). Temporal and spatial changes in the gene expression were not detected in the microarray because of the fact that the RNA samples were pooled from root and leaf tissues at different time points of salt stress in WT and *SaVHAc1*-rice lines. However, differences in their transcript accumulation were captured in the (semi)quantitative RT-PCR (Figure 7). Although the expression of cation transporter (*OsCTP*) and exchanger (*OsNHX1*) genes were immediately affected at the early stage of salt stress, their subsequent up-regulation was observed in the *SaVHAc1*-rice under stress, which further explained that a proton electrochemical gradient was generated by the expression of *SaVHAc1*. Higher accumulation of *OsCTP* (a novel cation transporter in rice with homology to *ChaC* associated with $Ca2^+/H^+$ transport in *Esc-*

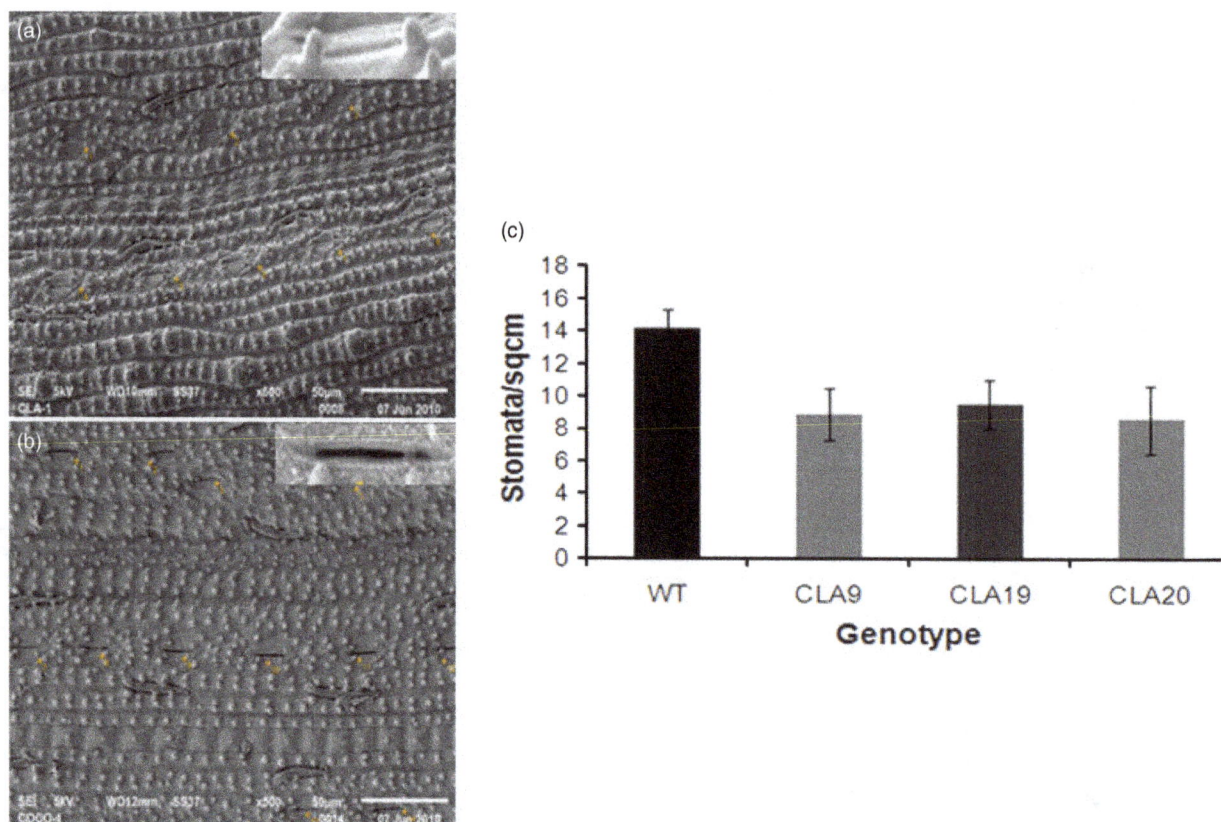

Figure 9 Scanning electron microscope picture of a leaf showing closed stomata (arrow marked and numbered) in *SaVHAc1*-rice (a) compared to open-type stomata under salt stress in WT rice (b). The *SaVHAc1*-rice has less stomata/sqcm in leaf surface than the WT under salinity (c). A 50× magnified picture of a representative stoma is shown as an inset on the top right corner of Panel a and b). Error bars represent standard error of means.

herichia coli) in the *SaVHAc1*-rice may have contributed to the increased Ca^{2+} transport. The high expression of *SaVHAc1*, even without salt stress and subsequent up-regulation of other genes including cation transporter genes, which may have been energized by the *SaVHAc1* expression as a component of coordinated regulation under salinity stress, led us to the presumption that the *SaVHAc1* lines maintained an anticipatory preparedness to adapt to salt stress.

SaVHAc1 and stomata closure

As an adaptation/avoidance strategy, plants are known to close the stomata under dehydration stress to save water and maintain turgor (Skirycz and Inze, 2010). The possible involvement of ABA in the regulatory pathway leading to induction of V-ATPase activity has been established (Tsiantis *et al.*, 1996). Overaccumulation of V-ATPase in *SaVHAc1*-rice might play an important role in the coregulation of ABA signalling induced by salt stress. ABA treatment mimics NaCl treatment of plants, and only the c-subunit of the V-ATPase has been shown to be transcriptionally regulated by NaCl in plants (Tsiantis *et al.*, 1996).

Stomata opening is induced by increased turgor pressure as a result of K^+ and anion influx energized by H^+-ATPase. Osmotic stress, at the early stage of salt stress first experienced by the roots, releases ABA signalling molecules, which under the overexpression of vacuolar ATPase could induce stomata closure (Allen *et al.*, 2000). The loss of turgidity as a result of K^+ and water efflux from the guard cell may have contributed to the stomata closure in *SaVHAc1*-rice plants (Figure 9). H^+ is known as secondary messenger in hormone action of plants. The

increase in cytoplasmic pH (by the expression of vacuolar ATPase) in the *SaVHAc1* in coordination with ABA molecules may have resulted in the opening of K^+ (out) channel and closing of K^+ (in) channel. Further, the increase in Ca^{2+}_{Cyt} in the *SaVHAc1*-rice possibly brought about a reduction in the K^+ (in) activity of the plasma membrane, leading to the reduction in guard cell turgidity and ultimately stomata closure. Although the Ca^{2+}- and H^+-mediated K^+ efflux are independent of each other, it is possible both the mechanisms are in operation to maintain a balance of the Ca^{2+}_{Cyt} and alkalinization (Swamy, 1999). Also, the kinetics of expression changes of the ABA signalling genes (*OsPLDα1* and *OsGPA1*) in the *SaVHAc1*-rice vis-à-vis WT plants provided clues to the closure of stomata in the leaves of *SaVHAc1*-rice. The expression of these two positive regulators of stomata ABA signalling was higher in *SaVHAc1*-rice as compared with the WT (Figure 8) without and with salt stress. Similarly, the expression of *OsSORK* (rice homologue of Arabidopsis *GORK*) was 2.8- and 1.9-fold higher in *SaVHAc1*-rice than WT with and without salt stress, respectively. Although the mechanism of reduced stomata density in *SaVHAc1*-rice remains to be investigated, the earlier observation that *det3* mutants were defective in guard cell signalling or movement could explain the apparent anatomical adjustments of *SaVHAc1*-rice to adapt to the stress condition. Further, the reduced stomata density in the *SaVHAc1*-rice could contribute to the slow/lower rate of water loss from *SaVHAc1*-rice leaves for which they maintained a higher RWC under salt stress as compared to the WT leaves. Higher RWC is one of the important factors contributing to tolerance to physiological drought that is experienced by the plant at the initial stage of salt stress as a result of low osmotic potential caused by the accumulation of soluble salts.

The time point analysis of *SaVHAc1* as well as other related genes further established that *SaVHAc1*-rice tolerated salt stress by an early stage priming and preconditioning response (Harb et al., 2010), thus maintaining a higher RWC and chlorophyll (thereby photosynthesis), protecting cytosol from ion toxicity, and early stomata closure. Present results suggested a clearly defined functional role of *SaVHAc1* with implications in the modification of a signalling pathway in the salt stress response by driving the expression of other genes involved in ion homoeostasis, solute accumulation, and ABA signalling. This study including that of others as described earlier again substantiates that grass halophytes, in addition to their usefulness to understand the gene regulation mechanism of their natural salinity-adaptability, could be effectively used as a source for mining and bioprospecting superior alleles for the manipulation of salt tolerance in monocot field crops such as rice.

Experimental procedures

Cloning of *SaVHAc1*, sequence analysis, and construction of binary vector

An expressed sequence tag (EST#617) of *Spartina alterniflora* (GenBank Acc. No EH277293; Baisakh et al., 2008) showed similarity to plant vacuolar H+-ATPase subunit c1 when its nucleotide and deduced protein sequences were blasted against the non-redundant nucleotide and protein database using BLASTN and BLASTP interface, respectively. The sequence alignment of this EST (hereinafter referred to as *SaVHAc1*) was carried out with orthologs from other plants (see Figure 1 for details) using CLUSTALW.

The complete open reading frame of *SaVHAc1* was amplified from the *S. alterniflora* root cDNA library (Baisakh et al., 2008) by PCR using primers *SaVHAc1* Fwd: 5′- ggaagatctatgtcgtcgacgttcag -3′ and *SaVHAc1* Rev: 5′- gggtwaccctaatctgcacggac -3′ containing the *Bgl*II and *Bst*EII restriction endonuclease (RE) recognition sites (underlined), respectively. The PCR product and pCAMBIA1301 (CAMBIA, Australia) was digested using the same REs and ligated to yield the binary plasmid *p35S:SaVHAc1* (Figure 2a). The identity and orientation of *p35S:SaVHAc1* was confirmed by restriction digestion and sequencing. The plasmid was mobilized into *Agrobacterium tumefaciens* LBA4404 using the freeze-thaw method (An et al., 1998).

Rice transformation

Agrobacterium tumefaciens transformation of rice cultivar 'Cocodrie' was performed following the method described earlier (Rao et al., 2009). LBA4404/*p35S:SaVHAc1* was precultured overnight at 28 °C in Luria-Bertani (LB) broth with rifampicin (20 μg/mL), spectinomycin (100 μg/mL), streptomycin (50 μg/mL) and kanamycin (50 μg/mL) under constant shaking at 200 rpm. The precultured bacteria were subcultured in fresh LB with the same antibiotics and grown for 24 h. Bacteria cells were resuspended in the MS (Murashige and Skoog, 1962) liquid medium supplemented with 2 mg/L 2,4-D and 100 μM acetosyringone (MSco) to a final titre of $A_{600} = 1.0$ for transformation.

Three–four-week-old seed-derived rice embryogenic calli were vacuum-infiltrated (0.4–0.6 atm) with the bacterial suspension for 10 min and co-cultivated for 3 days in solid MSco medium at 25 °C in the dark for 3 days. Embryogenic callus development, and selection and regeneration of the putative transgenic calli was performed following the method described earlier (Baisakh et al., 2001).

Molecular analysis of plants expressing *SaVHAc1*

Polymerase chain reaction

Total genomic DNA was isolated from leaf tissues using a modified CTAB method (Murray and Thompson, 1980). One hundred ng of DNA was subject to PCR analysis for selectable marker gene (*hph*) and target gene (*SaVHAc1*) using gene-specific primers (5′–3′) as follows: *HPH* F- tacttctacacagccatc, *HPH* R- tatgtcctgcgggtaaat; *SaVHAc1* F- aggagggtgtaccattcgtcaatg, *SaVHAc1* R- ccaggctcgtagagaataccattg.

Southern blot analysis

Southern blot analysis was performed following Baisakh et al. (2006a). Ten micrograms of genomic DNA were digested with a single cutter *Sst* I (for copy number analysis) and *Bgl*II and *Bst*EII (for releasing the full length *SaVHAc1*), electrophoresed on a 1% (w/v) TAE agarose gel, and transferred under alkaline denaturing conditions to Hybond N+ nylon membrane (GE Healthcare, Piscataway, NJ). A PCR generated 200-bp fragment of *SaVHAc1*, radiolabeled using (α-^{32}P)dCTP and the Rediprime labelling system (GE Healthcare), was used as the hybridization probe. Hybridization and follow-up membrane washing, and exposure to X-ray Hyperfilm™ MP (GE Healthcare) was performed as per Baisakh et al. (2006a).

Mapping the *SaVHAc1* integration site in rice

The insertion site of *SaVHAc1* was mapped by isolating the T-DNA-rice genome flanking sequences using TAIL-PCR technique as described by Liu et al. (1995). The gene-specific primers designed from 35S promoter and nosT sequences were used in combination with AD1 and AD4, and the nested primers were designed from the left and right border sequences of the T-DNA. Three nested gene-specific reverse primers and one of the four arbitrary degenerate (AD1-4) primers were used in successive rounds of TAIL-PCR cycling. The primary PCR product was diluted 40-fold and used in the secondary reaction, while the latter was diluted 5-fold for the tertiary reaction. The products of the primary, secondary and tertiary reactions were analysed on a 1.5% agarose gel. Fragments exhibiting a difference in size consistent with nested gene-specific primer positions were cloned using pGEMT-easy vector (Promega, Madison, WI) and sequenced as described earlier (Baisakh et al., 2006b). The genomic sequences flanking *SaVHAc1* were blasted against and mapped to the reference rice genome using megaBLASTN interface (http://blast.ncbi.nlm.nih.gov/Blast.cgi).

Salt stress treatment

Initially, cut leaves of three independent PCR positive *SaVHAc1*-rice plants were floated in 40 mL of water with 100 and 200 mM NaCl in plastic deep dishes and kept under continuous light for 72 h before scoring for chlorophyll bleach (Sanan-Mishra et al., 2005).

Four-week-old seedlings of homozygous *SaVHAc1*-rice (CLA9, CLA19, and CLA20) and WT rice were subject to salt stress (150 mM NaCl) under hydroponics in Yoshida's nutrient solution (Yoshida et al., 1976) following the method described earlier (Batlang et al., 2012). Twenty plants were included in each of the four replications. After 1 week of stress, shoot and root length, shoot and root fresh and dry weight, chlorophyll concentration and tissue ion concentrations were measured. After a week of stress, the *SaVHAc1*-rice and a few surviving WT plants (those not completely dead) were transferred to fresh nutrient solution without salt, and a week-old recovered plants were grown to maturity in the greenhouse maintained at 29/21 °C day/night temperature regime under natural day light condition. At reproductive stage, salt stress was imposed as described by Batlang *et al.* (2012). Ten *SaVHAc1* and WT plants during panicle initiation stage (90–95 days-old) were subject to 150 mM NaCl in the irrigation water continuously for 1 week before the pots were submerged up to the soil level in a deep plastic tray with salt-free water for 2 days and then irrigated normally until maturity. Data were recorded for number of seeds per plant and single panicle (primary tiller) grain yield.

Expression analysis

Microarray analysis

Total RNA was extracted from salt stressed and unstressed WT and transgenic plants at different time points (24, 48, 72 h, and 4 days of recovery) using RNeasy plant minikit (Qiagen, Valencia, CA). For microarray experiment, total RNA quality was checked in gel as well as through Agilent 2100 bioanalyzer. The RNA samples from different time points except recovery point were pooled for WT and *SaVHAc1* plants for each of the four biological replications. The four samples were named WT-Control (WT-C), WT-Stress (WT-S), *SaVHAc1*-rice-Control (*SaVHAc1*-rice-C) and *SaVHAc1*-rice-stress (*SaVHAc1*-rice-S).

Target preparation, that is, first-strand synthesis, second-strand synthesis, aRNA purification, dye labelling, and hybridization and washing, was performed as per the protocol described earlier (Edwards *et al.*, 2008; Data S1). Sixty-mer oligonucleotides designed from the rice unigenes and synthesized by Operon technologies were printed on slides in Galbraith laboratory (http://ag.arizona.edu/~dgalbrai) for use on microarray chips (45K). The GPR result file was analysed in R software (http://www.R-project.org; Data S1).

Enrichment of functional Gene Ontology categories (GO; http://www.geneontology.org) for genotype, environment and their interaction was tested using the GOEAST package (Zheng and Wang, 2008) with a Fisher's exact test, which considered topology of the GO relationships (Alexa *et al.*, 2006), followed by a Yekutieli adjustment for multiple comparisons (Yekutieli and Benjamini, 1999). Results were considered significant at 0.05 after adjustment. The set of oligos that responded to the environment were enriched for the biological processes related to biotic stimuli and stress responses.

(Semi)quantitative reverse transcription polymerase chain reaction (Sq/qRT-PCR)

The total RNA was extracted from 100 mg of freshly collected leaf and root tissues of *SaVHAc1*-rice and WT rice at 0 h (control), 2, 24, 48, 72 h and 4 days of recovery following salt stress. Two micrograms of total RNA were subject to sqRT-PCR of *SaVHAc1* gene and other genes that were up/down-regulated in the microarray data, employing a single-step RT-PCR kit (Qiagen, Valencia, CA). The products were resolved in 1.5% TAE agarose gel, visualized under UV transilluminator in a Kodak 200 gel doc apparatus (Carestream Health, Inc., Roches-

ter, NY). Rice actin gene 1 (*OsAct1*) was used as an internal control for the template validation. The primer sequences used in the study are provided in Table S2.

The qRT-PCR was carried out using the same RNA samples that were used for sqRT-PCR as per the method described (Baisakh *et al.*, 2008). Essentially, 1 μg total RNA was reverse transcribed using iScript 1st strand cDNA synthesis kit (Bio-rad, Carlsbad, CA). PCR was performed in triplicate (biological replicate) with two independent cDNA preparations (technical replicate) using SYBR green master mix (Bio-rad, Hercules, CA), 2 μL cDNA and 3.25 pmol of each gene-specific primer in a MyiQ Real-Time PCR detection system (Bio-rad, Hercules, CA). The relative expression ratio was calculated using the $2^{-\Delta\Delta Ct}$ method (Baisakh *et al.*, 2008) with rice elongation factor 1α gene (*OsEF1α*) as the reference gene.

Estimation of chlorophyll concentration

Total chlorophyll was extracted from one fully expanded leaf per plant (three plants from each of the *SaVHAc1*-rice and WT plants following salt stress) with 80% acetone twice. The chlorophyll a and b concentration was measured spectrophotometrically following the method described by Lichtenthaler (1987) to determine the extent of bleaching and chlorophyll loss. Total chlorophyll was also estimated from the SPAD502 meter (Konica Minolta Sensing, Inc., Ramsey, NJ) reading as described earlier (Monje and Bugbee, 1992).

Estimation of tissue ion concentration

Leaf and root tissues were harvested from 1-month-old seedlings of unstressed (control) and salt-treated *SaVHAc1*-rice and WT, and oven-dried at 80 °C for 48 h. Five hundred mg of dried tissues was extracted with HNO_3 digestion. The Na^+, K^+, Ca^{2+} and Mg^{2+} concentrations were measured through inductively coupled plasma-mass spectrometry (ICP-MS, Perkin-Elmer Plasma 400 emission spectrometer) in an in-house plant and soil testing laboratory.

Growth parameters study

Data were collected on different vegetative growth parameters (root and shoot length, root and shoot dry weight) of the *SaVHAc1*-rice lines vis-à-vis WT plants one week after salt stress was imposed. The single panicle grain yield from the primary tiller was recorded on the salt-stressed greenhouse-grown plants at maturity. Data were analysed for ANOVA with statistical analysis software SAS 9.1.3 (SAS Institute Inc, 2004).

Relative water content

The RWC of the leaves was determined following Slatyer (1967). Middle sections of second-youngest fully expanded leaves were collected and wrapped from three different WT and *SaVHAc1*-rice plants after a week of salt stress and weighed [fresh weight (FW)]. The leaf pieces were immersed in dH_2O placed in dark at 4 °C overnight and weighed after brief blot-drying [turgid weight (TW)]. Then, the pieces were dried at 60 °C for 24 h and weighed [dry weight (DW)]. RWC was estimated in percentage of the water content at a given time and tissue as related to the water content at full turgor using the formula:

$$RWC(\%) = (FW - DW)/(TW - DW) \times 100$$

Scanning electron microscopy (SEM)

Leaf samples were collected from *SaVHAc1*-rice and WT before and 24 h after salt stress and were fixed with glutaraldehyde buffer followed by gradual alcohol dehydration. The leaves were then critical point dried under liquid CO_2 and the probe surface sputter-coated with an electric-conducting gold layer before imaging with SEM (Cambridge S-260) at 5 kV.

Acknowledgements

The technical assistance of Greg Ford, SRRC, USDA-ARS is duly acknowledged. This work was financially supported by USDA-CSREES. This manuscript is approved for publication by the Director of Louisiana Agricultural Experiment Station as MS#2011-306-6572.

References

Alexa, A., Rahnenfuhrer, J. and Lengauer, T. (2006) Improved scoring of functional groups from gene expression data by decorrelating GO graph structure. *Bioinformat.* **22**, 1600–1607.

Allen, G.J., Chu, S.P., Schumacher, K., Shimazaki, C.T., Vafeados, D., Kemper, A., Hawke, S.D., Tallman, G., Tsien, R.Y., Harper, J.F., Chory, J. and Schroeder, J.I. (2000) Alteration of Stimulus-specific guard cell calcium oscillations and stomatal closing in *Arabidopsis det3* mutant. *Science*, **289**, 2338–2342.

An, G., Ebert, P.R, Mitra, A. and Ha, S.B. (1998) Binary vectors. In *Plant Molecular Biology* (Gelvin, S.B. and Shilperoort, R.A., eds), pp. 1–19. Dordrecht, the Netherlands: Kluwer Academic Publishers.

Apse, M.P., Aharon, G.S., Snedden, W.A. and Blumwald, E. (1999) Salt tolerance conferred by over expression of a vacuolar Na^+/H^+ antiporter in *Arabidopsis*. *Science*, **285**, 1256–1258.

Baisakh, N., Datta, K., Oliva, N., Ona, I., Rao, G.J.N., Mew, T.W. and Datta, S.K. (2001) Rapid development of homozygous transgenic rice using anther culture harboring rice *chitinase* gene for enhanced sheath blight resistance. *Plant Biotechnol.* **18**, 101–108.

Baisakh, N., Subudhi, P. and Parami, N. (2006a) cDNA-AFLP analysis reveals differential gene expression in response to salt stress in a halophyte *Spartina alterniflora* Loisel. *Plant Sci.* **170**, 1141–1149.

Baisakh, N., Rehana, S., Rai, M., Oliva, N., Tan, J., Mackill, D., Khush, G.S., Datta, K. and Datta, S.K. (2006b) Marker-free transgenic (MFT) near-isogenic introgression lines (NILs) of 'golden' indica rice (cv IR64) with accumulation of provitaminA in the endosperm tissue. *Plant Biotechnol. J.* **4**, 467–475.

Baisakh, N., Subudhi, P. and Varadwaj, P. (2008) Primary responses to salt stress in a halophyte *Spartina alterniflora* (Loisel). *Funct. Integr. Genomics*, **8**, 287–300.

Barkla, B.J., Zingarelli, L., Blumwald, E. and Smith, J.A.C. (1995) Tonoplast Na^+/H^+ antiport activity and its energization by the vacuolar H^+-ATPase in the halophytic plant *Mesembryanthemum crystallinum* L. *Plant Physiol.* **109**, 549–556.

Batlang, U., Baisakh, N., Ambavaram, M.M.R. and Pereira, A. (2012) Phenotypic and physiological evaluation of rice drought and salinity stress responses. In *Methods Molecular Biology* (Yang, Y., ed.), New York, NY: Humana Press (in press).

Chen, X., Kanopkorn, T., Zeng, Q., Wilkins, T.A. and Wood, A.J. (2002) Characterization of the V-type H+—ATPase in the resurrection plant *Tortula ruralis*: accumulation and polysomal recruitment of the proteolipid c subunit in response to salt stress. *J. Exp. Bot.* **53**, 225–232.

Dietz, K.J. and Arbinger, B. (1996) cDNA sequence and expression of SuE of the vacuolar H^+-ATPase in the inducible crassulacean acid metabolism plant *Mesembryanthemum crystallinum*. *Biochim. et Biophys. Acta*, **1281**, 134–138.

Dietz, K.J., Tavakoli, N., Kluge, C., Mimura, T., Sharma, S.S., Harris, G.C., Chardonnens, A.N. and Golldack, D. (2001) Significance of the V-type

ATPase for the adaptation to stressful growth conditions and its regulation on the molecular and biochemical level. *J. Exp. Bot.* **52**, 1969–1980.

Edwards, J.D., Janda, J., Sweeney, M.T., Gaikwad, A.B., Liu, B., Leung, H. and Galbraith, D.W. (2008) Development and evaluation of a high-throughput, low-cost genotyping platform based on oligonucleotide microarrays in rice. *Plant Methods*, **4**, 13.

Flowers, T.J. (2004) Improving crop salt tolerance. *J. Exp. Bot.* **55**, 307–319.

Gaxiola, R.A., Fink, G.R. and Hirschi, K.D. (2002) Genetic manipulation of vacuolar proton pumps and transporters. *Plant Physiol.* **129**, 967–973.

Gaxiola, R.A., Palmgren, M.G. and Schumacher, K. (2007) Plant proton pumps. *FEBS Lett.* **581**, 2204–2214.

Golldack, D. and Deitz, K.J. (2000) Salt-Induced expression of the vacuolar H^+-ATPase in the common ice plant is developmentally controlled and tissue specific. *Plant Physiol.* **125**, 1643–1654.

Harb, A., Krishnan, A., Ambavaram, M.M.R. and Pereira, A. (2010) Molecular and physiological analysis of drought stress in *Arabidopsis* reveals early responses leading to acclimation in plant growth. *Plant Physiol.* **154**, 1254–1271.

Hasegawa, P.M., Bressan, R.A., Zhu, J.K. and Bohnert, J.H. (2000) Plant cellular and molecular responses to high salinity. *Annu. Rev. Plant Physiol. Plant Mol. Biol.* **51**, 463–499.

Hirschi, K.D., Zhen, R.G., Cunningham, K.W., Rea, P.A. and Fink, G.R. (1996) CAX1, an H^+/Ca^{2+} antiporter from *Arabidopsis*. *Proc. Natl. Acad. Sci. USA*, **93**, 8782–8786.

Kirsch, M., An, Z., Viereck, R., Löw, R. and Rausch, T. (1996) Salt stress induces an increased expression of V-type H(+)-ATPase in mature sugar beet leaves. *Plant Mol. Biol.* **32**, 543–547.

Lehr, A., Kirsch, M., Viereck, R., Schiemann, J. and Rausch, T. (1999) cDNA and genomic cloning of sugar beet V-type H^+-ATPase subunit A and c isoforms: evidence for coordinate expression during plant development and coordinate induction in response to high salinity. *Plant Mol. Biol.* **39**, 463–475.

Lichtenthaler, H.K. (1987) Chlorophyll and carotenoids pigments of photosynthetic biomembranes. *Methods Enzymol.* **148**, 350–382.

Liu, Y.G., Mitsukawa, N., Oosumi, T. and Whittier, R.F. (1995) Efficient isolation and mapping of *Arabidopsis thaliana* T-DNA insert junctions by thermal asymmetric interlaced PCR. *Plant J.* **8**, 457–463.

Löw, R., Rockel, B., Kirsch, M., Ratajczak, R., Hortensteiner, S., Martinoia, E., Lüttge, U. and Rausch, T. (1996) Early salt stress effects on the differential expression of vacuolar H(+)-ATPase genes in roots and leaves of *Mesembryanthemum crystallinum*. *Plant Physiol.* **110**, 259–265.

Maathuis, F.J., Filatov, V., Herzyk, P., Krijger, G.C., Axelsen, K.B., Chen, S., Green, B.J., Li, Y., Madagan, K.L., Sánchez-Fernández, R., Forde, B.G., Palmgren, M.G., Rea, P.A., Williams, L.E., Sanders, D. and Amtmann, A. (2003) Transcriptome analysis of root transporters reveals participation of multiple gene families in the response to cation stress. *Plant J.* **35**, 675–692.

Michelet, B. and Boutry, M. (1995) The Plasma membrane H+-ATPase A highly regulated enzyme with multiple physiological functions. *Plant Physiol.* **108**, 1–6.

Monje, O.A. and Bugbee, B. (1992) Inherent limitations of nondestructive chlorophyll meters: a comparison of two types of meters. *HortSci.* **27**, 69–71.

Munns, R. and Tester, M. (2008) Mechanisms of salinity tolerance. *Annu. Rev. Plant Biol.* **59**, 651–681.

Murashige, T. and Skoog, F. (1962) A revised medium for rapid growth and bio-assays with tobacco tissue cultures. *Physiol. Plant.* **15**, 473–497.

Murray, M.G. and Thompson, W.F. (1980) Rapid isolation of high molecular weight plant DNA. *Nucl. Acids Res.* **8**, 4321–4325.

Nilson, S.E. and Assmann, S.M. (2010) The alpha-subunit of the *Arabidopsis* heterotrimeric G protein, GPA1, is a regulator of transpiration efficiency. *Plant Physiol.* **152**, 2067–2077.

Niu, X., Bressan, R.A., Hasegawa, P.M. and Pardo, J.M. (1995) Ion homeostasis in NaCl stress environments. *Plant Physiol.* **109**, 735–742.

Padmanaban, S., Lin, X., Perera, I., Kawamura, Y. and Sze, H. (2004) Differential expression of vacuolar H^+-ATPase subunit c genes in tissues active in membrane trafficking and their roles in plant growth as revealed by RNAi. *Plant Physiol.* **134**, 1514–1626.

Palmgren, M.G. (1998) Proton gradients and plant growth: role of the plasma membrane H^+-ATPase. *Adv. Bot. Res.* **28**, 1–70.

Rao, M.V.R., Behera, K.S., Baisakh, N., Datta, S.K. and Rao, G.J.N. (2009) Transgenic *indica* rice cultivar 'Swarna' expressing a potato chymotrypsin inhibitor *pin2* gene show enhanced levels of resistance to yellow stem borer. *Plant Cell Tiss. Organ Cult.* **99**, 277–285.

Ratajczak, R., Richter, J. and Lüttge, U. (1994) Adaptation of the tonoplast V-type H$^+$-ATPase of *Mesembryanthemum crystallinum* to salt stress, C$_3$-CAM transition and plant age. *Plant Cell Environ.* **17**, 1101–1112.

Sanan-Mishra, N., Pham, X.H., Sopory, S.K. and Tuteja, N. (2005) Pea DNA helicase 45 overexpression in tobacco confers high salinity tolerance without affecting yield. *Proc. Natl Acad. Sci. USA*, **102**, 509–514.

SAS Institute Inc. (2004) *SAS OnlineDoc® 9.1.3.* Cary, NC: SAS Institute Inc.

Schumacher, K., Vafeados, D., McCarthy, M., Sze, H., Wilkins, T. and Chory, J. (1999) The *Arabidopsis* det3 mutant reveals a central role for the vacuolar H$^+$-ATPase in plant growth and development. *Genes Dev.* **13**, 3259–3270.

Senthilkumar, P., Jithesh, M.N., Parani, M., Rajalakshmi, S., Praseetha, K. and Parida, A. (2005) Salt stress effects on the accumulation of vacuolar H+-ATPase subunit c transcripts in wild rice, *Porteresia coarctata* (Roxb.) Tateoka. *Curr. Sci.* **89**, 1386–1394.

Serrano, R. (1989) Structure and function of plasma membrane ATPase. *Annu. Rev. Plant Physiol. Plant Mol. Biol.* **40**, 61–94.

Skirycz, A. and Inze, D. (2010) More from less: plant growth under limited water. *Curr. Opini. Biotechnol.* **21**, 1–7.

Slatyer, R.O. (1967) *Plant–Water Relationships.* London/New York: Academic Press. pp. 121–126.

Sussman, M.R. (1994) Molecular analysis of proteins in the plasma membrane. *Annu. Rev. Plant Physiol. Plant Mol. Biol.* **45**, 211–234.

Swamy, S.K. (1999) Drought signaling in plants. *Resonance*, **4**, 34–44.

Sze, H., Ward, J.M. and Lai, S. (1992) Vacuolar-H$^+$-ATPases from plants. *J. Bioenerg. Biomemb.* **24**, 371–381.

Sze, H., Li, X. and Palmgren, M.G. (1999) Energization of plant cell membranes by H$^+$ pumping ATPases: regulation and biosynthesis. *Plant Cell*, **11**, 677–690.

Tester, M. and Bacic, A. (2005) Abiotic stress tolerance in grasses. From model plants to crop plants. *Plant Physiol.* **137**, 791–793.

Thomson, M., Ocampo, M., Egdane, J., Rahman, M.A., Sajise, A.G., Adorada, D.L., Timimbang-Raiz, E., Blumwald, E., Seraj, Z.I., Singh, R.K., Gregorio, G.B. and Ismail, A.M. (2010) Characterizing the Saltol quantitative trait locus for salinity tolerance in rice. *Rice*, **3**, 148–160.

Tsiantis, M.S., Bartholomew, D.M. and Smith, J.A. (1996) Salt regulation of transcript levels for the c subunit of a leaf vacuolar H$^+$-ATPase in the halophyte *Mesembryanthemum crystallinum. Plant J.* **9**, 729–736.

Tyagi, W., Rajagopal, D., Singla-pareek, S.L., Reddy, M.K. and Sopory, S.K. (2005) Cloning and regulation of a stress-regulated *Pennisetum glaucum* Vacuolar ATPase c gene and characterization of its promoter that is expressed in shoot hairs and floral organs. *Plant Cell Physiol.* **46**, 1411–1422.

Winicov, I. (1998) New molecular approaches to improving salt tolerance in crop plants. *Annals of Bot.* **82**, 703–710.

Yekutieli, D. and Benjamini, Y. (1999) Resampling-based false discovery rate controlling multiple test procedures for correlated test statistics. *J. Stat. Plan. Inf.* **82**, 171–196.

Yoshida, S., Forno, D.A., Cock, J.H. and Gomez, K.A. (1976) *Laboratory manual for physiological studies of rice.* Philippines: IRRI, 83.

Zheng, Q. and Wang, X.J. (2008) GOEAST: a web-based software toolkit for Gene Ontology enrichment analysis. *Nucl. Acids Res.* **36**, 358–363.

A late embryogenesis abundant protein HVA1 regulated by an inducible promoter enhances root growth and abiotic stress tolerance in rice without yield penalty

Yi-Shih Chen[1,2], Shuen-Fang Lo[1,3], Peng-Kai Sun[1], Chung-An Lu[2], Tuan-Hua D. Ho[3,4,5,]* and Su-May Yu[1,3,5,]*

[1]Institute of Molecular Biology, Academia Sinica, Taipei, Taiwan

[2]Department of Life Sciences, National Central University, Jhongli City, Taiwan

[3]Agricultural Biotechnology Center, National Chung Hsing University, Taichung, Taiwan

[4]Institute of Plant and Microbial Biology, Academia Sinica, Taipei, Taiwan

[5]Department of Life Sciences, National Chung Hsing University, Taichung, Taiwan

*Correspondence

email sumay@imb.sinica.edu.tw

email tho@gate.sinica.edu.tw;
ho@biology2.wustl.edu

Accession number: HVA1 (CAA31853);
CRL1 (AB200234); CYCB1.1 (AK111939);
CYCB2.1 (AK070211); CYCB2.2
(AK070518); MSU_Gene_ID: PIN1b
(LOC_Os02g50960); PIN2
(LOC_Os06g44970); PIN9
(LOC_Os01g58860).

Keywords: rice, barley HVA1, stress-
inducible composite promoter, ABA,
lateral and primary root growth,
abiotic stress tolerance.

Summary

Regulation of root architecture is essential for maintaining plant growth under adverse environment. A synthetic abscisic acid (ABA)/stress-inducible promoter was designed to control the expression of a late embryogenesis abundant protein (HVA1) in transgenic rice. The background of HVA1 is low but highly inducible by ABA, salt, dehydration and cold. HVA1 was highly accumulated in root apical meristem (RAM) and lateral root primordia (LRP) after ABA/stress treatments, leading to enhanced root system expansion. Water-use efficiency (WUE) and biomass also increased in transgenic rice, likely due to the maintenance of normal cell functions and metabolic activities conferred by HVA1 which is capable of stabilizing proteins, under osmotic stress. HVA1 promotes lateral root (LR) initiation, elongation and emergence and primary root (PR) elongation via an auxin-dependent process, particularly by intensifying asymmetrical accumulation of auxin in LRP founder cells and RAM, even under ABA/stress-suppressive conditions. We demonstrate a successful application of an inducible promoter in regulating the spatial and temporal expression of HVA1 for improving root architecture and multiple stress tolerance without yield penalty.

Introduction

Plant root architecture is essential for water and nutrient uptake, anchorage and interactions with microbes in the soil, functions that impact growth rate, yield and abiotic stress tolerance. Root branching is enhanced by both water and nutrient deficiencies, and such root developmental plasticity offers one of the major acclimation strategies for plants to adapt to limited supplies of water and nutrients in the soil (Lopez-Bucio et al., 2003; Potters et al., 2007). Understanding the mechanism regulating root architecture is of great agronomic importance; for example, the majority of drought-resistant rice varieties have a deeper and more highly branched root system than drought-sensitive varieties (Price et al., 1997; Uga et al., 2013). However, detailed mechanisms underlying how environmental cues regulate root architecture remain mostly unclear.

Studies in Arabidopsis have significantly advanced our knowledge of mechanisms controlling root development (Lavenus et al., 2013; Malamy, 2005; Peret et al., 2009; Potters et al., 2007); however, similar studies in cereals are relatively limited (Coudert et al., 2010). Unlike Arabidopsis, in which the PR iteratively branches to generate several orders of lateral roots, the cereals have several types of branched roots including shoot-born crown roots and root-born lateral roots. Although several homologous genes that play similar roles regulating root formation between Arabidopsis and cereals have been identified, distinct hormonal and developmental pathways have also been identified in cereals (Orman-Ligeza et al., 2013). Auxin acts as a common integrator for many endogenous and environmental signals regulating LR development in both dicots and monocots (Lavenus et al., 2013; Orman-Ligeza et al., 2013). In Arabidopsis, ABA stimulates main root elongation in response to drought and osmotic stresses, yet ABA and auxin signals seem to act antagonistically during LR initiation, with ABA acting as a repressor while auxin as a promoting agent (De Smet et al., 2006). However, low ABA concentrations have also been shown to be promotive in root development in rice and Arabidopsis (Xu et al., 2013). Both auxin and ABA play essential roles regulating root development, but how they interplay to regulate root development in response to abiotic stress is unclear neither.

The current global climate changes tend to shift weather to more extreme variations, which aggravate the world crop productivity that has already plateaued (Lucas, 2003). Rice is the major dietary staple for nearly half of humanity. However, of the 130 million hectares of rice lands, close to 30% is affected by drought (Tuong and Bouman, 2003), 20% by salinity (Negrao et al., 2011) and 10% by low temperature (Sthapit and Witcombe, 1998). The development of new strategies for breeding rice with tolerance to multiple stresses that still maintain high productivity remains an important subject of research. Over-accumulation of osmoprotectants or proteins has been widely pursued as effective strategies in improving plant stress tolerances (Ahuja et al., 2010; Bartels, 2001; Ho and Wu, 2004; Yamaguchi-Shinozaki and Shinozaki, 2006). However, under normal environmental conditions, constitutive overproduction of these compounds or proteins consumes extra energy and often results in plant growth retardation and yield penalty (Dubouzet et al., 2003; Hsieh et al., 2002; Kasuga et al., 1999). To minimize such negative effect while still improving stress tolerance of plants, there is promise in inducible promoters that control the accumulation of an osmoprotectant or a protein, ensuring that stress tolerance mechanisms are initiated only under stress conditions (Garg et al., 2002; Kasuga et al., 1999; Lee et al., 2003; Morran et al., 2011).

Late embryogenesis abundant (LEA) proteins are highly accumulated in seeds at the onset of desiccation and in plant vegetative tissues in response to water deficit (Dure, 1992). HVA1 is a group 3 small LEA protein specifically expressed in barley seeds during late stages of seed development undergoing desiccation and is rapidly induced in young barley seedlings, especially in roots, by abiotic stresses (Hong et al., 1992). Expression of the rice HVA1 homologue, LEA3, is also highly induced in roots of rice seedlings in response to ABA and salt (Moons et al., 1997). HVA1 contains an 11-amino acid consensus motif that is repeated nine times, potentially forming an amphipathic α-helical dimer suitable for accommodating positively and negatively charged ions, and has therefore been proposed to function in sequestering ions (Dure, 1993). Most LEA proteins belong to a class of proteins called hydrophilins that are proposed to exist as unstructured proteins in aqueous solutions and thus function in protecting cells from damages caused by water limitation (Olvera-Carrillo et al., 2011). Many in vitro experiments have demonstrated that LEA could prevent the inactivation of enzymes and aggregation of proteins or maintain the functional membrane structure under dehydration and freeze–thaw conditions (Battaglia and Covarrubias, 2013). Overexpression of a soybean LEA3 confers high salt and extreme temperature tolerance in E. coli (Liu et al., 2010). A shrimp LEA3-like protein confers enhanced viability in Drosophila melanogaster cells after desiccation and hyperosmotic stress (Marunde et al., 2013). Ectopic expression of HVA1 also protects plants against drought, salt and osmotic stresses in various monocot and dicot plant species (Baena-Gonzalez et al., 2007; Battaglia and Covarrubias, 2013; Nguyen and Sticklen, 2013; Xu et al., 1996); however, the mechanism related to HVA1 conferring stress tolerance in transgenic plants is unknown.

Many ABA-inducible promoters contain a conserved ABA response element (ABRE) with an ACGT core and a coupling element (CE) forming an ABA response complex (ABRC) (Shen et al., 1993). Single and multiple copies of ABRC1 from the HVA22 promoter have been used to manipulate stress-inducible gene expression and stress tolerance in transgenic rice and tomato (Garg et al., 2002; Lee et al., 2003), but high background of transgene expression is present under nonstress conditions (Garg et al., 2002).

In this study, an ABRC-based synthetic composite promoter 3xABRC321 was designed with minimal background activity but capable of conferring spatial and temporal activation of transgene expression under stress conditions. Ectopic expression of 3xABRC321:HVA1 led to a significant increase in the expansion of primary and branch root systems in response to ABA/stress and thus enhanced tolerance to drought, salt and cold stresses and WUE in transgenic rice. HVA1 was found to promote auxin accumulation in meristematic tissues that initiate LR and PR growth through an auxin-dependent process despite the presence of ABA, indicating that the ABA-mediated environmental signal crosstalks with the auxin-mediated developmental signal to regulate root architecture.

Results

The synthetic promoter *ABRC321* has low background activity but is highly inducible by ABA and abiotic stresses in transgenic rice

ABRC321 was generated by fusion of CE and ABRE from HVA1 and HVA22 promoters (Shen et al., 1996) (Figure 1a). The reporter GUS was fused downstream of 1–3 tandem repeats of ABRC321 promoter (Figure S1a) and used for rice stable transformation. The GUS activity in leaves and roots controlled by 3xABRC321 had relatively low background but was highly inducible by ABA (Figure S1c).

The 3xABRC321 promoter was then used to control the expression of HVA1 (Figure S1b) in transgenic rice. The expression of 3xABRC321:HVA1 was induced in leaves and roots of transgenic seedlings, with levels generally higher in roots, and the expression of endogenous LEA3 was also induced in roots of wild-type (WT) rice (Figure 1b). The accumulation of HVA1 in seedlings was induced and increased with time up to 20 h under ABA, NaCl, cold and dehydration treatments (Figure 1c), but decreased gradually after relief from these stresses (Figure 1d). The half-life of HVA1 was estimated to be 78.7 and 157.8 h, in contrast to that of tubulin to be 60.1 and 46.3 h, in the absence and presence of ABA, respectively (Figure 1e). These studies indicate that both enhanced mRNA transcription and protein stability of HVA1 may account for the high-level accumulation of HVA1 in response to ABA and abiotic stresses.

3xABRC321:HVA1 enhances root growth in transgenic rice under low ABA concentrations and osmotic stress

Rice contains seminal, crown and LRs (Figure S2). Crown roots of transgenic rice were longer, and LR numbers were greater in transgenic lines than in WT even without ABA and were further increased by 0.1 μM ABA in transgenic lines (Figure 2a,b). Although LRs became shorter and thicker and LR density decreased with ABA concentrations exceeding 0.5 μM, all roots in transgenic plants were still longer and LR number and density greater than WT at all ABA concentrations (Figure 2a,b).

Transgenic seedlings were also treated with 400 mM sorbitol, which mimics osmotic stress. LR number and density of crown and seminal root in transgenic lines were increased with sorbitol (Figure 3a). All roots in transgenic lines were longer and had higher LR density than those of WT with sorbitol and in transgenic lines were all longer and LR density higher with sorbitol (Figure 3b). Transgenic lines also have higher dry and fresh weights in shoots and roots than WT with sorbitol (Figure 3c).

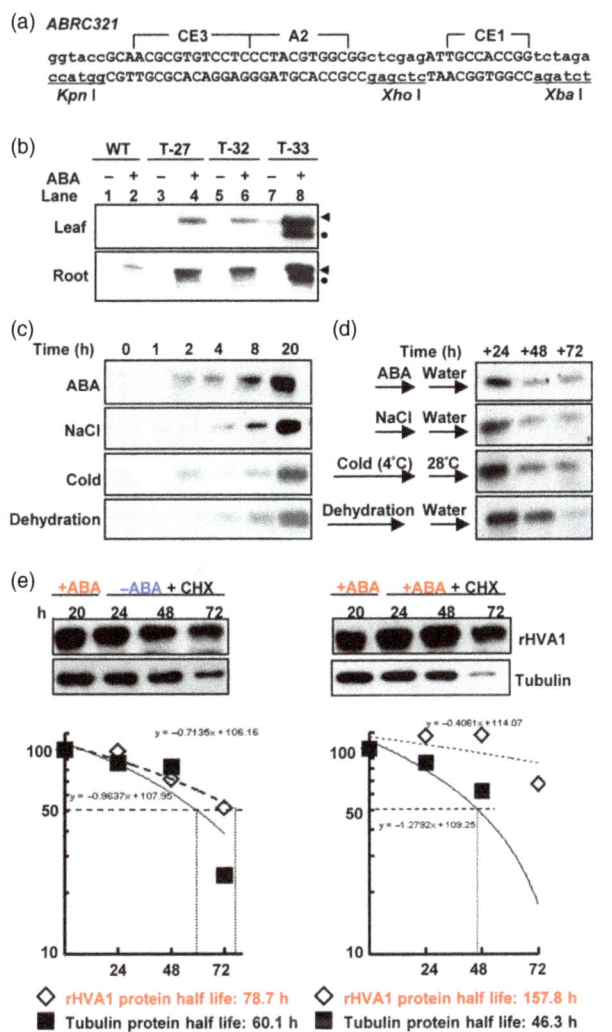

(a) ABRC321

(b)

(c)

(d)

(e)

Figure 1 *3xABRC321-HVA1* is highly inducible by abiotic stress, and the half-life of HVA1 is enhanced by ABA in transgenic rice. Ten-day-old seedlings of Tainung 67 transgenic lines were treated with various abiotic stresses. Total proteins were extracted from leaves and subjected to protein gel blot analysis using the antibarley HVA1 and anti-α-tubulin antibodies. (a) Nucleotide sequence of *ABRC321*. (b) Seedlings were treated with or without 10 μM ABA for 24 h. Triangle indicates the 27-kD HVA1, and dot indicates a smaller protein of 22 kD which was likely the degradation form of HVA1. (c) Seedlings were treated with 10 μM ABA, 200 mM NaCl or dehydration (slow air dry) at 28 °C or with cold (4 °C) for various lengths of time. (d) Seedlings were treated with various stresses for 20 h and then transferred to water at 28 °C for indicated time. Five seedlings were used for each treatment at each time point in (b)–(d). (e) Seedlings were incubated with 10 μM ABA for 20 h to allow the accumulation of HVA1, and then, seedlings were transferred to freshwater containing 200 μM cycloheximide (in 100% EtOH) with or without 10 μM ABA for up to 72 h. Levels of HVA1 shown in the protein gel blot were quantified densitometrically. The relative HVA1 and α-tubulin levels were then determined by a linear regression analysis, and the graph was plotted using a linear regression algorithm in Excel. The half-life of HVA1 and tubulin was estimated using equations derived from the linear regression analysis.

To determine whether HVA1 promotes LR initiation and/or elongation, roots of seedlings treated with 0.2 μM ABA were examined. The number of LRPs was reduced but that of LRs

increased in root regions 5.5–7.0 cm and 3.0–4.5 cm from root tips in transgenic lines as compared with WT, indicating that the majority of LRPs initiated were elongated in transgenic lines (Figure 4a,b). The number of LRPs and elongated LRs were both greater in region 0–1.5 cm from the root tip in transgenic lines than in WT (Figure 4a,b).

3xABRC321:HVA1 confers abiotic stress tolerance and promotes WUE in transgenic rice without apparent yield penalty

The tolerance to various abiotic stresses in transgenic rice was tested. With prior acclimation with 150 mM NaCl for 1 week, Tainung 67 transgenic lines displayed higher tolerance to 250 mM NaCl than WT, with survival rates 65–100% for five transgenic lines and only 5% for WT (Figure 5a). Kitaake was more sensitive to drought than Tainung 67 and used in the drought experiment. Although all plants displayed symptoms of leaf rolling and drying under drought, transgenic leaves recovered better than WT after rewatering, with survival rates 58–100% for six transgenic lines and only 25% for WT (Figure 5b). At 4 °C for 3 days, leaves of all plants rolled and dried, but leaves of Tainung 67 and Kitaake transgenic lines opened rapidly and continued to grow at 28 °C (Figure 5c).

Transgenic Tainung 67 and Kitaake expressing HVA1 were treated with sorbitol to induce root growth, and WUEs were determined. Although transgenic lines consumed higher amounts of water, they produced even more biomass compared with WT and thus had higher WUEs (Figure 6a). The performance of transgenic plants was further evaluated in fields. For three separate growing seasons, grain yields in transgenic lines were virtually the same as WT in irrigated fields (Figures 6b and S3). Kitaake was more sensitive to drought than Tainung 67, and therefore, only Kitaake was evaluated in nonirrigated fields. The result indicates that the grain yield in transgenic Kitaake was generally higher than WT in nonirrigated fields (Figures 6b and S3).

3xABRC321:HVA1 has similar tissue-specific and ABA-inducible expression patterns to the endogenous *LEA3* in transgenic rice roots

The expression pattern of HVA1 and the rice LEA3 in WT and transgenic roots was examined by immunocytochemistry assays using antibodies that recognize HVA1 and LEA3. The accumulation of LEA3 in WT and HVA1 in transgenic rice was barely detectable in roots without ABA, but was significantly increased in endodermis and pericycle (En/P) with ABA induction (Figure 7a and Figure S4a). LEA3 and HVA1 were also detected in LRP and in stele and RAM, and levels enhanced by ABA (Figure 7a and Figure S4a). The similarity in expression patterns between HVA1 and LEA3 is likely due to sharing of three cis-acting elements in promoters (Figure S4b). HVA1 was also detected in the cortex and exodermis overlaying the developing LRP in WT and transgenic lines treated with ABA if the antibody fluorescence signals were enhanced (Figure 7b and Figure S5).

The expression pattern of *3xABRC321:GUS* was examined in transgenic roots. GUS was barely detectable without ABA, but was detected at high levels mainly in En/P, founder cells of LRP and RAM after ABA induction (Figure 7c,d). GUS was also detected in the cortex and exodermis overlaying the developing LRP, and LR growth emerged through the

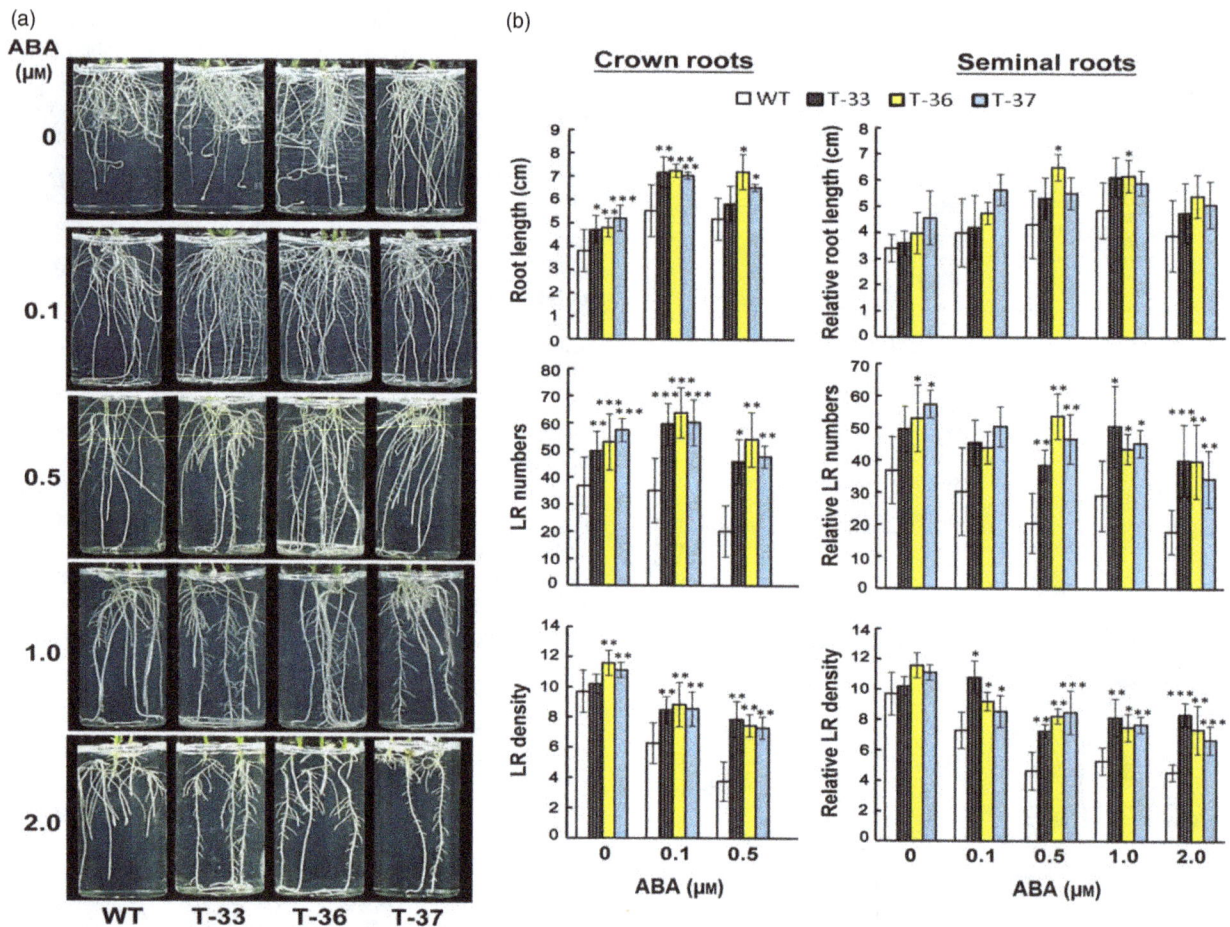

Figure 2 *3xABRC321:HVA1* enhances root growth in transgenic rice under low ABA concentrations. Three-day-old seedlings of three Tainung 67 (T) transgenic lines expressing *3xABRC321:HVA1* were transferred to MS medium containing various concentrations of ABA for 14 days. (a) Root morphology. (b) Relative root length, LR number and LR density of seminal root (right panel) and crown root (left panel) were determined. Error bars represent SD (n=9). Significance levels with the *t*-test: *$P < 0.05$, **$P < 0.01$, ***$P < 0.001$. Relative root length and LR number of seminal roots mean length of roots elongated and root developed, respectively, after transfer to medium containing various concentrations of ABA.

centre of the GUS-expressing cortex–exodermis region (Figure 7c, d).

3xABRC321:HVA1 and ABA promote root growth in transgenic rice through an auxin-dependent process

The amount and direction of auxin flow in the root determine both the positioning and frequency of LR initiation (Himanen *et al.*, 2002; Orman-Ligeza *et al.*, 2013). As HVA1 and low concentrations of ABA promote the initiation of LRP (Figure 4), we first determined whether auxin is necessary for the HVA1- and ABA-induced root growth. Seedlings were treated with ABA plus or minus the polar auxin transport inhibitor N-(1-naphthyl) phthalamic acid (NPA) (Reed *et al.*, 1998). As compared with the WT, without any treatment, crown roots and LRs were more abundant in transgenic lines (Figure 8a), and with 0.2 μM ABA, seminal and crown roots were significantly longer and LRs were more abundant in transgenic lines (Figure 8b). With NPA, the growth of all types of roots was inhibited (Figure 8c). With both ABA and NPA, the growth of crown roots and LRs was still inhibited, but that of seminal roots continued to extend, with length twice longer than that with NPA alone in transgenic lines (Figure 8d).

PIN-FORMED (PIN) proteins are auxin efflux transporters, and auxin gradient in roots, which depends on the PIN-driven polar auxin transport, is required for the establishment of LRP in both dicots and monocots (Benkova *et al.*, 2003; Petrasek and Friml, 2009). The expression of PINs and a few other proteins that are known to be related to root growth was then examined. The accumulation of *PIN1b*, *PIN2* and *PIN9* mRNAs was higher in transgenic roots than in WT and was all further enhanced by ABA (Figure 8e). Levels of mRNA of three cell cycle regulatory genes, *CYCB1.1*, *CYCB2.1* and *CYCB2.2*, and the auxin-inducible *Crown Rootless 1* (*CRL1*) essential for crown root and LR growth in rice (Inukai *et al.*, 2005) were also higher and further enhanced by ABA in roots of transgenic lines (Figure 8e).

3xABRC321:HVA1 and ABA enhance asymmetrical localization of auxin in LRP founder cells and auxin accumulation in RAM in transgenic rice

DR5:GUS is a widely used marker for estimating endogenous auxin levels (Ulmasov *et al.*, 1997). To determine whether the local distribution of auxin is affected by ectopic expression of HVA1 or by ABA, WT rice and transgenic rice carrying *3XA-*

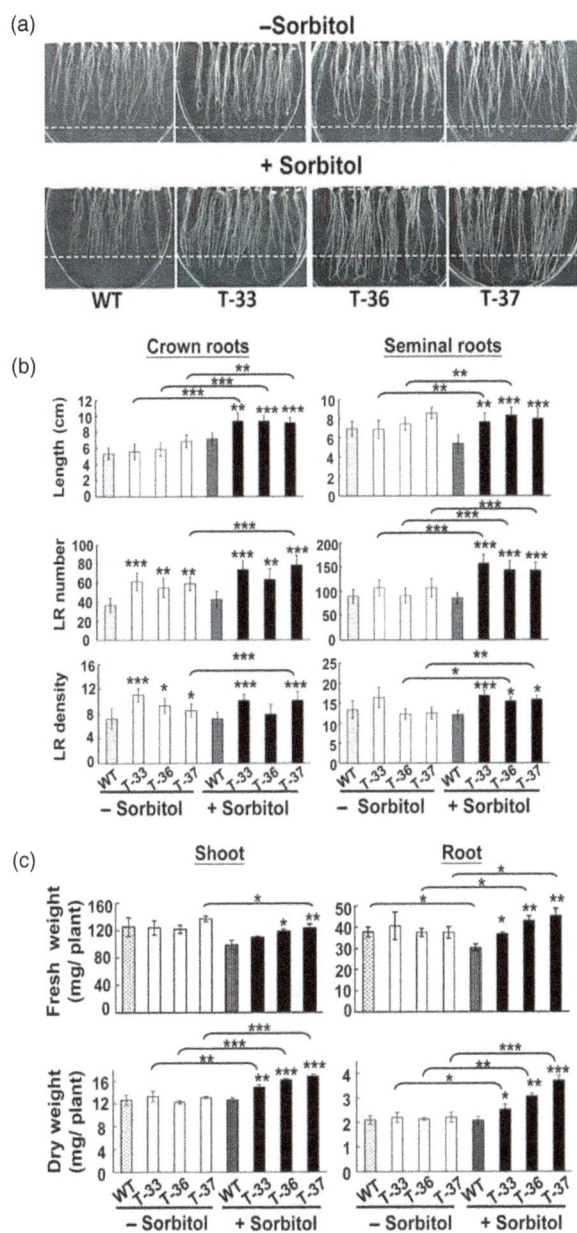

(a) −Sorbitol / + Sorbitol

WT T-33 T-36 T-37

(b) Crown roots Seminal roots

(c) Shoot Root

Figure 3 *3xABRC321:HVA1* enhances root growth in transgenic rice under osmotic stress. Three-day-old seedlings of three Tainung 67 (T) transgenic lines expressing *3xABRC321:HVA1* were transferred to MS medium containing 400 mM sorbitol for 11 days. (a) Root morphology. (b) Root length, LR number and LR density of crown roots (left panel) and seminal roots (right panel) were determined. The white dashed line indicates the position of root tips of most plants in WT. (c) Dry and fresh weights of shoots and roots of seedlings in (b) were determined. Error bars represent SD (*n* = 12). Significance levels with the *t*-test: *$P < 0.05$, **$P < 0.01$, ***$P < 0.001$.

BRC321:HVA1 were further transformed with the *DR5:GUS* construct, generating transgenic lines *DR5:GUS* and *HVA1 DR5: GUS*. In crown roots without ABA, GUS expression in both lines was visible along the En/P and at the founder cells of LRP and LR and in RAM (Figures 9a,b, S6a and S7). With ABA, GUS expression in line *DR5:GUS* was reduced or became hardly detectable, but was enhanced in line *HVA1 DR5:GUS* (Figures 9a,

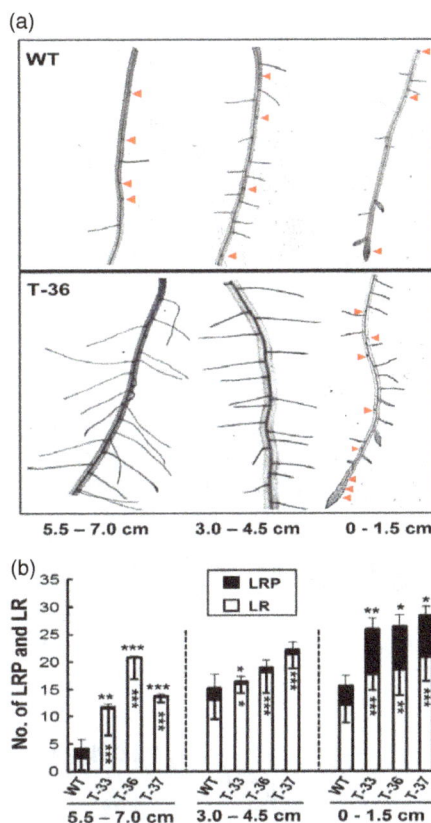

(a) WT / T-36

5.5 – 7.0 cm   3.0 – 4.5 cm   0 – 1.5 cm

(b) No. of LRP and LR — ■ LRP □ LR

5.5 – 7.0 cm   3.0 – 4.5 cm   0 – 1.5 cm

Figure 4 *3xABRC321:HVA1* enhances root initiation and elongation in transgenic rice treated with ABA. Three-day-old seedlings of three Tainung 67 (T) transgenic lines expressing *3xABRC321:HVA1* were transferred to MS medium containing 0.2 μM ABA. (a) Crown roots were treated with 75% ethanol, and LRP and extended LR in three segments (0–1.5 cm, 3.0–4.5 cm and 5.5–7.0 cm from the root tip) were examined. Upper panel: wild type. Lower panel: transgenic line T-36. Arrowhead indicates LRP. (b) The number of LRP and extended LR in (a) was quantified. Error bars represent SD (*n* = 12). Significance levels with the *t*-test: *$P < 0.05$, **$P < 0.01$, ***$P < 0.001$.

b, S6b and S7). The asymmetrical accumulation of GUS in LR founder cells was significantly reduced in line *DR5:GUS*, but was particularly high in line *HVA1 DR5:GUS* with ABA (Figure 9c and Figure S8).

Discussion

A major challenge in modern agricultural biotechnology is designing a transformation cassette enabling the precise temporal and spatial expression of transgene expression at appropriate levels to facilitate growth and enhance abiotic stress tolerance in crops. In this study, we demonstrated that *3xABRC321* functions as a molecular switch to keep low basal level of HVA1 under normal growth conditions, but confers a high level of expression upon induction by ABA, drought, salt and cold stresses in transgenic rice. Moreover, the expression of *3xABRC321:HVA1* successfully improves root architecture and enhances tolerance to various abiotic stresses and WUE in transgenic rice with no yield penalty in irrigated fields and higher yield in nonirrigated field as compared with the WT (Figure 6b).

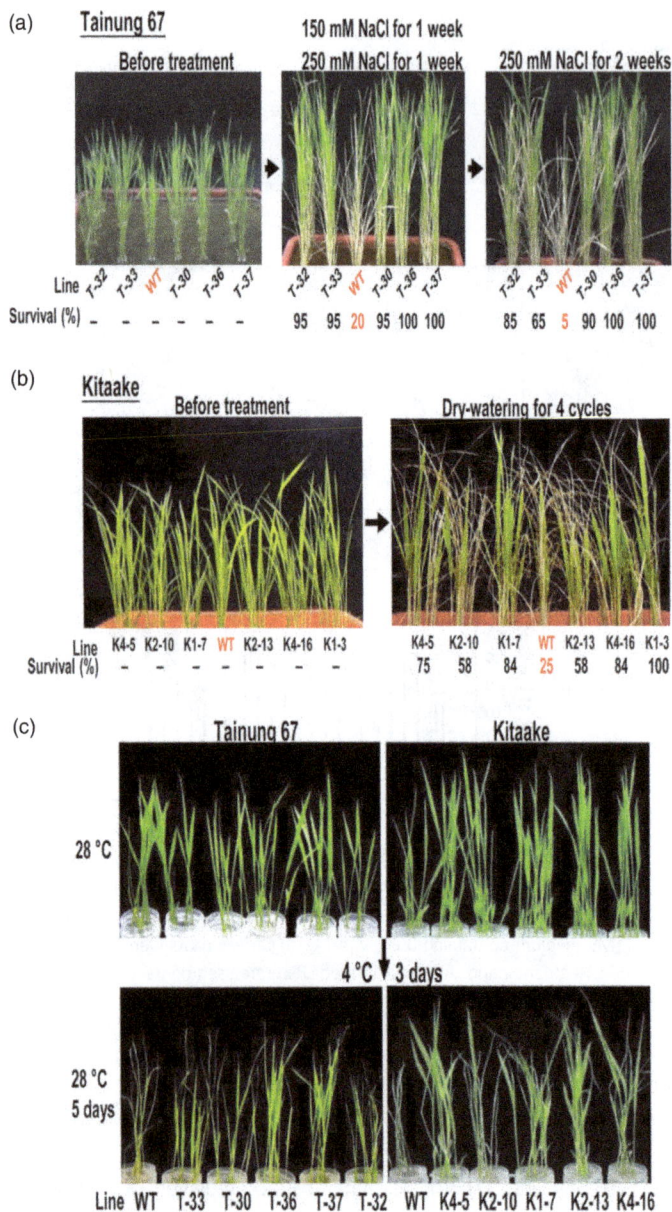

Figure 5 *3xABRC321:HVA1* confers abiotic stress tolerance. (a) Ten-day-old seedlings of Tainung 67 transgenic lines grown in soil were treated with 150 mM NaCl for 1 week and then 250 mM NaCl for 3 weeks. (b) Ten-day-old seedlings of Kitaake transgenic lines grown in soil were subjected to four cycles of complete dry (dehydration) and rewatering (wetting) treatments. Twenty seedlings per line were used for each treatment, and survival rates were determined by counting number of plants that grew when experiments were terminated in (a) and (b). (c) Ten-day-old seedlings of Tainung 67 (left panel) and Kitaake (right panel) transgenic lines grown in hydroponic solution at 28 °C were transferred to 4 °C incubator for 3 days and then shifted back to 28 °C for another 5 days. Survival rates were determined by counting number of plants that grew when experiments were terminated. Twenty seedlings per line were used for each treatment.

Transgenic rice and bent grass overexpressing HVA1 maintain higher leaf relative water content (RWC) and exhibit significantly less leaf wilting than WT and bent grass under drought stress (Babu *et al.*, 2004; Fu *et al.*, 2007). Transgenic mulberry overexpressing HVA1 also shows higher plasma membrane stability than wild type under salinity or water stress (Lal *et al.*, 2008). We observed enhanced expression of three cell cycle regulatory genes by ABA in roots of HVA1 transgenic lines (Figure 8e). These studies suggest that HVA1 transgenic rice may have greater water-holding capacity and RWC, thus be able to maintain photosynthesis and/or metabolic functions under drought conditions. All these metabolic activities eventually led to more active transpiration and growth, contributing to increased total biomass under water stress conditions.

The increase in root growth in HVA1 transgenic rice treated with ABA was resulted from the enhancement of LR initiation and elongation (Figure 4), and the process was auxin dependent based on two observations. First, the HVA1- and ABA-promoted

root elongation and LR growth in transgenic rice were inhibited by the auxin transport inhibitor NPA (Figure 8a,b,c,d). Second, the expression of genes encoding several auxin transporters, auxin-dependent cell cycle regulators and positive regulator of crown and LR growth in rice was all increased in transgenic rice and levels further enhanced by ABA (Figure 8e). In the presence of ABA, seminal roots overexpressing HVA1 were less sensitive to NPA inhibition (Figure 8d), suggesting that HVA1 promotes seminal root elongation even with reduced endogenous auxin levels.

In Arabidopsis, auxin is considered a key inducer, while ABA acts as an repressor of LR development (Peret *et al.*, 2009). However, physiological concentrations of ABA seem to induce LRP, as an ABA-insensitive mutant, *abi3*, has a reduced LRP formation in the presence of auxin (Brady *et al.*, 2003). In maize, ABA accumulates rapidly in roots under water stress, and an increase in ABA levels at the root tip is necessary for maintaining PR elongation at low water potentials (Yamaguchi and Sharp,

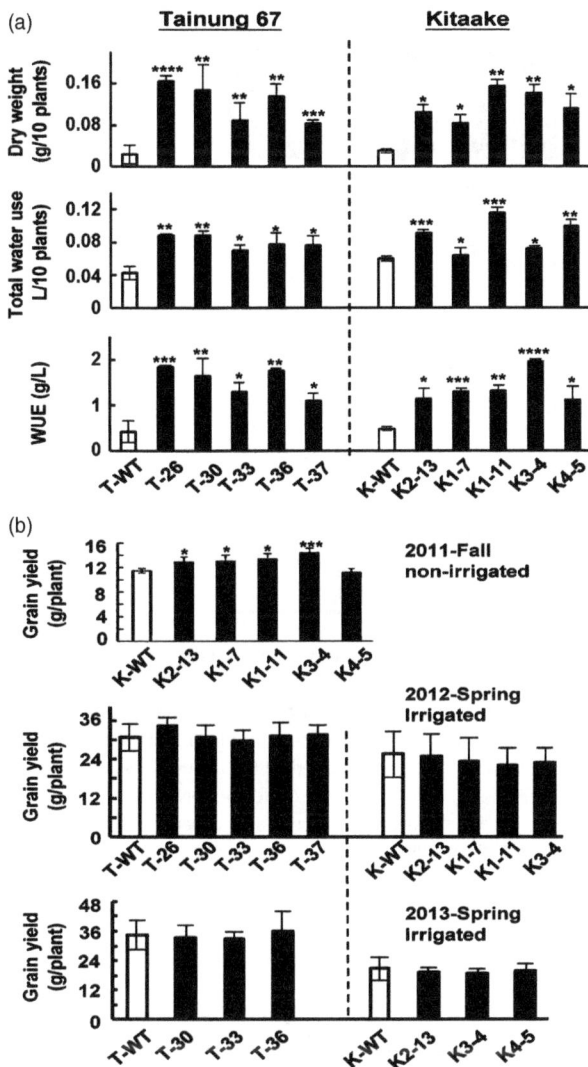

Figure 6 *3xABRC321:HVA1* promotes WUE in transgenic rice without apparent yield penalty. (a) Three-day-old seedlings of Tainung 67 (T) and Kitaake (K) transgenic plants were treated with two cycles of 250 mM sorbitol (3 days) and water (5 days) to induce root growth. Seedlings were then grown in Yoshida solution and transferred to the same amount of fresh solution every 2 days. Total water consumption was measured up to 19 days. Whole-plant dry weights, with two sets of same numbers ($n = 12$) of plants, before and after 19 days were measured. Upper panel: increased plant dry weights. Middle panel: total water use. Lower panel: WUE determined by dividing total increased plant dry weight by total water use. Error bars represent SD. (b) Tainung 67 (T) and Kitaake transgenic lines were grown in nonirrigated fields in the fall (August to December) of 2011 and in irrigated fields in the spring (March to July) of 2012 and 2013. Grain yield was determined after harvest. Error bars represent SD ($n = 10$). Significance levels with the t-test: *$P < 0.05$, **$P < 0.01$, ***$P < 0.001$, ****$P < 0.0001$.

2010). In rice, the effect of ABA on root growth appears to be concentration dependent. A low concentration of ABA, 0.1 μM, has been shown to induce root development with enhanced PR elongation and higher root hair density in rice and Arabidopsis (Xu et al., 2013). We also found that 0.1 μM ABA enhanced PR elongation and LR growth in both WT and HVA1 transgenic line (Figure 2). Interestingly, 0.2 μM ABA repressed PR elongation and LR growth in WT but enhanced those in HVA1 transgenic line

(Figure 4 and compare Figure 8a with Figure 8b). ABA concentrations higher than 0.5 μM inhibited crown root and LR growth, but less inhibitory to seminal root elongation, and enhanced root thickness in both WT and HVA1 transgenic rice (Figure 2). Hence, ABA treatment may mimic low water potentials or osmotic stress in dehydrated soils, with low concentrations inducing entire root system for increasing root surface for more efficient water uptake, while high concentrations repressing branched roots and promoting root thickness and thus facilitating downwards root growth.

The growth of LRs and PRs was increased in transgenic rice without any hormone or stress treatment (Figure 2). Neither typical auxin response element is found in *ABRC321* nor treatment of 2,4-D or IAA activates *3xABRC321:GUS* expression in transgenic rice. It is likely that the endogenous ABA activates *3xABRC321* to confer low level of HVA1 expression in roots to initiate LRP and PR growth. Exogenous application of low concentrations of ABA or sorbitol elevated the expression of *3xABRC321:HVA1*, leading to more significant increases in root growth.

For a long time, LEAs have represented an enigma in plant biology. The lack of similarity to other proteins of known functions and their high structural flexibility has hampered the research progress regarding the mechanism by which LEAs impact plant growth under abiotic stresses. LEA3 has been shown to be necessary for lateral root growth in Arabidopsis (Salleh et al., 2012). In the present study, we observed that HVA1 promotes the accumulation of auxin in LRP and RAM in the presence of low concentrations of ABA. The auxin-responsive reporter *DR5:GUS* revealed that the asymmetrical accumulation of the auxin maxima at the founder cells of LRP and in RAM was suppressed by 0.2 μM ABA in transgenic line *DR5:GUS* but enhanced in line *HVA1 DR5:GUS* (Figure 9). This result is consistent with the induction of PR and LR growth by 0.2 μM ABA in HVA1 transgenic lines but repression in WT (Figure 8b). The expression of *3xABRC321:GUS* and *DR5:GUS* was partially co-localized in tissues essential for LR and PR growth (Figures 7 and 9), indicating that the ABA-induced *3xABRC321:HVA1* expression plays a positive role in determining the positioning and frequency of LR initiation by altering the amount and/or transport direction of auxin.

The asymmetric accumulation of auxin during plant development could be regulated at the level of biosynthesis, degradation, metabolism or transport (Petrasek and Friml, 2009). PINs are localized in the plasma membrane (Benkova et al., 2003). As aforementioned, HVA1/LEA3 may stabilize proteins or maintain the functional membrane structure under abiotic stresses. Consequently, HVA1 may stabilize PINs in the presence of ABA or under abiotic stress. Alternatively, HVA1 may modulate the auxin homeostasis, for example biosynthesis, metabolism and transport, by stabling important proteins involved in these processes. This notion is supported by the result showing that the accumulation of *PIN1b*, *PIN2* and *PIN9* mRNAs was higher in transgenic roots than in WT, and levels of all of these mRNAs were further enhanced by ABA (Figure 8e).

It is interesting to note that *LEA3*, *3xABRC321:HVA1* and *3xABRC321:GUS* were all expressed in cortical and exodermal cells overlaying the developing LRP (Figure 7b,c,d), indicating that a local inductive signal was transmitted to cells involved in the early commitment stage of LR development. These cells seem to precisely guide and facilitate the LR to push through several cell layers and emerge out of the root surface. These studies

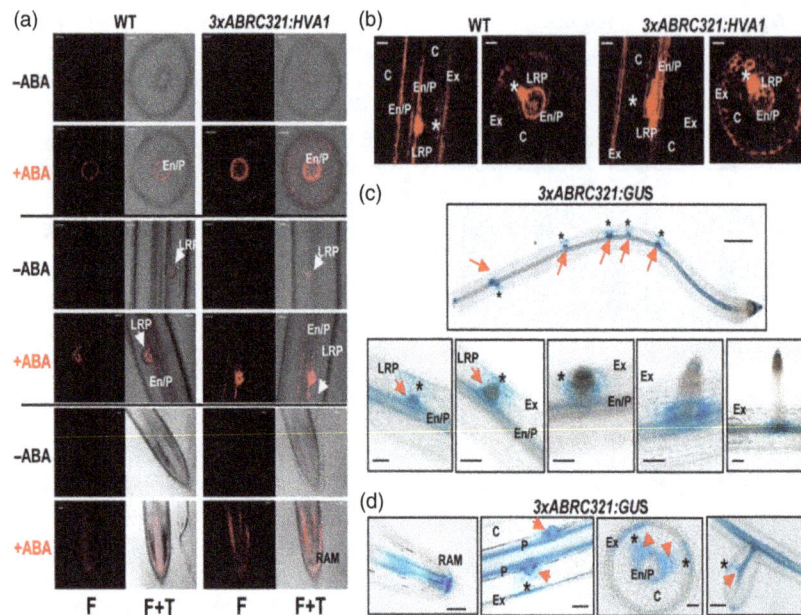

Figure 7 *3xABRC321:HVA1* has similar tissue-specific and ABA-inducible expression patterns to the endogenous *Lea3* in transgenic rice roots. (a) Ten-day-old seedlings of *3xABRC321:HVA1* transgenic line (T-33) were treated with or without 10 μM ABA for 16 h, and roots were sectioned for immunohistological fluorescence assay of LEA3 and HVA1. F: image of fluorescence field. F+T: composite images of fluorescence and transmission fields. For quantitative data of image intensity, see also Figure S4a. (b) Fluorescence images of root longitudinal and cross sections of WT and transgenic lines as prepared in (a) showing expression of LEA3 and HVA1 in cortex and exodermal cells overlaying the LRP as indicated by the asterisk. The image of fluorescence has been enhanced. For original images, please see also Figure S5. (c) Three-day-old seedlings of transgenic lines were treated with 0.2 μM ABA for 11 days, and roots were stained for GUS expression. (d) Roots in (c) were sectioned and stained for GUS expression. Asterisk indicates that GUS was also expressed in cortex and exodermal cells overlaying the LRP in (c) and (d). Arrows indicate LRP. Scale bars: 100 μm. Abbreviation: C: cortex; En: endodermis; Ex: exodermis; P, pericycle.

demonstrated the multiple roles of HVA1 and ABA in promoting LR initiation, elongation and emergence, likely through distinct regulatory pathways.

In summary, our studies demonstrate that *3xABRC321* is an ideal promoter for introducing HVA1 into breeding programs to improve cereals and possibly dicot crops for improvement of tolerance to a wide array of abiotic stresses without an apparent yield penalty. Future studies on mechanisms that regulate the crosstalk between ABA and auxin signals and the function of HVA1/LEA3 in regulating auxin accumulation and/or transport and root growth should facilitate engineering of root architecture for sustainable crop production under stressful environments.

Experimental procedures

Plant materials

Rice cultivars *Oryza sativa* L. cv Tainung 67 and *Oryza sativa* L. cv Kitaake were used for all experiments. Plasmid was introduced into *Agrobacterium tumefaciens* strain EHA101, and rice transformation was performed as described (Chen *et al.*, 2002). Homozygous transgenic lines were used in all experiments. For observation of root growth, transgenic rice seeds were germinated on the surface of half-strength MS medium without sugar but with or without ABA or sorbitol. For stress treatment, transgenic seeds were germinated in water for 5 days and transferred to soil unless otherwise indicated. For hydroponic culture, Yoshida solution was used. Seedlings were normally grown in 28 °C incubator with 12-h daily light.

Plasmids

Plasmid pAHC18 contains the luciferase (*Luc*) cDNA fused between the *Ubi* promoter and the *Nos* terminator (Bruce *et al.*, 1989). Plasmid MP64 contains the barley *Amy64* minimal promoter (−60 relative to the transcription start site) and its 5′ untranslated region (+57 relative to the transcription start site), *HVA22* intron1–exon2–intron2, the *GUS* coding region and the *HVA22* 3′ untranslated region (Shen and Ho, 1995). Plasmid QS115 contains a copy of *HVA22* ABRC1 fused upstream of the *Amy64* minimal promoter in plasmid MP64 (Shen and Ho, 1995).

Plasmid construction

Two 56-bp complementary oligonucleotides containing the CE3 and A2 elements from the *HVA1* promoter and the CE1 element from the *HVA22* promoter (Shen *et al.*, 1996) and restriction sites *Kpn*I, *Xho*I and *Xba*I were synthesized, annealed and designated as *ABRC321* (Figure 1a). *ABRC321* was self-ligated in two or three copies in correct orientations. For expression of GUS under the control of *ABRC321*, 1–3 copies of *ABRC321* were inserted into the *Xba*I site in MP64, so that the *ABRC321* was fused upstream of the barley *Amy64* minimal promoter, generating constructs *1xABRC321-GUS*, *2xABRC321-GUS* and *3xABRC321-GUS* (Figure S1a). For expression of *HVA1* under the control of *3xABRC321*, *GUS* cDNA in construct *3xABRC321-GUS* was replaced with *HVA1* cDNA, generating construct *3xABRC321-HVA1* (Figure S1b).

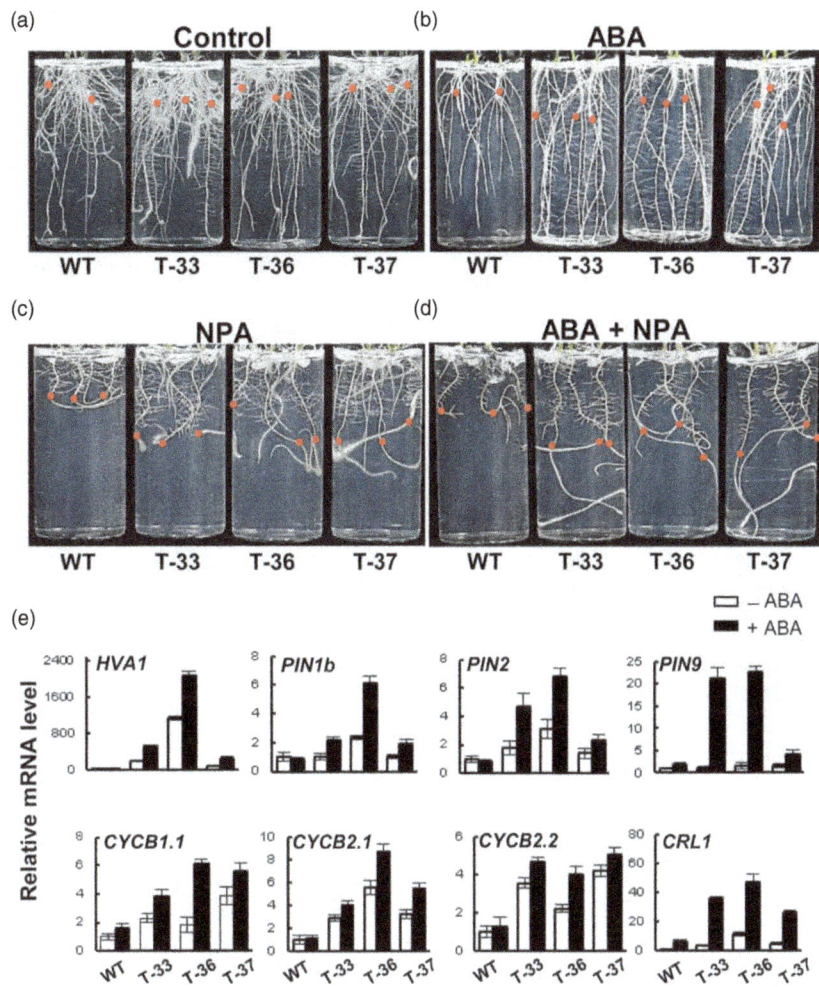

Figure 8 *3xABRC321:HVA1* and ABA promote root growth in transgenic rice through an auxin-dependent process. Three-day-old seedlings of *3xABRC321:HVA1* transgenic line (T-33) were transferred to MS medium containing ABA and/or NPA for 11 days, and root morphology was examined. (a) Medium only, (b) 0.2 μM ABA only, (c) 1 μM NPA only and (d) 0.2 μM ABA plus 1 μM NPA. Red dots indicate starting points of root growth after transferring to medium containing ABA and/or NPA. (e) mRNAs from roots of seedlings prepared in (b) were subjected to quantitative real-time RT-PCR analysis of indicated genes.

GUS activity staining

Sections of leaf and root from 10-day-old seedlings were cut with Microslicers DTK-1000 (TED PELLA, Inc., Redding, CA), incubated in water containing or lacking 10 μM ABA at 28 °C for 24 h and subjected to histochemical staining with a buffer (0.1 M NaPO$_4$, pH 7.0, 10 mM EDTA, 0.1% Triton X-100, 0.5 mM potassium ferricyanide, pH 7.0, and 1 mM X-glucuronide) at 37 °C as described (Jefferson, 1987). After GUS staining, leave samples were incubated in 70% ethanol at 65 °C for 1 h to remove chlorophyll.

Immunohistological fluorescence staining

Tissue localization of HVA1 was examined by modification of a described method (Long *et al.*, 2006). Rice roots were fixed with 2% paraformaldehyde (w/v) in 0.1 M NaPO$_4$ buffer, pH 7.0, and then embedded in 5% agar. Sections were sliced to 30 μm thickness using Microslicers DTK-1000 and incubated in PBS buffer containing 0.3% (v/v) Triton X-100 (PBS-T). The nonspecific reaction was blocked with 5% (w/v) bovine serum albumin in PBS-T. Samples were then incubated with purified rabbit antibarley HVA1 polyclonal antibodies and subsequently with secondary antibodies (Alexa Fluor 555 goat anti-rabbit IgG; Molecular Probes). Samples were examined with a laser scanning confocal microscope (LSM510 META; Zeiss).

Measurement of roots

Rice branched roots could be classified into three types (Supplementary Fig. 3a): seminal roots, also called radical, are the first root emerging during germination; crown roots emerge from the node of coleoptile; and lateral roots branch from the above two types of roots and can bear additional large or small lateral roots until the fifth order of branching. The number of each type of roots was measured by simple counting. Root length was measured, the number of roots was counted, and lateral root density was determined by dividing lateral root number by root length. Error bars represent SD ($n = 12$). Significance levels with the *t*-test are as follows: * $P < 0.05$, ** $P < 0.01$, *** $P < 0.001$.

Dehydration of seedlings

For dehydration tests, 10-day-old seedlings were planted in small pots containing 330 g of soil, grown in a 28 °C chamber and slowly air-dried for 6 days. Pots were then transferred to a bigger container and rewater until soil was completely submerged in water.

Field trial

To evaluate grain yield and biomass production in the field, 25-day-old seedlings were transplanted to a field with 25 × 25 cm of space between each plant. The irrigated field was flooded with

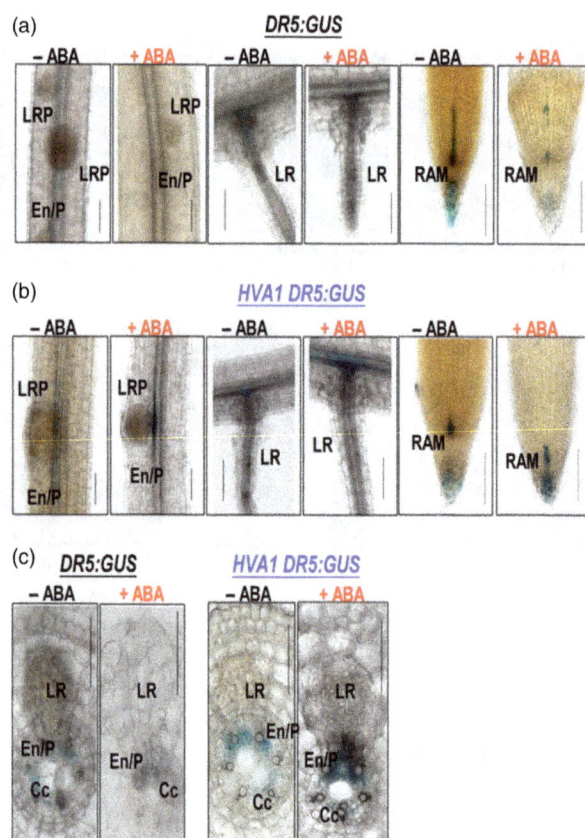

Figure 9 *3xABRC321:HVA1* and ABA enhance asymmetrical localization of auxin in LRP founder cells and auxin accumulation in RAM in transgenic rice. Three-day-old seedlings of transgenic lines *DR5:GUS* and *HVA1 DR5: GUS* were transferred to MS medium with (+) or without (−) 0.2 μm ABA for 11 days, and roots were stained for GUS expression in LR founder cells and RAM, treated with ethanol and examined under a microscope. (a) GUS expression in line *DR5:GUS* was suppressed by ABA. (b) GUS expression in line *HVA1 DR5:GUS* was enhanced or maintained with ABA treatment. For more images of LR founder cells and RAM stained with GUS, see also Figures S6 and S7. (c) Asymmetrical localization of GUS was suppressed in line *DR5:GUS* but enhanced or maintained in line *HVA1 DR5:GUS* with ABA treatment. Scale bar: 100 μm. For images of entire root cross sections, see also Figure S8. Abbreviation: C: cortex; Cc: companion cells; En: endodermis; Ex: exodermis; P, pericycle.

1–5 cm of water (soil water content 37%, v/v) until the end of active tillering stage (30–40 days after transplanting). Water was then drained (soil water content 27%, v/v) for 10–15 days at late tillering stage and flooded again with 3–10 cm of water until the milky stage. Soil in the nonirrigated field was kept moist (soil water content 20~25%, v/v) by intermittent irrigation during the entire planting period. Soil water content was measured using a Theta probe and meter (models ML2x and HH1, Delta-T devices, Cambridge, UK) (Ji *et al.*, 2012). Seeds were harvested, dried and yield determined.

Primer

Nucleotide sequences of primers are listed in Table S1.

Acknowledgements

We thank AndreAna Pena for critical reviewing of the manuscript and Kuo-Wei Lee, Chun-Hsien Lu, Lin-Chih Yu, Ding-Hua Lee and Fang-Min Wu for technical assistance. This work was supported by grants (94F005-1 and 94S1502 to Su-May Yu and 94F005-2 to Tuan-Hua David Ho) from Academia Sinica and NSC101-2321-B001-035 (to Su-May Yu) and NSC102-2321-B-001-048- from the National Science Council of the Republic of China and in part by the Ministry of Education, R.O.C. under the ATU plan to Su-May Yu and Tuan-Hua David Ho.

References

Ahuja, I., de Vos, R.C., Bones, A.M. and Hall, R.D. (2010) Plant molecular stress responses face climate change. *Trends Plant Sci.* **15**, 664–674.

Babu, R.C., Zhang, J., Blum, A., Ho, D.T.-H., Wu, R. and Nguyen, H.T. (2004) HVA, a LEA gene from Barly confers dehydration tolerance in transgenic rice (Oryza sativa L.) via cell membrane protection. *Plant Sci.* **166**, 855–862.

Baena-Gonzalez, E., Rolland, F., Thevelein, J.M. and Sheen, J. (2007) A central integrator of transcription networks in plant stress and energy signalling. *Nature* **448**, 938–942.

Bartels, D. (2001) Targeting detoxification pathways: an efficient approach to obtain plants with multiple stress tolerance? *Trends Plant Sci.* **6**, 284–286.

Battaglia, M. and Covarrubias, A.A. (2013) Late Embryogenesis Abundant (LEA) proteins in legumes. *Front. Plant Sci.* **4**, 190.

Benkova, E., Michniewicz, M., Sauer, M., Teichmann, T., Seifertova, D., Jurgens, G. and Friml, J. (2003) Local, efflux-dependent auxin gradients as a common module for plant organ formation. *Cell* **115**, 591–602.

Brady, S.M., Sarkar, S.F., Bonetta, D. and McCourt, P. (2003) The ABSCISIC ACID INSENSITIVE 3 (ABI3) gene is modulated by farnesylation and is involved in auxin signaling and lateral root development in Arabidopsis. *Plant J.* **34**, 67–75.

Bruce, W.B., Christensen, A.H., Klein, T., Fromm, M. and Quail, P.H. (1989) Photoregulation of a phytochrome gene promoter from oat transferred into rice by particle bombardment. *Proc. Natl Acad. Sci. USA* **86**, 9692–9696.

Chen, P.-W., Lu, C.-A., Yu, T.-S., Tseng, T.-H., Wang, C.-S. and Yu, S.-M. (2002) Rice alpha-amylase transcriptional enhancers direct multiple mode regulation of promoters in transgenic rice. *J. Biol. Chem.* **277**, 13641–13649.

Coudert, Y., Perin, C., Courtois, B., Khong, N.G. and Gantet, P. (2010) Genetic control of root development in rice, the model cereal. *Trends Plant Sci.* **15**, 219–226.

De Smet, I., Zhang, H., Inze, D. and Beeckman, T. (2006) A novel role for abscisic acid emerges from underground. *Trends Plant Sci.* **11**, 434–439.

Dubouzet, J.G., Sakuma, Y., Ito, Y., Kasuga, M., Dubouzet, E.G., Miura, S., Seki, M., Shinozaki, K. and Yamaguchi-Shinozaki, K. (2003) OsDREB genes in rice, Oryza sativa L., encode transcription activators that function in drought-, high-salt- and cold-responsive gene expression. *Plant J* **33**, 751–763.

Dure, L. (1992) The LEA proteins of higher plants. In *Control of Plant Gene Expression*, (Verma, D.P.S., ed), pp. 325–335. Boca Raton, FL: CRC Press.

Dure, L. 3rd. (1993) A repeating 11-mer amino acid motif and plant desiccation. *Plant J.* **3**, 363–369.

Fu, D., Huang, B., Xiao, Y., Muthukrishnan, S. and Liang, G.H. (2007) Overexpression of barley hva1 gene in creeping bentgrass for improving drought tolerance. *Plant Cell Rep.* **26**, 467–477.

Garg, A.K., Kim, J.K., Owens, T.G., Ranwala, A.P., Choi, Y.D., Kochian, L.V. and Wu, R.J. (2002) Trehalose accumulation in rice plants confers high tolerance levels to different abiotic stresses. *Proc. Natl Acad. Sci. USA* **99**, 15898–15903.

Himanen, K., Boucheron, E., Vanneste, S., de Almeida Engler, J., Inze, D. and Beeckman, T. (2002) Auxin-mediated cell cycle activation during early lateral root initiation. *Plant Cell* **14**, 2339–2351.

Ho, T.-H.D. and Wu, R. (2004) Genetic engineering for enhancing plant productivity and stress tolerance. In *Physiology and Biotechology Integration for Plant Breeding*, (Nguyen, H.T. and Blum, A., eds), pp. 489–502. New York, NY: Marcel Dekker Inc.

Hong, B., Barg, R. and Ho, T.-H. (1992) Developmental and organ-specific expression of an ABA- and stress-induced protein in barley. *Plant Mol. Biol.* **18**, 663–674.

Hsieh, T.H., Lee, J.T., Charng, Y.Y. and Chan, M.T. (2002) Tomato plants ectopically expressing Arabidopsis CBF1 show enhanced resistance to water deficit stress. *Plant Physiol.* **130**, 618–626.

Inukai, Y., Sakamoto, T., Ueguchi-Tanaka, M., Shibata, Y., Gomi, K., Umemura, I., Hasegawa, Y., Ashikari, M., Kitano, H. and Matsuoka, M. (2005) Crown rootless1, which is essential for crown root formation in rice, is a target of an AUXIN RESPONSE FACTOR in auxin signaling. *Plant Cell* **17**, 1387–1396.

Jefferson, R.A. (1987) Assaying chimeric genes in plants from gene fusion system. *Plant Mol. Biol. Report* **5**, 387–405.

Ji, K., Wang, Y., Sun, W., Lou, Q., Mei, H., Shen, S. and Chen, H. (2012) Drought-responsive mechanisms in rice genotypes with contrasting drought tolerance during reproductive stage. *J. Plant Physiol.* **169**, 336–344.

Kasuga, M., Liu, Q., Miura, S., Yamaguchi-Shinozaki, K. and Shinozaki, K. (1999) Improving plant drought, salt, and freezing tolerance by gene transfer of a single stress-inducible transcription factor. *Nat. Biotechnol.* **17**, 287–291.

Lal, S., Gulyani, V. and Khurana, P. (2008) Overexpression of HVA1 gene from barley generates tolerance to salinity and water stress in transgenic mulberry (Morus indica). *Transgenic Res.* **17**, 651–663.

Lavenus, J., Goh, T., Roberts, I., Guyomarc'h, S., Lucas, M., De Smet, I., Fukaki, H., Beeckman, T., Bennett, M. and Laplaze, L. (2013) Lateral root development in Arabidopsis: fifty shades of auxin. *Trends Plant Sci.* **18**, 450–458.

Lee, J.T., Prasad, V., Yang, P.T., Wu, J.F., Ho, T.H.D., Charng, Y.Y. and Chan, M.T. (2003) Expression of Arabidopsis CBF1 regulated by an ABA/stress inducible promoter in transgenic tomato confers stress tolerance without affecting yield. *Plant, Cell Environ.* **26**, 1181–1190.

Liu, Y., Zheng, Y., Zhang, Y., Wang, W. and Li, R. (2010) Soybean PM2 protein (LEA3) confers the tolerance of Escherichia coli and stabilization of enzyme activity under diverse stresses. *Curr. Microbiol.* **60**, 373–378.

Long, S.P., Zhu, X.G., Naidu, S.L. and Ort, D.R. (2006) Can improvement in photosynthesis increase crop yields? *Plant, Cell Environ.* **29**, 315–330.

Lopez-Bucio, J., Cruz-Ramirez, A. and Herrera-Estrella, L. (2003) The role of nutrient availability in regulating root architecture. *Curr. Opin. Plant Biol.* **6**, 280–287.

Lucas, J. (2003) *World food security at risk as crop yields plateau.* http://www.trust.org/item/20131119230754-aynqh/?source=spotlight.

Malamy, J.E. (2005) Intrinsic and environmental response pathways that regulate root system architecture. *Plant, Cell Environ.* **28**, 67–77.

Marunde, M.R., Samarajeewa, D.A., Anderson, J., Li, S., Hand, S.C. and Menze, M.A. (2013) Improved tolerance to salt and water stress in Drosophila melanogaster cells conferred by late embryogenesis abundant protein. *J. Insect Physiol.* **59**, 377–386.

Moons, A., De Keyser, A. and Van Montagu, M. (1997) A group 3 LEA cDNA of rice, responsive to abscisic acid, but not to jasmonic acid, shows variety-specific differences in salt stress response. *Gene* **191**, 197–204.

Morran, S., Eini, O., Pyvovarenko, T., Parent, B., Singh, R., Ismagul, A., Eliby, S., Shirley, N., Langridge, P. and Lopato, S. (2011) Improvement of stress tolerance of wheat and barley by modulation of expression of DREB/CBF factors. *Plant Biotechnol. J.* **9**, 230–249.

Negrao, S., Courtois, B., Ahmadi, N., Abreu, I., Saibo, N. and Oliveira, M. (2011) Recent updates on salinity stress in rice: from physiological to molecular responses. *Crit. Rev. Plant Sci.* **30**, 329–377.

Nguyen, T. and Sticklen, M.. (2013) Barley HVA1 gene confers drought and salt tolerance in transgenic maize (Zea Mays L.). *Adv. Crop. Sci. Tech.* **1**, 105. doi:10.4172/acst.1000105.

Olvera-Carrillo, Y., Luis Reyes, J. and Covarrubias, A.A. (2011) Late embryogenesis abundant proteins: versatile players in the plant adaptation to water limiting environments. *Plant Signal. Behav.* **6**, 586–589.

Orman-Ligeza, B., Parizot, B., Gantet, P.P., Beeckman, T., Bennett, M.J. and Draye, X. (2013) Post-embryonic root organogenesis in cereals: branching out from model plants. *Trends Plant Sci.* **18**, 459–467.

Peret, B., De Rybel, B., Casimiro, I., Benkova, E., Swarup, R., Laplaze, L., Beeckman, T. and Bennett, M.J. (2009) Arabidopsis lateral root development: an emerging story. *Trends Plant Sci.* **14**, 399–408.

Petrasek, J. and Friml, J. (2009) Auxin transport routes in plant development. *Development* **136**, 2675–2688.

Potters, G., Pasternak, T.P., Guisez, Y., Palme, K.J. and Jansen, M.A. (2007) Stress-induced morphogenic responses: growing out of trouble? *Trends Plant Sci.* **12**, 98–105.

Price, A.H., Tomos, A.D. and Virk, D.S. (1997) Genetic dissection of root growth in rice (Oryza sativa L.).1. a hydrophonic screen. *Theor. Appl. Genet.* **95**, 132–142.

Reed, R.C., Brady, S.R. and Muday, G.K. (1998) Inhibition of auxin movement from the shoot into the root inhibits lateral root development in Arabidopsis. *Plant Physiol.* **118**, 1369–1378.

Salleh, F.M., Evans, K., Goodall, B., Machin, H., Mowla, S.B., Mur, L.A., Runions, J., Theodoulou, F.L., Foyer, C.H. and Rogers, H.J. (2012) A novel function for a redox-related LEA protein (SAG21/AtLEA5) in root development and biotic stress responses. *Plant, Cell Environ.* **35**, 418–429.

Shen, Q. and Ho, T.H. (1995) Functional dissection of an abscisic acid (ABA)-inducible gene reveals two independent ABA-responsive complexes each containing a G-box and a novel cis-acting element. *Plant Cell* **7**, 295–307.

Shen, Q., Uknes, S.J. and Ho, T.H. (1993) Hormone response complex in a novel abscisic acid and cycloheximide-inducible barley gene. *J. Biol. Chem.* **268**, 23652–23660.

Shen, Q., Zhang, P. and Ho, T.H. (1996) Modular nature of abscisic acid (ABA) response complexes: composite promoter units that are necessary and sufficient for ABA induction of gene expression in barley. *Plant Cell* **8**, 1107–1119.

Sthapit, B. and Witcombe, J. (1998) Inheritance of tolerance to chilling stress in rice during germination and plumule greening. *Crop Sci.* **38**, 660–665.

Tuong, T.P. and Bouman, B.A.M. (2003) Rice production in water-scarce environments. In *Water Productivity in Agriculture: Limits and Opportunities for Improvement* (Kijne, J.W., Barker, R. and Molden, D., eds), pp. 53–67. UK: CABI Publishing.

Uga, Y., Sugimoto, K., Ogawa, S., Rane, J., Ishitani, M., Hara, N., Kitomi, Y., Inukai, Y., Ono, K., Kanno, N., Inoue, H., Takehisa, H., Motoyama, R., Nagamura, Y., Wu, J., Matsumoto, T., Takai, T., Okuno, K. and Yano, M. (2013) Control of root system architecture by DEEPER ROOTING 1 increases rice yield under drought conditions. *Nat. Genet.* **45**, 1097–1102.

Ulmasov, T., Murfett, J., Hagen, G. and Guilfoyle, T.J. (1997) Aux/IAA proteins repress expression of reporter genes containing natural and highly active synthetic auxin response elements. *Plant Cell* **9**, 1963–1971.

Xu, D., Duan, X., Wang, B., Hong, B., Ho, T. and Wu, R. (1996) Expression of a Late Embryogenesis Abundant Protein Gene, HVA1, from Barley Confers Tolerance to Water Deficit and Salt Stress in Transgenic Rice. *Plant Physiol.* **110**, 249–257.

Xu, W., Jia, L., Shi, W., Liang, J., Zhou, F., Li, Q. and Zhang, J. (2013) Abscisic acid accumulation modulates auxin transport in the root tip to enhance proton secretion for maintaining root growth under moderate water stress. *New Phytol.* **197**, 139–150.

Yamaguchi, M. and Sharp, R.E. (2010) Complexity and coordination of root growth at low water potentials: recent advances from transcriptomic and proteomic analyses. *Plant, Cell Environ.* **33**, 590–603.

Yamaguchi-Shinozaki, K. and Shinozaki, K. (2006) Transcriptional Regulatory Networks in Cellular Responses and Tolerance to Dehydration and Cold Stresses. *Annu. Rev. Plant Biol.* **57**, 781–803.

3

Agronomic nitrogen-use efficiency of rice can be increased by driving *OsNRT2.1* expression with the *OsNAR2.1* promoter

Jingguang Chen[1,2], Yong Zhang[1,2], Yawen Tan[1,2], Min Zhang[2], Longlong Zhu[1], Guohua Xu[1,2] and Xiaorong Fan[1,2,*]

[1]*State Key Laboratory of Crop Genetics and Germplasm Enhancement, Ministry of Agriculture, Nanjing Agricultural University, Nanjing, China*
[2]*Key Laboratory of Plant Nutrition and Fertilization in Low-Middle Reaches of the Yangtze River, Ministry of Agriculture, Nanjing Agricultural University, Nanjing, China*

*Correspondence

email xiaorongfan@njau.edu.cn

Summary

The importance of the nitrate (NO_3^-) transporter for yield and nitrogen-use efficiency (NUE) in rice was previously demonstrated using map-based cloning. In this study, we enhanced the expression of the *OsNRT2.1* gene, which encodes a high-affinity NO_3^- transporter, using a ubiquitin (*Ubi*) promoter and the NO_3^--inducible promoter of the *OsNAR2.1* gene to drive *OsNRT2.1* expression in transgenic rice plants. Transgenic lines expressing *pUbi:OsNRT2.1* or *pOsNAR2.1:OsNRT2.1* constructs exhibited the increased total biomass including yields of approximately 21% and 38% compared with wild-type (WT) plants. The agricultural NUE (ANUE) of the *pUbi:OsNRT2.1* lines decreased to 83% of that of the WT plants, while the ANUE of the *pOsNAR2.1:OsNRT2.1* lines increased to 128% of that of the WT plants. The dry matter transfer into grain decreased by 68% in the *pUbi:OsNRT2.1* lines and increased by 46% in the *pOsNAR2.1:OsNRT2.1* lines relative to the WT. The expression of *OsNRT2.1* in shoot and grain showed that *Ubi* enhanced *OsNRT2.1* expression by 7.5-fold averagely and *OsNAR2.1* promoters increased by about 80% higher than the WT. Interestingly, we found that the *OsNAR2.1* was expressed higher in all the organs of *pUbi:OsNRT2.1* lines; however, for *pOsNAR2.1:OsNRT2.1* lines, *OsNAR2.1* expression was only increased in root, leaf sheaths and internodes. We show that increased expression of *OsNRT2.1*, especially driven by *OsNAR2.1* promoter, can improve the yield and NUE in rice.

Keywords: *OsNAR2.1* promoter, *Oryza sativa*, *OsNRT2.1*, agronomic nitrogen-use efficiency.

Introduction

Rice (*Oryza sativa* L.) is not only a major staple food crop for a large part of the world population but also an important model monocot plant species for research because of its small genome size and the availability of the complete rice genome sequence (Feng *et al.*, 2002; Sasaki *et al.*, 2002). Nitrogen (N) nutrition affects all levels of plant function from metabolism to resource allocation, growth and development (Crawford, 1995; Scheible *et al.*, 1997, 2004; Stitt, 1999). The most abundant source for N acquisition by plant roots is nitrate (NO_3^-), which is present in naturally aerobic soils due to intensive nitrification from applied organic and fertilizer N. In contrast, ammonium (NH_4^+) is the main form of available N in flooded paddy soils due to the anaerobic soil conditions (Sasakawa and Yamamoto, 1978).

NO_3^- serves as a nutrient and as a signal that induces changes in the growth and gene expression (Coruzzi and Bush, 2001; Coruzzi and Zhou, 2001; Crawford and Forde, 2002; Crawford and Glass, 1998; Kirk and Kronzucker, 2005; Kronzucker *et al.*, 2000; Wang *et al.*, 2000; Zhang and Forde, 2000). Two different NO_3^- uptake systems in plants, the high- and low-affinity NO_3^- uptake systems designated as HATS and LATS, respectively, are regulated by NO_3^- supply and enable plants to cope with high or low NO_3^- concentrations in soils (Fan *et al.*, 2005).

Some high-affinity NO_3^- transporters belonging to the NRT2 family have been shown to require a partner protein, NAR2, for their function (Xu *et al.*, 2012). Quesada *et al.* (1994) identified the *CrNar2* gene, which encodes a small protein of approx-

imately 200 amino acid residues and which has no known transport activity, but is required for complementation of NO_3^- transport in *Chlamydomonas reinhardtii* mutants defective in uptake. In *Arabidopsis*, Okamoto *et al.* (2006) showed that both constitutive and NO_3^--inducible HATS, but not LATS, depended on the expression of the NAR2-type gene, for example *Arabidopsis AtNRT3.1*. Orsel *et al.* (2006) used yeast split-ubiquitin and oocyte expression systems to show that *AtNAR2.1* (*AtNRT3.1*) and *AtNRT2.1* interacted to produce a functional HATS. Yong *et al.* (2010) showed that the NRT2.1 and NAR2.1 polypeptides interact directly at the plasma membrane to constitute an oligomer that may act as the functional unit for high-affinity NO_3^- influx in *Arabidopsis* roots. In rice, the *OsNRT2.1*, *OsNRT2.2* and *OsNRT2.3a* gene products were similarly shown to require the protein encoded by OsNAR2.1 for NO_3^- uptake (Feng *et al.*, 2011; Liu *et al.*, 2014; Yan *et al.*, 2011), and their interaction at the protein level was demonstrated using a yeast two-hybrid assay and by Western blotting (Liu *et al.*, 2014; Yan *et al.*, 2011).

Rice seedling growth was improved slightly by increased *OsNRT2.1* expression, but N uptake remained unaffected (Katayama *et al.*, 2009), probably due to the absence of the interaction with OsNAR2.1, which is required for the functional NO_3^- transport (Feng *et al.*, 2011; Yan *et al.*, 2011).

In this study, we transformed the open reading frame (ORF) of the *OsNRT2.1* gene into rice with the expression driven by the *OsNAR2.1* promoter to modify the coexpression of the *OsNRT2.1* and *OsNAR2.1* genes in rice plants and to investigate the biological function of their coexpression *in vivo*. Transgenic lines expressing the *OsNRT2.1* gene under the control of the *OsNAR2.1* promoter exhibited greatly increased the growth, biomass and yield compared with transgenic lines expressing *OsNRT2.1* under a ubiquitin promoter. We analysed *OsNRT2.1* and *OsNAR2.1* expression patterns during the whole-plant growth and show that modification of the ratio of *OsNRT2.1* to *OsNAR2.1* expression in stems altered the rice growth and agricultural N-use efficiency (ANUE).

Results

Generation of transgenic rice plants expressing *pUbi:OsNRT2.1* and *pOsNAR2.1:OsNRT2.1* constructs and field analysis of traits

The ubiquitin promoter (pUbi) has been used as a strong promoter in a variety of applications in gene transfer studies and was shown to drive gene expression most actively in rapidly dividing cells (Cornejo *et al.*, 1993). Overexpression of just the *OsNRT2.1* gene in rice was previously shown to not increase NO_3^- uptake (Katayama *et al.*, 2009).

We introduced *pUbi:OsNRT2.1* (Figure S1a) and *pOsNAR2.1:OsNRT2.1* (Figure S1b) expression constructs into Wuyunjing 7 (WYJ7), a rice cultivar that produces high yields in Jiangsu Province, using *Agrobacterium tumefaciens*-mediated transformation. We generated 23 lines exhibiting increased *OsNRT2.1* expression, including 12 *pUbi:OsNRT2.1* lines and 11 *pOsNAR2.1:OsNRT2.1* lines (Figure S2).

We analysed the grain yield and biomass of transgenic lines in the T0 and T1 generations. Relative to the wild-type (WT) plants, the biomass, including the grain yield, of the 12 *pUbi:OsNRT2.1* lines increased by approximately 21.8% (Figure S2e) and 20.9% (Figure S3a) in T0 and T1 plants, respectively, but the grain yield decreased approximately 18.4% (Figure S2c) and 16.6% (Figure S3a) in T0 and T1 plants, respectively. Relative to the WT, the biomass, including the grain yield, of the 11 *pOsNAR2.1:OsNRT2.1* lines increased by average values of 32.2% (Figure S2f) and 27.1% (Figure S3b) in T0 and T1 plants, respectively, and the grain yield increased by average values of 30.7% (Figure S2d) and 28.1% (Figure S3b) in T0 and T1 plants, respectively. Based on the Southern blot analysis of T1 plants (Figure S4) and RNA expression data for the T0 generation (Figure S2a,b), we selected three independent *pUbi:OsNRT2.1* T1 lines OE1–2, OE2–5 and OE3–4 [renamed as OE1, OE2 and OE3 (Figure 1a)] and three independent *pOsNAR2.1:OsNRT2.1* T1 lines O6–4, O7–6 and O8–3 [renamed as O6, O7 and O8 (Figure 1b)].

Agricultural traits of these six lines were investigated in the field in the T1 through T4 generations, with a particular focus on the T3 generation. *OsNRT2.1* expression in roots was enhanced four- to sevenfold in the OE1, OE2 and OE3 lines but only 2.5- to threefold in the O6, O7 and O8 lines relative to the WT. In culms, *OsNRT2.1* expression was increased approximately sixfold in the OE lines and approximately threefold in the O lines. In leaf blades, however, only the OE lines exhibited increased *OsNRT2.1* expression (four- to sevenfold) compared with the WT, and no change in the expression was observed in the O lines (Figure 1c,

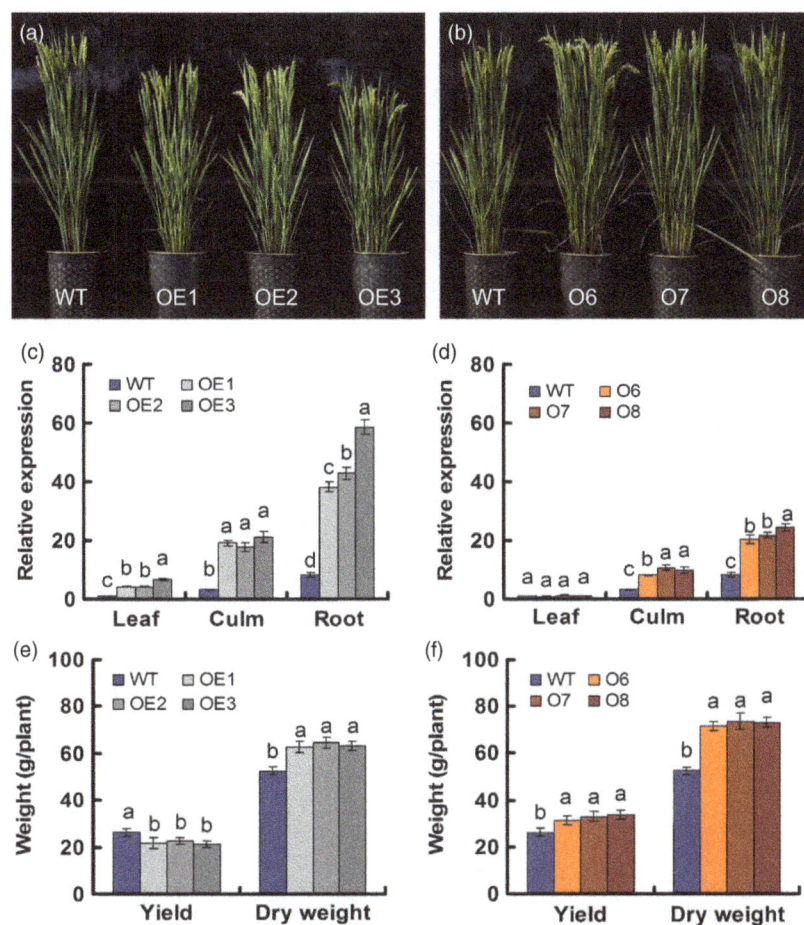

Figure 1 Characterization of transgenic lines. (a) Gross morphology of *pUbi:OsNRT2.1* transgenic lines (OE1, OE2 and OE3) and the wild-type (WT). (b) Gross morphology of *pOsNAR2.1:OsNRT2.1* transgenic lines (O6, O7 and O8) and the WT. (c, d) Real-time quantitative RT-PCR analysis of endogenous *OsNRT2.1* expression in various transgenic lines and WT plants. (c) *pUbi:OsNRT2.1* transgenic lines (OE1, OE2 and OE3) and the WT, and (d) *pOsNAR2.1:OsNRT2.1* transgenic lines (O6, O7 and O8) and the WT. RNA was extracted from leaf blade I, culm and root. (e, f) Grain yield and dry weight per plant for transgenic and WT plants grown in the field. Dry weight mean values are for all aboveground biomass, including the grain yield. (e) *pUbi:OsNRT2.1* transgenic lines and WT. (f) *pOsNAR2.1:OsNRT2.1* transgenic lines and WT. Statistical analysis was performed on data derived from the T3 generation. Error bars: SE (*n* = 3). Significant differences between the transgenic lines and WT are indicated by different letters (*P* < 0.05, one-way ANOVA).

Table 1 Comparison of the grain yield, dry weight and agronomic nitrogen-use efficiency (ANUE) between the wild-type (WT) and transgenic lines in the T1–T3 generations

		pUBi:OsNRT2.1			pOsNAR2.1:OsNRT2.1		
	WT	OE1	OE2	OE3	O6	O7	O8
Grain yield (kg/m^2)							
T1	0.52 b	0.42 c	0.44 c	0.43 c	0.69 a	0.69 a	0.71 a
T2	0.66 b	0.54 c	0.56 c	0.54 c	0.89 a	0.91 a	0.93 a
T3	0.70 b	0.58 c	0.60 c	0.57 c	0.94 a	0.98 a	1.00 a
Dry weight (kg/m^2)							
T1	1.05 c	1.31 b	1.29 b	1.31 b	1.43 a	1.45 a	1.47 a
T2	1.27 c	1.55 b	1.61 b	1.58 b	1.77 a	1.83 a	1.77 a
T3	1.40 c	1.67 b	1.73 b	1.69 b	1.91 a	1.96 a	1.95 a
ANUE (g/g)							
T1	15.48 b	12.43 c	12.94 c	12.56 c	19.64 a	19.63 a	19.86 a
T2	20.28 b	16.46 c	17.02 c	16.25 c	26.41 a	26.71 a	27.50 a
T3	21.33 b	17.42 c	18.41 c	17.01 c	26.17 a	27.86 a	28.12 a

Dry weight mean values are for all aboveground biomass, including the grain yield. For each mean, $n = 3$. Significant differences between the transgenic lines and WT are indicated by different letters ($P < 0.05$, one-way ANOVA).

d). The field data showed that both the OE and O lines exhibited increased growth and biomass, but only the O lines produced higher yields than the WT (Figure 1e,f).

Based on the agricultural traits of the T1–T4 generation plants in the field, the total aboveground biomass including the grain yield increased by 21% for the *pUbi:OsNRT2.1* lines and by 38% for the *pOsNAR2.1:OsNRT2.1* lines, while the biomass without grain yield increased by 190% for the *pUbi:OsNRT2.1* lines and by 160% for the *pOsNAR2.1:OsNRT2.1* lines. The grain yields of the *pUbi:OsNRT2.1* lines decreased over the three successive generations (Table 1), but the yields of the *pOsNAR2.1:OsNRT2.1* lines increased significantly from the T1 to T3 generation (Table 1). The yields of the O lines were enhanced by approximately 33% in T1 plants grown at Ledong and by 34%–42% in

the T2 and T3 generations grown at Nanjing relative to the WT, while the OE lines exhibited lower yields than the WT by approximately 17% in all three generations (Table 1). We also analysed the yield and the biomass of the WT and T4 generation transgenic plants at Nanjing under low (180 kg N/ha) and normal N (300 kg N/ha) supplies. At the level of 180 kg N/ha, compared with WT, the yield of OE lines was reduced by 17%, and the biomass increased by 14%, while the yield and biomass of O lines were increased by 25% and 27% (Figure S5a), respectively. At the level of 300 kg N/ha, the yield of OE lines was reduced by 16%, and the biomass increased by 12%, as for O lines the yield and biomass were increased by 21% and 22%, respectively, compared with WT (Figure S5b).

The total tiller number per plant in the T3 generation at the harvest stage increased 27.1% on average for both *pOsNAR2.1: OsNRT2.1* and *pUbi:OsNRT2.1* transgenic plants relative to the WT with no difference between the transgenic lines (Table 2); however, the grain number per panicle differed significantly between the OE and O lines (Table 2). The grain number per panicle increased approximately 15% in the O lines; the panicle length increased in the O lines approximately 12%, and the seed setting rate increased in the O lines by 14% relative to the WT (Table 2). The grain yields of the O lines increased by 24.2% relative to the WT (Table 2).

Nitrogen-use efficiency of transgenic lines

Because the biomass and yields increased in the *pOsNAR2.1: OsNRT2.1* transgenic plants, we also analysed ANUE in T1–T4 generations of transgenic plants, N recovery efficiency (NRE), physiological N-use efficiency (PNUE) and N harvest index (NHI) traits at the harvest stage in T3 generation transgenic lines to determine whether N use was altered in these plants, as modified the calculation method of the reference in Zhang *et al.* (2009). The ANUE of the O lines was enhanced by approximately 33% in T1 plants grown at Ledong and by 34%–42% in the T2 and T3 generations grown at Nanjing relative to the WT, while the OE lines exhibited lower ANUE than the WT by approximately 17% in all three generations (Table 1). In T4 plants at Nanjing, at the level of 180 kg N/ha, compared with WT, the ANUE of OE lines was

Table 2 Comparison of agronomic traits between the wild-type (WT) and transgenic lines

Genotype	WT	pUBi:OsNRT2.1			pOsNAR2.1:OsNRT2.1		
		OE1	OE2	OE3	O6	O7	O8
Plant height (cm)	83.21 b	80.18 c	79.25 c	76.25 d	87.27 a	86.85 a	88.69 a
Total tiller number per plant	20.26 b	25.57 a	23.24 a	24.81 a	25.83 a	26.84 a	24.41 a
Panicle length (cm)	13.19 b	12.48 c	11.48 c	11.12 c	14.40 a	14.13 a	14.48 a
Grain weight (g/panicle)	2.22 d	1.16 e	0.96 f	1.23 e	3.87 a	2.74 c	3.01 b,c
Seed setting rate (%)	70.45 b	59.79 c	57.82 c	61.94 c	80.46 a	75.92 a	78.95 a
Grain number per panicle	132.58 b	105.67 c	97.25 c	101.61c	154.50 a	166.25a	149.75 a
1000-grain weight (g)	25.24 a	24.89 a	24.39 a	24.45 a	25.28 a	25.67 a	25.89 a
Grain yield (g/plant)	26.21 b	21.61 c	22.63 c	21.19 c	31.17 a	32.81 a	33.64 a

Statistical analysis was performed on data derived from the T3 generation. Significant differences between the transgenic lines and WT are indicated by different letters ($P < 0.05$, one-way ANOVA, $n = 3$).

reduced by 22% and the ANUE of O lines was increased by 33%, and at the level of 300 kg N/ha, the ANUE of OE lines was reduced by 17% and the ANUE of O lines was increased by 28% (Figure S5c). In the OE lines, the NRE increased to approximately 115% of the WT, and the PNUE and NHI were reduced to approximately 71% of the WT values. In the O lines, the ANUE increased to approximately 128% of the WT, the NRE increased to approximately 136% of the WT, and the PNUE and NHI were not significantly different from WT values (Table 4).

We sampled shoot tissues at the anthesis stage (60 days after transplanting) and the mature stage (90 days after transplanting) to determine the total N content. At the anthesis stage, total N was concentrated mainly in the culm with no difference between the OE and O lines, but with an increase of approximately 27% relative to the WT. In leaves, the total N content was the same in the O and WT lines, but was approximately 33% higher in the OE lines. The total N content in the grain was the same in all lines (Figure 2a). At the mature stage, total N was concentrated mainly in the grain, with the N content decreased by approximately 10% in the OE lines and increased by approximately 38% in the O lines relative to the WT (Figure 2b).

Translocation of dry matter and N in transgenic lines

We investigated the dry matter and N translocation (NT) in rice plants by determining dry matter at anthesis (DMA), dry matter at maturity (DMM), total N accumulation at anthesis (TNAA) and total N accumulation at maturity (TNAM). For the OE lines, the DMA, the DMM, the TNAA and the TNAM increased by approximately 27%, 21%, 25% and 21%, respectively, relative to the WT. For the O lines, the DMA, the DMM, the TNAA and the TNAM increased by approximately 46%, 38%, 15% and 27%, respectively, relative to the WT (Table 3).

We also investigated the dry matter translocation (DMT), the DMT efficiency (DMTE), the contribution of preanthesis assimilates to grain yield (CPAY) and the harvest index (HI), based on the calculation method of the reference in Ntanos and Koutroubas (2002). For the OE lines, the DMT, DMTE, CPAY and HI decreased by approximately 68%, 75%, 61% and 31%, respectively, relative to the WT. For the O lines, the DMT increased by approximately 46%, while the DMTE, CPAY and HI did not differ between the O lines and the WT (Table 4).

We investigated postanthesis N uptake (PANU), NT, N translocation efficiency (NTE) and the contribution of preanthesis N to grain N accumulation (CPNGN), as modified the calculation method of the reference in Ntanos and Koutroubas (2002) and Zhang *et al.* (2009). The PANU and CPNGN did not differ between the OE lines and the WT, but the NT and the NTE decreased by approximately 16% and 32%, respectively, in the OE lines relative to the WT. The NTE did not differ between the O lines and the WT, while the PANU and NT increased by

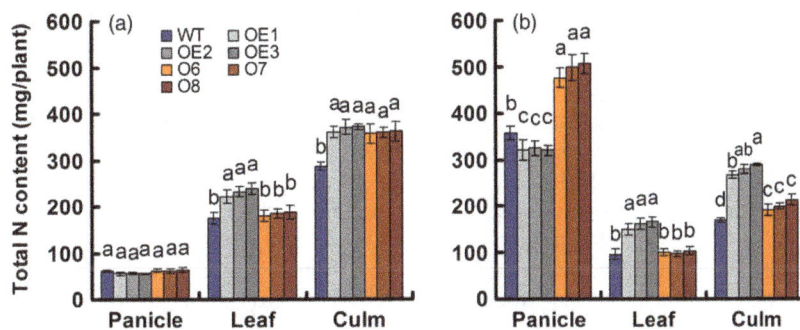

Figure 2 N content in various parts of the wild-type (WT) and transgenic plants at two growth stages. (a) Sixty days after transplant, anthesis stage. (b) Ninety days after transplant, maturity stage. Error bars: SE ($n = 3$). Statistical analysis was performed on data derived from the T3 generation. Significant differences between the transgenic lines and WT are indicated by different letters ($P < 0.05$, one-way ANOVA).

Table 3 Comparison of dry matter accumulation and N content between the wild-type (WT) and transgenic lines

Dry matter and nitrogen components	WT	pUBi:OsNRT2.1			pOsNAR2.1:OsNRT2.1		
		OE1	OE2	OE3	O6	O7	O8
Dry matter at anthesis (kg/m²)	0.90 c	1.14 b	1.15 b	1.14 b	1.30 a	1.31 a	1.35 a
Dry matter at maturity (kg/m²)	1.40 c	1.67 b	1.78 b	1.69 b	1.91 a	1.96 a	1.95 a
Total nitrogen accumulation at anthesis (g/m²)	13.98 c	17.02 a	17.62 a	17.83 a	16.02 b	16.20 b	16.38 b
Total nitrogen accumulation at maturity (g/m²)	16.62 b	19.68 a	20.45 a	20.67 a	20.47 a	21.22 a	21.98 a
Grain nitrogen accumulation at maturity (g/m²)	9.56 b	8.56 c	8.67 c	8.54 c	12.69 a	13.30 a	13.53 a

Statistical analysis was performed on data derived from the T3 generation. For each mean, $n = 3$. Significant differences between the transgenic lines and WT are indicated by different letters ($P < 0.05$, one-way ANOVA).

Table 4 Comparison of N-use efficiency, dry matter transport efficiency and N transport efficiency between the wild-type (WT) and transgenic rice lines

	WT	pUBi:OsNRT2.1			pOsNAR2.1:OsNRT2.1		
		OE1	OE2	OE3	O6	O7	O8
N recovery efficiency (%)	39.06 c	44.59 b	45.40 b	45.64 b	52.13 a	53.29 a	53.68 a
Physiological N-use efficiency (g/g)	54.55 a	39.96 b	40.56 b	37.26 b	50.10 a	51.49 a	52.40 a
N harvest index (%)	59.49 a	43.52 b	42.39 b	41.31 b	61.98 a	62.68 a	61.56 a
Dry matter transfer (g/m^2)	198.95 c	72.25 d	51.03 e	67.74 d	301.22 a	278.87 b	293.48 a,b
Dry matter transfer efficiency (%)	22.10 a	6.32 b	4.45 c	5.95 b	23.23 a	21.22 a	21.78 a
Contribution of preanthesis assimilates to grain yield (%)	28.45 a	12.53 b	8.45 c	11.98 b	30.10 a	28.39 a	29.22 a
Harvest index (%)	49.93 a	34.46 b	34.96 b	33.47 b	49.20 a	50.34 a	51.46 a
Postanthesis N uptake (g/m^2)	2.64 c	2.66 c	2.83 c	2.84 c	4.45 b	5.03 a	5.40 a
N translocation (g/m^2)	6.91 b	5.91 c	5.84 c	5.70 c	8.24 a	8.28 a	7.93 a
N translocation efficiency (%)	49.45 a	34.69 b	33.14 b	31.97 b	51.42 a	51.10 a	48.42 a
Contribution of preanthesis N to grain N accumulation (%)	72.34 a	68.95 a	68.36 a	69.76 a	61.93 b	62.21 b	58.62 b

Statistical analysis was performed on data derived from the T3 generation. Methods of calculations in Table S4. For each mean, $n = 3$. Significant differences between the transgenic lines and WT are indicated by different letters ($P < 0.05$, one-way ANOVA).

approximately 87% and 18%, respectively, and the CPNGN decreased by approximately 16% in the O lines relative to the WT (Table 4).

Expression patterns of *OsNRT2.1* and *OsNAR2.1* in different organs of the WT and transgenic lines

Rice was previously shown to have a two-component NO_3^- uptake system consisting of *OsNRT2.1* and *OsNAR2.1*, similar to the system in *Arabidopsis* (Feng *et al.*, 2011; Liu *et al.*, 2014; Yan *et al.*, 2011). We analysed the *OsNRT2.1* and *OsNAR2.1* expression patterns in the WT and transgenic lines during the filling stage. The detail about RNA samples was described in Figure S6 and Experimental procedures. The *OsNRT2.1* expression pattern in the WT showed that *OsNRT2.1* gene expressed most in root, secondly in leaf sheaths, thirdly in leaf blades and internodes and lest in grain including seed, palea and lemma (Table S5, Figure 3a). As for *OsNAR2.1*, it was expressed also most in root, secondly in leaf sheaths, thirdly in internodes and lest in grain and leaf blades (Table S5, Figure 3b). The coexpression pattern of *OsNRT2.1* and *OsNAR2.1* happened in root, leaf sheaths, internodes and grain but not in leaf blades (Table S5, Figure S7).

Compared with WT, the *OsNRT2.1* expression increased by about 7.5-fold averagely in all organs of OE lines including root. The increase pattern of *OsNRT2.1* in OE lines showed the similar trade as the native expression of *OsNRT2.1* in the WT that was most in root, secondly in leaf sheaths, thirdly in leaf blades and internodes and lest in grain (Table S5, Figure 3a). It was very interesting that we found that in OE lines, the *OsNAR2.1* was also increased with the pattern as most in root, secondly in leaf sheaths, thirdly in internodes, fourthly in leaf blades and lest in grain (Table S5, Figure 3b). The coexpression increase pattern of *OsNRT2.1* and *OsNAR2.1* occurred in all organs of OE lines (Table S5, Figure S7).

Compared with WT, the *OsNRT2.1* expression was not changed in grain and leaf blades in O lines and increased in leaf sheaths, internodes and root significantly with a same pattern as WT, which is most in root, secondly in leaf sheaths, thirdly in internodes, fourthly in leaf blades and lest in grain (Table S5, Figure 3a). For *OsNAR2.1* expression in O lines, it was also not increased in grain and leaf blades only increased in leaf sheaths,

internodes and root significantly with a same pattern as WT, which was most in root, secondly in leaf sheaths, thirdly in internodes and lest in grain and leaf blades (Table S5, Figure 3b). The coexpression increase pattern of *OsNRT2.1* and *OsNAR2.1* occurred in leaf sheaths, internodes and root of O lines (Table S5, Figure S7).

Expression patterns of *OsNRT2.1* and *OsNAR2.1* in different growth stages of the WT and transgenic lines

In this study, we found that the *OsNRT2.1* and *OsNAR2.1* mRNA levels in the culms including the leaf sheath and internode (Figure S8) were significantly higher in all of the transgenic plants than in the WT plants (Figure 4a,b). *OsNRT2.1* expression was 3–20-fold higher in the OE lines than in the WT, but was only 31%–45% higher in the O lines than in the WT (Figure 4a). *OsNAR2.1* expression was two- to ninefold higher in the OE lines than in the WT and was one- to eightfold higher in the O lines than in the WT (Figure 4b). Throughout the experimental growth period, *OsNRT2.1* expression was significantly higher in the culms of the OE lines than the O lines, but no significant difference in *OsNAR2.1* expression was observed between the OE and O transgenic lines.

During the entire experimental growth period, no significant differences in the *OsNRT2.1* and *OsNAR2.1* expression were found between the leaf blade I of the O lines and WT plants, but the expression levels of both *OsNRT2.1* and *OsNAR2.1* were up-regulated significantly in the OE plants relative to the WT (Figure S9).

Growth rate in transgenic lines

N transport and the growth of rice biomass are closely related, and *OsNRT2.1* overexpression was previously shown to affect the rice growth (Katayama *et al.*, 2009). In this study, the OE and O lines began to show significantly higher biomass than WT plants at 45 days after transplanting and had accumulated 21% and 38% more biomass at 90 days (Figure 4c). The growth rates of the OE and O lines reached peak values at 60 days and were higher than those of the WT plants (Figure 4d). The growth rates of the OE and O lines were approximately 25% and 58% greater, respectively, than the WT. The growth rates of the transgenic and

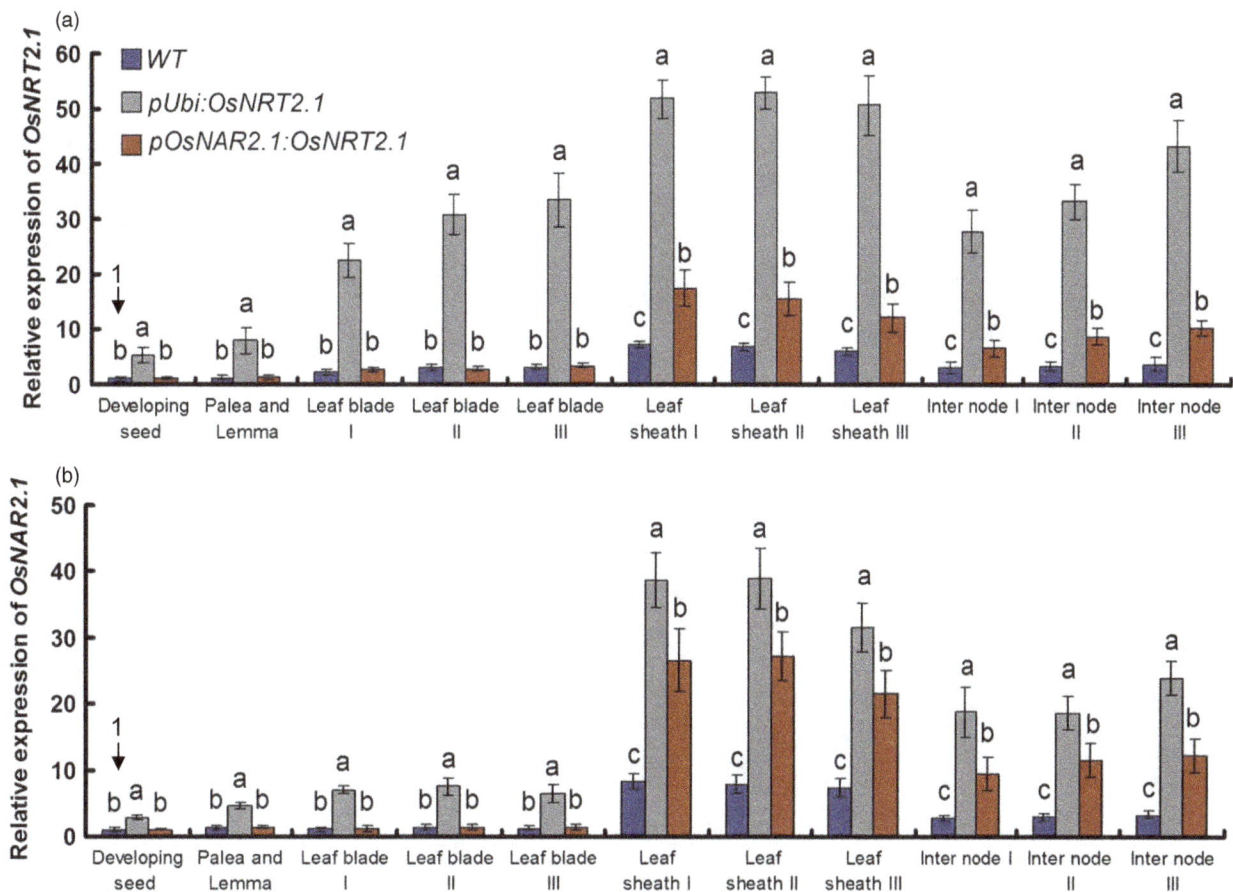

Figure 3 Expression pattern of *OsNRT2.1* and *OsNAR2.1*. Relative expression of (a) *OsNRT2.1* and (b) *OsNAR2.1* in various organs at 14 days after pollination. *pUbi:OsNRT2.1* represents the average of OE1, OE2 and OE3. *pOsNAR2.1:OsNRT2.1* represents the average of O6, O7 and O8. Statistical analysis was performed on data derived from the T4 generation. We defined that developing seed of the wild-type (WT) expression was set equal to 1. Error bars: SE (*n* = 3). Significant differences between the transgenic lines and WT are indicated by different letters (*P* < 0.05, one-way ANOVA).

WT plants were identical after 75 days during the grain filling stage (Figure 4d).

The coexpression of *OsNRT2.1* and *OsNAR2.1* in the WT and transgenic plants

The expression pattern of *OsNRT2.1* and *OsNAR2.1* in different organs showed that there exists a strong coexpression pattern of these two genes in rice plants (Figure S7). The coexpression pattern of *OsNRT2.1* and *OsNAR2.1* was altered very much in OE lines compared with O and WT lines (Figure S7). The expression ratio of *OsNRT2.1* and *OsNAR2.1* is 5.4 : 1 in the OE organs and 3.6 : 1 in the O lines compared with 3.9 : 1 in the WT organs (Figure S7). Furthermore, we specially investigated the ratio of *OsNAR2.1* to *OsNRT2.1* expression in root as 6.3 : 1 in the OE lines, 4.1 : 1 in the O lines and 4.2 : 1 in the WT plants, with no significant differences between the O lines and WT plants (Table S5).

The culm is important for N storage and translocation in rice shoots. In rice shoot, *OsNRT2.1* and *OsNAR2.1* expression was expressed most in leaf sheaths of culm (Figure 3). Our expression data also confirmed that *OsNRT2.1* and *OsNAR2.1* expression in the culm could play a key role in NO_3^- remobilization. To further study the possible relationship between *OsNRT2.1* and *OsNAR2.1* expression and rice growth, we compared the ratio of *OsNRT2.1* and *OsNAR2.1* expression in

rice plants. The expression ratio was approximately 11.3 : 1 in the OE lines and approximately 4.7 : 1 in the O lines compared with approximately 7.2 : 1 in the WT plants (Figure 5). We also investigated the ratio of *OsNAR2.1* to *OsNRT2.1* expression in leaf blade I. The expression ratio was 7.3 : 1 in the OE lines, 4 : 1 in the O lines and 5.2 : 1 in the WT plants, with no significant differences between the O lines and WT plants (Figure S10). The ratio of *OsNAR2.1* to *OsNRT2.1* expression correlated with the grain yield.

Discussion

N nutrition affects all levels of plant function, from metabolism to resource allocation, growth and development (Crawford, 1995; Scheible *et al.*, 1997, 2004; Stitt, 1999). As one form of available N nutrient to plants, NO_3^- is taken up in the roots by active transport processes and stored in vacuoles in rice shoots (Fan *et al.*, 2007; Li *et al.*, 2008). In rice, OsNAR2.1 acts as a partner protein with OsNRT2.1 in the uptake and transport of NO_3^- (Liu *et al.*, 2014; Tang *et al.*, 2012; Yan *et al.*, 2011). *OsNAR2.1* gene expression was shown to be up-regulated by NO_3^- and down-regulated by NH_4^+ (Feng *et al.*, 2011; Nazoa *et al.*, 2003; Zhuo *et al.*, 1999).

Rooke *et al.* (2000) reported that the maize Ubi-1 promoter had strong activity in young, metabolically active tissues and in

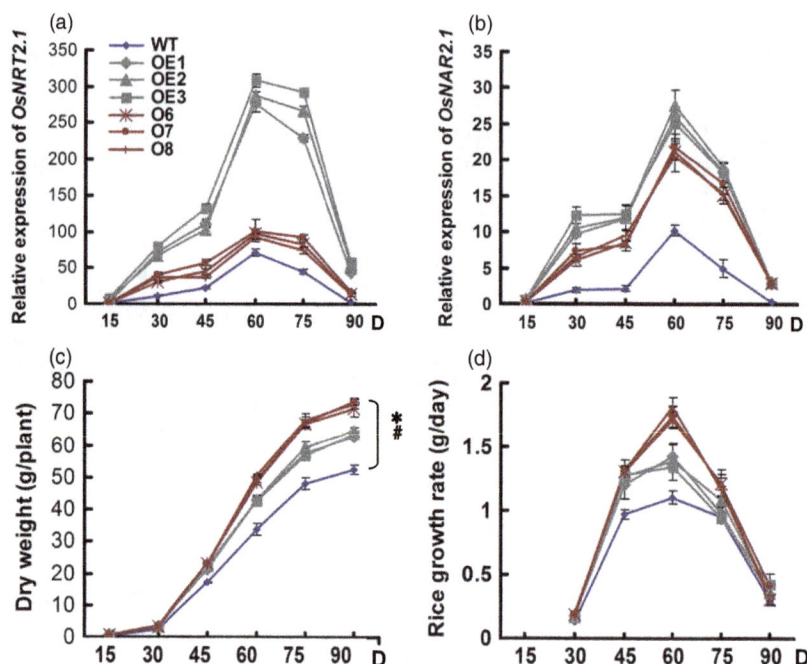

Figure 4 Growth status of the wild-type (WT) and transgenic lines during the experimental growth period. (a) Changes in *OsNRT2.1* expression over the experimental growth period. (b) Changes in *OsNAR2.1* expression over the experimental growth period. RNA was extracted from culms. (c) Dry weight. Dry weight mean values are for all aboveground biomass, including the grain yield. (d) Growth rate. Samples were collected at 15-day intervals after seedlings were transplanted to the field. Statistical analysis was performed on data derived from the T3 generation. Error bars: SE (n = 3). D in *x*-axis means the day after transplanting. The asterisk at the end of time course indicates their statistically significant differences among plants, and #statistically significant differences during the growth stages (P < 0.05, ANCOVA).

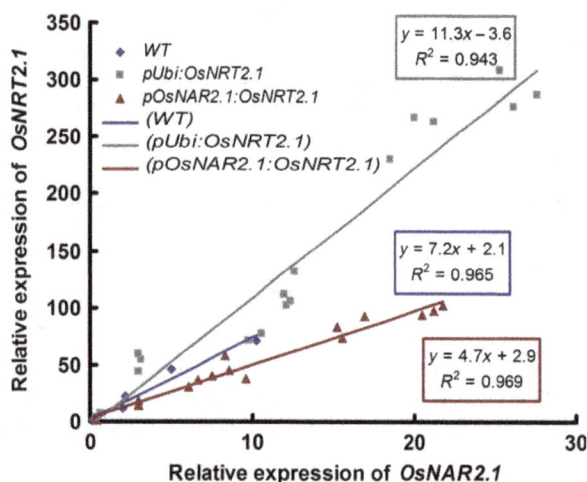

Figure 5 Ratios of *OsNRT2.1* to *OsNAR2.1* expression in culms of the wild-type (WT) and transgenic lines over the course of the study. The ratios of *OsNRT2.1* and *OsNAR2.1* expression during different periods in the culms of *pUbi:OsNRT2.1* lines (OE1, OE2 and OE3), *pOsNAR2.1:OsNRT2.1* lines (O6, O7 and O8) and WT were presented.

pollen grains. Furthermore, Cornejo *et al.* (1993) performed histochemical localization of Ubi-GUS activity and showed that the Ubi promoter was most active in rapidly dividing cells; however, Chen *et al.* (2012) reported that the Ubi promoter drove strong *OsPIN2* expression in all tissues. Chen *et al.* (2015) reported that ectopic expression of the *WOX11* gene driven by the promoter of the *OsHAK16* gene, which encodes a potassium (K) transporter that is induced by low K levels, led to an extensive root system, adventitious roots and increased tiller numbers in rice. In contrast, *WOX11* overexpression driven by the Ubi promoter induced ectopic crown roots in rice and failed to present any similar super growth phenotype in field (Zhao *et al.*, 2009) as described by Chen *et al.* (2015). These

results suggested that the use of a specific inducible promoter-driven gene function could be a good strategy for plant breeding.

In this study, *OsNRT2.1* expression was up-regulated significantly in both the aboveground and underground parts of *pUbi:OsNRT2.1* transgenic plants relative to the WT (Figure 1c), while *OsNRT2.1* expression in *pOsNAR2.1:OsNRT2.1* transgenic plants was increased significantly only in the roots and culms and not enhanced significantly in the leaves (Figure 1d). Specific induction of expression by the *OsNAR2.1* promoter in rice roots and culms based on GUS fusion data has been reported previously (Feng *et al.*, 2011); therefore, we investigated the effects of tissue-specific induction of *OsNRT2.1* expression in roots and culms on plant growth and nitrogen-use efficiency (NUE).

Effect of *pOsNAR2.1:OsNRT2.1* expression on NUE in transgenic rice

N redistribution during the reproductive stage was shown to vary significantly among cultivars and under various N management strategies (Souza *et al.*, 1998). Mae and Ohira (1981) reported that a major proportion of N was redistributed from vegetative organs to panicles during grain filling, 64% of which was derived from leaf blades and 36% from culms. The NTE values of the WT, *pUbi:OsNRT2.1* and *pOsNAR2.1:OsNRT2.1* plants were averagely 49.5%, 33.4% and 50.3%, indicating that N transfer from the shoots into grain was significantly less in *pUbi:OsNRT2.1* transgenic plants than in the WT or *pOsNAR2.1:OsNRT2.1* plants (Table 4). This lower level of N transfer from vegetative organs to grain during grain filling in *pUbi:OsNRT2.1* plants affected spike formation and final grain yield compared with the WT and *pOsNAR2.1:OsNRT2.1* plants (Table 1). The DMTE values for WT, *pUbi:OsNRT2.1* and *pOsNAR2.1:OsNRT2.1* plants were 22.1%, 5.5% and 22.1%, averagely (Table 4) demonstrating that markedly less dry matter was transferred into grain yield in the *pUbi:OsNRT2.1* lines. These data confirmed that the transport of N and biomass during the transition from the flowering to harvest stages affected the final yield and NUE of rice (Zhang *et al.*, 2009) and also indicated that the Ubi promoter decreased N and biomass translocation, while the *OsNAR2.1* promoter did not.

In both types of *OsNRT2.1* overexpression line, NT was reduced during the reproductive stage and NUE was reduced before flowering. The CPAY average values of the WT, *pUbi: OsNRT2.1* and *pOsNAR2.1:OsNRT2.1* plants were 28.5%, 11% and 34.9%, respectively. The CPAY of the *pOsNAR2.1:OsNRT2.1* plants was higher than that of the WT plants that had higher CPAY than the *pUbi:OsNRT2.1* plants (Table 4). The HI was much lower for the *pUbi:OsNRT2.1* plants than for the WT or *pOsNAR2.1:OsNRT2.1* plants (Table 4), indicating that the Ubi promoter affected NO_3^- uptake and N use before the flowering stage and that levels of *OsNRT2.1* overexpression in rice that were excessive did not benefit N use during either the vegetative or reproductive stages.

The coexpression pattern of *OsNRT2.1* and *OsNAR2.1* is an important factor controlling N transport in rice

How to assess the effect of NO_3^- transporter expression on rice NUE is a key question for rice breeding. The NO_3^- transporter, OsNRT1.1B, was shown to improve the NUE of rice by approximately 30% (Hu *et al.*, 2015), while our data showed that not the higher expression level of NO_3^- transporter was relative to the higher yield and NUE of rice (Tables 1 and 4, and Figure 4). After determining the expression levels of *OsNRT2.1* and its partner gene, *OsNAR2.1*, we calculated the coexpression ratio of these genes in rice plants.

The coexpression pattern of *OsNRT2.1* and *OsNAR2.1* happened in the WT and transgenic plants (Figures 3 and 4, Table S5). However, the coexpression pattern of *OsNRT2.1* and *OsNAR2.1* was changed in OE lines compared with O and WT lines (Figure S7), which suggested that *OsNRT2.1* driven by different promoters had a different coexpression patterns with *OsNAR2.1*. But it is still not clear that why increasing *OsNRT2.1* expression would induce *OsNAR2.1* expression and what mechanism exists behind the coexpression pattern of *OsNRT2.1* and *OsNAR2.1* in the gene regulation.

However, the ratio changes of *OsNRT2.1* to *OsNAR2.1* expression may be a clue for the explanation of the rice growth and nitrogen-use difference in the WT and transgenic lines. The ratio changes of *OsNRT2.1* to *OsNAR2.1* expression in different organs were increased significantly in *pUbi:OsNRT2.1* lines compared with WT and *pOsNAR2.1:OsNRT2.1* lines (Figure S7). Also during the growth stages, the ratio of *OsNRT2.1* to *OsNAR2.1* expression in culm (including the internode and leaf sheath) was increased in *pUbi:OsNRT2.1* lines compared with WT and the *pOsNAR2.1: OsNRT2.1* lines (Figure 5). These data indicated that the interaction between *OsNRT2.1* and *OsNAR2.1* in *pUbi:OsNRT2.1* plants differed from WT and that in the *pOsNAR2.1:OsNRT2.1* lines. Furthermore in culms, *pOsNAR2.1:OsNRT2.1* lines showed a lower expression ratio of these two genes, in which more OsNAR2.1 protein may be available to interact with OsNRT2.1 protein. Therefore, the efficiency of OsNRT2.1 function in rice plants should differ between the two types of transgenic plants resulting in different rice yield and NUE phenotypes. On the other hand, the higher expression of *OsNRT2.1* and *OsNAR2.1* in all the organs of pUbi:*OsNRT2.1* than WT may cause some disadvantages to plants such as high cost for mRNA synthesis or disturbing of nitrogen transport in the leaf blades. All possibilities remain to be confirmed by further analysis.

In this study, we showed that the rice yield and NUE could be improved by increasing *OsNRT2.1* expression, especially in

combination with a lower expression ratio with its partner gene *OsNAR2.1*, which encodes a high-affinity NO_3^- transporter.

Experimental procedures

Construction of vectors and rice transformation

We amplified the *OsNRT2.1/OsNRT2.2* ORF sequence, which is identical for both genes, from cDNA isolated from *Oryza sativa* L. ssp. Japonica cv. Nipponbare using the primers listed in Table S1. We amplified the *OsNAR2.1* and ubiquitin promoters from the *pOsNAR2.1(1698bp):GUS* (Feng *et al.*, 2011) and *pUbi:OsPIN2* (Chen *et al.*, 2012) constructs, respectively, using the primers listed in Table S2. The PCR products were cloned into the pMD19-T vector (TaKaRa Biotechnology, Dalian, China) and confirmed by restriction enzyme digestion and DNA sequencing. The *pUbi:OsNRT2.1* and *pOsNAR2.1:OsNRT2.1* vectors were constructed as shown in Figure S1. These constructs were introduced into *Agrobacterium tumefaciens* strain EHA105 by electroporation and then transformed into rice as described previously (Tang *et al.*, 2012).

Southern blot analysis

Transgene copy number was determined by the Southern blot analysis following procedures described previously (Jia *et al.*, 2011). Briefly, genomic DNA was extracted from the leaves of wild type (WT) and digested with HindIII and EcoRI restriction enzymes. The digested DNA was separated on a 1% (w/v) agarose gel, transferred to a Hybond-N+ nylon membrane and hybridized with hygromycin-resistant gene.

Biomass, total nitrogen (N) measurement and calculation of NUE

WT and transgenic rice plants were harvested at 9:00 a.m. and heated at 105 °C for 30 min. Panicles, leaves and culms were then dried at 75 °C for 3 days. Dry weights were recorded as biomass values. Samples collected at 15-day intervals from WT and transgenic lines grown in soil in pots were used to calculate whole-plant biomass values.

Total N content was measured using the Kjeldahl method (Li *et al.*, 2006). The total dry weight (biomass) was estimated as the sum of weights of all plant parts. Total N accumulation was estimated as the sum of the N contents of all plant parts. Agronomic NUE (ANUE, g/g) was calculated as (grain yield − grain yield of zero N plot)/N supply; NRE (%) was calculated as (total N accumulation at maturity for N-treated plot − total N accumulation at maturity of zero N plot)/N supply; physiological NUE (PNUE, g/g) was calculated as (grain yield − grain yield of zero N plot)/total N accumulation at maturity; and the NHI (%) was calculated as (grain N accumulation at maturity/total N accumulation at maturity). Dry matter and NT and translocation efficiency method for the calculation of the reference of Ntanos and Koutroubas (2002) and Zhang *et al.* (2009). Dry matter translocation (DMT, g/m²) was calculated as dry matter at anthesis − (dry matter at maturity − grain yield); dry matter translocation efficiency (DMTE, %) was calculated as (dry matter translocation/dry matter at anthesis) × 100%; the CPAY (%) was calculated as (dry matter translocation/grain yield) × 100%; the HI (%) was calculated as (grain yield/dry matter at maturity) × 100%; PANU (g/m²) was calculated as total N accumulation at maturity − total N accumulation at anthesis; NT (g/m²) was calculated as total N accumulation at anthesis − (total N accumulation at maturity − grain N accumulation at maturity); NTE (%) was calculated as (N translocation/ total N accumulation at anthesis) × 100%; and the CPNGN (%) was calculated as (N translocation/grain N accumulation at maturity) × 100% (Table S4).

The growth conditions for T0 to T4 transgenic plants

T0, T2, T3 and T4 generation plants were grown in plots at the Nanjing Agricultural University in Nanjing, Jiangsu (Figure S11). T1 generation plants were grown in Sanya, Hainan. Jiangsu is in a subtropical monsoon climate zone. Chemical properties of the soils in the plots at the Nanjing Agricultural University included organic matter, 11.56 g/kg; total N content, 0.91 g/kg; available P content, 18.91 mg/kg; exchangeable K, 185.67 mg/kg; and pH 6.5. Basal applications of 30 kg P/ha ($Ca(H_2PO_4)_2$) and 60 kg/K ha (KCl) were made to all plots 3 days before transplanting. N fertilizer accounted for 40%, 30% and 40% of the total N fertilizer was applied prior to transplanting, at tillering, just before the heading stage, respectively.

The field experiments for yield harvest

T0–T4 generation seedlings were planted in the same experiment site in Nanjing, except T1 in Sanya. Seed generation transgenic lines and WT were surface-sterilized with 10% (v : v) hydrogen peroxide (H_2O_2) for 30 min and rinsed thoroughly with deionized water. The transgenic seeds were soaked in water containing 25 mg/L hygromycin, and the WT seeds were soaked in water. After 3 days, the sterilized seeds were sown evenly in wet soil. The similar seedlings were transplanted to field plots after 3 weeks of germination.

T1–T3 plants were planted in plots fertilized at a rate of 300 kg N/ha as urea and in plots without N fertilization. Plots were 2 × 2.5 m in size with the seedlings planted in a 10 × 10 array. Plants at the edges of all four sides of each plot were removed at maturity to avoid the influence of edge effects. Four points, each containing four seedlings, totally 16 seedlings, were selected randomly within the remaining centre 8 × 8 array of plants, and samples were collected (Khuram et al., 2013; Ookawa et al., 2010; Pan et al., 2013; Srikanth et al., 2016). Yield and biomass values determined from these four points in each plot were used to calculate the yield per hectare and biomass of each line, and three random plots for each line were designed in the experiment (Figure S11).

T3 generation plants were sampled at 15-day intervals for the determination of the grain yield, biomass and N content. The growth rate was the dry weight of the weight increase in the unit time after seedlings were transplanted to the plots.

T4 generation plants were planted in a plot fertilized at a rate of 0, 180 and 300 kg N/ha as urea. Same random field plots with three replicates were designed as T1–T3 plants for yield, and biomass values determined from these four points were used to calculate the yield and biomass per plant and ANUE of each line.

mRNA sampling and qRT-PCR assay

To investigate the expression pattern in plant organs, we sampled mRNA for seeds, palea and lemma, leaf blade I, leaf blade II, leaf blade III, leaf sheath I, leaf sheath II, leaf sheath III, internode I, internode II, internode III and newly developed root (3 cm from root tips) at the grain filling stage (described in Figure S6). Tracking rice in the whole growth period of gene expression in T3 generation, we sampled mRNA from culms including leaf sheath and internode I (described in Figure S8) at 15, 30, 45, 60, 75 and 90 days after transplanting.

Total RNAs were prepared from the various tissues of the WT and transgenic plants using TRIzol reagent (Vazyme Biotech Co., Ltd, http://www.vazyme.com). Real-time PCR was carried as described before (Li et al., 2014). All primers used for qRT-PCR are listed in Table S3.

Statistical analysis

Data were analysed by Tukey's test of one-way analysis of variance (ANOVA), except that analysis of covariate (ANCOVA) was used in the biomass and growth rate during growth stages (Figure 4a,b). Different letters on the histograms or after mean values indicate statistically significant differences at $P < 0.05$ between the transgenic plants and WT (one-way ANOVA). The asterisk at the end of time course indicates their statistically significant differences among plants, and #statistically significant differences during the growth stages at $P < 0.05$ (ANCOVA). All statistical evaluations were conducted using the IBM SPSS Statistics version 20 software (SPSS Inc., Chicago, IL).

Acknowledgements

This work was supported by the National Natural Science Foundation (No. 31372122, 31172013) and Nanjing 321 Talents Program. The English in this document has been checked by two native speakers of English, and for a certificate, please see: http://www.textcheck.com/certificate/SoEldm.

References

Chen, Y., Fan, X., Song, W., Zhang, Y. and Xu, G. (2012) Over-expression of OsPIN2 leads to increased tiller numbers, angle and shorter plant height through suppression of OsLAZY1. Plant Biotechnol. J. 10, 139–149.

Chen, G., Feng, H., Hu, Q., Qu, H., Chen, A., Yu, L. and Xu, G. (2015) Improving rice tolerance to potassium deficiency by enhancing OsHAK16p:WOX11-controlled root development. Plant Biotechnol. J. 13, 833–848.

Cornejo, M.J., Luth, D., Blankenship, K.M., Anderson, O.D. and Blechl, A.E. (1993) Activity of a maize ubiquitin promoter in transgenic rice. Plant Mol. Biol. 23, 567–581.

Coruzzi, G. and Bush, D.R. (2001) Nitrogen and carbon nutrient and metabolite signaling in plants. Plant Physiol. 125, 61–64.

Coruzzi, G.M. and Zhou, L. (2001) Carbon and nitrogen sensing and signaling in plants: emerging 'matrix effects'. Curr. Opin. Plant Biol. 4, 247–253.

Crawford, N.M. (1995) Nitrate: nutrient and signal for plant growth. Plant Cell, 7, 859–868.

Crawford, N.M. and Forde, B.G. (2002) Molecular and developmental biology of inorganic nitrogen nutrition. In The Arabidopsis Book (Meyerowitz, E.M., ed.), pp. 1–25. Rockville, MD: American Society of Plant Biologists.

Crawford, N.M. and Glass, A.D.M. (1998) Molecular and physiological aspects of nitrate uptake in plants. Trends Plant Sci. 3, 389–395.

Fan, X., Shen, Q., Ma, Z., Zhu, H., Yin, X. and Miller, A.J. (2005) A comparison of nitrate transport in four different rice (Oryza sativa L.) cultivars. Sci. China C Life Sci. 48, 897–911.

Fan, X., Jia, L., Li, Y., Smith, S.J., Miller, A.J. and Shen, Q. (2007) Comparing nitrate storage and remobilization in two rice cultivars that differ in their nitrogen use efficiency. J. Exp. Bot. 58, 1729–1740.

Feng, Q., Zhang, Y., Hao, P., Wang, S., Fu, G., Huang, Y., Li, Y. et al. (2002) Sequence and analysis of rice chromosome 4. Nature, 420, 316–320.

Feng, H., Yan, M., Fan, X., Li, B., Shen, Q., Miller, A.J. and Xu, G. (2011) Spatial expression and regulation of rice high-affinity nitrate transporters by nitrogen and carbon status. J. Exp. Bot. 62, 2319–2332.

Hu, B., Wang, W., Ou, S., Tang, J., Li, H., Che, R., Zhang, Z. et al. (2015) Variation in NRT1.1B contributes to nitrate-use divergence between rice subspecies. Nat. Genet. 47, 834–838.

Jia, H., Ren, H., Gu, M., Zhao, J., Sun, S., Zhang, X., Chen, J. et al. (2011) The phosphate transporter gene OsPht1;8 is involved in phosphate homeostasis in rice. Plant Physiol. 156, 1164–1175.

Katayama, H., Mori, M., Kawamura, Y., Tanaka, T., Mori, M. and Hasegawa, H. (2009) Production and characterization of transgenic rice plants carrying a high-affinity nitrate transporter gene (OsNRT2.1). Breed. Sci. 59, 237–243.

Khuram, M., Asif, I., Muhammad, H., Faisal, Z., Siddiqui, M.H., Mohsin, A.U., Bakht, H.F.S.G. et al. (2013) Impact of nitrogen and phosphorus on the growth, yield and quality of maize (Zea mays L.) fodder in Pakistan. Philipp. J. Crop Sci. 38, 43–46.

Kirk, G.J.D. and Kronzucker, H.J. (2005) The potential for nitrification and nitrate uptake in the rhizosphere of wetland plants: a modelling study. Ann. Bot. 96, 639–646.

Kronzucker, H.J., Glass, A.D.M., Siddiqi, M.Y. and Kirk, G.J.D. (2000) Comparative kinetic analysis of ammonium and nitrate acquisition by tropical lowland rice: implications for rice cultivation and yield potential. New Phytol. 145, 471–476.

Li, B., Xin, W., Sun, S., Shen, Q. and Xu, G. (2006) Physiological and molecular responses of nitrogen-starved rice plants to re-supply of different nitrogen sources. Plant Soil, 287, 145–159.

Li, Y.L., Fan, X.R. and Shen, Q.R. (2008) The relationship between rhizosphere nitrification and nitrogen-use efficiency in rice plants. Plant Cell Environ. 31, 73–85.

Li, Y., Gu, M., Zhang, X., Zhang, J., Fan, H., Li, P., Li, Z. et al. (2014) Engineering a sensitive visual tracking reporter system for real-time monitoring phosphorus deficiency in tobacco. Plant Biotechnol. J. 12, 674–684.

Liu, X., Huang, D., Tao, J., Miller, A.J., Fan, X. and Xu, G. (2014) Identification and functional assay of the interaction motifs in the partner protein OsNAR2.1 of the two-component system for high-affinity nitrate transport. New Phytol. 204, 74–80.

Mae, T. and Ohira, K. (1981) The remobilization of nitrogen related to leaf growth and senescence in rice plants (Oryza sativa L.). Plant Cell Physiol. 22, 1067–1074.

Nazoa, P., Vidmar, J.J., Tranbarger, T.J., Mouline, K., Damiani, I., Tillard, P., Zhuo, D. et al. (2003) Regulation of the nitrate transporter gene AtNRT2.1 in Arabidopsis thaliana: responses to nitrate, amino acids, and developmental stage. Plant Mol. Biol. 52, 689–703.

Ntanos, D.A. and Koutroubas, S.D. (2002) Dry matter and N accumulation and translocation for Indica and Japonica rice under Mediterranean conditions. Field. Crop. Res. 74, 93–101.

Okamoto, M., Kumar, A., Li, W., Wang, Y., Siddiqi, M.Y., Crawford, N.M. and Glass, A.D. (2006) High-affinity nitrate transport in roots of Arabidopsis depends on expression of the NAR2-like gene AtNRT3.1. Plant Physiol. 140, 1036–1046.

Ookawa, T., Hobo, T., Yano, M., Murata, K., Ando, T., Miura, H., Asano, K. et al. (2010) New approach for rice improvement using a pleiotropic QTL gene for lodging resistance and yield. Nat. Commun. 1, 132.

Orsel, M., Chopin, F., Leleu, O., Smith, S.J., Krapp, A., Daniel-Vedele, F. and Miller, A.J. (2006) Characterization of a two-component high-affinity nitrate uptake system in Arabidopsis. Physiology and protein–protein interaction. Plant Physiol. 142, 1304–1317.

Pan, S., Rasul, F., Li, W., Tian, H., Mo, Z., Duan, M. and Tang, X. (2013) Roles of plant growth regulators on yield, grain qualities and antioxidant enzyme activities in super hybrid rice (Oryza sativa L.). Rice, 6, 9.

Quesada, A., Galvan, A. and Fernandez, E. (1994) Identification of nitrate transporter genes in Chlamydomonas reinhardtii. Plant J. 5, 407–419.

Rooke, L., Byrne, D. and Salgueiro, S. (2000) Marker gene expression driven by the maize ubiquitin promoter in transgenic wheat. Ann. Appl. Biol. 136, 167–172.

Sasakawa, H. and Yamamoto, Y. (1978) Comparison of the uptake of nitrate and ammonium by rice seedlings. Plant Physiol. 62, 665–669.

Sasaki, T., Matsumoto, T., Yamamoto, K. et al. (2002) The genome sequence and structure of rice chromosome 1. Nature, 420, 312–316.

Scheible, W.R., Gonzalez-Fontes, A., Lauerer, M., Muller-Rober, B., Caboche, M. and Stitt, M. (1997) Nitrate acts as a signal to induce organic acid metabolism and repress starch metabolism in tobacco. Plant Cell, 9, 783–798.

Scheible, W.R., Morcuende, R., Czechowski, T., Fritz, C., Osuna, D., Palacios-Rojas, N., Schindelasch, D. et al. (2004) Genome-wide reprogramming of primary and secondary metabolism, protein synthesis, cellular growth processes, and the regulatory infrastructure of Arabidopsis in response to nitrogen. Plant Physiol. 136, 2483–2499.

Souza, S.R., Stark, E.M.L.M. and Fernandes, M.S. (1998) Nitrogen remobilization during the reproductive period in two Brazilian rice varieties. J. Plant Nutr. 21, 2049–2063.

Srikanth, B., Subhakara, R.I., Surekha, K., Subrahmanyam, D., Voleti, S.R. and Neeraja, C.N. (2016) Enhanced expression of OsSPL14 gene and its association with yield components in rice (Oryza sativa) under low nitrogen conditions. Gene, 576, 441–450.

Stitt, M. (1999) Nitrate regulation of metabolism and growth. Curr. Opin. Plant Biol. 2, 178–186.

Tang, Z., Fan, X., Li, Q., Feng, H., Miller, A.J., Shen, Q. and Xu, G. (2012) Knockdown of a rice stelar nitrate transporter alters long-distance translocation but not root influx. Plant Physiol. 160, 2052–2063.

Wang, R., Guegler, K., LaBrie, S.T. and Crawford, N.M. (2000) Genomic analysis of a nutrient response in Arabidopsis reveals diverse expression patterns and novel metabolic and potential regulatory genes induced by nitrate. Plant Cell, 12, 1491–1509.

Xu, G., Fan, X. and Miller, A.J. (2012) Plant nitrogen assimilation and use efficiency. Annu. Rev. Plant Biol. 63, 153–182.

Yan, M., Fan, X., Feng, H., Miller, A.J., Sheng, Q. and Xu, G. (2011) Rice OsNAR2.1 interacts with OsNRT2.1, OsNRT2.2 and OsNRT2.3a nitrate transporters to provide uptake over high and low concentration ranges. Plant, Cell Environ. 34, 1360–1372.

Yong, Z., Kotur, Z. and Glass, A.D. (2010) Characterization of an intact two-component high-affinity nitrate transporter from Arabidopsis roots. Plant J. 63, 739–748.

Zhang, H. and Forde, B.G. (2000) Regulation of Arabidopsis root development by nitrate availability. J. Exp. Bot. 51, 51–59.

Zhang, Y.L., Fan, J.B., Wang, D.S. and Shen, Q.R. (2009) Genotypic differences in grain yield and physiological nitrogen use efficiency among rice cultivars. Pedosphere, 19, 681–691.

Zhao, Y., Hu, Y., Dai, M., Huang, L. and Zhou, D.X. (2009) The WUSCHEL-related homeobox gene WOX11 is required to activate shoot-borne crown root development in rice. Plant Cell, 21, 736–748.

Zhuo, D., Okamoto, M., Vidmar, J.J. and Glass, A.D. (1999) Regulation of a putative high-affinity nitrate transporter (Nrt2;1At) in roots of Arabidopsis thaliana. Plant J. 17, 563–568.

4

Overexpression of *Arabidopsis* and Rice stress genes' inducible transcription factor confers drought and salinity tolerance to rice

Karabi Datta[1], Niranjan Baisakh[2], Moumita Ganguly[1], Sellapan Krishnan[3], Kazuko Yamaguchi Shinozaki[4] and Swapan K. Datta[1,5]

[1]*Department of Botany, University of Calcutta, Kolkata, India*
[2]*School of Plant, Environmental and Soil Sciences, Louisiana State University Agriculture Center, Baton Rouge, LA, USA*
[3]*Department of Botany, Goa University, Goa, India*
[4]*Graduate School of Agricultural and Life Sciences, University of Tokyo, Yayoi, Bunkyo-Ku, Tokyo, Japan*
[5]*Indian Council of Agricultural Research, Krishi Bhawan, New Delhi, India*

*Correspondence
email
swpndatta@yahoo.com

Keywords: transcription factor, stress-inducible promoter, osmotic stress tolerance, rice, transformation.

Summary

Rice yield is greatly affected by environmental stresses such as drought and salinity. In response to the challenge of producing rice plants tolerant to these stresses, we introduced cDNA encoding the transcription factors *DREB1A* and *DREB1B* under the control of the stress inducible *rd29* promoter. Two different indica rice cultivars were used, BR29, an improved commercially cultivated variety from Bangladesh and IR68899B, an IRRI bred maintainer line for hybrid rice. *Agrobacterium* mediated transformation of BR29 was done independently with *DREB1A* isolated from rice and *Arabidopsis* and *DREB1B* isolated from rice, whereas biolistic transformation was done with rice- *DREB1B* in the case of IR68899B. Initial genetic integration was confirmed by PCR and Southern blot analysis. Salinity tolerance was assayed in very young seedlings. Drought stress tests were found to be more reliable when they were carried out at the pre-flowering booting stage. RNA gel blot analysis as well as quantitative PCR analysis was performed to estimate the transcription level under stressed and unstressed conditions. Agronomic performance studies were done with stressed and unstressed plants to compare the yield losses due to dehydration and salt loading stresses. Noticeably enhanced tolerance to dehydration was observed in the plants transformed with *DREB1A* isolated from *Arabidopsis* while *DREB1B* was found to be more effective for salt tolerance.

Introduction

Environmental stresses have great negative effects on plant growth and productivity. Unfavorable growing environments routinely cause major reduction of the yield potential of crops. Abiotic stresses such as drought, low temperature and saline soils account for more losses of agricultural production than any other factors. Hence, it is a major challenge to sustain or to improve crop yields to feed the rapidly increasing world population.

Plants respond to different stresses by a number of biochemical and physiological adaptations that involve function of many genes. However, the number of genes that regulate the process and the way they are coordinated is not yet clearly understood. Although, different plant species have different thresholds for stress tolerance, most cultivated crop plants are highly sensitive when they are exposed to long periods of stress. Osmotic stress particularly water deficit in plants occur when water loss due to transpiration exceeds the supply of water from the soil. Prolonged water shortages affect metabolic activities and eventually result in severe reductions in plant productivity. Improvement of water stress tolerance in crop plants has a significant impact on agricultural productivity.

Breeding for osmotic stress tolerance in crop plants utilizes existing genetic stocks, which are limited. However, a transgenic approach may offer an effective alternative way of improving tolerance to dehydration by incorporating genes from any source with a better understanding of the mechanism involved. Various approaches have been used to produce transgenic plants with increased tolerance to osmotic stress. These include overproduction of enzymes responsible for biosynthesis of osmolytes, late-embryogenesis-abundant proteins and detoxification enzymes, which are involved in reducing the reactive oxygen species (Bohnert and Jensen, 1996). Each of these strategies involved transfer of a gene expressing a single specific stress protective protein. However, it is well established that plant tolerance to dehydration stress is mediated by a number of physiological and biochemical process which means a multigene trait. The activation of such genes must involve a distinctive set of transcription factors. By over-expressing transcription factor gene(s), it may be possible to change or to increase the level of expression of several downstream target genes responsible for dehydration tolerance at the same time (Varshney et al., 2011).

Signal transduction during plant stress responses has been studied in various plants from a variety of angles to identify stress responsive genes and their regulatory mechanisms.

A large number of genes are regulated when plants are exposed to osmotic stress. The transcription factor *DREB1A* interacts with cis-acting DRE (dehydration responsive element) and regulates expression of many stress tolerance genes under drought, high salinity, and cold stress in *Arabidopsis* (Liu et al., 1998). In plant, majority of the *DREB1* subfamily members are expressed under low temperature and increased level of drought or salt (Dubouzet et al., 2003; Haake et al., 2002; Huang et al., 2007; Xiao et al., 2006). It has been reported that homologous or heterologous expression of *DREB1*s can confer tolerance to multiple abiotic stresses as shown in rice (Wang et al., 2008), peanut (Bhatnagar-Mathur et al., 2007) and other plants (Varshney et al., 2011).

Over-expression of *DREB1A* cDNA under the control of CaMV35S constitutive promoter in transgenic *Arabidopsis* plants activated the expression of many stress tolerance genes and resulted in increased tolerance to drought, salt loading and freezing (Gilmour et al., 2000; Liu et al., 1998). It has been reported that the over-expression of cDNA encoding *DREB1A* induced the expression of many stress tolerance genes such as *rd29A, kin1, Cor 6.6, Cor 15a, rd17, erd10, erd1* and *P5CS* in *Arabidopsis* (Kasuga et al., 1999). Similarly, over-expression of *DREB1* genes can increase the expression of their direct downstream genes with DRE-cis element, such as RD29A, COR15A, ERD10, COR47 and GoLS2 (Maruyama et al., 2004).

The strong constitutive *CaMV35S* promoter driving expression of *DREB1A* resulted in growth abnormalities under normal conditions. Use of the stress-inducible promoters such as rRab16A, 4XABRE, 2XABRC are effectively expressed in rice (Ganguly et al., 2011). Similarly, *rd29A* promoter to drive the expression of *DREB1A* provided a better stress tolerance of the transgenic *Arabidopsis* under stress conditions with minimal effects on plant growth than when the expression of the same gene was driven by the 35S promoter (Kasuga et al., 1999). Similar results were observed in rice plants over-expressing *OsDREB1A, OsDREB1B* or *OsDREB1F*, cloned from rice (Ito et al., 2006; Wang et al., 2008). Rice, OsDREB1-type proteins show high homology to *Arabidopsis* DREB1A proteins (Dubouzet et al., 2003).

Abiotic stresses like drought and salinity are major environmental constraints of rice production in non-irrigated rice areas. Of the world's 130 million hectares of rice cultivated land, an estimated 20% are periodically subject to drought conditions, 30% contain enough salt to limit cultivation and reduce yield, and 10% occasionally experience low temperature (15 °C or below) (Lane, 2002). Conventional breeding approach has been used to exploit natural genetic variation in improving rice varieties. But until now, rice plants except for a few rice cultivars that are well adapted to water stress conditions, do not show osmotic adjustment, the most effective component of abiotic stress. Several stress-related genes have been cloned and introduced in rice to enhance tolerance to drought and salinity stress (Datta et al., 2008).

In the research reported here, we introduced the transcription factor *DREB1A* and *DREB1B* cDNA driven by *rd29A* promoter into two different indica rice cultivars. *DREB1A* cDNA has been isolated from both *Arabidopsis* and rice plants. Over-expression of *DREB1A* cDNA isolated from *Arabidopsis* performed better in rice than the same isolated from rice when the plants are exposed to artificial drought stress in pots, whereas plants with inserted *DREB1B* cDNA from rice showed better performance in highly saline conditions.

Results

Transformation

Plants were transformed with a G29 AHS vector expressing the *DREB1A* cDNA isolated from either rice or *Arabidopsis* or *DREB1B* cDNA isolated from rice with the *rd29* promoter (Figure 1). Twenty six antibiotic resistant (hygromycin) BR29 plants carrying rice *DREB1A* (OsDREB1A), fourteen antibiotic resistant BR29 plants carrying rice *DREB1B* (OsDREB1B) and 23 hygromycin resistant BR29 plants with *Arabidopsis* DREB1A (AtDREB1A) were generated using the vacuum infiltration method for *Agrobacterium* mediated transformation. Twenty antibiotic resistant IR68899B plants carrying rice *DREB1B* (OsDREB1B) were obtained by biolistic transformation (Table 1). From the generated plants, 24 primary independent transgenics from variety BR29 and four primary independent transgenics from IR68899B were selected by PCR and DNA-blot analysis. Growth and fertility of the plants were compared with wild-type of their respective cultivars. Most of these independent primary transgenics showed a normal phenotype and were fertile. No significant difference was found in growth and fertility status of the primary transgenic plants with their respective wild type based on selected fast growing *in vitro* regenerants.

Progeny

The T_0 plants after self pollination gave segregating T_1 progeny. PCR analysis was done for all the T_1 progeny to find out the segregation ratio and to identify the positive plants containing gene of interest. Most of the progeny lines showed the Mendelian segregation ratio of 3:1 (Figure 2). Southern blot analysis was done with the identified positive plants from different lines. The Southern-positive plants of BR29, transformed by *Agrobacterium* mediated transformation, showed a very simple pattern of integration with one to four copies of the transgene. Figure 3 represents one Southern blot showing integration of *AtDREB1A* in positive progeny of different lines of BR29 selected by PCR analysis (line BRSH29, BRSH24, and BRSH22) at the T_1 generation. All the plants showed the integrated 0.8 Kb fragment of *DREB1* gene.

Drought tolerance at vegetative stage

Water stress tolerance experiments were conducted on homozygous plants at the T_3 generation. To study drought tolerance, three sets of experiments were performed for three phases of plant growth. For the vegetative stage drought tolerance experiment, 6-week-old non-transformed wild type and transgenic plants grown in soil were subjected to 21 days of continuous drought stress i.e. watering was stopped completely. After

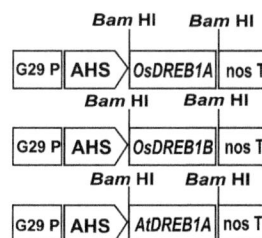

Figure 1 Partial maps of the plasmid vectors of *DREB* genes driven by rd29 promoter used for Biolistic or *Agrobacterium*-mediated transformation in indica rice.

Table 1 Transgenic indica rice developed with *DREB1* genes

Cultivar	Gene of interest	Method	Promoter	No. of plants GGH	No. of Independent. Transgenic events
BR29	OsDREB1A	Agrobacterium	rd29A	26	12
	OsDREB1B	Agrobacterium	rd29A	14	5
	AtDREB1A	Agrobacterium	rd29A	23	5
IR68899B	OsDREB1B	Biolistic	rd29A	20	4

GGH, grown in greenhouse.

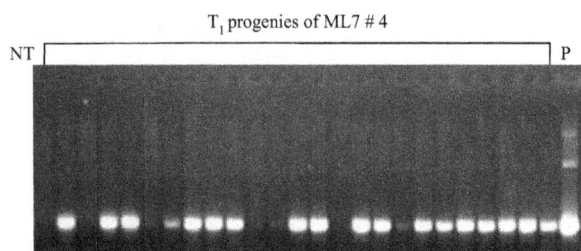

Figure 2 PCR analysis showing the segregating *OsDREB1B* gene in T_1 transgenics of rice cv. IR68899B developed through biolistic transformation.

drought treatments, a variation of response was found in individual lines. Non transgenic wild type plants exhibited stress symptoms like wilting and drought induced leaf rolling within 7 days, whereas the transgenic lines did not show much difference within this short period of water stress. After drought stress and subsequent watering, recovery of the plants was observed depending on the damage caused by the stress. Sometimes new tillers came out and ultimately this stress at the vegetative stage did not cause much damage to the crop yield. Figure 4a–d represent differential phenotypic expression due to the water stress treatment and recovery of plants after watering in the case of BR29 (BRSH29 homozygous progeny). It has been observed that the lines with inserted *AtDREB1A* gene may be more tolerant to drought stress than the plants carrying *OsDREB1A* or *OsDREB1B*.

Stomatal behavior

Experiments were carried out to understand the stomatal behavior of transgenic and wild type plants during a period of drought. Water was withdrawn from both transgenic and wild type plants for 6 days in greenhouse conditions and stomatal response was observed. The results showed that in wild type rice plants almost all the stomata were open; however in transgenic rice (BRSH29) most of the stomata were closed (Figure 5). The stomatal closure in transgenic plants during drought conditions may be due to the expression of the transgene regulating the functioning of guard cells in rice plants.

Expression analysis of the transgenics

The expression of *rd29A DREB1* gene during drought stress at the vegetative stage was detected by RNA gel blot analysis. RNA was extracted from the leaves of transgenic plants and their respective wild type plants grown under normal condition and also after exposure to drought stress for 6 days. The induced expression of the inserted gene has been detected by the strong expression of mRNA extracted from leaves of dehydration stressed transgenic plants with *AtDREB1A* gene controlled by *rd29* stress inducible promoter. Figure 6 represents mRNA expression of three different plants from different lines during each day after exposing to water stress.

Quantitative RT-PCR (qRT-PCR) analyses showed 10.8- to 14.4-fold induction of AtDREB1A mRNA levels in the drought stressed transgenic lines compared to that of well watered controlled transgenic lines (Figure 7). The fold-induction levels varied widely in the transgenics, the maximum induction being recorded in BRSH 22-22-25-7, which was 14.4-fold upon 6 days of drought treatment.

Drought tolerance at flowering stage

To study drought tolerance at the pre-flowering stage, 10 week old plants in pots just before the emergence of flowers i.e. the booting stage, were subjected to drought stress for 7 days. After 7 days the soil became too hard. More damage was observed in the flowering of the wild type than in the different variable response in the transgenic plants at the pre-flowering stage water stress situation, which was also reflected in the crop yield. In Table 2 the effect of water stress on crop yield in some selected transgenic BR29 plants (homozygous progenies of BRSH22 and BRSH29) at the pre-flowering stage has been shown. Wild type BR29 severely affected by 7 days drought exposure at the pre-flowering stage whereas transgenics were less affected. The grain yield per plant was affected by water stress in all the plants which were compared with that of their normal watered plants of the same line but damage in yield

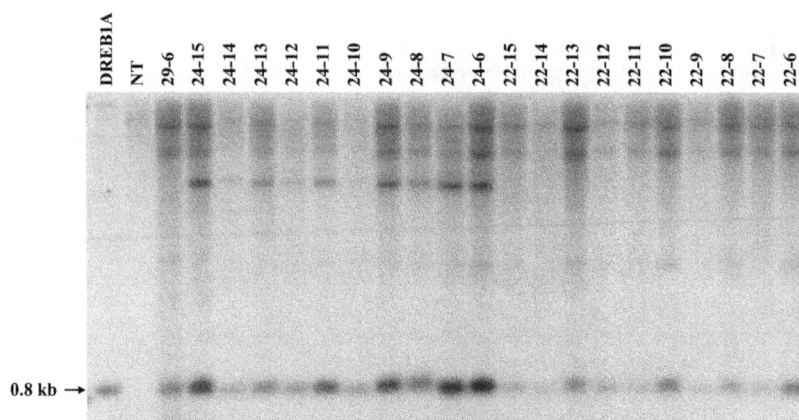

Figure 3 Southern blot showing the stable integration of 0.8 kbp of *DREB1A* in progenies of different lines of BR29.

Figure 4 Differential phenotypic expression due to water stress at vegetative stage (BRSH29 progeny). (a) Showing levels of tolerance after 14 days (b) selected transgenic plants surviving in hard soil after 21 days without water. (c) Quick recovery of water stressed transgenic plants after watering (d) recovery of water stressed plants after watering, two plants from right are non transgenic wild type.

was more in case of wild type (7.84 g/plant) as compared to that of transgenics (14.20–17.81 g/plant).

Tolerance to salt

The technique for screening salt tolerance is based on the ability of seedlings to grow in high saline conditions. To examine the tolerance level of the transgenic plants, homozygous progeny of BRSH29 and BRSH22 containing *AtDREB1A* and progeny of IR68899B (designated as ML7) containing *OsDREB1B,* to salt stress, 8 day old seedlings were subjected to salt stress in hydroponic nutrient solution. Fourteen days of continuous salt stress with EC 12 salt level (with renewal of the nutrient solution every 7 days) the transgenic plants showed a moderately tolerant nature (Table 3). After 14-days in EC 12 the same plants were subjected to EC18 salt level for another 14 days. This stringent salt pressure killed almost all the plants excepting only a few salt exposed IR68899B (ML7) plants with *DREB1B* gene inserted (Figure 8). Five plants from the surviving line ML-4-17 were grown to soil after the experiment. Those plants showed much better seed setting than the non-transgenic wild type stressed plant but less seed setting than those plants which

were grown in normal condition without salt stress from seeding stage (Table 4).

Discussion

Plant responses to water and salt stresses have much similarity in most metabolic processes. The high saline condition reduces the ability of plants to take up water. The water stress effect caused by drought on growth is similar to the osmotic effect of salt on the initial phase. Thus, any improvement in drought resistance may make a plant more adapted to saline soil.

Many plant genes have been reported to respond to abiotic stresses like drought, high salt condition and low temperature, and proteins encoded by these genes are expected to enhance the tolerance level of plants to these stresses. The analyses of these stress proteins and their corresponding genes provide us with the possibility of developing stress resistant plants.

In this present study, we were able to generate transgenic rice plants in which the *DREB1A* or *DREB1B* cDNAs were introduced to overexpress the DREB protein. The transgenic plants

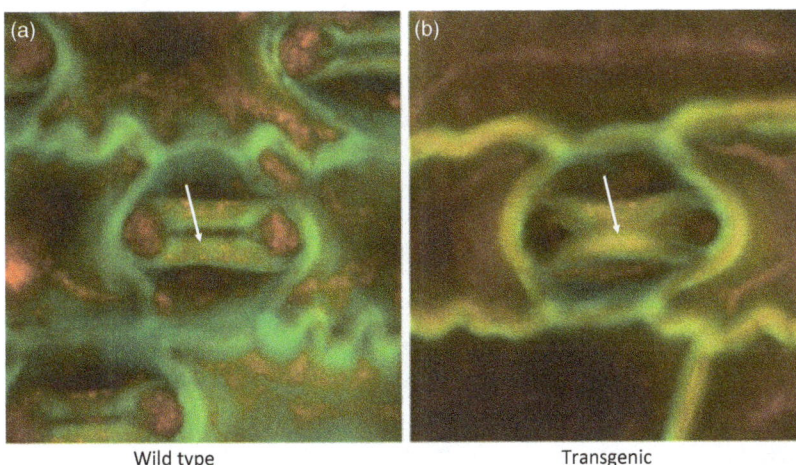

Wild type   Transgenic

Figure 5 Laser confocal micrographs of epidermal peels of transgenic and wild type rice plants under 6 days of drought condition stained with acridine orange. Wild type and transgenic (a) shows the stomata opening (wild type) and (b) closer of stomata (transgenic BRSH29 progeny).

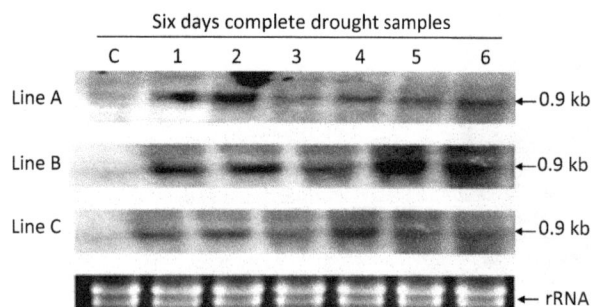

Figure 6 RNA gel blot showing gene expression in different lines under drought stress condition. Line A: BRSH29-22-2-19-4; Line B: BRSH29-22-2-16-15; Line C: BRSH22-22-25-7-13; Days 1–6 represents each day after exposing to drought stress. Letter 'C' indicates well watered controlled transgenic plants.

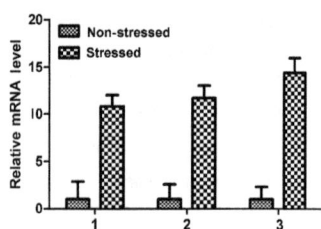

Figure 7 qRT-PCR analyses showing the fold induction of AtDREB1A expression in response to drought stress for 6 days and well watered controlled transgenic plants of three different transgenic lines 1:BRSH29-22-2-19-4; 2:BRSH29-22-2-16-15; 3:BRSH22-22-25-7-13. The expression of gene determined using *tubulin* gene as an internal control. Results represent means ± SE, based on three replicates.

showed enhanced tolerance to drought and salinity because of overexpression of this single transcription factor governed by *rd29* stress inducible promoter. The stress response study of the

exposed plants at their different growth and developmental phases was based on the phenotypic changes due to stress and their ultimate effect on grain yield.

Homozygous lines with the *DREB1* gene have been developed and the phenotypic expression study was conducted with the over-expressed promising lines as has been selected by molecular analysis. It was observed that DREB1A transgenic rice plants remained green for longer time during drought stress condition when compared to DREB1B transgenic rice and the non-transgenic wild type. It has also been reported earlier that DREB1A *Arabidopsis* plants had significantly improved freezing and drought tolerance and downstream gene expression level in DREB1A were significantly higher than that in DREB1B *Arabidopsis* plants (Novillo *et al.*, 2007; Tong *et al.*, 2009).

Expression of *DREB1A cDNA* encoded by *rd29* inducible promoter in our study showed no significant phenotypic change in growth of transgenic rice plant. The stress inducible promoter appears to minimize the negative effects of the transgene on plant growth has been reported in *Arabidopsis* (Kasuga *et al.*, 1999).

Stomata closure to reduce water loss is an important feature to improve tolerance of plants under water limitation. Genetic determinants governing the stomatal function and consequent improvement of plant performance under water limitation have been identified (Pardo, 2010). Several transcription factors have been known which are involved in the regulation to the signaling network that controls stomatal movements in *Arabidopsis*. Transgenic plants expressing the transcription factor *MYB44* showed ABA-induced stomata closure response in transgenic plants (Jung *et al.*, 2008). Overexpression of *ARAG1*, a transcription factor gene, in rice plant may promote the synthesis of higher level of ABA in transgenic plants (Zhao *et al.*, 2010). The internal increase in concentration of phytohormone ABA may eventually lead to early stomata closer in transgenic plants which prevent major water loss (Apel and Hirt, 2004). Interestingly, in the present study, stomata closure in transgenic plants

Table 2 Effect of water stress on agronomic trait in selected BR29 transgenic plants with At*DREB1A* gene at T_3 generation (pre flowering stage: 7 days of water stress)

Plant line	Plant No.	No. reproductive tillers	No. filled grain	No. unfilled grain	Ratio (f/uf)	Grain yield/plant (g)
BRSh-22-24-6-11	6	10	670	488	1.37	16.20
	7	10	611	516	1.18	14.36
	9	7	601	283	2.15	14.20
	Watered*	9	937	126	7.43	22.59
BRSh-29-22-2-16	10	8	614	501	1.22	14.52
	13	8	618	349	1.77	15.22
	15	9	739	362	2.04	17.81
	Watered*	8	767	142	5.40	20.17
BRSh-29-22-2-19	4	6	687	279	2.46	16.16
	8	8	695	303	2.29	15.88
	12	6	541	294	1.8	13.20
	Watered*	7	681	119	5.7	16.35
BRSh-22-22-25-6	9	8	685	522	1.31	16.63
	Watered*	6	797	137	5.81	18.84
BRSh-22-22-25-7	13	9	583	533	1.09	17.14
	Watered*	7	809	127	6.37	19.40
Non transgenic control	Treated	7	339	525	0.64	7.84
	Watered*	6	754	98	7.69	18.56

*Data scored based on average five plants of the same transgenic line.

Table 3 Response of transgenic lines transformed with *DREB1* genes to salinity

| Rice line | Gene used | EC-rating | |
		EC12*	EC18*
BR29-22-22-25-6	At*DREB1A*	5	9
BR29-22-22-25-7	At*DREB1A*	5	9
BR29-22-22-25-10	At*DREB1A*	**3**	9
BR29-22-22-25-11	At*DREB1A*	5	9
BR29-22-22-25-13	At*DREB1A*	5	9
BR29-29-22-25-14	At*DREB1A*	5	9
BR29-29-22-25-16	At*DREB1A*	5	9
BR29-29-22-25-5	At*DREB1A*	5	9
ML7-4-13	Os*DREB1B*	5	7
ML7-4-16	Os*DREB1B*	**3**	9
ML7-4-17	Os*DREB1B*	**3**	**5**
ML7-4-18	Os*DREB1B*	**3**	7
BR29 (control)	Check	9	9
IR29 (control)	Sensitive check	9	9
ML7 (control)	Check	9	9

ML7 = IR6899B (Indica maintainer line).

*Data scored after 2 and 4 weeks (at EC12, 2 week & EC18 2 week);

9 rating represent maximum negative effect.

Bold EC 3 means tolerant and bold 5 indicates moderately tolerant.

during drought conditions may be due to the expression of the transcription factor transgene in rice plants as shown in Figure 5.

In rice plants, water stress at the vegetative stage does not have much effect on the grain yield as rice plants at the tillering stage can recover very quickly and new tillers can replace the damaged tillers. But in the pre-flowering stage and post flowering stage dehydration affects proper flowering and seed development. It was shown that the non-transgenic wild type plants under water stress produced more unfilled grains than filled grains and consequently grain yield is very low, whereas yield

Figure 8 Effect of 14 days continuous salt stress with EC18; non-transgenic wild plants (left five rows), Transgenic IR68899B with Os*DREB1B* (b) (right five rows). EC has been measured in desi Siemens per meter (dS/m).

loss in the case of the transgenic plants under water stress is much less. Water stress after flowering affects the grain filling of rice, thus causing severe reduction of the grain yield.

Transgenic plants with the *DREB1* gene showed more tolerance to salt stress when compared with their respective wild type plants. All the transgenic plants grown at the EC12 salt level for fourteen days were moderately tolerant [3–5 standard evaluation score (SES)] compared with their wild type (9 SES). The same plants after growing in nutrient solution with salinity level EC12 for 2 weeks were again grown in salinity level EC18 most of the transgenic plants died (i.e. SES 9), excepting only one line transformed with *DREB1B* gene which survived. The tolerance level of wild type variety BR29 and line IR68899B seems to be the same as is shown by the data scored after 2 weeks (SES 9), all died at salinity level EC12. It might be possible that for salinity tolerance, *DREB1B* gene may be more effective than the *DREB1A*.

Here we have demonstrated the enhancement of drought and salt tolerance with a single transcription factor gene encoded by a stress inducible promoter. It is known that different independent regulatory systems are involved in abiotic stress responsive gene expression. A better understanding of the physiological and biochemical basis of stress tolerance is still needed.

Experimental protocol

Plasmid DNA and bacterial strain

The plasmids pG29 AHS + Os*DREB1A*, pG29 AHS + Os*DREB1B* and pG29 AHS + At*DREB1A* were introduced into *Agrobacterium tumefaciens* strain EHA105. These plasmids contain marker genes for neomycin phosphotransferase (*nptII*), hygromycin phosphotransferase (*hph*) and the stress responsive gene Os*DREB1A*, Os*DREB1B* and At*DREB1A* controlled by *rd29* stress inducible promoter (Kasuga *et al.*, 1999). The bacteria were grown for 48 h in LB medium containing rifampicin (20 µg/mL) and kanamycin (50 µg/mL).

Transformation

Two different rice cultivars BR29 and IR68899B were chosen for rice transformation. Embryogenic calli were obtained from 9 to 14 day-old immature embryos of BR29. Three-to-four-week-old actively growing embryogenic calli was taken as explants for *Agrobacterium* mediated transformation whereas for IR68899B, 10 day old immature embryos were used for biolistic transformation.

For *Agrobacterium*-mediated transformation we followed the method described by Datta *et al.*, 2000;. Vacuum infiltration was applied and calli were co-cultivated with *Agrobacterium* in 200 µM acetosyringone containing medium for 3 days. After 3 days of cocultivation, the *Agrobacterium* was removed by washing and the plant tissue was allowed to grow in medium containing 250 mg/L cefotaxime (to inhibit the growth of *Agrobacterium*) and 50 mg/L hygromycin (selective agent for transformation). After four cycles of selection the embryogenic calli were allowed to regenerate in regeneration medium. Regenerated plants were transferred to pots for natural growth. For biolistic transformation of IR68899B we followed the method described in earlier paper (Datta *et al.*, 1998). Selected regenerated putative transgenic plants were grown in pots in greenhouse.

Table 4 Effect of salinity stress on yield performance of salinity stress survived plants (IR68899B)

Plant no.	Reproductive tillers	No. of filled grains	No. of unfilled spikelets	No. of total spikelets	Filled: unfilled	Grain yield per plant (g)
ML7-4-17-1	10	594	408	1002	1.45	12.14
ML7-4-17-2	5	409	169	578	2.42	8.67
ML7-4-17-3	9	355	135	490	2.63	7.10
ML7-4-17-4	10	581	278	859	2.09	10.79
ML7-4-17-5	11	602	338	940	1.78	11.96
*IR68899B (without stress)	10.4	1017	156	1173	6.51	19.95
IR68899B (stressed control)	5	365	257	622	1.42	7.3

*Average of five plants grown under normal condition up to maturity.

Southern blot analysis

Ten microgram of DNA was digested with *BamHI* restriction endonuclease (Invitrogen, Carlsbad, CA). The digested DNA samples were separated by electrophoresis on 1% (W/V) TAE-agarose gel. Southern membrane transfer, hybridization and autoradiography were done as previously described (Datta *et al.*, 1998). The *BamHI* fragments containing the ~0.8 Kb DREB genes of *DREB1A* and *DREB1B* were radio labeled with '-α32P-dCTP and used correspondingly as hybridization probe.

RNA gel blot analysis

Total RNA was isolated by the GITC (Guanidine isothiocyanate) method (Chomczynski and Sacchi, 1987). Equal amounts of total RNA (20 μg) were electrophoresed on 1.4% MOPS-Formaldehyde gel, transferred to nylon membrane (Hybond-N; Amersham place, Buckinghamshire, UK) and baked by 80 °C for 2 h. PCR amplified product of *DREB* gene were labeled by random prime method (Radiprime; Amersham) and used as a probe. Radio-labeled (50 μci α32P-dCTP; Perkin-Elmer, Wellesley, MA) probes were denatured and hybridized to the membrane at 65 °C in a hybridization buffer (6× SSC, 5% Dextran sulphate, 0.05 M Sodium phosphate pH 7.2, 5× Denhardt's solution, 0.5% SDS, 0.0025 M EDTA and 100 μg/mL Salmon sperm DNA) for 12–16 h and washed for 15 min each with 1× SSC, 0.5% SDS and 0.5× SSC, 0.5% SDS. The hybrdization signal was observed on Kodak X ray film s after 2–3 days exposure (Chomczynski and Sacchi, 1987).

Quantitative RT-PCR

The qRT-PCR was performed after 6 days of drought stress with gene specific primers and the cycle was as follows: 95° C for 30 s, 60 °C for 30 s and 72 °C for 30 s. The procedure was according to the manufacturer's instructions (CFX 96 Real time system; Biorad, Hercules, CA). The amplification of *tubulin* gene was used as internal control to normalize all data. To validate the qRT-PCR results, the experiments were repeated three times. The mean values for the expression levels of the genes were calculated from three independent experiments.

Drought stress treatment

The screening for drought tolerance is based on the phenotypic changes and survival of plants due to the water stress condition and ultimately its effect on grain yield. It was performed at three phases of plant growth with three different sets of plants. For the vegetative phase watering of 6-week-old plants was stopped for 21-days or as long as the wild type plants have severe damage. In the case of pre-flowering stage screening, water supply was stopped for 7-days at the booting stage. For the flowering stage screening, plants were not supplied with water at the start of the flowering stage for 7-days. In all the three different sets of experiment, visual phenotype changes, recovery of the plants and its effect on grain yield were observed.

Stomata studies

Water was withdrawn from transgenic and wild type rice plants of five different transgenic plants of BRSH29 for 6 days and leaf materials were collected and used for stomata studies. Epidermal peels were obtained from rice leaves 1-cm-long pieces of the leaves were scraped on the abaxial sides to remove most of the cells above the adaxial epidermis, then the isolated adaxial epidermis was stained with 0.1% acridine orange for 15 min (dye was prepared by dissolving 0.1 g of acridine orange in 0.05 M phosphate buffer), washed thoroughly in distilled water, mounted with dilute glycerin, observed and photographed using laser confocal microscope (LSM 510 with Carl Zeiss Axioplan-2, Göttingen, Germany).

Screening for salinity tolerance

Salinity tolerance screening was done at the seedling stage. Four day old seedlings were placed on styrofoam seedling float with nylon net bottom on distilled water. The radicles were inserted through the nylon mesh. After 3 days when the seedlings were well established the water was replaced with salinized nutrient solution. Initial salinity was EC = 12 dS/m by adding NaCl to the nutrient solution. Renewal of the solution was done every 7-days and pH at 5.0. The first scoring was taken after 14-days at EC = 12 dS/m. After 14-days, the salinity level was increased to EC = 18 dS/m. The second scoring was taken after 14-days at EC = 18 dS/m. The modified SES in rating the visual symptom of salt toxicity was used (Gregorio *et al.*, 1997).

Acknowledgements

The Financial support from Department of Biotechnology (DBT), Goverment of India in the form of DBT Programme Support is thankfully acknowledged. The authors are grateful to Professor Malcolm Elliott, University of Leicester, UK for editorial assistance.

References

Apel, K. and Hirt, H. (2004) Reactive oxygen species: metabolism, oxidative stress and signal transduction. *Ann. Rev. Plant Biol.* **55**, 373–379.

Bhatnagar-Mathur, P., Devi, M.J., Reddy, D.S., Lavanya, M., Vadez, V., Serraj, R., Yamaguchi-Shinozaki, K. and Sharma, K.K. (2007) Stress-inducible expression of *AtDREB1A* on transgenic peanut (*Arachis hypogea* L.) increases transpiration efficiency under water-limiting conditions. *Plant Cell Rep.* **26**, 2071–2082.

Bohnert, H.J. and Jensen, R.G. (1996) Strategies for engineering water-stress tolerance in plants. *Trends Biotechnol.* **14**, 89–97.

Chomczynski, P. and Sacchi, N. (1987) Single step method of RNA isolation by acid guanidinium thiocyanate-phenol-chloroform extraction. *Anal. Biochem.* **162**, 156–159.

Datta, K., Vasquez, A., Tu, J., Torrizo, L., Alam, M.F., Oliva, N., Abrigo, E., Khush, G.S. and Datta, S.K. (1998) Constitutive and tissue-specific differential expression of *crylA(b)* gene in transgenic rice plants conferring resistance to rice insect pest. *Theor. Appl. Genet.* **97**, 20–30.

Datta, K., Koukolíková-Nicola, Z., Baisakh, N., Oliva, N. and Datta, S.K. (2000) Agrobacterium-mediated engineering for sheath blight resistance of indica rice cultivars from different ecosystems. *Theor. Appl. Genet.* **100**, 832–839.

Datta, K., Tuteja, N. and Datta, S.K. (2008) Transgenic research on plant abiotic stress and nutrition improvement of food derived from plants. In *A Transgenic Approach in Plant Biochemistry and Physiology* (Rivera-Dominguez, M., Rojas, R.T. and Tiznado-Hernandez, M.E., eds), pp. 181–215, Trivandrum: Signpost.

Dubouzet, J.G., Sakuma, Y., Ito, Y., Kasuga, M., Dubouzet, E.G., Miura, S., Seki, M., Shinozaki, K. and Yamaguchi-Shinozaki, K. (2003) *OsDREB* genes in rice, *Oryza sativa* L., encode transcription activators that function in drought-, high-salt and cold-responsive gene expression. *Plant J.* **33**, 751–763.

Ganguly, M., Roychoudhury, A., Sarkar, SN., Sengupta, D.N., Datta, S.K. and Datta, K. (2011) Inducibility of three salinity/abscisic acid regulated promoters in transgenic rice with *gusA* reporter gene. *Plant Cell Rep.* **30**, 1617–1625.

Gilmour, S.J., Sebolt, A.M., Salazar, M.P., Everard, J.D. and Thomashow, M.F. (2000) Overexpression of the Arabidopsis CBF3 transcriptional activator mimics multiple biochemical changes associated with cold acclimation. *Plant Physiol.* **124**, 1854–1865.

Gregorio, G.B., Senadhira, D. and Mendoza, R.D. (1997) The screening rice for salinity tolerance. IRRI Discussion Paper Series No. 22

Haake, V., Cook, D., Riechmann, J.L., Pineda, O., Thomashow, M.F. and Zhang, J.Z. (2002) Transcription factor *CBF4* is a regulator of drought adaptation in *Arabidopsis*. *Plant Physiol.* **130**, 639–648.

Huang, B., Jin, L.G. and Liu, J.Y. (2007) Molecular cloning and functional characterization of a *DREB1/CBF*-like gene (*GhDREB1L*) from cotton. *Sci. China Ser. C.* **50**, 7–14.

Ito, Y., Katsura, K., Maruyama, K., Taji, T., Kobayashi, M., Seki, M., Shinozaki, K. and Yamaguchi-Shinozaki, K. (2006) Functional analysis of rice *DREB1/CBF*-type transcription factors involved in cold-responsive gene expression in transgenic rice. *Plant Cell Physiol.* **47**, 141–153.

Jung, C., Seo, J.S., Han, S.W., Koo, Y.J., Kim, C.H., Song, S.I., Nahm, B.H., Choi, Y.D. and Cheong, J.J. (2008) Overexpression of AtMYB44 enhances stomatal closure to confer abiotic stress tolerance in *Arabidopsis*. *Plant Physiol.* **146**, 623–635.

Kasuga, M., Liu, Q., Miura, S., Yamaguchi-Shinozaki, K. and Shinozaki, K (1999) Improving plant drought, salt and freezing tolerance by gene transfer of a single stress-inducible transcription factor. *Nature Biotechnol.* **17**, 287–291.

Lane, P. (2002) C.U. engineers new rice strain. 26 November 2002. Cornell Daily Sun. http://www.cornelldailysun.com/articles/7100/ (accessed 15 March 2004).

Liu, Q., Kasuga, M., Sakuma, Y., Abe, H., Miura, S., Yamaguchi-Shinozaki, K. and Shinozaki, K. (1998) Two transcription factors, DREB1 and DREB2, with an EREBP/AP2 DNA binding domain separate two cellular signal transduction pathways in drought-and low-temperature-responsive gene expression, respectively, in Arabidopsis. *Plant Cell*, **10**, 1391–1406.

Maruyama, K., Sakamura, Y., Kasuga, M., Ito, Y., Seki, M., Goda, H., Shimada, Y., Yoshida, S., Shinozaki, K. and Yamaguchi-Shinozaki, K. (2004) Identification of cold-inducible downstream genes of *Arabidopsis DREB1A/CBF3* transcriptional factor using two microarray systems. *Plant J.* **38**, 982–993.

Novillo, F., Medina, J. and Salinas, J. (2007) *Arabidopsis CBF1* and *CBF3* have a different function than *CBF2* in cold acclimation and define different gene classes in the *CBF* regulation. *Proc. Natl Acad. Sci. U.S.A.* **104**, 21002–21007.

Pardo, J.M. (2010) Biotechnology of water and salinity stress tolerance. *Curr. Opin. Biotechnol.* **21**, 185–196.

Tong, Z., Hong, B., Yang, Y., Li, Q., Ma, N., Ma, C. and Gao, J. (2009) Overexpression of two chrysanthemum DgDREB1 group genes causing delayed flowering or dwarfism in *Arabidopsis*. *Plant Mol. Biol.* **71**, 115–129.

Varshney, R. K., Bansal, K.C., Aggarwal, P.K., Datta, S.K. and Craufurd, P.Q. (2011) Agricultural biotechnology for crop improvement in a variable climate: hope or hype? *Trends Plant Sci.* **16**, 363–371.

Wang, Q.Y., Guan, Y.C., Wu, Y.R., Chen, H.L., Chen, F. and Chu, C.C. (2008) Overexpression of a rice *OsDREB1F* gene increase salt, drought and low temperature tolerance in both *Arabidopsis* and rice. *Plant Mol. Biol.* **67**, 589–602.

Xiao, H., Siddiqua, M., Braybrook, S. and Nassuth, A. (2006) Three grape *CBF/DREB1* genes respond to low temperature, drought and abscisic acid. *Plant Cell Environ.* **29**, 1410–1421.

Zhao, L., Hu, Y., Chong, K. and Wang, T. (2010) *ARAG1*, an ABA-responsive *DREB* gene, plays a role in seed germination and drought tolerance of rice. *Ann. Bot.* **105**, 3401–3409.

5

Identification and characterization of *OsEBS*, a gene involved in enhanced plant biomass and spikelet number in rice

Xianxin Dong[1], Xiaoyan Wang[1], Liangsheng Zhang[1,3], Zhengting Yang[1], Xiaoyun Xin[1], Shuang Wu[2], Chuanqing Sun[2], Jianxiang Liu[1], Jinshui Yang[1] and Xiaojin Luo[1,*]

[1]State Key Laboratory of Genetic Engineering, Institute of Genetics, Institute of Plant Biology, School of Life Sciences, Fudan University, Shanghai, China
[2]The Department of Plant Genetics and Breeding, China Agricultural University, Beijing, China
[3]Department of Biology, The Pennsylvania State University, University Park, PA, USA

*Correspondence
email luoxj@fudan.
edu.cn
Accession numbers: Sequence data from this article can be found in the GenBank/EMBL database under the following accession numbers: JX162210; JX162211; LOC_Os05g51360; LOC_Os05g51380; LOC_Os05g51390; X67711.2; NM_001056254; *OsEBS*: JN008171; *OsActin1*: Os03g0718100; *AtActin2*: AT3g18780.

Keywords: biomass, spikelet number, *OsEBS*, map-based cloning, cell number, common wild rice (*Oryza rufipogon* Griff.).

Summary

Common wild rice (*Oryza rufipogon* Griff.) is an important genetic reservoir for rice improvement. We investigated a quantitative trait locus (QTL), *qGP5-1*, which is related to plant height, leaf size and panicle architecture, using a set of introgression lines of *O. rufipogon* in the background of the Indica cultivar Guichao2 (*Oryza sativa* L.). We cloned and characterized *qGP5-1* and confirmed that the newly identified gene *OsEBS* (enhancing biomass and spikelet number) increased plant height, leaf size and spikelet number per panicle, leading to an increase in total grain yield per plant. Our results showed that the increased size of vegetative organs in *OsEBS*-expressed plants was enormously caused by increasing cell number. Sequence alignment showed that OsEBS protein contains a region with high similarity to the N-terminal conserved ATPase domain of Hsp70, but it lacks the C-terminal regions of the peptide-binding domain and the C-terminal lid. More results indicated that *OsEBS* gene did not have typical characteristics of Hsp70 in this study. Furthermore, Arabidopsis (*Arabidopsis thaliana*) transformed with *OsEBS* showed a similar phenotype to *OsEBS*-transgenic rice, indicating a conserved function of *OsEBS* among plant species. Together, we report the cloning and characterization of *OsEBS*, a new QTL that controls rice biomass and spikelet number, through map-based cloning, and it may have utility in improving grain yield in rice.

Introduction

Rice (*Oryza sativa* L.) is one of the most important staple food worldwide. To keep pace with the increasing food demands of a growing human population, rice breeding faces more complex requirements. However, rice breeding programmes are approaching a plateau because of the limited genetic variability of available parental materials (Tanksley and McCouch, 1997). Wild species are important reservoirs of useful genes for agronomic traits, as they have accumulated abundant genetic diversity. Of the 21 wild species of rice (Vaughan *et al.*, 2003), common wild rice (*O. rufipogon* Griff.) is generally recognized as a putative ancestor of Asian cultivated rice (*O. sativa* L.) (Chang, 1976; Khush, 1997). Many quantitative trait loci (QTLs) mapping studies have shown that *O. rufipogon* could serve as an important source of beneficial agronomic traits such as yield regulation (Luo *et al.*, 2011; Marri *et al.*, 2005; Moncada *et al.*, 2001; Septiningsih *et al.*, 2003a; Tan *et al.*, 2007; Thomson *et al.*, 2003; Tian *et al.*, 2006; Xiao *et al.*, 1998; Xie *et al.*, 2006, 2008; Xue *et al.*, 2008), grain quality (Septiningsih *et al.*, 2003b; Yuan *et al.*, 2010), abiotic stress tolerance (Koseki *et al.*, 2010; Nguyen *et al.*, 2003; Tian *et al.*, 2011; Zhang *et al.*, 2006) and biotic stress tolerance (Huang *et al.*, 2001, 2008; Tan *et al.*, 2004; Yang *et al.*, 2002). As those useful genes are identified, the utilization of wild rice

resources for crop breeding may pave the way for a new green revolution.

Among the agronomic traits of rice, yield is the most important one. It possesses three main components: number of tillers (panicles) per plant, number of grains per panicle and 1000-grain weight. These yield traits are usually governed by a number of QTLs. A few genes corresponding to yield QTL, such as *MOC1*, *Gn1a*, *GS3*, *LOG*, *GW2*, *qSW5 (GW5)*, *GIF1*, *Ghd7*, *GW8* and *DST*, have been fine-mapped and cloned (Ashikari *et al.*, 2005; Fan *et al.*, 2006; Kurakawa *et al.*, 2007; Li *et al.*, 2003, 2013; Shomura *et al.*, 2008; Song *et al.*, 2007; Wang *et al.*, 2008, 2012; Weng *et al.*, 2008). Furthermore, yield is a complex quantitative trait; besides the three main components, there are also some other traits relevant to rice yield, such as plant height, plant architecture, biomass and seed-setting rate (Jiao *et al.*, 2010; Jin *et al.*, 2008; Miura *et al.*, 2010; Thangasamy *et al.*, 2011; Xue *et al.*, 2008). Therefore, in the process of improving rice yield trait, multiple agronomic traits should be taken into consideration.

Heat-shock protein 70 (Hsp70) family plays an important role as molecular chaperones in protein folding and transport, broadly and highly conserved across prokaryotes and eukaryotes (Craig *et al.*, 1994; Lindquist and Craig, 1988). Structurally, Hsp70 with the molecular weight 70 kDa can be divided into two main domains: an approximately 45-kDa N-terminal ATPase domain (Flaherty et al., 1990) and an approximately 25-kDa C-terminal peptide recognition and binding domain (Zhu et al., 1996), both are highly conserved. Whereas some plant Hsp70s also possess a variable sequence in the N- and C-terminal domain, which is related to organelles transit and localization of the proteins. According to phylogenetic analyses, plant Hsp70s can be classified into four major cellular compartments with their conserved C terminus (Boorstein et al., 1994): cytosol, endoplasmic reticulum (ER), mitochondria and chloroplasts (Sung et al., 2002). In addition to the character of proteins, heat-shock element (HSE) has been recognized as an essential element in the regulatory region of many Hsp70 genes. Its consensus sequence is CT-GAA-TTC-AG, placing upstream of TATA box (Pelham, 1982). When the organisms are exposed to elevated temperatures, dehydration/drought, oxidants, viral infections and other biotic or abiotic stress, heat-shock factor (HSF) assembles into a trimer, binds to HSE in the heat-shock gene promoter, then activates the transcription of a series of heat-shock proteins, ultimately prevents protein misfolding and protects the cells and organisms from severe damage (Morimoto, 1993).

In the current study, we cloned a gene, OsEBS (enhancing biomass and spikelet number), from the QTL qGP5-1. This QTL was mapped from BIL112, a large-panicle introgression line constructed using an accession of Dongxiang wild rice as the donor parent and Guichao2 as the recurrent parent (Luo et al., 2011). Our results demonstrated that the OsEBS gene increased plant biomass caused by more cell number and led to an average of 37.62% increase in total grain yield per plant. The purpose of this study was to elucidate the effect of OsEBS on rice important agronomic traits, including plant height, biomass and spikelet numbers per panicle.

Results

Fine mapping of QTL qGP5-1

Previously, we developed an introgression line (IL) population comprising 265 lines carrying variant introgressed segments of Dongxiang common wild rice (O. rufipogon Griff.) collected from Dongxiang County, Jiangxi Province, China. This population was in the background of the Indica (O. sativa L. ssp. Indica) cultivar Guichao2 (Luo et al., 2009). BIL112, a large-panicle IL (Figure 1a, b) from the above IL population, was backcrossed to the recurrent parent Guichao2 to produce a BCF_2 population. Using the F_2 population, a major QTL for number of spikelets per panicle, qGP5-1, was mapped on chromosome 5. The O. rufipogon allele at qGP5-1 contributed to increased number of spikelet per panicle by 11% (Luo et al., 2011). In our previous study, we mapped a QTL for number of spikelet per panicle near the qGP5-1 region on the long arm of chromosome 5 using the IL population described above (Luo et al., 2009), suggesting that they may be the same QTL. Consequently, qGP5-1 was selected as the target for map-based cloning.

We used BIL112 to develop a set of near isogenic lines (NILs) of the target QTL and mapped qGP5-1 between the simple sequence repeat (SSR) markers RM3068 and RM5818 using a chromosome fragment substitution analysis, as described previously (Figure 1c,d) (He et al., 2006). An NIL homozygous for the target QTL was backcrossed with Guichao2 to generate a segregating population (NIL-F_2). A total of 3265 NIL-F_2 progeny were subjected to SSR analysis with ten markers between RM3068 and RM5818 (Figure 1d). Further high-resolution map-

ping of qGP5-1 was carried out by progeny testing of homozygous recombinant plants from the NIL-F_2 progeny and newly developed markers between J9-4 and RM19202 (Figure 1e). Using the same procedure, we localized qGP5-1 on a high-resolution linkage map using 14 $F_{2:3}$ families of recombinant plants (BC_3F_4) and narrowed the qGP5-1 locus to a 27-kb region between markers JS-2 and RM19202 (for mapping primers, see Table S1).

According to the fine mapping of qGP5-1, the significant spikelet-enhancing effect in the NILs resulted from introgression of the genomic region between markers JS-2 and RM19202 of Dongxiang wild rice into the genetic background of Guichao2. To identify the candidate gene(s), we sequenced a 26.3-kb genomic region from Dongxiang wild rice and a 24.6-kb genomic region from Guichao2 between the markers JS-2 and RM19202 (for sequencing primers, see Table S1). The sequences were submitted to GenBank with accession numbers of JX162210 and JX162211.

Referring to the annotation information for Nipponbare genomic sequences, we found three hypothetical genes (LOC_Os05g51360, LOC_Os05g51380 and LOC_Os05g51390) in this fragment (Figure 2a). To tentatively identify putative candidate gene(s) for the QTL qGP5-1, sequence consistency and expression levels of these three genes were analysed. Sequence alignment between the two mapping parents revealed that there was no different nucleotide within LOC_Os05g51390. However, as for LOC_Os05g51360 and LOC_Os05g51380, the alignment showed sequence differences. There were two single-nucleotide polymorphisms and a three-base deletion in the genomic sequence of LOC_Os05g51380 as compared between Dongxiang wild rice and Guichao2 (Figure 2a). These changes resulted in the loss of one amino acid, leucine, from the predicted sequence of the gene product in Dongxiang wild rice. Comparatively speaking, much severe sequence discrepancy existed in LOC_Os05g51360 between the two mapping parents. The variant in Guichao2 had a 939-bp deletion, consisting of a 379-bp deletion from the region upstream of the start codon and a 560-bp deletion from the downstream region (Figure 2b), and this deletion may absolutely abolish the expression of this gene. The mRNA abundance of these three genes was then quantified by quantitative real-time PCR (qRT-PCR) analysis. As a result of the sequence deletion, LOC_Os05g51360 was not expressed at all in Guichao2 (Figure 2c). Only mild change existed in the level of LOC_Os05g51380 and LOC_Os05g51390 transcripts between Dongxiang wild rice and Guichao2 (Figure S1). LOC_Os05g51360 in Dongxiang wild rice contained two exons (Figure 1d) and encoded a protein of 437 amino acids, and this protein showed 54%–57% similarity to the ATPase domain of OsBiPs and AtBiPs, which are a kind of Hsp70 located in ER (Figure S2). Therefore, our results strongly indicated that LOC_Os05g51360, designated as OsEBS, was a good candidate for the QTL qGP5-1.

Complementary test of OsEBS

To investigate whether OsEBS was responsible for the phenotypic changes, a 2.4-kb genomic fragment of OsEBS, including the 1.1-kb upstream regulatory sequence and the 1.3-kb coding region from BIL112, was cloned into the vector pCAMBIA1304 (Figure S3a). We introduced this plasmid into the gene-loss recurrent parent, Guichao2. Seven independent plantlets were confirmed as positive transformants by PCR analysis. To further confirm the authenticity of the transgenic

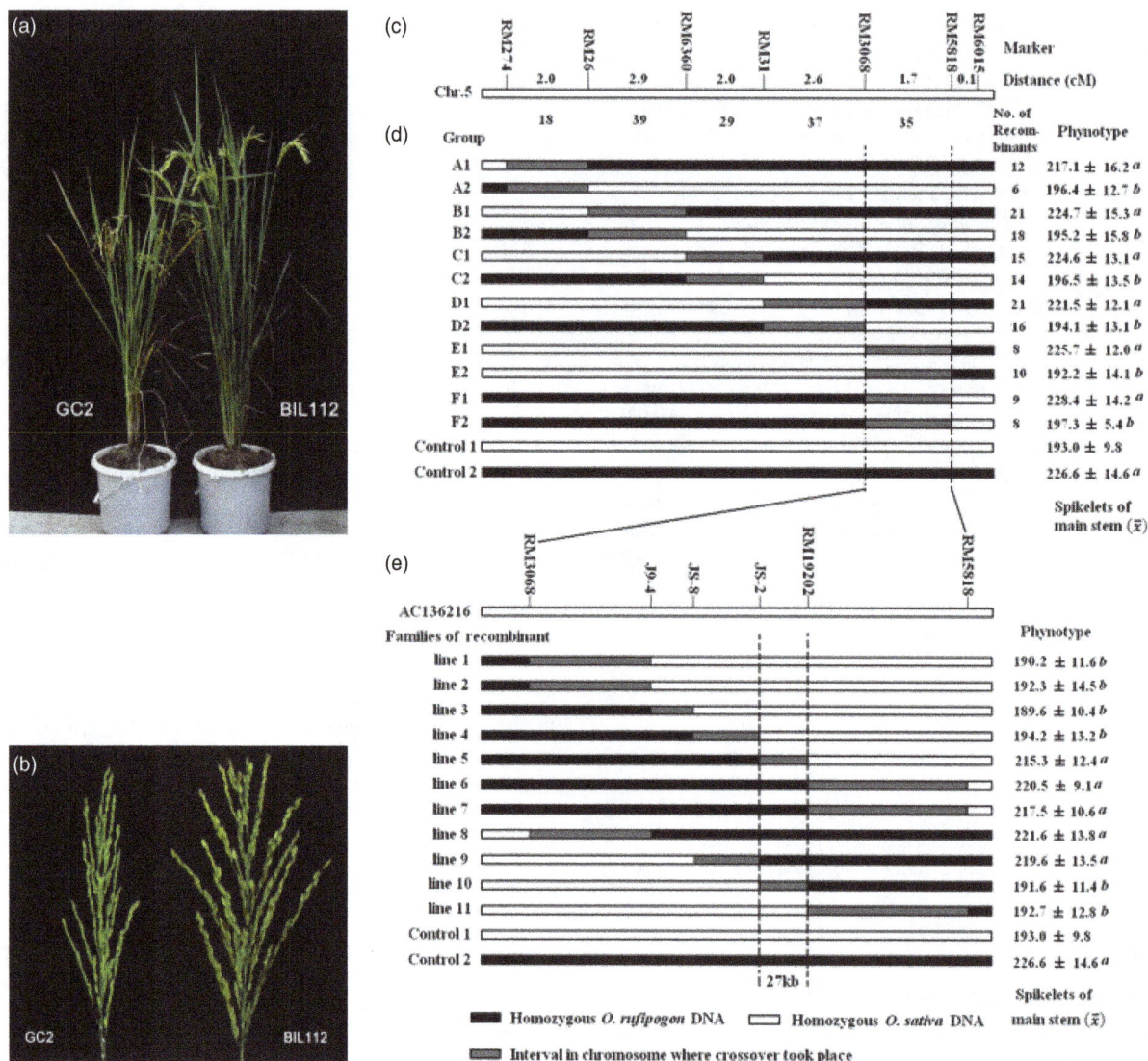

Figure 1 Substitution mapping of *qGP5-1*. (a) Introgression line plant (BIL112; right) with *qGP5-1* and recurrent parent plant (GC2; left) under natural long-day (NLD) conditions at maturity. (b) Inflorescences of BIL112 (right) and GC2 (left). (c) High-resolution linkage map of the *qGP5-1* region produced with 3265 F$_2$ plants. Number of recombinants between adjacent markers is indicated under linkage map. (d) Progeny testing of homozygous recombinants delimited the *qGP5-1* locus to the region between markers RM3068 and RM5818. Recombinants (158 in total) formed 12 groups based on genotypes. Number of recombinants in each group was marked on the right. Phenotypic difference in number of spikelets per panicle, compared with control 1, is shown for each group. (e) Fine mapping of *qGP5-1*. 11 recombinants between markers RM3068 and RM5818 are shown at left. Open bar shows part of BAC clone AC136216. On right, phenotypic difference in number of spikelets per panicle, compared with control 1, is shown for each recombinant family. 'a', Mean phenotypic value of recombinant is significantly different from that of control 1 at $P < 0.001$; 'b', mean phenotypic value of recombinant is not significantly different from that of the control 1 at $P > 0.001$.

plants, the expression level of *OsEBS* was detected by qRT-PCR analysis. This gene was expressed in transgenic plants (T$_2$) but not in the control Guichao2 (Figure S3b). All seven lines containing *OsEBS* showed greater plant height and more spikelets per panicle than Guichao2, with a phenotype similar to that of BIL112. These results confirmed that *OsEBS* was integrated into the rice genome and transmitted genetically to the next generation. It also provided further evidence that *OsEBS* was the gene responsible for the spikelet-enhancing QTL *qGP5-1*. We isolated and cloned its sequence including the 2500-bp promoter region and its open reading frame (ORF) from genomic DNA of BIL112. Sequence data for this gene have been deposited at GenBank under accession number JN008171.

Characteristics of *OsEBS*-transgenic rice plants

OsEBS-transgenic plants and BIL112 (collectively referred to as *OsEBS*-positive plants) showed significantly different phenotypes to that of the control Guichao2. There were no obvious phenotypic differences at the seedling stage; however, the differences became evident as the plants grew. At the ten-leaf stage, transgenic lines showed greater plant height than control plants (Figure 3a). At the flowering stage, the heading time was delayed by approximately 7 days in *OsEBS*-positive plants. BIL112 and transgenic plants had greater biomass than Guichao2 at maturity (Figure 3b). The *OsEBS*-positive plants had larger leaves, greater plant height, longer internodes, greater culm diameter and more spikelets on the main panicle (Figure 3c–e, Table 1). To

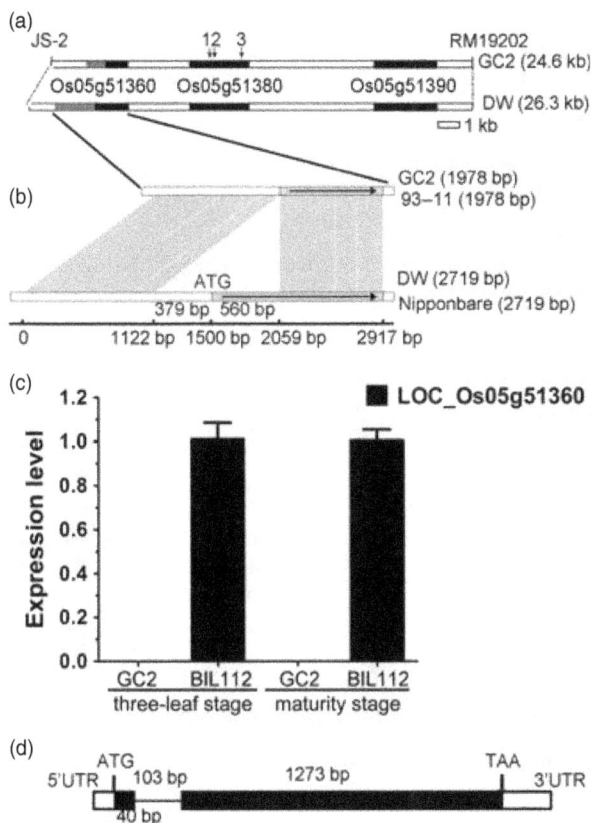

Figure 2 Identification of *OsEBS* gene associated with mapped QTL, *qGP5-1*. (a) Schematic diagram of sequence comparison in QTL region between Dongxiang wild rice (DW) and Guichao2 (GC2). LOC_Os05g51360, LOC_Os05g51380 and LOC_Os05g51390 are three hypothetical genes based on annotation information from Nipponbare genomic sequence. Arrows 1, 2 and 3 indicate sequence differences in LOC_Os05g51380 gene between Dongxiang wild rice and Guichao2. (b) Sequence alignment of LOC_Os05g51360 gene between Guichao2 (or 93–11) and Dongxiang wild rice (or Nipponbare). Horizontal arrows in shaded rectangles indicate polarity and position of LOC_Os05g51360 gene copies. (c) qRT-PCR analysis of expression level of Os05g51360 in Guichao2 and BIL112. Values are means ± SD; *n* = 3 individuals and show expression level relative to that of *OsActin1*. (d) Exon/intron structure of *OsEBS*. Black bars represent exons.

validate the enhancing effects of *OsEBS* on plant biomass, we measured fresh weight and dry weight of Guichao2, BIL112 and B102 plants throughout the growth period (Figure 4). *OsEBS*-positive plants showed greater biomass than Guichao2.

At maturity, we compared some important agronomic traits among Guichao2, BIL112 and two transgenic lines (B102 and B103). Each line consisted of 12 individual plants. There was almost no difference in main panicle length and the number of panicles per plant (Table 2). However, relative to control plants, *OsEBS*-positive plants showed increased plant height (BIL112, 16.49%; B102, 14.35%; B103, 14.25%) and bigger biomass (BIL112, 35.38%; B102, 36.08%; B103, 17.56%). We also counted the number of spikelets per panicle; the *OsEBS*-positive plants produced more spikelets than control plants (BIL112, 37.11%; B102, 19.68%; B103, 34.44%) (Table 2) and produced an average of 905, 1164, 1257 and 1142 grains per plant, respectively. These results suggested that *OsEBS* greatly affected spikelet number. However, the average 1000-grain weights

showed only slight differences (Table 2). The mean grain yield per plant was 20.29, 27.36, 29.01 and 27.4 g in Guichao2, BIL112, B102 and B103, respectively. These results indicate that BIL112, B102 and B103 showed an increase in grain yield per plant over the control of 34.84%, 42.97% and 35.04%, respectively (Table 2).

In summary, expression of *OsEBS* led to increased internode length, larger leaf size, greater number of spikelets per panicle and greater number of grains per plant. These changes resulted in a yield-increased phenotype. Therefore, we concluded that *OsEBS* encodes a protein regulating increased rice yield via increasing plant biomass and spikelet number.

OsEBS increases cell number in vegetative organs

To examine the cellular basis of larger individual plants, we prepared histological sections of roots, leaves and stems from Guichao2, BIL112 and the transformant B102. In terms of root anatomical structure, the thicker roots were due to increased numbers of cortical parenchyma cells around the root pericycle. The cell size did not differ significantly among the different plant lines (Figure 5a, Table S2). Analyses of transverse sections of stems showed that there were 34 vascular bundles in BIL112, 33 in B102 and 32 in Guichao2; the greater number of vascular bundles and longer distances between them accounted for the larger stem diameter (Figure 5b, Figure S4, Table S2). Analyses of longitudinal stem sections revealed that the thicker stem wall was because of an increase in the number of cell layers in the ground tissue (Figure 5c, Table S2). We assumed that the differences in leaf blades would be attributed to the greater number of vascular bundles, as indicated in the transverse sections. There were 43 vascular bundles in BIL112, 44 in B102 and only 39 in Guichao2 (Figure 5d, Figure S5). However, there was no distinguishable difference in cell size between *OsEBS*-positive and *OsEBS*-negative plants (Table S2). These results showed that the increased size of vegetative organs in individual *OsEBS*-positive plants was mainly because of increased cell number, rather than increased cell size.

Expression pattern of *OsEBS* in rice

To investigate the expression profile of *OsEBS*, we searched the CREP rice gene expression database (crep.ncpgr.cn), which contains a vast amount of microarray data covering the entire life cycle of rice. *OsEBS* was almost undetectable at a trace expression level in many tissues at different growth stages (Table S3).

To further confirm the expression pattern of *OsEBS*, we investigated its expression level in BIL112 and the transformant B102 using qRT-PCR. *OsEBS* was expressed at basal levels in the callus, root and leaf of plants at different growth stages and in panicles in mature plants. It was not expressed in the stem at any growth stage. Its expression was higher in leaves than in roots at the tiller stage and in panicles at the heading stage (Figure 6).

Characterization of *OsEBS* gene

Amino acid sequence alignment analysis of OsEBS indicated that it was homologous to BiP (luminal binding protein) (Table S4), which is a type of Hsp70 located in ER. Compared with the domain architecture of Hsp70, OsEBS contained only an ATPase domain with a molecular weight of 47 kD and lacked the peptide-binding domain and the C-terminal ER retention signal (Figure 7a).

Figure 3 Performance of *OsEBS*-transgenic line grown under natural long-day conditions. (a) Transgenic plant (B102, T$_2$ line; right) and control plant (GC2, recurrent parent; left) at ten-leaf stage. (b) Transgenic plant (B102, T$_2$ line; right) and control plant (GC2, recurrent parent; left) at maturity. (c) Flag leaf of transgenic B102 (right) and control GC2 (left) at maturity. (d) Internodes, main panicles and main culm of transgenic B102 (right) and control GC2 (left) at maturity. (e) Inflorescences of transgenic B102 (right) and control GC2 (left) at maturity.

Table 1 Growth traits of IL, transgenic plants and recurrent parent (control) at heading stage

	GC2	BIL112	B102	B103
Flag leaf				
Leaf length (cm)	28.83 ± 3.12	36.69 ± 3.64**	35.11 ± 3.64**	34.14 ± 3.64**
Leaf width (cm)	1.38 ± 0.07	1.55 ± 0.10**	1.53 ± 0.10**	1.54 ± 0.10**
Second leaf				
Leaf length (cm)	37.52 ± 3.16	44.96 ± 3.58**	44.98 ± 3.56**	45.15 ± 3.56**
Leaf width (cm)	1.27 ± 0.06	1.34 ± 0.06**	1.33 ± 0.06**	1.31 ± 0.06*
Third leaf				
Leaf length (cm)	45.40 ± 2.86	45.45 ± 2.77	43.79 ± 2.10	44.77 ± 2.46
Leaf width (cm)	1.14 ± 0.05	1.18 ± 0.04	1.16 ± 0.05	1.15 ± 0.05
Plant height (cm)	94.76 ± 2.5	106.47 ± 2.91**	111.71 ± 2.91**	109.80 ± 2.91**
Internode length (cm)				
Uppermost internode	31.77 ± 1.98	34.58 ± 2.21**	35.71 ± 2.21**	34.96 ± 2.21**
Second internode	17.61 ± 1.96	18.07 ± 1.00	20.55 ± 1.12**	20.10 ± 1.12**
Third internode	12.39 ± 1.93	14.12 ± 1.20**	16.61 ± 1.20**	15.49 ± 1.22**
Fourth internode	7.22 ± 1.03	9.65 ± 1.87**	10.49 ± 1.87**	10.14 ± 1.88**
Basal internode	4.52 ± 0.86	4.95 ± 1.49	4.79 ± 1.22	4.92 ± 0.79
Third internode diameter (mm)	4.65 ± 0.19	5.17 ± 0.42**	5.61 ± 0.42**	5.52 ± 0.42**

GC2, recurrent parent Guichao2 (control); BIL112, introgression line (IL); B102 and B103, *OsEBS*-positive transgenic plant lines.

Values are means (n = 12). Two-tailed Student's t-test was used to test difference between two means: *$0.05 < P < 0.01$; **$P < 0.01$.

The sampling plants in this experiment were grown in Shanghai test field (31°11′N, 121°29′E).

The C-terminal domain is related to protein subcellular localization in some plant HSP70s (Boorstein *et al.*, 1994; Guy and Li, 1998; Munro and Pelham, 1987). Therefore, the lack of C-terminal ER retention domain in OsEBS may affect its subcellular localization. To test this, the OsEBS-GFP fusion protein driven by the 35S promoter was expressed in Arabidopsis root and tobacco leaf epidermal cells (Figure 7b,c). It was found that OsEBS-GFP signal was not co-localized with the ER protein marker (AtWAK2-mCherry-HDEL), indicating that OsEBS may involve different function compared with BiP.

Many heat-responsive Hsp70 genes have HSE *cis*-element on their promoter regions. We searched for HSE in the *OsEBS* promoter

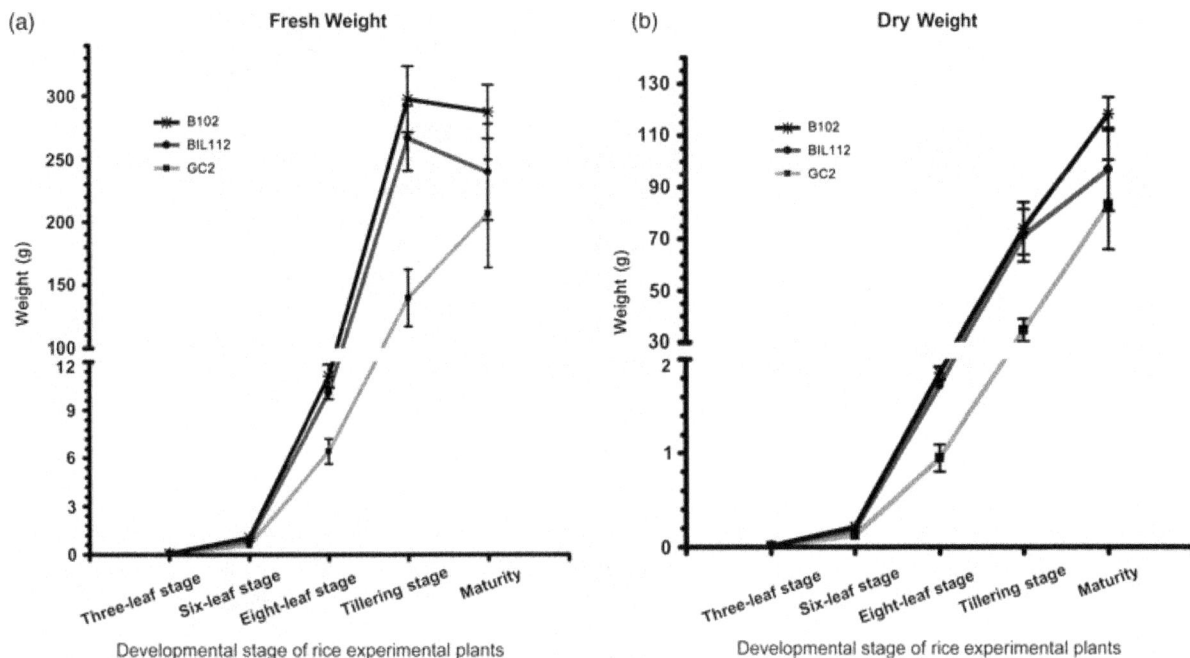

Figure 4 Comparison of biomass among IL, transgenic plants and recurrent parents (control). (a) Fresh weight of B102, BIL112 and GC2 plants during growth period. Whole individual plant without roots was immediately weighed after harvest. Values are means (n = 12). (b) Dry weight of B102, BIL112 and GC2 plants during growth period. Whole individual plant without roots was weighed after drying at 80 °C for 48 h. Values are means (n = 12).

using the Web tools at PLACE (www.dna.affrc.go.jp/PLACE/signalscan.html) and PLantCARE (bioinformatics.psb.ugent.be/webtools/plantcare/html/) and found no HSE in the 2-kb upstream sequence of *OsEBS* promoter. To know whether the expression of *OsEBS* responds to heat shock, we treated rice plants at 42 °C for 2 h. While two heat-responsive Hsp70 genes (X67711.2 and NM_001056254) were induced by heat shock (Han *et al.*, 2009; Wang and Fang, 1996), *OsEBS* gene was even down-regulated by heat stress (Figure 7d).

Characterization of transgenic Arabidopsis plants

As the transgenic rice plants expressing *OsEBS* showed a phenotype of increased biomass and spikelet number, we were curious about the function of *OsEBS* in Arabidopsis. Therefore, we proceeded detailed phenotypic analysis of Arabidopsis plants transformed with CaMV 35S:*OsEBS*. We randomly chose two transgenic lines, namely EBS-1 and EBS-2, and their expression levels of *OsEBS* were confirmed by qRT-PCR detection (Figure 8a). Compared with Col-0 plants, the transformants had larger leaves at the vegetative stage (Figure 8b,c), particularly, the most obvious difference of leaf area belonged to the fifth to eleventh leaf (Table S5). Arabidopsis overexpressing *OsEBS* showed about a 4-day delay in bolting (Figure 8d). Ultimately, the bigger leaf area and slightly longer growth period caused the biomass increase in transgenic plants in comparison with Col-0. To validate this conclusion, we measured the fresh weight of Col-0, EBS-1 and EBS-2 from 20 to 40 days, transgenic plants showed greater biomass than wild type (Figure 8e). Observations of histological sections of the leaves showed that the large-leaf phenotype of the Arabidopsis transformants was because of increased cell number (Figure 5f), as in rice leaves. The above phenotype of transgenic Arabidopsis plants was similar to the *OsEBS* phenotype in rice plants, suggesting that *OsEBS* performed

a similar function in Arabidopsis and rice, namely involved in plant growth regulation.

Furthermore, we used OsEBS protein sequence as a query to identify a number of homologous genes in diverse plant species using BLAST analysis, the subject proteins of BLAST derived from seven monocots, 18 dicots, two Gymnospermae and one Pteridophyta. Among these, we only found the truncated Hsp70 protein like OsEBS in some monocot, including *Oryza sativa*, *Aegilops tauschii*, *Brachypodium distachyon*, *Sorghum bicolor* (Table S6), and there was no domain-loss Hsp70 protein in Arabidopsis and other dicots. Although *OsEBS* was not widespread in all plants, our result showed that it had a conserved function in rice and Arabidopsis.

Discussion

OsEBS, a newly identified gene from common wild rice, enhances rice biomass and spikelet number

The exploitation of favourable genes from wild rice might further improve tolerance to biotic and abiotic stress, yield and other important agronomic traits for rice variety (Luo *et al.*, 2011). Previous studies have confirmed that Dongxiang common wild rice (*O. rufipogon*) possesses a large number of trait-enhancing genes (He *et al.*, 2006; Li *et al.*, 2002; Luo *et al.*, 2009; Tian *et al.*, 2006; Zha *et al.*, 2009). In this study, we have successfully cloned a QTL, *qGP5-1*, from Dongxiang common wild rice. Through the transgenic complementary test, we confirmed that the gene *OsEBS* was responsible for *qGP5-1* and related to the function of improving grain yield via enhancing rice biomass and spikelet number.

Panicle architecture is one of important traits for rice plant type because it can affect directly grain yield as a factor of sink size (Zhou *et al.*, 2009). To increase the value of each panicle

Table 2 Comparison of agronomic traits among IL, transgenic plants and recurrent parent (control)

Plant line	Plant height (cm)	Biomass (g)	Main panicle length (cm)	Panicles per plant	Spikelets per panicle	Grains per plant	Weight per 1000 grains (g)	Grain yield per plant (g)	Increase (%)
GC2	93.80 ± 4.65	31.60 ± 8.60	22.18 ± 1.38	6.75 ± 1.42	134.08 ± 28.61	905.08 ± 290.66	22.42 ± 1.72	20.29 ± 0.50	—
BIL112	109.27 ± 4.85**	42.78 ± 13.45*	21.50 ± 0.99	6.33 ± 1.55	183.84 ± 37.05**	1164.33 ± 327.97*	23.50 ± 1.31	27.36 ± 0.43*	34.84
B102	107.26 ± 3.39**	43.00 ± 12.99**	21.19 ± 0.90	7.83 ± 2.28	160.47 ± 29.45*	1257.00 ± 495.18**	23.08 ± 1.31	29.01 ± 0.64**	42.97
B103	107.17 ± 5.60**	37.15 ± 8.36*	21.32 ± 0.77	6.33 ± 1.49	180.26 ± 31.57**	1141.66 ± 295.21*	24.00 ± 1.59*	27.40 ± 0.47*	35.04

GC2, recurrent parent Guichao2 (control); BIL112, introgression line (IL); B102 and B103, OsEBS-positive transgenic plant lines.

Values are means ($n = 12$). Two-tailed Student's t-test was used to test difference between two means: *$0.05 < P < 0.01$; **$P < 0.01$.

The sampling plants in this experiment were grown in Sanya test field (18°14′N, 109°31′E).

characteristic would be an effective strategy for improving grain yield. Some QTLs for spikelet number have been identified during the past decade. *DEP1*, which encodes a truncated phosphatidylethanolamine-binding protein-like domain protein, is a locus that enhances meristematic activity, resulting in an increased number of grains per panicle and, consequently, increased grain yield. Mutants with knockout *DEP1* (NIL-*dep1*) showed a greater number of cells across the longitudinal axis of the plant compared with NIL-*DEP1* plants (Huang et al., 2009; Zhou et al., 2009). In this study, expression of *OsEBS* in rice increased plant height (Figure 3a,b) and the number of spikelets per panicle (Figure 3e), which resulted in a 34.84% increase in grain yield (Table 2). Our current results showed that the *OsEBS* gene increased the number of cells in some vegetative organs (Figure 5; Table S2). This finding is consistent with the results of previous research by Huang et al. and suggests that the function of *OsEBS* may be related to cell division. Further research is required to elucidate the exact molecular mechanism of *OsEBS*.

Yield traits are always controlled by pleiotropism genes. *OsSPL14* controls shoot branching at the vegetative and reproductive stage and, hence, regulates the number of tillers and spikelets (Jiao et al., 2010; Miura et al., 2010). *Ghd7* has major effects on several traits in rice, including the number of grains per panicle, plant height and heading date (Xue et al., 2008). Our study demonstrated that the biomasses of BIL112 (42.78 g) and the transgenic lines B102 (43.00 g) and B103 (37.15 g) were greater than that of Guichao2 (31.60 g), and the slightly longer growth duration and increase in biomass caused the greater yield potential (Ying et al., 1998; Zhang et al., 2009). What is more, the mean spikelet numbers in BIL112, B102 and B103 (183.84, 160.47 and 180.26, respectively) were greater than that in Guichao2 (134.08). Consequently, the mean grain yield per plant was estimated at 27.36 g for BIL112, 29.01 g for B102, 27.40 g for B103 and 20.29 g for Guichao2 (Table 2). These results confirmed that *OsEBS* had the potential to increase the yield of rice plants grown under normal cultivation conditions.

OsEBS, responsible to QTL *qGP5-1*, derived from Dongxiang wild rice, was observed in Nipponbare (*O.sativa*, japonica) as a full size, but in Guichao2 and 93-11 (*O.sativa*, indica) as a truncated one. We further sequenced a set of indica/japonica varieties and wild rices, and the results (Table S7) showed that *OsEBS* gene in most indica varieties was a truncated one and in most japonica varieties was a full-size one, whereas exceptions also exited in both japonica and indica. The distribution of *OsEBS* gene type in wild rice was random, and there were truncated genes and full-size genes in both *O. rufipogon* and *O. nivara*. This result indicated that the *OsEBS* gene differentiation occurred before common wild rice evolved into cultivated varieties.

As to the spikelet number, we calculated the spikelet number per panicle of two indica varieties (93-11 and Teqing 2) and two japonica varieties (Nipponbare and Wuyujing 3) planted in Shanghai test fields. The statistical data showed that 93-11 and Teqing 2 with truncated *OsEBS* gene were 186 ± 9.9 and 126.7 ± 9.4, respectively, while the Nipponbare and Wuyujing 3 with full-length *OsEBS* gene were 98.3 ± 8.3 and 133.6 ± 7.6, respectively; the varieties with full-length *OsEBS* gene did not always have a more spikelet number than these with truncated gene. This result indicated that spikelet number, as a quantitative trait, was governed by several genes and mainly affected by varied genetic background, so one single gene, like *OsEBS*, may not determine spikelet number in all rice varieties.

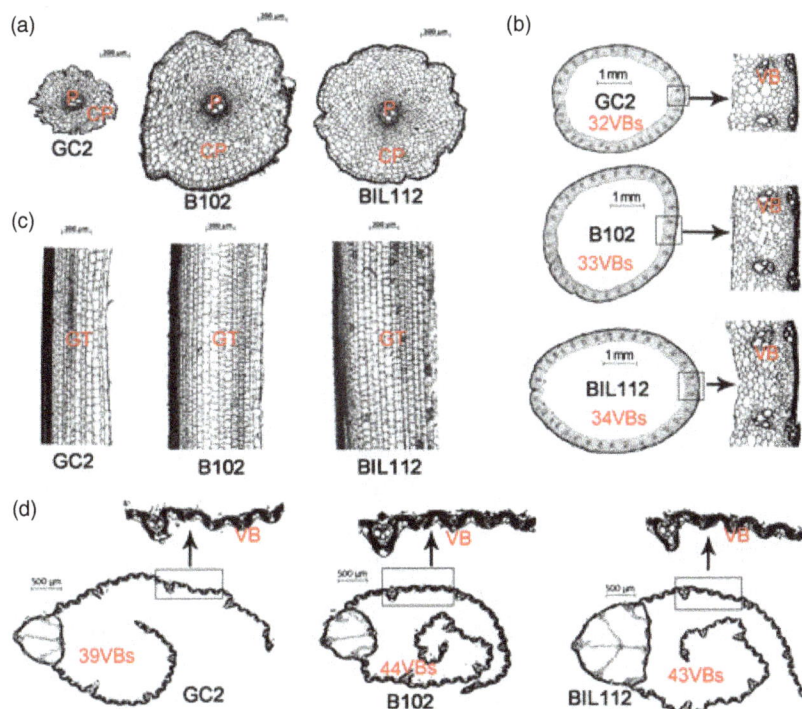

Figure 5 Internal structures of root, stem and leaf of IL, transgenic plants and recurrent parent. (a) Transverse sections of adventitious roots at tiller stage. Same positions are shown in B102, BIL112 and GC2 plants. (b) Transverse sections of stems at heading stage. Sections were from approximately 10 cm above uppermost nodes from main culms of plants. The clearer picture is showed in Figure S4. (c) Longitudinal sections of stems at heading stage. Sections were from approximately 10 cm above uppermost nodes from main culms of plants. (d) Transverse sections of middle part of flag leaf blades at heading stage. The clearer picture is showed in Figure S5. Scale bars are shown for each figure. P, pericycle; CP, cortical parenchyma; VB, vascular bundles; GT, ground tissue.

Domain loss of OsEBS leads to a different function from that of canonical chaperones

Sequence alignment analyses showed that OsEBS encodes a conserved ATPase domain, analogous to BiP, a kind of Hsp70 localized to ER (Figure 7a, Figure S2, Table S4). As we all know, during heat stress, unfolded proteins accumulate within the ER, and plant cells increase the expression of chaperones to overcome the abundance of unfolded proteins and to alleviate ER stress, among which, BiP is one of the proteins that are involved in quality control and unfolded protein response (UPR) (Koizumi, 1996). For example, AtBiP2 is required for the transport and secretion of proteins in the ER and has been used as a common marker for UPR activation in Arabidopsis (Koizumi et al., 2001; Martinez and Chrispeels, 2003). However, OsEBS was not localized to the ER (Figure 7b,c). The different organelle localization indicated that OsEBS may have lost the function as a chaperone in ER during UPR.

Structurally, OsEBS is markedly shorter than BiP as a result of the loss of the peptide-binding domain and the C terminus. The peptide-binding domain, which contains a groove with affinity for newly synthesized or unfolding polypeptides (Fourie et al., 1994), interacts with the ATPase domain (Davis et al., 1999) and acts as a molecular chaperone in protein folding and transport (Lindquist and Craig, 1988). The loss of peptide-binding domain in OsEBS may lead to the loss of function in binding newly synthesized and unfolding polypeptides.

Cytosolic chaperones are usually induced by heat stress. Our data also showed that the promoter of OsEBS lacked HSEs, and the expression of OsEBS decreased, rather than increased, under

heat-shock treatment (Figure 7d). These results pointed out that OsEBS lacked the canonical characters of heat-inducible Hsp70.

The transgenic plants of OsEBS-expressed rice and Arabidopsis showed bigger leaf size and more biomass (Figures 3 and 8): this implies that OsEBS gene had a conserved biological function among different plant species. The only left ATPase domain of OsEBS potentially refers the conserved function in rice and Arabidopsis. Considering the increased cell number in OsEBS-positive plants (Figures 5 and 8f), we infer that the ATPase domain is involved in the control of cell division. In both prokaryotes and eukaryotes, proteins with an ATPase domain are reported to relate to cell division processes. Such proteins include MinD and axonemal dynein in bacteria (Gibbons and Rowe, 1965; Goehring and Beckwith, 2005; Karki and Holzbaur, 1999; Motallebi-Veshareh et al., 1990), KaiC in Synechococcus elongatus (Dong et al., 2010), and members of the AAA (ATPase associated with different cellular activities)-ATPase protein family (Lupas and Martin, 2002). CDC48/P47 (cell division cycle), a highly abundant type II AAA-ATPase widely found in yeast, mammals and plants (Peters et al., 1990), is involved in cell cycle control (Moir et al., 1982) and cell proliferation (Egerton and Samelson, 1994). Hsp70 was predicted to have a similar ATPase subdomain and three-dimensional structures to those of the prokaryotic cell cycle proteins MreB, FtsA and StbA (Bork et al., 1992). This suggests that the ATPase domain of OsEBS may function similarly to cell cycle proteins on cell number regulation. Further research is required to explore this possibility.

Greater cell number results in larger vegetative organs and greater biomass production in rice (Sairam, 1994); our results in the present study support this conclusion, which is consistent

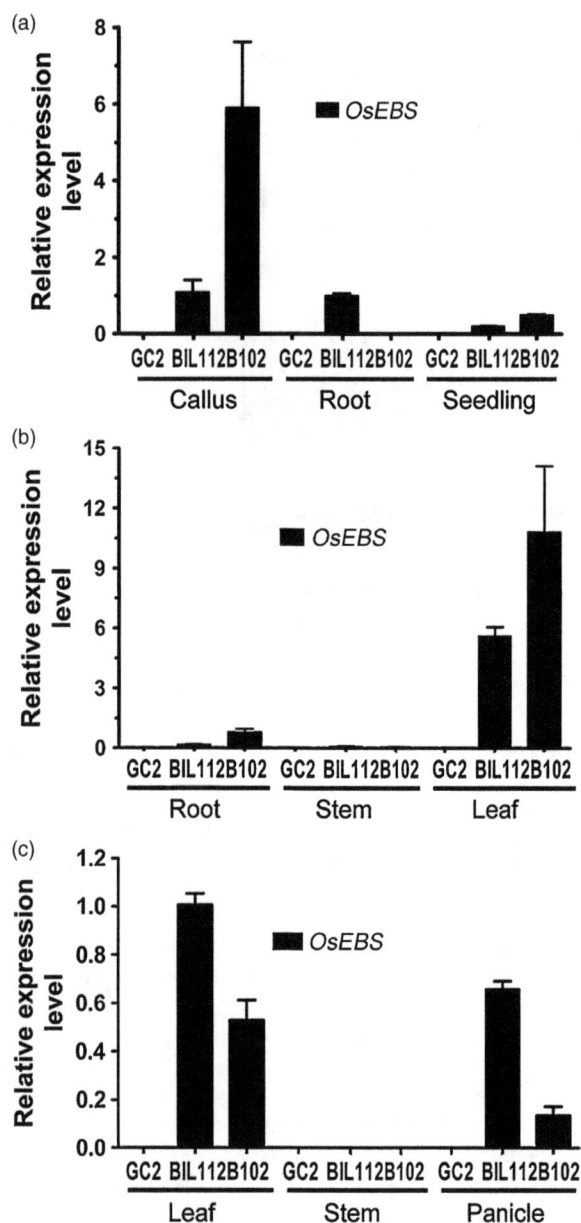

Figure 6 Expression pattern of *OsEBS* in rice tissues at different growth stages. (a) qRT-PCR analysis of expression of rice *OsEBS* in callus, root and seedling at 3-leaf stage. (b) qRT-PCR analysis of expression of rice *OsEBS* in root, stem and leaf at tillering stage. (c) qRT-PCR analysis of expression of rice *OsEBS* in stem, leaf and panicle at maturity. Values are means ± SD, $n = 3$ individuals and show expression level of *OsEBS* relative to that of *OsActin1*.

plant growth and plant yield (Kant and Rothstein, 2009; Kant et al., 2009), was suppressed in *OsEBS*-positive plants. The results above are consistent with the phenotypes of bigger individual and more yields in the OsEBS-positive plants. To confirm these results, we performed qRT-PCR analysis to examine the expression of these genes at three-leaf stage (Table S8). Unfortunately, other hitherto known component involved in *OsEBS* regulation network is still missing, and the identification of more new genes regulating spikelet initiation and development is required to better understand the molecular mechanism of *OsEBS* in rice spikelet development.

Experimental procedures

Plant materials and growth conditions

Plants or seeds were photographed with a Nikon E995 digital camera. The rice mapping population and transgenic plants were grown under normal conditions in test fields in Shanghai (31°11′ N, 121°29′E) and Sanya (18°14′N, 109°31′E), and the other rice materials for expression analysis were grown in a greenhouse at 28 °C under a 14-h-light/10-h-dark photoperiod. All the wild-type and transgenic Arabidopsis lines used in this study were in the Col-0 background and grown at 21 ± 1 °C under 60% relative humidity, with an approximately 100 µmol/m^2/s light intensity under a 16-h-light/8-h-dark photoperiod.

Rice transformation

The binary plasmid vector pCAMBIA1304 (Center for the Application of Molecular Biology of International Agriculture) was constructed by inserting the promoter and coding region of the *OsEBS* gene from BIL112 between Hind III and Spe I restriction sites in a sense orientation. The construct was transformed into Guichao2 as previously described (Qi et al., 2011).

Phenotype analysis of transgenic rice

Seeds from T$_3$ lines and equivalent control plants were collected and germinated by soaking in water for 2 days at 37 °C. Germinating seeds were sown in pots and grown in the greenhouse as described above. At the five-leaf stage, 20 seedlings of each T$_3$ line and equivalent control were transplanted to larger pots under the same growing conditions. At maturity, the phenotypic characteristics were measured. The data of phenotypic traits are expressed as means ± standard deviation (SD) calculated in an analysis of variance (t-test: $P < 0.05$).

To evaluate the biomass difference, we sampled ten individual plants of one transgenic line B102 and two control lines, Guichao2 and BIL112, every three weeks through the whole growth period. Only the aerial parts except roots of plant individual were weighed for the fresh weight and dry weight before and after oven drying samples at 80 °C for 48 h.

Sample paraffin sectioning and microscopy

Plant materials were fixed in FAA (50% ethanol, 5% glacial acetic acid and 5% formaldehyde) for 16 h, dehydrated in an ethanol series and embedded in paraffin. Tissue sections (10 mm thick) were cut with a rotary microtome (LEICA RM2235) and then mounted on glass slides. Sections were photographed under an inverted optical microscope (ZEISS Ser. No. 1 03 11 4545).

RNA extraction and qRT-PCR

Plants were cultivated under controlled environmental conditions. Fresh materials were harvested and immediately frozen in liquid

with the results of our study (Figure 5). According to this phenotype, we speculated that *OsEBS* functions in cell proliferation and probably differentiation. To test this hypothesis, we investigated differentially expressed genes between *OsEBS*-positive seedling and control seedling at the six-leaf stage using microarray analysis. The microarray results showed that some genes involved in the response to auxin were differentially expressed in positive plants, and the higher transcript level of auxin growth promoter genes promoted plant growth and obtained more biomass (Table S8). In particular, *OsSAUR39* (a small auxin-up RNA gene), which acts as a negative regulator of

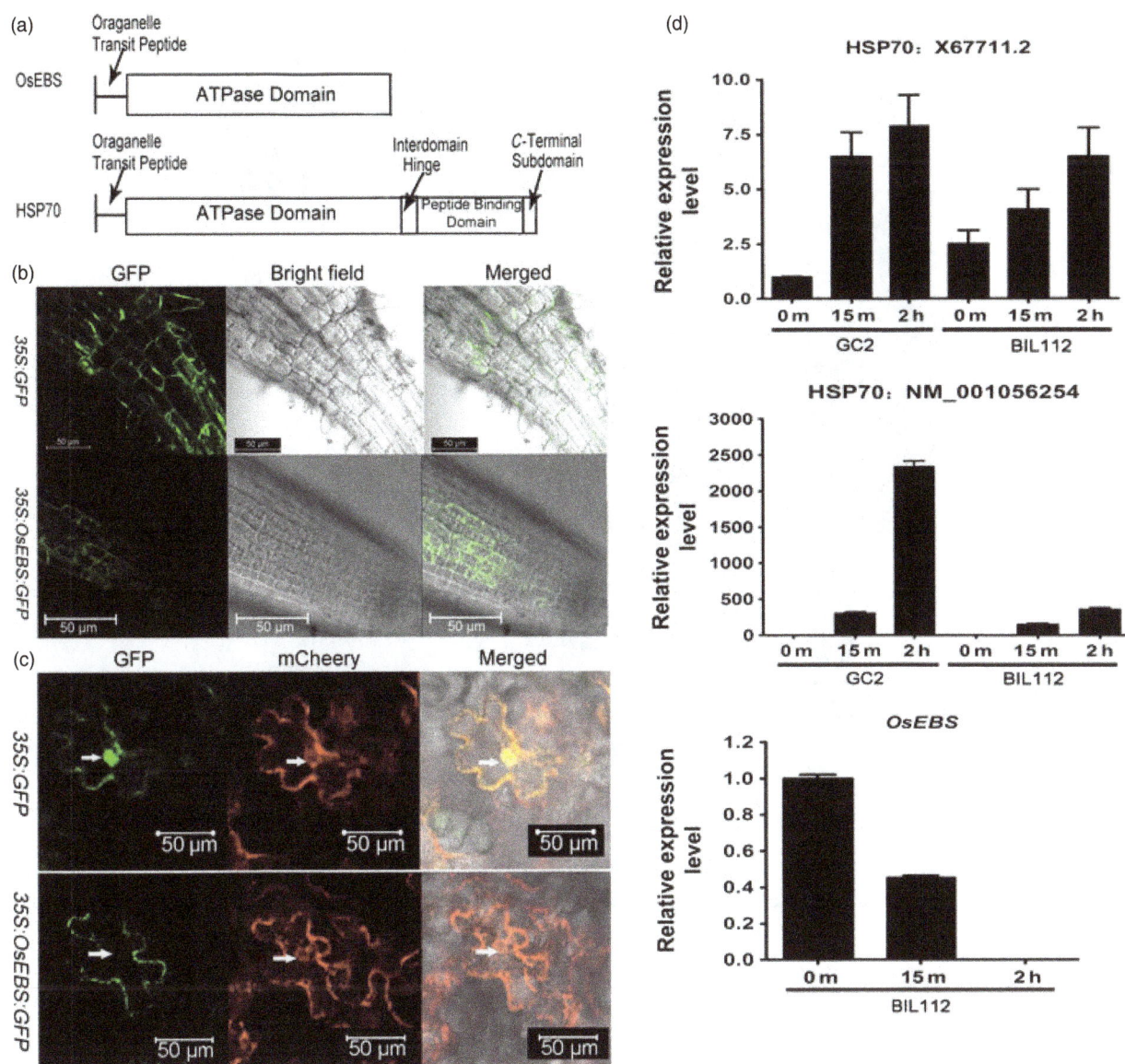

Figure 7 Characteristics of *OsEBS* gene. (a) Schematic representation of OsEBS and Hsp70 protein domains. (b) Subcellular localization of 35S:GFP (top) and 35S:OsEBS:GFP fusion proteins (bottom) in Arabidopsis roots. (c) Subcellular localization of 35S::GFP (top) and 35S:OsEBS:GFP fusion proteins (bottom) co-transformed with the *ERmarker-mCheery* in tobacco leaf epidermal cells. mCherry panel shows the ER localization, and arrows indicate the position of nuclei. (d) Expression pattern of *OsEBS* under heat-shock treatment. Seedlings (15 days old) of GC2 and BIL112 plants were subjected to a heat-shock treatment and sampled at 0 and 15 min, and 2 h. Expressions of heat-responsive genes X67711.2 and ABE95267 served as positive controls. Values are means ± SD, $n = 3$ individuals, and show expression level of three genes relative to that of *OsActin1*.

nitrogen and then stored at −80 °C. Total RNAs were extracted using the RNAprep Plant kit (TIANGEN, Beijing, China). First-strand cDNA was generated using Primescript RT reagent (Perfect Real Time; TaKaRa, Otsu, Shiga, Japan). The primers used for qRT-PCR are shown in Table S1.

In this experiment, qRT-PCR analysis of *OsEBS* was carried out using a Bio-Rad iCycler IQ Real-Time PCR Detection System (Bio-Rad) and a FastStart TaqMan® Probe Master kit (Roche) (Horst and Peterhansel, 2007). The probes and primers were designed using the Assay Design Center at the UPL Website (www.universalprobelibrary.com). For qRT-PCR analysis of other genes, we used the Primescript RT reagent (Perfect Real Time) kit (TaKaRa). Melting curves and standard curves were calculated and analysed for detected genes and *Actin*.

Subcellular localization assay and Arabidopsis transformation

To investigate the cellular localization of OsEBS, the full ORF of *OsEBS* was amplified from BIL112 using reverse-transcribed cDNA as the template and OsEBS-GFP primers (Table S1). The PCR product was subcloned into the *Bgl* II and *Spe* I sites of the pCAMBIA1304 vector and fused with *GFP* in-frame under the control of the cauliflower mosaic virus *35S* promoter. The resulting constructs as well as the vector control were introduced into *Agrobacterium tumefaciens* strain GV3101. Arabidopsis plants were transformed via the floral-dipping method (Clough and Bent, 1998). Putative transgenic plants were initially selected with 50 mg/L hygromycin, and the transgenic

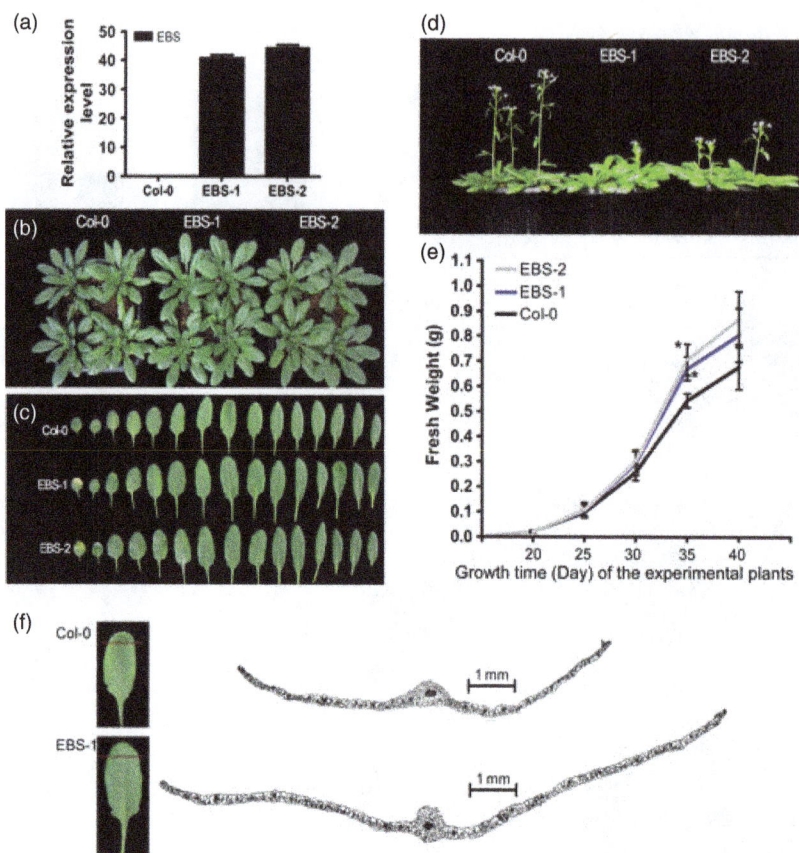

Figure 8 Characterization of Col-0 transformed with *35S:OsEBS*. (a) qRT-PCR analysis of *OsEBS* expression in rosette leaves of 20 days old. EBS-1 and EBS-2: two transgenic plants. Values are means ± SD, $n = 3$ individuals, and show expression level of OsEBS relative to that of *AtActin2*. (b) 35-day-old plants of transgenic lines (EBS-1 and EBS-2) and wild type. (c) Leaves of 35-day-old plants from transgenic lines (EBS-1 and EBS-2) and wild type. Detached leaves were counted from base of plant (oldest). (d) 41-day-old plants of transgenic lines (EBS-1 and EBS-2) and wild type. (e) Comparison of biomass among transgenic plants and Col-0 (control). The chart showed the fresh weight change in EBS-1, EBS-2 and Col-0 during vegetative growth stage. Values are means ($n = 15$), two-tailed Student's *t*-test was used to test difference between two means: *$0.05 < P < 0.01$; **$P < 0.01$. (f) Transverse sections of the eighth rosette leaf showed in the right, and red square marked the slice position. Scale bars are shown on figures.

lines were confirmed by PCR analysis with the primers HPT-JC (Table S1).

To verify whether OsEBS localized in ER or not, *35S:GFP, 35S: OsEBS:GFP* and *AtWAK2(signal peptide)-mCherry-HDEL* fusion constructs were introduced to *Agrobacterium* strain GV3101 through the freeze–thaw method, *Agrobacterium* infiltration method was applied to transfer the constructs to tobacco (*Nicotiana benthamiana*) leaf cells. The infiltrated plants were grown in greenhouse for an additional two days, and subcelluar localization of each fusion proteins was observed under confocal fluorescence microscopy (Zeiss LSM A710). The co-localization ER marker was as follows: CDC-959 for ER marker (Nelson *et al.*, 2007).

Microarray analysis

Control and transgenic seedlings were grown to the six-leaf stage in greenhouse under a 14/10-h (light/dark) photoperiod at 28 °C. RNA samples were extracted using an RNeasy Mini Kit (Qiagen) following the manufacturer's instructions. To determine integration of RNA, the RNA integrity number was determined using a bioanalyser (Agilent Technologies, Santa Clara, CA). Microarray analyses and data processing were carried out as previously described (Qi *et al.*, 2011). All microarray data from this work are available from the National Center for Biotechnology Information

(NCBI) Gene Expression Omnibus (GEO) (www.ncbi.nlm.nih.gov/geo/) under the series entry GSE46616.

Bioinformatic analysis

To fine mapping of *qGP5-1*, some new InDel molecular markers were designed from publicly available rice genome sequence, and the likelihood of detecting polymorphism between *Oryza rufipogon* Griff and Guicao2 was predicted by comparing sequence from the *japonica* cultivar, cv. Nipponbare (rgp.dna.affrc.go.jp/), and the *indica* cultivar, cv. 93-11 (rice.genomics.org.cn/), and the primers used in this study are all designed by Primer-BLAST (www.ncbi.nlm.nih.gov/tools/primer-blast/index.cgi?LINK_LOC=BlastHome).

To investigate *OsEBS* gene structure, the exon/intron boundary was investigated using GenScan (genes.mit.edu/GENSCAN.html). InterProScan (www.ebi.ac.uk/Tools/pfa/iprscan/) was used for domain prediction. BLAST (blast.ncbi.nlm.nih.gov/Blast.cgi) was used for sequence alignment.

Acknowledgements

This research was supported by grants from the Genetically Modified Organisms Breeding Major Projects (2013ZX08001004-009), the National Natural Science Foundation of China

(No.30900881), the Shanghai Natural Science Foundation (13ZR1402800).

References

Ashikari, M., Sakakibara, H., Lin, S.Y., Yamamoto, T., Takashi, T., Nishimura, A., Angeles, E.R., Qian, Q., Kitano, H. and Matsuoka, M. (2005) Cytokinin oxidase regulates rice grain production. *Science*, **309**, 741–745.

Boorstein, W.R., Ziegelhoffer, T. and Craig, E.A. (1994) Molecular evolution of the HSP70 multigene family. *J. Mol. Evol.* **38**, 1–17.

Bork, P., Sander, C. and Valencia, A. (1992) An ATPase domain common to prokaryotic cell cycle proteins, sugar kinases, actin, and hsp70 heat shock proteins. *Proc. Natl Acad. Sci. USA*, **89**, 7290.

Chang, T.T. (1976) Origin, evolution, cultivation, dissemination, and diversification of Asian and African rices. *Euphytica*, **25**, 425–441.

Clough, S.J. and Bent, A.F. (1998) Floral dip: a simplified method for Agrobacterium-mediated transformation of *Arabidopsis thaliana*. *Plant J.* **16**, 735–743.

Craig, E.A., Weissman, J.S. and Horwich, A.L. (1994) Heat-Shock proteins and molecular chaperones: mediators of protein conformation and turnover in the cell. *Cell*, **78**, 365–372.

Davis, J.E., Voisine, C. and Craig, E.A. (1999) Intragenic suppressors of Hsp70 mutants: Interplay between the ATPase- and peptide-binding domains. *Proc. Natl. Acad. Sci. U. S. A.* **96**, 9269–9276.

Dong, G., Yang, Q., Wang, Q., Kim, Y.I., Wood, T.L., Osteryoung, K.W., van Oudenaarden, A. and Golden, S.S. (2010) Elevated ATPase activity of KaiC applies a circadian checkpoint on cell division in Synechococcus elongatus. *Cell*, **140**, 529–539.

Egerton, M. and Samelson, L.E. (1994) Biochemical characterization of valosin-containing protein, a protein tyrosine kinase substrate in hematopoietic cells. *J. Biol. Chem.* **269**, 11435–11441.

Fan, C.H., Xing, Y.Z., Mao, H.L., Lu, T.T., Han, B., Xu, C.G., Li, X.H. and Zhang, Q.F. (2006) GS3, a major QTL for grain length and weight and minor QTL for grain width and thickness in rice, encodes a putative transmembrane protein. *Theor. Appl. Genet.* **112**, 1164–1171.

Flaherty, K.M., DeLuca-Flaherty, C. and McKay, D.B. (1990) Three-dimensional structure of the ATPase fragment of a 70K heat-shock cognate protein. *Nature*, **346**, 623–628.

Fourie, A.M., Sambrook, J.F. and Gething, M.J.H. (1994) Common and divergent peptide binding specificities of hsp70 molecular chaperones. *J. Biol. Chem.* **269**, 30470–30478.

Gibbons, I. and Rowe, A. (1965) Dynein: a protein with adenosine triphosphatase activity from cilia. *Science*, **149**, 424–426.

Goehring, N.W. and Beckwith, J. (2005) Diverse paths to midcell: assembly of the bacterial cell division machinery. *Curr. Biol.* **15**, 514–526.

Guy, C.L. and Li, Q.B. (1998) The organization and evolution of the spinach stress 70 molecular chaperone gene family. *Plant Cell*, **10**, 539–556.

Han, F., Chen, H., Li, X.J., Yang, M.F., Liu, G.S. and Shen, S.H. (2009) A comparative proteomic analysis of rice seedlings under various high-temperature stresses. *Biochim. Biophys. Acta*, **1794**, 1625–1634.

He, G.M., Luo, X.J., Tian, F., Li, K.G., Zhu, Z.F., Su, W., Qian, X.Y., Fu, Y.C., Wang, X.K., Sun, C.Q. and Yang, J.S. (2006) Haplotype variation in structure and expression of a gene cluster associated with a quantitative trait locus for improved yield in rice. *Genome Res.* **16**, 618–626.

Horst, I. and Peterhansel, C. (2007) Quantification of Zea mays mRNAs by Real-Time PCR Using the Universal ProbeLibrary. *Biochemica*, **1**, 8–10.

Huang, Z., He, G., Shu, L., Li, X. and Zhang, Q. (2001) Identification and mapping of two brown planthopper resistance genes in rice. *Theor. Appl. Genet.* **102**, 929–934.

Huang, C.L., Hwang, S.Y., Chiang, Y.C. and Lin, T.P. (2008) Molecular evolution of the Pi-ta gene resistant to rice blast in wild rice (*Oryza rufipogon*). *Genetics*, **179**, 1527.

Huang, X.Z., Qian, Q., Liu, Z.B., Sun, H.Y., He, S.Y., Luo, D., Xia, G.M., Chu, C.C., Li, J.Y. and Fu, X.D. (2009) Natural variation at the DEP1 locus enhances grain yield in rice. *Nat. Genet.* **41**, 494–497.

Jiao, Y.Q., Wang, Y.H., Xue, D.W., Wang, J., Yan, M.X., Liu, G.F., Dong, G.J., Zeng, D.L., Lu, Z.F., Zhu, X.D., Qian, Q. and Li, J.Y. (2010) Regulation of

OsSPL14 by OsmiR156 defines ideal plant architecture in rice. *Nat. Genet.* **42**, 541–546.

Jin, J., Huang, W., Gao, J.P., Yang, J., Shi, M., Zhu, M.Z., Luo, D. and Lin, H.X. (2008) Genetic control of rice plant architecture under domestication. *Nat. Genet.* **40**, 1365–1369.

Kant, S. and Rothstein, S. (2009) Auxin-responsive SAUR39 gene modulates auxin level in rice. *Plant Signal. Behav.* **4**, 1174–1175.

Kant, S., Bi, Y.M., Zhu, T. and Rothstein, S.J. (2009) SAUR39, a small auxin-up RNA gene, acts as a negative regulator of auxin synthesis and transport in rice. *Plant Physiol.* **151**, 691–701.

Karki, S. and Holzbaur, E.L.F. (1999) Cytoplasmic dynein and dynactin in cell division and intracellular transport. *Curr. Opin. Cell Biol.* **11**, 45–53.

Khush, G.S. (1997) Origin, dispersal, cultivation and variation of rice. *Plant Mol. Biol.* **35**, 25–34.

Koizumi, N. (1996) Isolation and responses to stress of a gene that encodes a luminal binding protein in Arabidopsis thaliana. *Plant Cell Physiol.* **37**, 862–865.

Koizumi, N., Martinez, I.M., Kimata, Y., Kohno, K., Sano, H. and Chrispeels, M.J. (2001) Molecular characterization of two arabidopsis Ire1 homologs, endoplasmic reticulum-located transmembrane protein kinases. *Plant Physiol.* **127**, 949–962.

Koseki, M., Kitazawa, N., Yonebayashi, S., Maehara, Y., Wang, Z.X. and Minobe, Y. (2010) Identification and fine mapping of a major quantitative trait locus originating from wild rice, controlling cold tolerance at the seedling stage. *Mol. Genet. Genomics*, **284**, 45–54.

Kurakawa, T., Ueda, N., Maekawa, M., Kobayashi, K., Kojima, M., Nagato, Y., Sakakibara, H. and Kyozuka, J. (2007) Direct control of shoot meristem activity by a cytokinin-activating enzyme. *Nature*, **445**, 652–655.

Li, D., Sun, C., Fu, Y., Li, C., Zhu, Z., Chen, L., Cai, H. and Wang, X. (2002) Identification and mapping of genes for improving yield from Chinese common wild rice (*O. rufipogon* Griff.) using advanced backcross QTL analysis. *Chin. Sci. Bull.* **47**, 1533–1537.

Li, X.Y., Qian, Q., Fu, Z.M., Wang, Y.H., Xiong, G.S., Zeng, D.L., Wang, X.Q., Liu, X.F., Teng, S., Hiroshi, F., Yuan, M., Luo, D., Han, B. and Li, J.Y. (2003) Control of tillering in rice. *Nature*, **422**, 618–621.

Li, S., Zhao, B., Yuan, D., Duan, M., Qian, Q., Tang, L., Wang, B., Liu, X., Zhang, J. and Wang, J. (2013) Rice zinc finger protein DST enhances grain production through controlling Gn1a/OsCKX2 expression. *Proc. Natl Acad. Sci. USA*, **110**, 3167–3172.

Lindquist, S. and Craig, E.A. (1988) The Heat-Shock proteins. *Annu. Rev. Genet.* **22**, 631–677.

Luo, X., Tian, F., Fu, Y., Yang, J. and Sun, C. (2009) Mapping quantitative trait loci influencing panicle-related traits from Chinese common wild rice (*Oryza rufipogon*) using introgression lines. *Plant Breed.* **128**, 559–567.

Luo, X.J., Wu, S., Tian, F., Xin, X.Y., Zha, X.J., Dong, X.X., Fu, Y.C., Wang, X.K., Yang, J.S. and Sun, C.Q. (2011) Identification of heterotic loci associated with yield-related traits in Chinese common wild rice (*Oryza rufipogon* Griff.). *Plant Sci.* **181**, 14–22.

Lupas, A.N. and Martin, J. (2002) AAA proteins. *Curr. Opin. Struct. Biol.* **12**, 746–753.

Marri, P.R., Sarla, N., Reddy, L.V. and Siddiq, E.A. (2005) Identification and mapping of yield and yield related QTLs from an Indian accession of *Oryza rufipogon*. *BMC Genet.* **6**, 33–46.

Martinez, I.M. and Chrispeels, M.J. (2003) Genomic analysis of the unfolded protein response in Arabidopsis shows its connection to important cellular processes. *Plant Cell*, **15**, 561–576.

Miura, K., Ikeda, M., Matsubara, A., Song, X.J., Ito, M., Asano, K., Matsuoka, M., Kitano, H. and Ashikari, M. (2010) OsSPL14 promotes panicle branching and higher grain productivity in rice. *Nat. Genet.* **42**, 545–549.

Moir, D., Stewart, S.E., Osmond, B.C. and Botstein, D. (1982) Cold-sensitive cell-division-cycle mutants of yeast: isolation, properties, and pseudoreversion studies. *Genetics*, **100**, 547–563.

Moncada, P., Martinez, C.P., Borrero, J., Chatel, M., Gauch, H., Guimaraes, E., Tohme, J. and McCouch, S.R. (2001) Quantitative trait loci for yield and yield components in an *Oryza sativa* x *Oryza rufipogon* BC2F2 population evaluated in an upland environment. *Theor. Appl. Genet.* **102**, 41–52.

Morimoto, R.I. (1993) Cells in stress - transcriptional activation of Heat-Shock genes. *Science*, **259**, 1409–1410.

Motallebi-Veshareh, M., Rouch, D. and Thomas, C. (1990) A family of ATPases involved in active partitioning of diverse bacterial plasmids. *Mol. Microbiol.* **4**, 1455–1463.

Munro, S. and Pelham, H.R.B. (1987) A C-terminal signal prevents secretion of luminal ER proteins. *Cell*, **48**, 899–907.

Nelson, B.K., Cai, X. and Nebenführ, A. (2007) A multicolored set of *in vivo* organelle markers for co-localization studies in Arabidopsis and other plants. *Plant J.* **51**, 1126–1136.

Nguyen, B.D., Brar, D.S., Bui, B.C., Nguyen, T.V., Pham, L.N. and Nguyen, H.T. (2003) Identification and mapping of the QTL for aluminum tolerance introgressed from the new source, *Oryza rufipogon Griff.*, into indica rice (*Oryza sativa L.*). *Theor. Appl. Genet.* **106**, 583–593.

Pelham, H.R.B. (1982) A regulatory upstream promoter element in the drosophila HSP70 Heat-Shock gene. *Cell*, **30**, 517–528.

Peters, J., Walsh, M. and Franke, W. (1990) An abundant and ubiquitous homo-oligomeric ring-shaped ATPase particle related to the putative vesicle fusion proteins Sec18p and NSF. *EMBO J.* **9**, 1757.

Qi, W.W., Sun, F., Wang, Q.J., Chen, M.L., Huang, Y.Q., Feng, Y.Q., Luo, X.J. and Yang, J.S. (2011) Rice ethylene-response AP2/ERF factor OsEATB restricts internode elongation by down-regulating a gibberellin biosynthetic gene. *Plant Physiol.* **157**, 216–228.

Sairam, R. (1994) Effects of homobrassinolide application on plant metabolism and grain yield under irrigated and moisture-stress conditions of two wheat varieties. *Plant Growth Regul.* **14**, 173–181.

Septiningsih, E.M., Prasetiyono, J., Lubis, E., Tai, T.H., Tjubaryat, T., Moeljopawiro, S. and McCouch, S.R. (2003a) Identification of quantitative trait loci for yield and yield components in an advanced backcross population derived from the *Oryza sativa* variety IR64 and the wild relative *O.rufipogon*. *Theor. Appl. Genet.* **107**, 1419–1432.

Septiningsih, E.M., Trijatmiko, K.R., Moeljopawiro, S. and McCouch, S.R. (2003b) Identification of quantitative trait loci for grain quality in an advanced backcross population derived from the *Oryza sativa* variety IR64 and the wild relative *O.rufipogon*. *Theor. Appl. Genet.* **107**, 1433–1441.

Shomura, A., Izawa, T., Ebana, K., Ebitani, T., Kanegae, H., Konishi, S. and Yano, M. (2008) Deletion in a gene associated with grain size increased yields during rice domestication. *Nat. Genet.* **40**, 1023–1028.

Song, X.J., Huang, W., Shi, M., Zhu, M.Z. and Lin, H.X. (2007) A QTL for rice grain width and weight encodes a previously unknown RING-type E3 ubiquitin ligase. *Nat. Genet.* **39**, 623–630.

Sung, D.Y., Kaplan, F. and Guy, C.L. (2002) Plant Hsp70 molecular chaperones: protein structure, gene family, expression and function. *Physiol. Plant.* **113**, 443–451.

Tan, G.X., Weng, Q.M., Ren, X., Huang, Z., Zhu, L.L. and He, G.C. (2004) Two whitebacked planthopper resistance genes in rice share the same loci with those for brown planthopper resistance. *Heredity*, **92**, 212–217.

Tan, L.B., Liu, F.X., Xue, W., Wang, G.J., Ye, S., Zhu, Z.F., Fu, Y.C., Wang, X.K. and Sun, C.Q. (2007) Development of *Oryza rufipogon* and *O. sativa* introgression lines and assessment for yield-related quantitative trait loci. *J. Integr. Plant Biol.* **49**, 871–884.

Tanksley, S.D. and McCouch, S.R. (1997) Seed banks and molecular maps: unlocking genetic potential from the wild. *Science*, **277**, 1063–1066.

Thangasamy, S., Guo, C.L., Chuang, M.H., Lai, M.H., Chen, J.C. and Jauh, G.Y. (2011) Rice SIZ1, a SUMO E3 ligase, controls spikelet fertility through regulation of anther dehiscence. *New Phytol.* **189**, 869–882.

Thomson, M.J., Tai, T.H., McClung, A.M., Lai, X.H., Hinga, M.E., Lobos, K.B., Xu, Y., Martinez, C.P. and McCouch, S.R. (2003) Mapping quantitative trait loci for yield, yield components and morphological traits in an advanced backcross population between *Oryza rufipogon* and the *Oryza sativa* cultivar Jefferson. *Theor. Appl. Genet.* **107**, 479–493.

Tian, F., Zhu, Z.F., Zhang, B.S., Tan, L.B., Fu, Y.C., Wang, X.K. and Sun, C.Q. (2006) Fine mapping of a quantitative trait locus for grain number per panicle from wild rice (*Oryza rufipogon Griff.*). *Theor. Appl. Genet.* **113**, 619–629.

Tian, L., Tan, L.B., Liu, F.X., Cai, H.W. and Sun, C.Q. (2011) Identification of quantitative trait loci associated with salt tolerance at seedling stage from *Oryza rufipogon*. *J. Genet. Genomics*, **38**, 593–601.

Vaughan, D.A., Morishima, H. and Kadowaki, K. (2003) Diversity in the *Oryza* genus. *Curr. Opin. Plant Biol.* **6**, 139–146.

Wang, Q. and Fang, R.X. (1996) Structure and expression of a rice hsp70 gene. *Sci. China. C-Life Sci.* **39**, 291–299.

Wang, E., Wang, J., Zhu, X.D., Hao, W., Wang, L.Y., Li, Q., Zhang, L.X., He, W., Lu, B.R., Lin, H.X., Ma, H., Zhang, G.Q. and He, Z.H. (2008) Control of rice grain-filling and yield by a gene with a potential signature of domestication. *Nat. Genet.* **40**, 1370–1374.

Wang, S., Wu, K., Yuan, Q., Liu, X., Liu, Z., Lin, X., Zeng, R., Zhu, H., Dong, G. and Qian, Q. (2012) Control of grain size, shape and quality by OsSPL16 in rice. *Nat. Genet.* **44**, 950–954.

Weng, J.F., Gu, S.H., Wan, X.Y., Gao, H., Guo, T., Su, N., Lei, C.L., Zhang, X., Cheng, Z.J., Guo, X.P., Wang, J.L., Jiang, L., Zhai, H.Q. and Wan, J.M. (2008) Isolation and initial characterization of GW5, a major QTL associated with rice grain width and weight. *Cell Res.* **18**, 1199–1209.

Xiao, J.H., Li, J.M., Grandillo, S., Ahn, S.N., Yuan, L.P., Tanksley, S.D. and McCouch, S.R. (1998) Identification of trait-improving quantitative trait loci alleles from a wild rice relative, *Oryza rufipogon*. *Genetics*, **150**, 899–909.

Xie, X.B., Song, M.H., Jin, F.X., Ahn, S.N., Suh, J.P., Hwang, H.G. and McCouch, S.R. (2006) Fine mapping of a grain weight quantitative trait locus on rice chromosome 8 using near-isogenic lines derived from a cross between *Oryza sativa* and *Oryza rufipogon*. *Theor. Appl. Genet.* **113**, 885–894.

Xie, X.B., Jin, F.X., Song, M.H., Suh, J.P., Hwang, H.G., Kim, Y.G., McCouch, S.R. and Ahn, S.N. (2008) Fine mapping of a yield-enhancing QTL cluster associated with transgressive variation in an *Oryza sativa* x *O.rufipogon* cross. *Theor. Appl. Genet.* **116**, 613–622.

Xue, W.Y., Xing, Y.Z., Weng, X.Y., Zhao, Y., Tang, W.J., Wang, L., Zhou, H.J., Yu, S.B., Xu, C.G., Li, X.H. and Zhang, Q.F. (2008) Natural variation in Ghd7 is an important regulator of heading date and yield potential in rice. *Nat. Genet.* **40**, 761–767.

Yang, H.Y., Ren, X.A., Weng, Q.M., Zhu, L.L. and He, G.G. (2002) Molecular mapping and genetic analysis of a rice brown planthopper (Nilaparvata lugens Stal) resistance gene. *Hereditas*, **136**, 39–43.

Ying, J., Peng, S., He, Q., Yang, H., Yang, C., Visperas, R.M. and Cassman, K.G. (1998) Comparison of high-yield rice in tropical and subtropical environments: I. Determinants of grain and dry matter yields. *Field Crops Res.* **57**, 71–84.

Yuan, P.R., Kim, H.J., Chen, Q.H., Ju, H.G., Ji, S.D. and Ahn, S.N. (2010) Mapping QTLs for grain quality using an introgression line population from a cross between *Oryza sativa* and *O. rufipogon*. *J. Crop. Sci. Biotechnol.* **13**, 205–212.

Zha, X.J., Luo, X.J., Qian, X.Y., He, G.M., Yang, M.F., Li, Y. and Yang, J.S. (2009) Over-expression of the rice LRK1 gene improves quantitative yield components. *Plant Biotechnol. J.* **7**, 611–620.

Zhang, X., Zhou, S., Fu, Y., Su, Z., Wang, X. and Sun, C. (2006) Identification of a drought tolerant introgression line derived from Dongxiang common wild rice (*O. rufipogon Griff.*). *Plant Mol. Biol.* **62**, 247–259.

Zhang, Y., Tang, Q., Zou, Y., Li, D., Qin, J., Yang, S., Chen, L., Xia, B. and Peng, S. (2009) Yield potential and radiation use efficiency of "super" hybrid rice grown under subtropical conditions. *Field Crops Res.* **114**, 91–98.

Zhou, Y., Zhu, J.Y., Li, Z.Y., Yi, C.D., Liu, J., Zhang, H.G., Tang, S.Z., Gu, M.H. and Liang, G.H. (2009) Deletion in a quantitative trait gene qPE9-1 associated with panicle erectness improves plant architecture during rice domestication. *Genetics*, **183**, 315–324.

Zhu, X., Zhao, X., Burkholder, W.F., Gragerov, A., Ogata, C.M., Gottesman, M.E. and Hendrickson, W.A. (1996) Structural analysis of substrate binding by the molecular chaperone DnaK. *Science*, **272**, 1606.

6

Population genomics identifies the origin and signatures of selection of Korean weedy rice

Qiang He[1], Kyu-Won Kim[1] and Yong-Jin Park[1,2,]*

[1]Department of Plant Resources, College of Industrial Science, Kongju National University, Yesan, 32439, Korea
[2]Center for crop genetic resource and breeding (CCGRB), Kongju National University, Cheonan, 31080, Republic of Korea

Summary

*Correspondence
email yjpark@kongju.
ac.kr

Keywords: genetic resources, natural selection, origin, weedy rice, whole-genome resequencing.

Weedy rice is the same biological species as cultivated rice (*Oryza sativa*); it is also a noxious weed infesting rice fields worldwide. Its formation and population-selective or -adaptive signatures are poorly understood. In this study, we investigated the phylogenetics, population structure and signatures of selection of Korean weedy rice by determining the whole genomes of 30 weedy rice, 30 landrace rice and ten wild rice samples. The phylogenetic tree and results of ancestry inference study clearly showed that the genetic distance of Korean weedy rice was far from the wild rice and near with cultivated rice. Furthermore, 537 genes showed evidence of recent positive or divergent selection, consistent with some adaptive traits. This study indicates that Korean weedy rice originated from hybridization of modern *indica/indica* or *japonica/japonica* rather than wild rice. Moreover, weedy rice is not only a notorious weed in rice fields, but also contains many untapped valuable traits or haplotypes that may be a useful genetic resource for improving cultivated rice.

Introduction

Weedy rice (*Oryza sativa* f. *spontanea* Rosh.) is one of the most notorious weeds in rice fields worldwide, causing both crop yield losses and degrading the quality of rice (Qiu *et al.*, 2014). It can be defined as any spontaneously and strongly shattering rice that occurs in cultivated rice fields, and it harbours phenotypes of both wild and domesticated rice. Weedy rice only occurs in or nearby rice fields and is characterized by its easy seed shattering, deep dormancy and red pericarp, which serve as keys to distinguish weedy rice from wild species and cultivated rice (Xia *et al.*, 2011). Weedy rice produces fewer grains per plant and competes aggressively with the cultivated rice. It is present worldwide, including South and North America, southern Europe and southern and southeast Asia (Noldin *et al.*, 1999; Sun *et al.*, 2013). In the last 20 years, weedy rice has become a serious agricultural issue because of direct seeding. In southern parts of the United States, weedy rice causes an annual loss of more than $50 million (Gealy *et al.*, 2002). In China, weedy rice affects more than three million ha and reduces the crop yield by about 3.4 billion kg (Liang and Qiang, 2011).

In Korea, a lot of weedy rice strains (*O. sativa* L., locally called 'Aengmi' and 'Share') have been collected from farmers' fields. The regional distribution and genetics of weedy rice have been extensively characterized (Hak-Soo and Mun-Hue, 1992; Heu, 1988). 'Share rice' in Korea occurs mainly on Kanghwa Island and is geographically isolated from other weedy subpopulations. 'Share rice' has almost all of the characteristics of 'Red rice', while its spikelet within one panicle always has a different ripening time and the culm length is always shorter than those of 'Red rice' and landraces (Chung and Park, 2010). Although many morphological traits have been studied in detail, the large variability in complex quantitative traits in weedy rice remains unexploited (Chung and Park, 2010). Weedy rice plants are disease-resistant and herbicide-resistant (Chen *et al.*, 2004; Olofsdotter *et al.*, 2000). The

seeds of weedy rice have high ability to persist in the soil (Delouche and Labrada, 2007). Therefore, it is difficult to obliterate weedy rice from rice fields. On the other hand, cultivated rice is susceptible to rice diseases such as rice blast disease, which leads to a 20%–30% loss of annual rice harvest (Kou and Wang, 2012). Weedy rice has attracted attention in the area of rice research because of its tolerance against biostress and abiostress (Chen *et al.*, 2004; Liu *et al.*, 2014; Lu *et al.*, 2014; Qiu *et al.*, 2014; Sun *et al.*, 2013). Because weedy rice in Korea has acclimatized and adapted to different growing environments, it is also thought to be tolerant to a wide range of adverse conditions. In addition, we postulated that weedy rice may have additional useful characteristics, either in stress tolerance or rice eating quality, that could be used to improve cultivated rice in the future.

In addition, there is an active area of research toward determining the origin of weedy rice as a means to keep it out of rice fields. The origin of weedy rice has been debated intensively over the past 40 years, and there are three main hypotheses: (i) weedy rice descended directly from cultivated rice, with some populations from *indica*-type cultivars (Londo and Schaal, 2007), some from *japonica*-type cultivars (Cao *et al.*, 2006, 2009; Vaughan *et al.*, 2008), and others from *indica* × *japonica* hybrids (Ishikawa *et al.*, 2005; Qiu *et al.*, 2014); (ii) weedy rice may have generated from the process of ongoing selection and adaptation of wild rice (*Oryza rufipogon* and *Oryza nivara* which is considered as annual *O. rufipogon*) (De Wet and Harlan, 1975; Kelly Vaughan *et al.*, 2001); and (iii) weedy rice may have originated from the hybridization between cultivated rice and its wild ancestor *O. rufipogon* (Londo and Schaal, 2007). A number of other hypotheses have also been proposed, including weedy rice originating from reversion of cultivated rice when domesticated rice was abandoned (Bres-Patry *et al.*, 2001) or forming by hybridization of its cultivated relatives (Ishikawa *et al.*, 2005; Reagon *et al.*, 2010; Xiong *et al.*, 2012). This is supported by observations that some weedy-type offspring could occur after inter-subspecies and inter-varietal hybridization in rice. Despite increasing focus on the evolutionary study of weedy rice, its origin is still unclear. A number of molecular markers have been applied to infer the population structure, genetic diversity and adaptive loci of weedy rice. However these molecular markers, either simple sequence repeats or restriction fragment length polymorphisms, which were frequently used in studies of weedy rice, only allowed scanning of the genome at very low density (Chung and Park, 2010; Gross et al., 2010; Kelly Vaughan et al., 2001; Londo and Schaal, 2007; Reagon et al., 2010; Sun et al., 2013; Thurber et al., 2010). It is easy to obtain biased conclusions with low-density molecular markers because the accuracy of population structure, genetic diversity and adaptive loci identities is dependent on the representativeness of the markers, and it is difficult to evaluate the representative efficiency of the markers for unknown populations.

The rapid increase in the density of molecular markers through next-generation sequencing (NGS) technology has facilitated building more solid hypotheses on genomics-based study of rice (He et al., 2015; Huang et al., 2010, 2012a,b; Xu et al., 2012). Qiu et al. (2014) used a whole-genome rese-quencing strategy for three weedy rice samples and suggested the origin of weedy rice from domesticated indica/japonica hybridization. However, the case study of these three weedy rice samples is not representative of all weedy rice populations, and the limited population size was not suitable to identify the adaptation loci for weedy rice.

Recently, a number of analytical methods have been performed to identify signals of recent positive selection or adaptation loci on a genome-wide scale using huge NGS data, such as the fixation index (F_{ST}) (Akey et al., 2002), nucleotide diversity (π) (Nei and Li, 1979), integrated haplotype score (iHS) (Voight et al., 2006), extended haplotype homozygosity (EHH) (Sabeti et al., 2002), cross-population EHH (XP-EHH) (Sabeti et al., 2007; Tang et al., 2007), composite likelihood ratio (CLR) (Nielsen et al., 2005; Williamson et al., 2007) and cross-population CLR (XP-CLR) (Chen et al., 2010). Most of these methods were first used in human populations and were later extended to animals and plants. All of the techniques significantly stimulate population genetic studies. In rice, Xu et al. (2012) detected many domes-ticated or selective sweeps by comparing the reduction in nucleotide diversity and F_{ST} between wild rice and cultivated rice. They found two well-known domesticated rice genes, prog1 (Jin et al., 2008) and sh4 (Li et al., 2006), among their candidate selective sweep regions. However, this approach has not been used for weedy rice with NGS data.

In this study, we resequenced the whole genomes of 30 Korean weedy rice individuals and 30 Korean landrace rice individuals. Through high-density single-nucleotide polymorphism (SNP) markers and accredited population size, we performed population studies of Korean weedy rice to obtain reasonable hypotheses for the origin of Korean weedy rice. Furthermore, we detected many adaptive loci in the weedy rice population that may be valuable resources for control and utilization of weedy rice in the future.

Results

Polymorphisms across the rice genome

From high-coverage whole-genome sequencing of 30 weedy rice and 30 landrace unrelated individuals (Table S1) we collected more than two billion clean reads aligned to 374 Mb of the O. sativa genome (http://rapdb.dna.affrc.go.jp/download/archive/irgsp1/IRGSP-1.0_genome.fasta.gz). Ninety-eight percent of reads were mapped to the reference with an average depth of 10× (Table S2). From these data, we detected more than eight million SNPs and one million InDels. For the ten wild rice plants we finally obtained 10.6 million SNPs and 1.3 million InDels (Table 1). It is reasonable that wild rice has much more variation than cultivated rice or weedy rice. In addition to SNP and Indel analysis, we studied the structure variations among different groups. We found 3957, 3754, and 3006 copy number variations among weedy rice, landrace rice and wild rice, respectively. A total of 19 391 inversions and 22 812 translocations were detected in weedy rice. While only 12 381 inversions and 19 767 transloca-tions were detected in landrace rice (Table 1). It is suggested that weedy rice have much more structure variations than landrace rice.

Most of the variations were rare (minor allele frequency [MAF] <0.05). To obtain the SNPs used for population studies, we excluded those missing in any of 70 accessions and those with an MAF of <0.05, because these variations may biased the popu-lation study. This yielded a final total of 2.2 million high-quality SNPs (Tables 1 and S8). To our knowledge, this represents the largest high-quality SNP data set for weedy rice reported to date. These data will be used to identify important adaptation loci of weedy or landrace rice, as well as for breeding. Indeed, using this data set we identified one novel allele on the badh2 gene, which can be used to improve cultivated rice (He and Park, 2015).

Population structure of weedy, landrace and wild rice

To understand the genetic population structure and relationships among the major groups of Korean landrace and weedy rice, we constructed a maximum likelihood (ML) tree and conducted population structure analysis based on the 2.25 million high-quality SNPs. The ML tree contained four major groups corre-sponding to O. rufipogon, O. nivara, O. japonica and O. indica. This was consistent with the results of Xu et al. (2012). However, the O. nivara in our study was separated from cultivated rice compared with the results reported by Xu et al. (2012). This may have been because Korean indica rice has a different genetic background from those of other regions. The 30 weedy rice can be clearly divided into two major groups, ind_weedy (9) and jap_weedy (21). Similar to weedy, landrace can also be separated into two groups, ind_landrace (5) and jap_landrace (25). More-over, we found that weedy rice could be divided into three major subgroups and some admixed individuals (Figure 1a). Because most people in Korea prefer japonica rice in their daily diet, it is reasonable that more than 76% of local rice samples belonged to the japonica group. Furthermore, we investigated the population structure of the 70 samples. We analysed the data by increasing K (number of populations) from 2 to 7 (Figure 1b). For K = 2, we found a division between japonica and others. It is suggested the Korea indica rice was more closely related to wild rice than japonica. For K = 4, ind_weedy (G1) was separated from ind_-landrace. For K = 5, one of the jap_weedy (G3) subgroups was clearly separated from jap_landrace, while another jap_weedy subgroup, G2, was always mixed with other jap_landrace. From K = 2 to 7, there were no patterns indicating that weedy rice formed directly from wild rice. Because the phylogenetic tree showed that ind_weedy was consistently farther from wild rice than ind_landrace (Figure 1a), and it did not contain any wild

Table 1 Summary of sequencing variations for 70 samples

Populations	Indels (M)	SNPs (M)	High-quality Indels (M)	High-quality SNPs (M)	CNV	Inversion	Translocation
Wild rice	1.33	10.63	0.914	6.96	3006	3666	24 357
Landrace rice	0.96	6.98	0.374	2.54	3754	12 381	19 767
Weedy rice	0.86	6.42	0.342	2.44	3957	19 391	22 812
Landrace_weedy	1.09	8.18	0.313	2.23	-	-	-
Total	1.65	13.25	0.338	2.25	-	-	-

CNV, copy number variation; SNP, single-nucleotide polymorphism.

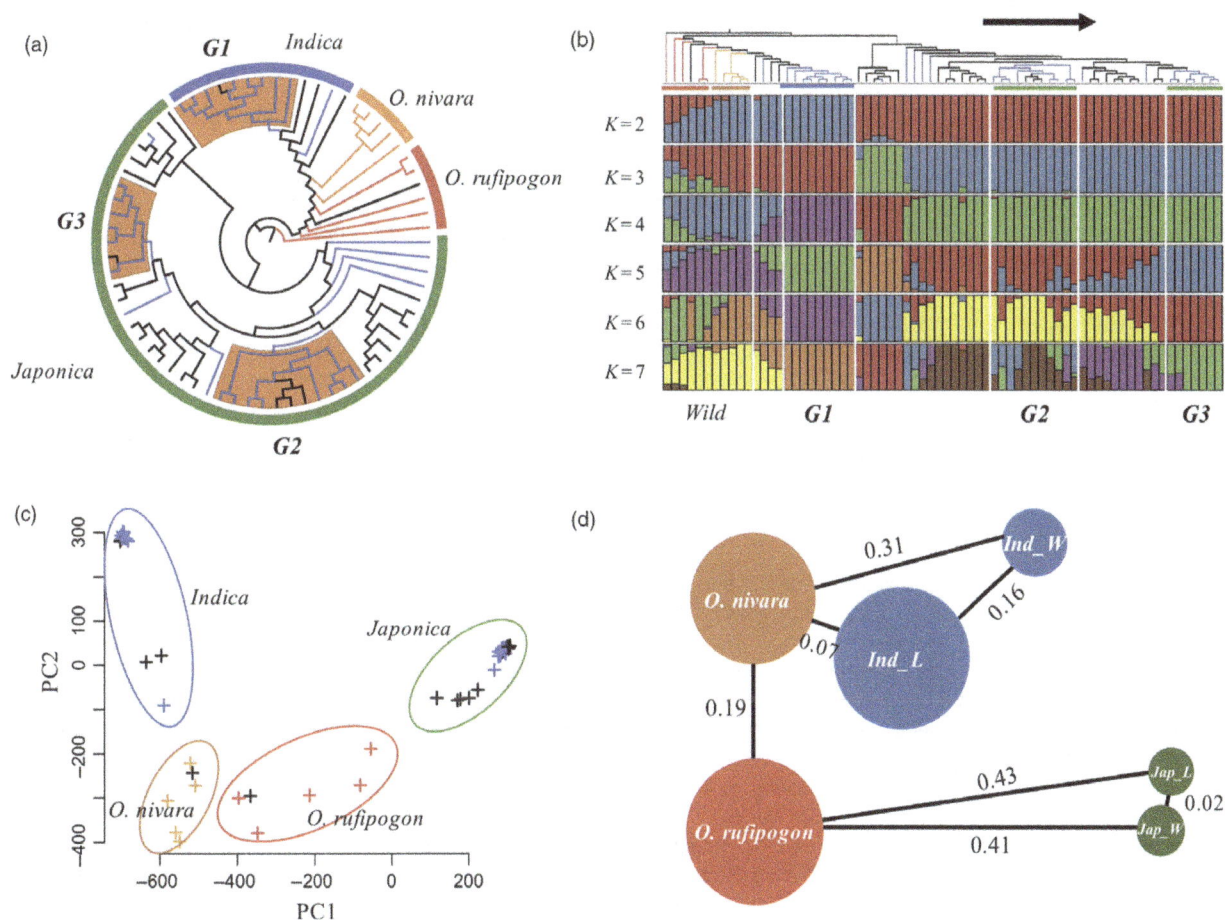

Figure 1 Population structure analysis of 30 weedy rice samples with 40 other *Oryza* species. (a) Maximum likelihood phylogenetic tree. Blue branches are weedy rice, black branches are landrace rice. (b) Maximum likelihood clustering with *K* ranging from 2 to 7. For each *K*, the different colour represents different populations. Each accession is represented by a vertical bar, and the length of each coloured segment in each vertical bar represents the proportion contributed by ancestral populations. (c) PCA plots. Blue dots are weedy rice, black dots are landrace rice. (d) Illustration of genetic diversity and population differentiation in wild rice and *japonica_landrace*, *japonica_weedy*, *indica_landrace* and *indica_weedy*. The sizes of the circles represent the levels of genetic diversity of groups, and the distances are F_{ST} values between different groups.

patterns from the *japonica* rice group in the structural study (Figure 1b), we assumed that Korean *ind_weedy* rice was formed from *indica* rice without admixture of patterns from other populations. For *K* = 5, the *japonica* group was separated into three subgroups, with many admixture patterns among the different subgroups. In comparison to *ind_weedy*, *jap_weedy* had more admixture patterns from among *jap_weedy* or between *jap_weedy* and *jap_landrace*, but not from *indica* or wild species. Therefore, we postulated that *jap_weedy* was derived from

japonica rice and that gene exchange occurred frequently between or among *japonica* rice. Meanwhile, the PCA plot showed clearly that weedy rice was clustered with landrace rice in both the *indica* and *japonica* groups (Figure 1c).

For all subgroups, we calculated the π values with a 100-kb sliding window, the common summary statistics for measuring genetics diversity in a population. The average diversity levels of *O. rufipogon* and *O. nivara* are markedly higher than those of *jap_landrace*, *jap_weedy*, and *ind_weedy*, but not *ind_landrace*

(Figure 1d). It is suggested that Korean *japonica* rice may have undergone a stronger reduction in effective population size than *indica* or weedy rice, while the *indica* landrace may not have encountered a strong bottleneck during domestication. The F_{ST} by the 100-kb sliding window was used to measure the level of population differentiation. The average F_{ST} between *O. rufipogon* and *O. nivara* (0.19) was much lower than that between wild rice and *O. sativa* [ruf-jap_weedy (0.41), ruf-jap_landrace (0.43), niv-ind_weedy (0.31). However, the F_{ST} between *O. nivara* and ind_landrace was only 0.07, which was much lower than that with ind_landrace-ind_weedy (0.16). This suggests that *indica* landrace is more closely related to *O. nivara* than the local weedy rice, possibly due to the materials of *O. nivara*. In the study by Xu et al. (2012), these five individuals were mixed with the *indica* varieties in the phylogenetic analysis, like in our study. The average F_{ST} between jap_landrace and jap_weedy was only 0.02 because of the complex and frequent gene exchange between these two subgroups (Figure 1d). This was confirmed by phylogenetic and population structure pattern analysis.

To estimate the linkage disequilibrium (LD) patterns in different rice groups, we calculated r^2 between pairs of SNPs using PLINK. We quantified the average extent of the genome-wide LD decay distance in wild rice, Korean landrace rice, and Korean weedy rice. These estimates were approximately 11, 72 and 80 kb in these three groups, respectively, where the r^2 dropped to half from the highest value (0.50, 0.53 and 0.66 respectively) (Figure S1). The LD decay of wild rice here was the same as reported previously. Because our weedy and landrace groups included *indica* and *japonica* subgroups, the LD decay was lower than *japonica* and larger than *indica*, as expected.

Recent positive and divergent selection

Population studies suggested that weedy rice formed from modern rice and not wild rice. LD patterns showed large differences in genome patterns between landrace and weedy rice, although the population study suggested that landrace rice and weedy rice undergo frequent gene exchange. To find the adaptive genome patterns of weedy rice, we employed five different methods to detect selective signatures between Korean landrace rice and weedy rice. Five distinct metrics of natural selection using a 100-kb sliding window or per site across the whole genome were used; that is, differentiation (F_{ST}), reduction in diversity (ROD), XP-CLR, iHS and XP-EHH. From these data, we classified the empirical top 2.5% of the windows or regions as 'selection outliers'. After annotation, we found 1068, 821, 552, 536 and 1329 genes over the outliers using ROD, F_{ST}, XP-EHH, iHS and XP-CLR respectively (Table S3). Most of the significantly selected genes (top 2.5%) occurred only one time among different selection approaches, suggesting that each statistic method will give a different view of selection. It is similar with previous study on *Populus trichocarpa*, that different selective forces are shaping different genomic regions (Evans et al., 2014). We also termed the regions in the top 2.5% for at least two of the selection scan metrics as candidate selection genes. Finally, 537 genes were detected as candidate selection genes (Table S4 and Figure 2a).

To explore the biological functions of 537 candidate selective genes, we performed a functional enrichment analysis to identify gene ontology terms. In total, 27 significantly enriched GO terms ($Q < 0.05$) were characterized in three major categories: biological process, cellular component and molecular function (Figure 2b and Table S5). By using the rice GO terms as background,

the candidate selected genes were enriched in response to stress, metabolic process and cellular process in biological process; extracellular region, membrane, cell and cell part in cellular component; organic acid transmembrane transporter activity, carboxylic acid transmembrane transporter activity and other 18 categories in molecular function (Figure 2b). Meanwhile, we performed KEGG pathway analysis for these 537 genes, which were enriched into 59 pathways, especially ascorbate and aldarate metabolism, RNA transport, histidine metabolism and Porphyrin and chlorophyll metabolism pathways (Figure 2c and Table S6).

Furthermore, we clustered those candidate selective genes against published database by using the PantGSEA. All genes were clustered into 90 independent experiment based data sets. We clustered these experiment-based data sets into three categories: plant development or growth-related categories, which are concentrated in tissue-specific expression genes; abiotic stress-related categories, including genes regulated by abiotic stress, such as cold stress, drought stress, salinity stress, Pi starvation stress, Cr stress, Fe stress, auxin stress etc.; and biotic stress-related categories, which were regulated by agrobacterium infection, blast fungus, bacterium infection in rice (Table S7). Among the 537 genes, 36% were related to plant development, 18% were related to abiotic stress, 17% were related to biotic stress and the remaining were still unknown in current rice database (Figure 3a). Some genes played different roles in different biological processes, in which 21 genes were clustered into three categories, and 66 genes shared in two different categories (Figure 3b).

To check the RNA expression pattern in specific tissues, we performed *k*-means co-expression clustering for these candidate selection genes. We downloaded the expression patterns of 11 different rice tissues from RiceXPro database (http://ricexpro.dna.affrc.go.jp). Three hundred and ninety-seven candidate selection genes were detected among this database. We clustered ten co-expression clusters (Figure S2 and Table S9) across all 11 tissues. Genes were defined as tissue-specific expressed genes if their tissue expression pattern deviation over twofold of standard deviation among the cluster. Cluster 10 was removed from the tissue-specific expression genes due to very low expression level. Finally, we found six tissue-specific expression clusters (cluster 1, 2, 3, 4, 8, 9) among the ten clusters (Figure S2). Cluster 1, 4, 8, 9 have high expression patterns in rice leaf, endosperm, embryo and root respectively. We found two pleiotropic drug resistance (PDR) family genes in cluster 9, *PDR8* and *PDR9* (Table S9). The *PDR9* in rice were significantly induced by jasmonates, which are involved in plant defence (Moons, 2008). We also found two receptor-like cytoplasmic kinase (RLCK) family gene cluster 1, which play roles in development and stress response in plants (Vij et al., 2008). It is suggested that genes in cluster 1, 4, 8, 9 may play important roles in weedy rice development and stress response. Cluster 2 and 3 show low expression levels in endosperm and leaf respectively. These genes in cluster 2 and 3 may not play important roles for rice endosperm and leaf development, but important for other rice tissue development.

In order to obtain a global view of gene functions in biotic stress, we used MapMan analysis, which groups candidate gene sets into hierarchical functional categories on the basis of putative involvement in biotic stress (Figure 3c). Four hundred and forty-eight genes out of 537 candidate genes were mapped to MapMan rice database. Among that, 107 genes were mapped to biotic stress pathway. Including five genes related to respiratory

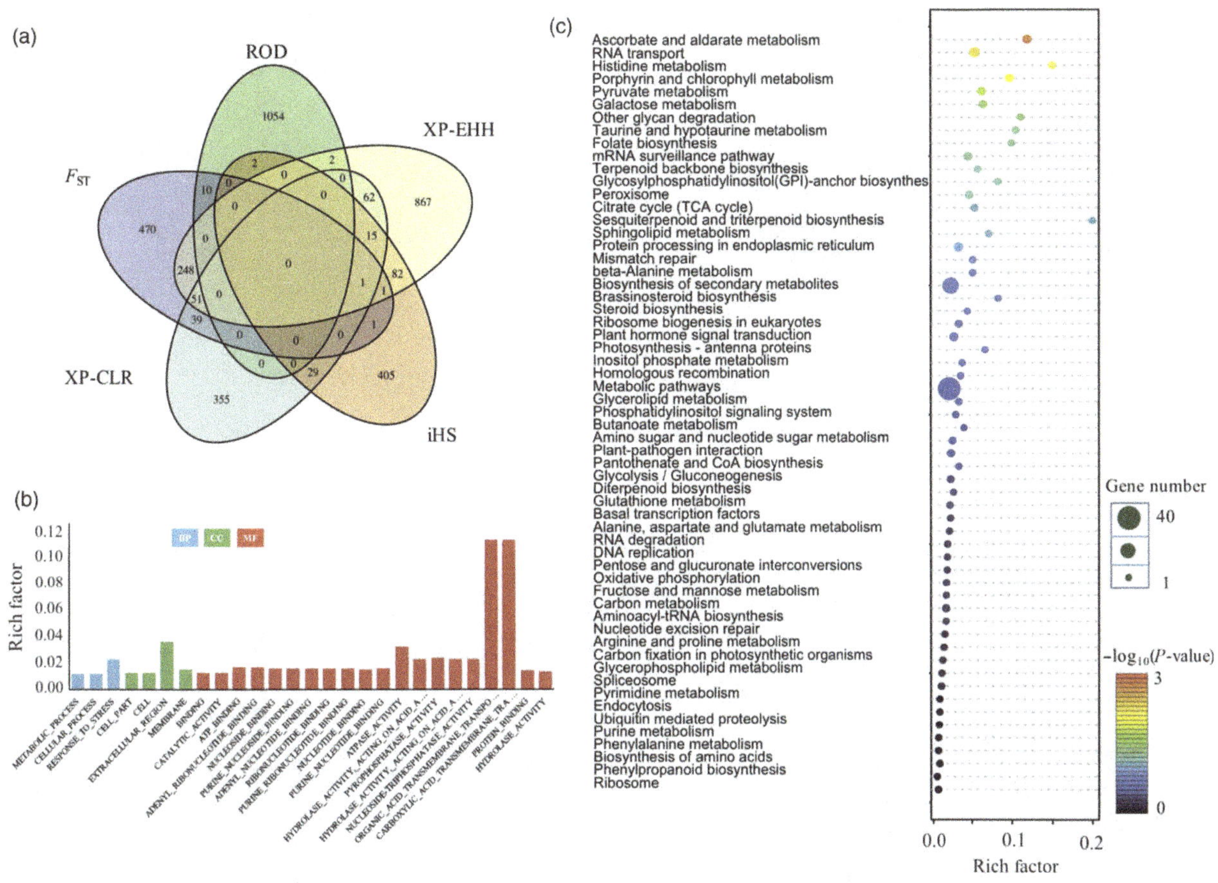

Figure 2 Unique and shared genomic regions (genes) among five selection scans. (a) Venn diagram of the number of genes throughout the genome in the top 2.5% for each selection scan. (b) GO enrichment of candidate selection genes. The blue bar indicates candidate selection genes, and the green bar shows the rice reference genome genes (c) KEGG pathway enrichment of candidate selection genes.

burst, 46 genes related to signalling, seven are transcription factors, eight secondary metabolites, two heat shock proteins, ten PR-proteins, five related to hormone signalling, nine cell wall genes, 19 proteolysis related genes and one beta glucanase. These annotations provided a valuable source of data for investigation of processes, function and pathways involved in the reaction to biotic stress in weedy rice. When we preformed the same strategy to overview the receptor-like kinases, 29 genes were enriched. Most of them (18) have LRR motifs, which were usually involved in protein–protein interactions and played an important role in regulating plant development and defence (Diévart and Clark, 2004). Six genes are receptor-like cytoplasmatic kinases, which may related to rice development or stress (Vij et al., 2008). Five genes are S-locus genes, which related to self-incompatibility responses of crucifers (Xing et al., 2013) (Figure 3d).

Discussion

Weedy rice is a noxious agriculture pest that has significantly reduced the grain yield of cultivated rice worldwide (He et al., 2014; Qiu et al., 2014). However, it has recently been realized that weedy rice possesses a number of adaptive traits lacking in cultivated rice, such as high reproduction and both biostress and abiostress tolerance (Green et al., 2001), so it may become an important resource for improving current cultivated rice. Knowledge regarding the origin, genetic diversity, structure and

adaptive genomic loci of weedy rice populations will facilitate the design of effective methods to control and use this weed.

Origin of Korean weedy rice

Most previous population genetics studies of weedy rice used limited polymorphism markers. Here, we used the whole-genome resequencing method and detected more than two million high-quality SNPs from 30 weedy rice, 30 landrace rice and ten wild rice samples. Using these high-density SNP makers (approximately 5.2 SNP/kb) will provide a more confident phylogenetic relationship and admixture pattern among these three rice populations.

The results of this study suggest that the weedy rice in Korea originated directly from japonica/japonica or indica/indica rice hybridization rather than from wild rice. Three major results support this speculation. (i) The phylogenetic tree based on 2.2 million SNPs indicated that 30 weedy rice samples clearly clustered into two major subgroups with japonica landrace and indica landrace rice, respectively. Weedy rice was more closely related to landrace than wild rice in either the japonica group or indica group, based on the evolutionary tree topology (Figure 1a). (ii) The results of the ancestry study of 70 samples showed that weedy rice divided into ind_weedy and jap_weedy subgroups and that landrace rice divided into ind_landrace and jap_landrace subgroups (Figure 1b). From $K = 3$ to $K = 6$, two major weedy rice subgroups (G1, G3) were separated from indica landrace and japonica landrace rice. Another main weedy subgroup, G2, had many admixture patterns from jap_landrace and other

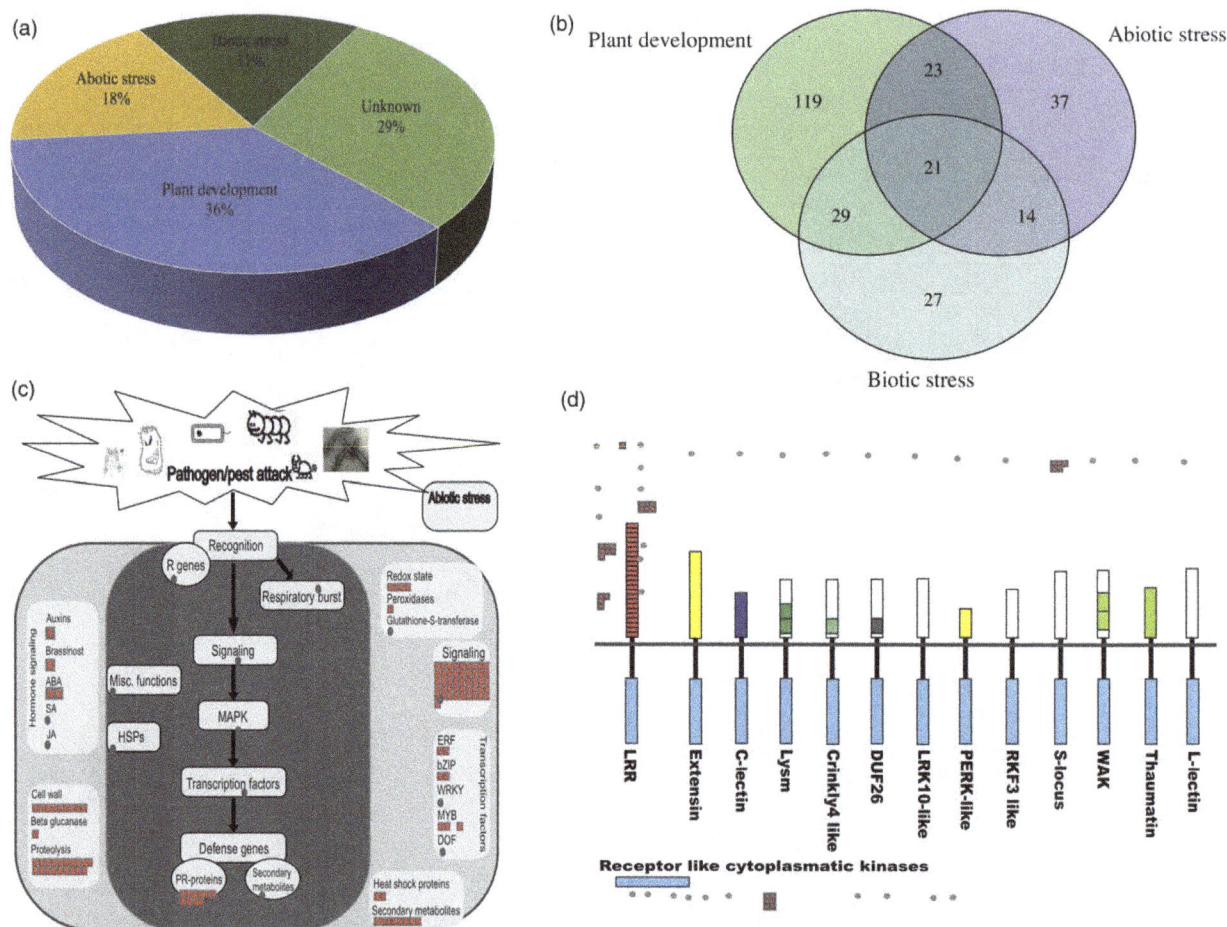

Figure 3 Candidate selected genes clustered based on public databases. (a) Functional category of candidate selected genes. (b) Venn diagram of three functional category. (c) The 'Biotic Stress overiew' MapMan pathway was used to visualize candidate genes. Red square are genes. (d) Genes involved in receptor-like kinases. Red square are genes.

jap_weedy. There were no wild rice patterns in weedy rice. Therefore, we propose that *ind_weedy* has only two genomic components: *ind_landrace* and *ind_weedy*. It was clearly demonstrated that *jap_weedy* had only two genomic components: *jap_landrace* and *jap_weedy*. (iii) The PCA study showed that weedy rice clearly clustered with landrace rice and not wild rice (Figure 1c).

Genomic adaptive of weedy rice

Determining the influences of positive and purifying selection, as well as neutral forces in shaping genetic variation is the primary goal of evolutionary biology (Evans *et al.*, 2014). Genetic diversity and phenotypic plasticity allow weeds to exploit novel and diverse opportunities as they occur in and infest agroecosystems (Dekker, 1997), there has long been an interest in understanding the genetic basis of weedy rice adaptation (Delouche and Labrada, 2007; He *et al.*, 2014; Qiu *et al.*, 2014). Understanding the adaptive genomic loci is a prerequisite for designing relevant strategies for effective control and management of different types of weeds in agroecosystems, as well as using these adaptive loci to improve cultivated rice. With the large genome-wide data sets for 70 rice variations discussed here, there is now an unprecedented opportunity for investigating issues such as the extent to which phenotypic differences among weedy rice and

cultivated rice (*O. sativa*) populations are driven by natural selection.

In this study, we concluded that weedy rice formed from cultivated rice. We performed positive selection or adaptive genomic loci identification using five different approaches between weedy and landrace rice. The ROD between two populations was used to detect reduced diversity at putatively neutral sites. F_{ST} was used to detect excess SNPs (or regions) with extreme population differentiation. XP-CLR, which is not sensitive to SNP ascertainment bias, was used to detect the selective sweeps based on multilocus allele frequency differentiation between two populations. XP-EHH and iHS were used to detect ongoing or nearly fixed selective sweeps by comparing haplotypes from two populations. A total of 537 genes were detected by at least two approaches, and this included many important genes; that is, the abiotic stress-related genes *SNAC3* on chromosome 1 (Fang *et al.*, 2015), *OsAsr1* on chromosome 2 (Vaidyanathan *et al.*, 1999), and *OsGIRL1* on chromosome 2 (Park *et al.*, 2014); the panicle morphology-related gene *OsRCN2* on chromosome 2 (Nakagawa *et al.*, 2002); biotic stress-related genes *PDR8* on chromosome 1 (Stein *et al.*, 2006) and *Snl6* on chromosome 1 (Bart *et al.*, 2010); the ion channel gene related to biotic and abiotic stresses *OsCNGC* on chromosome 12 (Nawaz *et al.*, 2014); the chlorophyll IIb biosynthesis-related gene *CAO* on chromosome 11 (Oster *et al.*, 2000); tiller development-related gene *SAD1* on chromosome 8 (Li *et al.*, 2015); and others (Table S4). It is well known that weedy rice always accompanied by red pericarp, while most cultivated rice have white pericarp, which mainly because of the 14 bp deletion of *Rc* gene on chromosome 7. *Rc* was already proved as the domesticated gene in rice domestication. But we did not find this gene among all the candidate selective genes. This is because we used all SNPs without Indels for selective signature detection. But selective sweeps always clustered or had selective blocks on chromosomes due to the linkage disequilibrium or other reasons. We mapped all 537 genes to the whole rice chromosomes. It is clearly shown that these candidate genes clustered on certain regions of rice chromosomes (Figure S3). When we checked the *Rc* region, two genes (Os07g0200700 and Os07g0222300) were found in 600 kb frank region of *Rc* (Figure S3). Around these two genes may have selective block and genetic linked with *Rc*. There's a big selective block from 8.7 to 13.6 M on chromosome 5 (Figure S3). However, we are did not find the popular domesticated genes in this region. We presume that this region may became hot research region for weedy rice study in the future. Although most of the candidate selection genes have not been cloned and their functions are not yet clear, most of these genes appear to be related to very important functions for weedy rice or cultivated rice. Os01g0191300 (SNAC3) (Fang *et al.*, 2015), which was detected using F_{ST} and XP-CLR, was recently confirmed to be related to drought and heat tolerance in rice, providing us a stronger reason to suggest that most of the candidate selected genes played important roles in rice or weedy rice evolution.

We noticed that the genes enriched in response to stress gene ontology terms, with rich factor 0.02 in biological process by using GO analysis (Figure 2b and Table S5). KEGG pathway enrichment suggested roles of these genes in histidine metabolism, which drives the reproduction and plant development-related pathway (Figure 2c and Table S6). These supported the idea that weedy rice has high reproduction (Song *et al.*, 2009) and high stress-resistant capacity (Chen *et al.*, 2004; Liu *et al.*, 2014; Lu *et al.*, 2014; Qiu *et al.*, 2014; Sun *et al.*, 2013). Meanwhile there are additional 26 GO terms ($Q < 0.05$) and 58 pathways were enriched among these 537 genes. It is suggested that the adaptive regions in weedy rice may not only just relate to stress or plant development but also relate to many other biological progress. The tissue-specific expression pattern study suggested that these candidate selection genes plays different roles at different tissues (Table S9 and Figure S2). The further study based on published databases suggested that most of the genes were related to the plant development and stress resistance. Except 29% unknown genes, others are all related to plant development, abiotic stress and biotic stress among 537 candidate genes (Figure 3a). Among those, 107 genes were enriched in biotic stress pathway, based on MapMan database (Figure 3c). These genes are distributed mostly in the sub-biological process response to biotic stresses, especially in signalling and proteolysis processes. Moreover, many of the candidate genes are LRR motif enriched gene, receptor-like cytoplasmatic kinases genes and S-locus genes (Figure 3d), which are related to plant development, defence and self-incompatibility based on previous studies (Diévart and Clark, 2004; Vij *et al.*, 2008; Xing *et al.*, 2013). In summary, these results suggested that weedy rice population significantly accumulated plant development and defence-related gene variations during the evolution. This may be the main reason why weedy rice exploit novel and diverse opportunities as they occur in and infest agroecosystems.

The results of this study using high-density polymorphism information clearly showed that Korean weedy rice formed by hybridization of *O. sativa* rather than wild rice. The weedy rice populations contain many genomic adaptive loci, which are different from other populations. These valuable candidate-selective genes or regions will make weedy rice a good resource to improve cultivated rice in the future, and the total 2.2 million high-quality SNPs generated here can be used for multiple objectives in the further rice genetics, genomics or evolution studies.

Experimental procedures

Plant materials

In 2010, we developed one core set for 4406 worldwide varieties, which collected from the National Genebank of the Rural Development Administration (RDA-Genebank, Republic of Korea) using the program PowerCore (Kim *et al.*, 2007; Zhao *et al.*, 2010). Among those, there are 30 Korean landrace rice and 30 Korean weedy rice accessions. These 60 available accessions were maintained by selfing at Kongju National University Experimental Farm. Each accession was transplanted in two rows, with 15 cm between plants and 30 cm between rows. Field management essentially followed normal agricultural practice. Total genomic DNA from a single plant of each accession was extracted from the leaf tissues by using the DNeasy Plant Mini Kit (Qiagen).

Whole-genome resequencing, SNP calling and structure variation calling

In previous study, we resequenced 137 core rice accessions (Kim *et al.*, 2016). Here, we use 60 of 137 sequence data and ten wild rice sequence data for this study. HiSeq 2500 were used for whole-genome resequencing of the 60 rice accessions (Kim *et al.*, 2016). Raw sequences were first processed to remove residual adapter sequences from the reads using Trimmomatic ver 0.36 (http://www.usadellab.org/cms/index.php?page=trimmomatic). Next, high-quality reads were aligned to the rice reference genome IRGSP-1.0 (http://rapdb.dna.affrc.go.jp/download/irgsp1.html) using the Burrows–Wheeler Aligner (BWA) (version 0.7.5a) with the default parameters (Li and Durbin, 2009). The reads will be removed if it did not meet BWA quality criteria or did not align to the reference genome. Duplicate reads were removed by using PICARD (version 1.88) (http://broadinstitute.github.io/picard/). Regional realignment and quality score recalibration were carried out using the Genome Analysis Toolkit (version 2.3.9 Lite) (McKenna *et al.*, 2010), and then variations were identified with ≥3× read depth coverage. Overall, the mapping depth was about 10× on average (Table S2). Sequence data for ten wild rice accessions (*O. rufipogon*, $n = 5$; *O. nivara*, $n = 5$) were downloaded from the National Centre for Biotechnology Information under accession number of SRA023116, and the same method, used for the 60 accessions, was used for SNP calling. The copy number variation study was performed by CNVnator (Abyzov *et al.*, 2011). Inversion and translocation were detected by using DELLY program (Rausch *et al.*, 2012).

Population structure and positive selection analysis

The population phylogenetic tree was constructed by the maximum likelihood method using RaxML (Stamatakis, 2014), and FigTree (Rambaut, 2007) was used to display the tree. We used a maximum likelihood method-based program, Frappe (Tang *et al.*, 2005), to generate the population structure. For positive selection, we used VCFtools (Danecek *et al.*, 2011) for nucleotide diversity and F_{ST} analysis. We performed XP-CLR using the XP-CLR program (http://genepath.med.harvard.edu/~reich). XP-EHH and iHS were performed by using selscan software (Szpiech and Hernandez, 2014).

Gene ontology (GO), Kyoto Encyclopedia of Genes and Genomes (KEGG) pathway

All candidate selective genes were used to GO and KEGG analysis. GO enrichment analysis was performed with PlantGSEA (Yi *et al.*, 2013) with the 'Oryza sativa' set as species background. KEGG pathway was performed by KOBAS (Xie *et al.*, 2011).

Acknowledgements

This work was carried out with the support of "Cooperative Research Program for Agriculture Science & Technology Development (Project No. PJ01116101)" Rural Development Administration, Republic of Korea. And also supported by Basic Science Research Program through the National Research Foundation of Korea(NRF) funded by the Ministry of Education (NRF-2014R1A1A2059399).

Author contributions

K.K.W and P.Y.J supervised the projects. K.K.W and P.Y.J contributed materials and analysis tools. H.Q, K.K.W and P.Y.J designed the research and wrote the manuscript.

Reference

Abyzov, A., Urban, A.E., Snyder, M. and Gerstein, M. (2011) CNVnator: an approach to discover, genotype, and characterize typical and atypical CNVs from family and population genome sequencing. *Genome Res.* **21**, 974–984.

Akey, J.M., Zhang, G., Zhang, K., Jin, L. and Shriver, M.D. (2002) Interrogating a high-density SNP map for signatures of natural selection. *Genome Res.* **12**, 1805–1814.

Bart, R.S., Chern, M., Vega-Sánchez, M.E., Canlas, P. and Ronald, P.C. (2010) Rice Snl6, a cinnamoyl-CoA reductase-like gene family member, is required for NH1-mediated immunity to Xanthomonas oryzae pv. oryzae. *PLoS Genet.* **6**, e1001123.

Bres-Patry, C., Bangratz, M. and Ghesquiere, A. (2001) Genetic diversity and population dynamics of weedy rice in Camargue area [France]. *Genet. Sel. Evol.* **33**, 425–440.

Cao, Q., Lu, B.-R., Xia, H., Rong, J., Sala, F., Spada, A. and Grassi, F. (2006) Genetic diversity and origin of weedy rice (Oryza sativa f. spontanea) populations found in north-eastern China revealed by simple sequence repeat (SSR) markers. *Ann. Bot.* **98**, 1241–1252.

Cao, Q.J., Xia, H., Yang, X. and Lu, B.R. (2009) Performance of hybrids between weedy rice and insect-resistant transgenic rice under field experiments: implication for environmental biosafety assessment. *J. Integr. Plant Biol.* **51**, 1138–1148.

Chen, L.J., Lee, D.S., Song, Z.P., Suh, H.S. and LU, B.R. (2004) Gene flow from cultivated rice (Oryza sativa) to its weedy and wild relatives. *Ann. Bot.* **93**, 67–73.

Chen, H., Patterson, N. and Reich, D. (2010) Population differentiation as a test for selective sweeps. *Genome Res.* **20**, 393–402.

Chung, J.W. and Park, Y.J. (2010) Population structure analysis reveals the maintenance of isolated sub-populations of weedy rice. *Weed Res.* **50**, 606–620.

Danecek, P., Auton, A., Abecasis, G., Albers, C.A., Banks, E., DePristo, M.A., Handsaker, R.E. *et al.* (2011) The variant call format and VCFtools. *Bioinformatics*, **27**, 2156–2158.

De Wet, J. and Harlan, J.R. (1975) Weeds and domesticates: evolution in the man-made habitat. *Econ. Bot.* **29**, 99–108.

Dekker, J. (1997) Weed diversity and weed management. *Weed Sci.* **45**, 357–363.

Delouche, J.C. and Labrada, R. (2007) *Weedy rices: origin, biology, ecology and control.* Food & Agriculture Org. **188**, 45–93.

Diévart, A. and Clark, S.E. (2004) LRR-containing receptors regulating plant development and defense. *Development*, **131**, 251–261.

Evans, L.M., Slavov, G.T., Rodgers-Melnick, E., Martin, J., Ranjan, P., Muchero, W., Brunner, A.M. *et al.* (2014) Population genomics of Populus trichocarpa identifies signatures of selection and adaptive trait associations. *Nat. Genet.* **46**, 1089–1096.

Fang, Y., Liao, K., Du, H., Xu, Y., Song, H., Li, X. and Xiong, L. (2015) A stress-responsive NAC transcription factor SNAC3 confers heat and drought tolerance through modulation of reactive oxygen species in rice. *J. Exp. Bot.* doi:10.1093/jxb/erv386.

Gealy, D.R., Tai, T.H. and Sneller, C.H. (2002) Identification of red rice, rice, and hybrid populations using microsatellite markers. *Weed Sci.* **50**, 333–339.

Green, J., Barker, J., Marshall, E., Froud-Williams, R., Peters, N., Arnold, G., Dawson, K. *et al.* (2001) Microsatellite analysis of the inbreeding grass weed Barren Brome (Anisantha sterilis) reveals genetic diversity at the within-and between-farm scales. *Mol. Ecol.* **10**, 1035–1045.

Gross, B.L., Reagon, M., Hsu, S.C., Caicedo, A.L., Jia, Y. and Olsen, K.M. (2010) Seeing red: the origin of grain pigmentation in US weedy rice. *Mol. Ecol.* **19**, 3380–3393.

Hak-Soo, S. and Mun-Hue, H. (1992) Collection and evaluation of Korean red rices I. Regional distribution and seed characteristics. *Korean J. Crop Sci.* **37**, 425–430.

He, Q. and Park, Y.-J. (2015) Discovery of a novel fragrant allele and development of functional markers for fragrance in rice. *Mol. Breed.* **35**, 1–10.

He, Z., Jiang, X., Ratnasekera, D., Grassi, F., Perera, U. and Lu, B.-R. (2014) Seed-mediated gene flow promotes genetic diversity of weedy rice within populations: implications for weed management. *PLoS ONE*, **9**, e112778.

He, Q., Yu, J., Kim, T.-S., Cho, Y.-H., Lee, Y.-S. and Park, Y.-J. (2015) Resequencing reveals different domestication rate for BADH1 and BADH2 in rice (Oryza sativa). *PLoS ONE*, **10**, e0134801.

Heu, M. (1988) Weed rice" Sharei" showing closer cross-affinity to Japonica type. *Rice Genet. Newsl.* **5**, 72–74.

Huang, X., Wei, X., Sang, T., Zhao, Q., Feng, Q., Zhao, Y., Li, C. *et al.* (2010) Genome-wide association studies of 14 agronomic traits in rice landraces. *Nat. Genet.* **42**, 961–967.

Huang, X., Kurata, N., Wei, X., Wang, Z.-X., Wang, A., Zhao, Q., Zhao, Y. *et al.* (2012a) A map of rice genome variation reveals the origin of cultivated rice. *Nature*, **490**, 497–501.

Huang, X., Zhao, Y., Wei, X., Li, C., Wang, A., Zhao, Q., Li, W. *et al.* (2012b) Genome-wide association study of flowering time and grain yield traits in a worldwide collection of rice germplasm. *Nat. Genet.* **44**, 32–39.

Ishikawa, R., Toki, N., Imai, K., Sato, Y., Yamagishi, H., Shimamoto, Y., Ueno, K. *et al.* (2005) Origin of weedy rice grown in Bhutan and the force of genetic diversity. *Genet. Resour. Crop Evol.* **52**, 395–403.

Jin, J., Huang, W., Gao, J.-P., Yang, J., Shi, M., Zhu, M.-Z., Luo, D. *et al.* (2008) Genetic control of rice plant architecture under domestication. *Nat. Genet.* **40**, 1365–1369.

Kelly Vaughan, L., Ottis, B.V., Prazak-Havey, A.M., Bormans, C.A., Sneller, C., Chandler, J.M. and Park, W.D. (2001) Is all red rice found in commercial rice really Oryza sativa? *Weed Sci.* **49**, 468–476.

Kim, K.-W., Chung, H.-K., Cho, G.-T., Ma, K.-H., Chandrabalan, D., Gwag, J.-G., Kim, T.-S. *et al.* (2007) PowerCore: a program applying the advanced M strategy with a heuristic search for establishing core sets. *Bioinformatics*, **23**, 2155–2162.

Kim, T.-S., He, Q., Kim, K.-W., Yoon, M.-Y., Ra, W.-H., Li, F.P., Tong, W. *et al.* (2016) Genome-wide resequencing of KRICE_CORE reveals their potential for future breeding, as well as functional and evolutionary studies in the post-genomic era. *BMC Genom.*, **17**, 1.

Kou, Y. and Wang, S. (2012) Toward an understanding of the molecular basis of quantitative disease resistance in rice. *J. Biotechnol.* **159**, 283–290.

Li, H. and Durbin, R. (2009) Fast and accurate short read alignment with Burrows-Wheeler transform. *Bioinformatics*, **25**, 1754–1760.

Li, C., Zhou, A. and Sang, T. (2006) Rice domestication by reducing shattering. *Science*, **311**, 1936–1939.

Li, W., Yoshida, A., Takahashi, M., Maekawa, M., Kojima, M., Sakakibara, H. and Kyozuka, J. (2015) SAD1, an RNA polymerase I subunit A34. 5 of rice, interacts with Mediator and controls various aspects of plant development. *Plant J.* **81**, 282–291.

Liang, D. and Qiang, S. (2011) Current situation and control strategy of weedy rice in China. *China Plant Prot.* **31**, 21–24.

Liu, Y., Jia, Y., Qi, X., Olsen, K., Caicedo, A. and Gealy, D. (2014) Insights into molecular mechanism of blast resistance in weedy rice. In *APS Annual Meeting.* pp. 67–70.

Londo, J. and Schaal, B. (2007) Origins and population genetics of weedy red rice in the USA. *Mol. Ecol.* **16**, 4523–4535.

Lu, Y.-L., Burgos, N.R., Wang, W.-X. and Yu, L.-Q. (2014) Transgene flow from Glufosinate-resistant rice to improved and weedy rice in China. *Rice Sci.* **21**, 271–281.

McKenna, A., Hanna, M., Banks, E., Sivachenko, A., Cibulskis, K., Kernytsky, A., Garimella, K. *et al.* (2010) The genome analysis toolkit: a MapReduce framework for analyzing next-generation DNA sequencing data. *Genome Res.* **20**, 1297–1303.

Moons, A. (2008) Transcriptional profiling of the PDR gene family in rice roots in response to plant growth regulators, redox perturbations and weak organic acid stresses. *Planta*, **229**, 53–71.

Nakagawa, M., Shimamoto, K. and Kyozuka, J. (2002) Overexpression of RCN1 and RCN2, rice TERMINAL FLOWER 1/CENTRORADIALIS homologs, confers delay of phase transition and altered panicle morphology in rice. *Plant J.* **29**, 743–750.

Nawaz, Z., Kakar, K.U., Saand, M.A. and Shu, Q.-Y. (2014) Cyclic nucleotide-gated ion channel gene family in rice, identification, characterization and experimental analysis of expression response to plant hormones, biotic and abiotic stresses. *BMC Genom.* **15**, 1.

Nei, M. and Li, W.-H. (1979) Mathematical model for studying genetic variation in terms of restriction endonucleases. *Proc. Natl Acad. Sci.* **76**, 5269–5273.

Nielsen, R., Williamson, S., Kim, Y., Hubisz, M.J., Clark, A.G. and Bustamante, C. (2005) Genomic scans for selective sweeps using SNP data. *Genome Res.* **15**, 1566–1575.

Noldin, J.A., Chandler, J.M. and McCauley, G.N. (1999) Red rice (Oryza sativa) biology. I. Characterization of red rice ecotypes. *Weed Technol.* **13**, 12–18.

Olofsdotter, M., Valverde, B. and Madsen, K.H. (2000) Herbicide resistant rice (Oryza sativa L.): global implications for weedy rice and weed management*. *Ann. Appl. Biol.* **137**, 279–295.

Oster, U., Tanaka, R., Tanaka, A. and Rüdiger, W. (2000) Cloning and functional expression of the gene encoding the key enzyme for chlorophyll b biosynthesis (CAO) from Arabidopsis thaliana. *Plant J.* **21**, 305–310.

Park, S., Moon, J.-C., Park, Y.C., Kim, J.-H., Kim, D.S. and Jang, C.S. (2014) Molecular dissection of the response of a rice leucine-rich repeat receptor-like kinase (LRR-RLK) gene to abiotic stresses. *J. Plant Physiol.* **171**, 1645–1653.

Qiu, J., Zhu, J., Fu, F., Ye, C.-Y., Wang, W., Mao, L., Lin, Z. *et al.* (2014) Genome re-sequencing suggested a weedy rice origin from domesticated indica-japonica hybridization: a case study from southern China. *Planta*, **240**, 1353–1363.

Rambaut, A. (2007) *FigTree, a graphical viewer of phylogenetic trees.*

Rausch, T., Zichner, T., Schlattl, A., Stütz, A.M., Benes, V. and Korbel, J.O. (2012) DELLY: structural variant discovery by integrated paired-end and split-read analysis. *Bioinformatics*, **28**, i333–i339.

Reagon, M., Thurber, C.S., Gross, B.L., Olsen, K.M., Jia, Y. and Caicedo, A.L. (2010) Genomic patterns of nucleotide diversity in divergent populations of US weedy rice. *BMC Evol. Biol.* **10**, 1.

Sabeti, P.C., Reich, D.E., Higgins, J.M., Levine, H.Z., Richter, D.J., Schaffner, S.F., Gabriel, S.B. *et al.* (2002) Detecting recent positive selection in the human genome from haplotype structure. *Nature*, **419**, 832–837.

Sabeti, P.C., Varilly, P., Fry, B., Lohmueller, J., Hostetter, E., Cotsapas, C., Xie, X. *et al.* (2007) Genome-wide detection and characterization of positive selection in human populations. *Nature*, **449**, 913–918.

Song, X., Liu, L., Wang, Z. and Qiang, S. (2009) Potential gene flow from transgenic rice (Oryza sativa L.) to different weedy rice (Oryza sativa f. spontanea) accessions based on reproductive compatibility. *Pest Manag. Sci.* **65**, 862–869.

Stamatakis, A. (2014) RAxML version 8: a tool for phylogenetic analysis and post-analysis of large phylogenies. *Bioinformatics*, **30**, 1312–1313.

Stein, M., Dittgen, J., Sánchez-Rodríguez, C., Hou, B.-H., Molina, A., Schulze-Lefert, P., Lipka, V. *et al.* (2006) Arabidopsis PEN3/PDR8, an ATP binding cassette transporter, contributes to nonhost resistance to inappropriate pathogens that enter by direct penetration. *Plant Cell*, **18**, 731–746.

Sun, J., Qian, Q., Ma, D.R., Xu, Z.J., Liu, D., Du, H.B. and Chen, W.F. (2013) Introgression and selection shaping the genome and adaptive loci of weedy rice in northern China. *New Phytol.* **197**, 290–299.

Szpiech, Z.A. and Hernandez, R.D. (2014) Selscan: an efficient multithreaded program to perform EHH-based scans for positive selection. *Mol. Biol. Evol.* **31**, 2824–2827.

Tang, H., Peng, J., Wang, P. and Risch, N.J. (2005) Estimation of individual admixture: analytical and study design considerations. *Genet. Epidemiol.* **28**, 289–301.

Tang, K., Thornton, K.R. and Stoneking, M. (2007) A new approach for using genome scans to detect recent positive selection in the human genome. *PLoS Biol.* **5**, e171.

Thurber, C.S., Reagon, M., Gross, B.L., Olsen, K.M., Jia, Y. and Caicedo, A.L. (2010) Molecular evolution of shattering loci in US weedy rice. *Mol. Ecol.* **19**, 3271–3284.

Vaidyanathan, R., Kuruvilla, S. and Thomas, G. (1999) Characterization and expression pattern of an abscisic acid and osmotic stress responsive gene from rice. *Plant Sci.* **140**, 21–30.

Vaughan, D.A., Lu, B.-R. and Tomooka, N. (2008) The evolving story of rice evolution. *Plant Sci.* **174**, 394–408.

Vij, S., Giri, J., Dansana, P.K., Kapoor, S. and Tyagi, A.K. (2008) The receptor-like cytoplasmic kinase (OsRLCK) gene family in rice: organization, phylogenetic relationship, and expression during development and stress. *Mol. Plant*, **1**, 732–750.

Voight, B.F., Kudaravalli, S., Wen, X. and Pritchard, J.K. (2006) A map of recent positive selection in the human genome. *PLoS Biol.* **4**, e72.

Williamson, S.H., Hubisz, M.J., Clark, A.G., Payseur, B.A., Bustamante, C.D. and Nielsen, R. (2007) Localizing recent adaptive evolution in the human genome. *PLoS Genet.* **3**, e90.

Xia, H.-B., Wang, W., Xia, H., Zhao, W. and Lu, B.-R. (2011) Conspecific crop-weed introgression influences evolution of weedy rice (Oryza sativa f. spontanea) across a geographical range. *PLoS ONE*, **6**, e16189.

Xie, C., Mao, X., Huang, J., Ding, Y., Wu, J., Dong, S., Kong, L. *et al.* (2011) KOBAS 2.0: a web server for annotation and identification of enriched pathways and diseases. *Nucleic Acids Res.* **39**, W316–W322.

Xing, S., Li, M. and Liu, P. (2013) Evolution of S-domain receptor-like kinases in land plants and origination of S-locus receptor kinases in Brassicaceae. *BMC Evol. Biol.* **13**, 69.

Xiong, H., Xu, H., Xu, Q., Zhu, Q., Gan, S., Feng, D., Zhang, X. *et al.* (2012) Origin and evolution of weedy rice revealed by inter-subspecific and inter-varietal hybridizations in rice. *Mol. Plant Breed.* **10**, 131–139.

Xu, X., Liu, X., Ge, S., Jensen, J.D., Hu, F., Li, X., Dong, Y. *et al.* (2012) Resequencing 50 accessions of cultivated and wild rice yields markers for identifying agronomically important genes. *Nat. Biotechnol.* **30**, 105–111.

Yi, X., Du, Z. and Su, Z. (2013) PlantGSEA: a gene set enrichment analysis toolkit for plant community. *Nucleic Acids Res.* **41**, W98–W103.

Zhao, W., Cho, G.-T., Ma, K.-H., Chung, J.-W., Gwag, J.-G. and Park, Y.-J. (2010) Development of an allele-mining set in rice using a heuristic algorithm and SSR genotype data with least redundancy for the post-genomic era. *Mol. Breed.* **26**, 639–651.

7

Transgenic rice with inducible ethylene production exhibits broad-spectrum disease resistance to the fungal pathogens *Magnaporthe oryzae* and *Rhizoctonia solani*

Emily E. Helliwell, Qin Wang and Yinong Yang*

Department of Plant Pathology and Huck Institutes of Life Sciences, Pennsylvania State University, University Park, PA, USA

*Correspondence
email yuy3@psu.edu

Summary

Rice blast (*Magnaporthe oryzae*) and sheath blight (*Rhizoctonia solani*) are the two most devastating diseases of rice (*Oryza sativa*), and have severe impacts on crop yield and grain quality. Recent evidence suggests that ethylene (ET) may play a more prominent role than salicylic acid and jasmonic acid in mediating rice disease resistance. In this study, we attempt to genetically manipulate endogenous ET levels in rice for enhancing resistance to rice blast and sheath blight diseases. Transgenic lines with inducible production of ET were generated by expressing the rice *ACS2* (1-aminocyclopropane-1-carboxylic acid synthase, a key enzyme of ET biosynthesis) transgene under control of a strong pathogen-inducible promoter. In comparison with the wild-type plant, the *OsACS2*-overexpression lines showed significantly increased levels of the *OsACS2* transcripts, endogenous ET and defence gene expression, especially in response to pathogen infection. More importantly, the transgenic lines exhibited increased resistance to a field isolate of *R. solani*, as well as different races of *M. oryzae*. Assessment of the growth rate, generational time and seed production revealed little or no differences between wild type and transgenic lines. These results suggest that pathogen-inducible production of ET in transgenic rice can enhance resistance to necrotrophic and hemibiotrophic fungal pathogens without negatively impacting crop productivity.

Keywords: ethylene, disease resistance, transgenic rice, rice blast, rice sheath blight.

Introduction

Rice (*Oryza sativa*) is of utmost importance to the human population, as more than half of the global population is dependent on the crop for the majority of its food requirements. As the human population is expected to rise to about 9 billion by the year 2050, rice crop yields will need to at least double by that time (Skamnioti and Gurr, 2009). One of the major limiting factors to yield is the occurrence of diseases caused by various fungal, bacterial and viral pathogens. Rice blast, caused by the ascomycete hemibiotrophic fungus *Magnaporthe oryzae*, is the most devastating rice disease in the world and often results in yield loss as high as 30% (Skamnioti and Gurr, 2009). Besides rice, *M. oryzae* can infect other grass species, such as perennial ryegrass (*Lolium perenne*) causing grey leaf spot, and wheat (*Triticum* spp.) causing wheat blast. Host resistance to *M. oryzae* is conferred by both race-specific resistance (*R*) genes, as well as by nonrace specific resistance quantitative trait loci (QTLs). *R* gene-mediated resistance frequently leads to a rapid and complete inhibition of the pathogen colonisation; however, this resistance is narrow spectrum, meaning that each *R* gene only recognises pathogen races that carry the corresponding aviru-lence (*Avr*) gene. As a result, *R* gene-mediated resistance is prone to breakdown due to point mutations, deletion and/or recombi-nation of *Avr* genes in the pathogen, which leads to disease susceptibility (Bonman, 1992; Dai *et al.*, 2010). Besides *R* gene-mediated resistance, another type of genetic resistance worth elucidating in plants is known as broad-spectrum resistance. Broad-spectrum resistance is defined as resistance that is effective against two or more pathogen species, and/or many different races within one pathogen species (Kou and Wang, 2010; Wisser *et al.*, 2005).

Sheath blight, caused by the fungus *Rhizoctonia solani* anas-tomosis group 1-IA, is the second-most devastating disease of rice, with yield loss between 10% and 25% (Banniza and Holderness, 2001). Unlike the case of *M. oryzae*, there are no known major *R* genes corresponding to *R. solani*, and resistance is conferred solely by the additive effect of nonrace-specific resistance QTL (Lee and Rush, 1983; Li *et al.*, 1995; Liu *et al.*, 2009; Pinson *et al.*, 2005). Resistance QTL is thought to confer a variety of traits, including components of basal resistance, developmental or morphological phenotypes that are not conducive to infection, production of antimicrobial compounds by the plant (phytoalexins and phytoanticipins), hormonal production and signalling, or other types of defence mechanisms (Poland *et al.*, 2008). Due to the specific and sometimes transient nature of *R*-gene-mediated resistance, along with the lack of known *R* genes for pathogens such as *R. solani*, it is important to study the more broad spectrum and quantitative types of host resistance.

The phytohormone ethylene (ET) is a small, gaseous molecule that plays numerous roles in plant growth, development and response to environmental stresses. Depending on the plant–pathogen combination and specific environmental conditions, ET may act as a positive or negative modulator of disease resistance (Broekaert *et al.*, 2006; Geraats *et al.*, 2003; Hoffman *et al.*, 1999; van Loon *et al.*, 2006). It is known that partial submer-gence of rice plants results in the biosynthesis and physical entrapment of ET within the hollow aerenchyma tissue (Steffens and Sauter, 2009). In paddy fields, enhanced resistance to *M. oryzae* infection was observed in rice plants grown under flood or anaerobic conditions (Lai *et al.*, 1999). Singh *et al.* (2004) proposed that the flood or hypoxia-induced ET biosynthesis in rice is critical for mediating horizontal resistance to blast infection. They demonstrated that application of AVG (aminoethoxyvinylglycine hydrochloride, an ET biosynthesis inhibitor) increased blast disease severity and negated the flood-induced resistance in rice plants. By contrast, application of ethephon (2-chloroethylphosphonic acid, an ET generator) significantly enhanced rice blast resistance in disease susceptible cultivars (Singh *et al.*, 2004). Exogenous ET is also known to induce pathogenesis-related (*PR*) genes such as *PR1*, *PR5* and *PR10* in rice plants (Agrawal *et al.*, 2003). Besides *M. oryzae*, ET has also been implicated in *Medicago* resistance to *R. solani*. Overexpression of an ET response factor (*MtERF1-1*) in the roots of *Medicago truncatula* resulted in enhanced resistance to *R. solani* AG8, whereas an ethylene-insensitive *ein2* (sickle) mutant was highly susceptible to the fungus (Anderson and Singh, 2011; Anderson *et al.*, 2010; Penmetsa *et al.*, 2008).

In the ET biosynthetic pathway, the conversion of S-adenosyl-L-methionine (AdoMet) to 1-aminocyclopropane-1-carboxylic acid (ACC) is the first committed step, which is catalysed by ACC synthase (ACS; Chae and Kieber, 2005). In higher plants, *ACS* genes are encoded by a multigene family, with nine members in *Arabidopsis* and six members in rice. Differential expression of individual *ACS* genes has been demonstrated in various plants to control specific aspects of plant development, maintenance and response to environmental cues. For example, *Arabidopsis ACS2* and *ACS6* are expressed after wounding and *Pseudomonas syringae* inoculation (Liang *et al.*, 1996). Tomato *ACS2*, *ACS4* and *ACS6* play differential roles in fruit development and ripening (Alexander and Grierson, 2002), and *ACS2* and *ACS6* are additionally induced by ozone stress (Moeder *et al.*, 2002). Maize *ACS6* is responsible for leaf senescence under normal and drought conditions (Young *et al.*, 2004). During the rice–*M. oryzae* interaction, endogenous ET levels increased within 48 h after inoculation with either avirulent or virulent isolates, with a significantly higher production of ET in the incompatible *Pii R* gene-mediated interaction (Iwai *et al.*, 2006). Whereas *OsACS3* and *OsACS4* were expressed constitutively, *OsACS1* and *OsACS2* were significantly induced upon *M. oryzae* infection, along with the induction of an ACC oxidase (ACO) gene, *OsACO7*. In a follow-up study, silencing of *OsACS2* and *OsACO7* by RNA interference (RNAi) resulted in increased susceptibility to rice blast (Seo *et al.*, 2010), suggesting that *OsACS2* and ET production play a positive role in rice resistance to *M. oryzae* infection.

In this study, we generated transgenic rice with inducible overproduction of ET by placing the ET biosynthetic gene *OsACS2* under the control of a strong, pathogen-inducible *PBZ1* promoter. Molecular, physiological and pathological analyses reveal that inducible overexpression of *OsACS2* in transgenic rice lines results in an induction of ET production and *PR* gene expression as well as enhanced resistance to both rice blast and sheath blight pathogens.

Results

Generation and verification of transgenic *OsACS2* overexpression lines

Agrobacterium-mediated transformation of the *PBZ1::OsACS2* construct yielded 22 independent T0 lines, with 55 plants total.

Five-week-old T0 plants were analysed for basal levels of *OsACS2* expression through quantitative real-time PCR (qRT-PCR), using a pair of *OsACS2*-specific primers. The resulting values were variable due to the nature of the stress-inducible *PBZ1* promoter. However, at least five independent lines showed a higher than threefold induction of *OsACS2* mRNA as compared with non-transformed cv. Kitaake lines (Figure 1b). The basal ET levels of these five transgenic lines were also higher than that of the wild-type plant (Figure 1c), which was consistent with the qRT-PCR data. Three independent lines (OX-7, OX-8 and OX-20) were chosen to advance to the T1 and T2 generations for further analysis based on the high-level inducibility of the *OsACS2* transgene and ET production.

Transgenic lines show inducible overexpression of *OsACS2* and increased production of ethylene

Upon comparison of the dH$_2$O- and benzothiodiazole (BTH)-treated plants, the dH$_2$O-treated OsACS2-OX lines OX-7 and OX-20 showed no significant change in the expression of *OsACS2* (Figure 2a), which can be explained by the lack of induction of *PBZ1* promoter under optimal growth conditions in absence of environmental stress. However, OX-8 showed a slight, but not

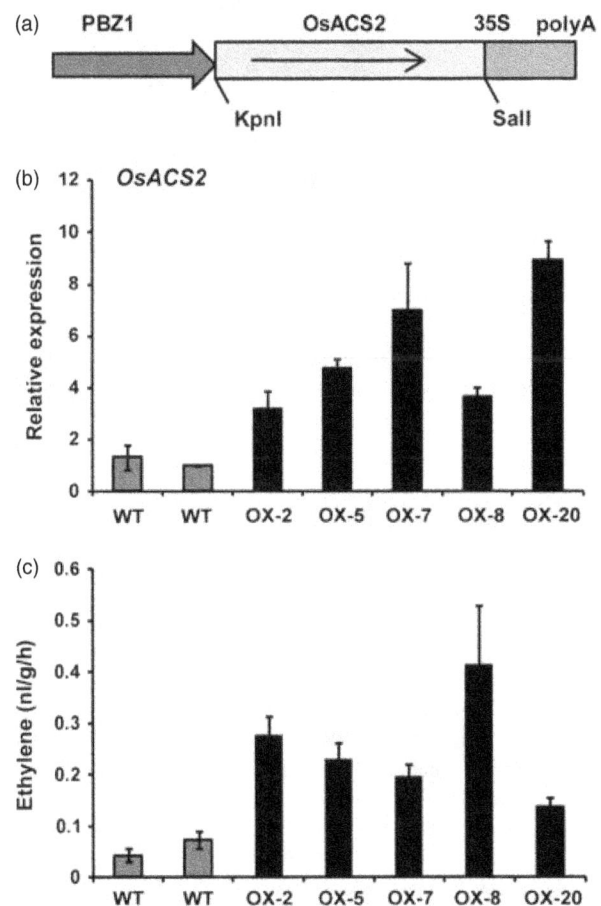

Figure 1 Generation and verification of *OsACS2*-overexpression lines. (a) The *PBZ1* promoter::*OsACS2* construct. (b) Relative quantity of *OsACS2* mRNA in leaves of wild type (WT) and T0 overexpression (OX) lines. (c) Ethylene production in foliar tissues of WT and T0 transgenic lines. The data were averaged from three leaf replicates with standard error. Experiments were conducted twice with similar results.

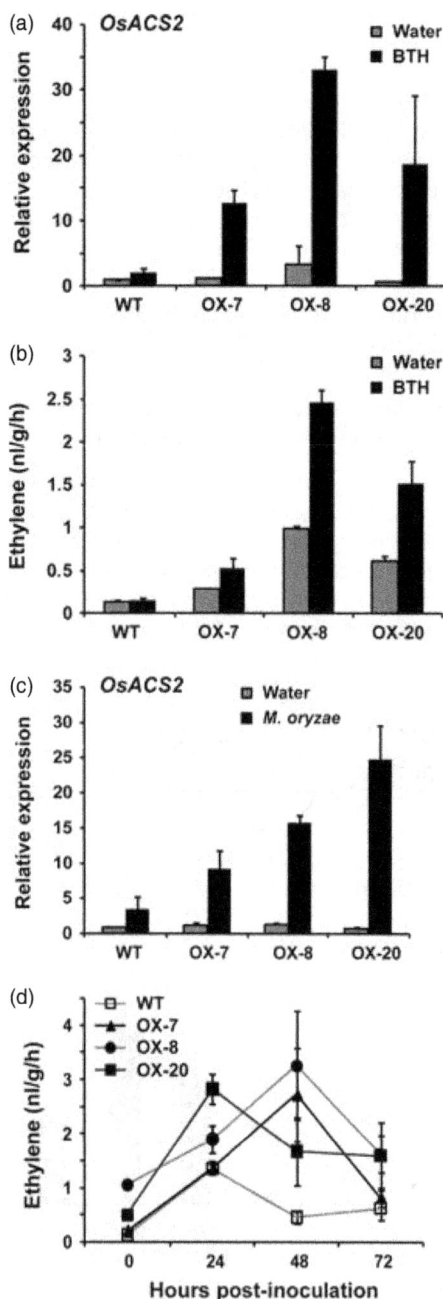

Figure 2 Induction of the *OsACS2* transgene and ethylene production in T2 homozygous lines. (a) Relative quantity of *OsACS2* mRNA in wild type (WT) and OsACS2-OX lines at 24 h post-treatment with either water or 0.25 mM benzothiodiazole (BTH). (b) Ethylene production in WT and OsACS2-OX lines at 24 h post-treatment with either water or 0.25 mM BTH. (c) Relative quantity of *OsACS2* mRNA in WT and OsACS2-OX lines at 72 h postinoculation with either water or *Magnaporthe oryzae* isolate IC17-18/1. (d) Ethylene production at 0, 24, 48 and 72 h postinoculation with *M. oryzae* isolate IC17-18/1. Experiments were conducted three times, and the data were averaged from three independent replicates with standard error.

significant increase of *OsACS2*. In contrast, the OsACS2-OX plants treated with BTH showed a drastic increase of *OsACS2* mRNA over the wild-type Kitaake ($F = 19.66$; $P = 0.018$). A second set of water- and BTH-treated lines were harvested and used to measure the ET levels (Figure 2b). Interestingly, all

water-treated OsACS2-OX lines showed significantly higher production of ET as compared with wild type (WT) Kitaake, despite lack of induction of *OsACS2* ($F = 103.99$; $P < 0.0001$). Basal ET levels of OX-8 were the highest among the OX and WT lines, which mirrored the slightly elevated levels of *OsACS2* mRNA of that same line. This pattern was amplified in the BTH-treated lines, as all OX lines showed a significant increase in ET production, with OX-8 containing the highest level of ET. The increase in both *OsACS2* expression and ET production in BTH-treated T2 lines shows a successful introduction of a functional and strongly inducible *OsACS2* transgene that is stably maintained through generations.

To examine the patterns of *OsACS2* expression and ET production in OsACS2-OX lines after pathogen challenge, WT cv. Kitaake and OsACS2-OX lines were inoculated with *M. oryzae* isolate IC17-18/1. *OsACS2* transcript levels were measured using qRT-PCR (Figure 2c). Similar to that of the BTH-treatment experiment, *OsACS2* was not significantly expressed in the OsACS2-OX lines as compared with the WT cv. Kitaake under basal conditions. At 72 h postinoculation, levels of *OsACS2* mRNA in the OsACS2-OX lines were significantly higher than that of the WT cv. Kitaake ($F = 14.12$; $P = 0.014$). This result indicates that the *OsACS2* transgene is induced to a higher degree than the endogenous *OsACS2* transcripts after *M. oryzae* infection. The production of ET in OsACS2-OX and WT cv. Kitaake lines was measured at 0, 24, 48 and 72 h postinoculation by *M. oryzae* isolate IC17-18/1 (Figure 2d). The OX-8 and OX-20 lines showed a marked increase in ET production over WT cv. Kitaake at all four time points; however, the kinetics of ET production were slightly different. The ET production of OX-20 peaked earlier (24 h postinoculation) than that of OX-8 (48 h postinoculation). The ET production of OX-7 was similar to that of WT cv. Kitaake, with the exception of 48 h postinoculation, in which OX-7 line produced a higher amount of ET than both cv. Kitaake and OX-20. Despite the differences in the kinetics of ET production, the OsACS2-OX lines produced significantly more ET than that of WT cv. Kitaake lines ($F = 10.79$; $P < 0.0001$) after inoculation with *M. oryzae*.

Increased expression of *PR* genes in *OsACS2* overexpression lines

Quantitative real-time PCR was used to measure the basal levels of *OsPR1b* and *OsPR5* transcripts in both 5-week-old T0 and 2-week-old T2 transgenic plants. *OsPR1b* was shown to be significantly induced in all three OsACS2-OX lines as compared with the nontransformed control, ranging from 10-fold to about 60-fold increase in both generations (Figure 3a). Likewise, all three OsACS2-OX lines also showed significantly higher expression of *OsPR5* in both generations, however, to a smaller degree than that of *OsPR1b* with fold increases from 2.0 to 7.9 in the T0 generation, and 2.3–4.2 in the T2 generation (Figure 3b). These results show that OsACS2-OX lines display higher basal expression of *PR* genes, which may imply higher levels of host resistance.

OsACS2 overexpression lines show increased resistance to different races of *Magnaporthe oryzae*

To evaluate the disease resistance of transgenic rice lines, two isolates of *M. oryzae* with differing degrees of virulence were spray-inoculated onto OsACS2-OX lines and wild-type Kitaake. Disease severity was assessed through counting the number of lesions per leaf (an indicator of disease incidence) as well as

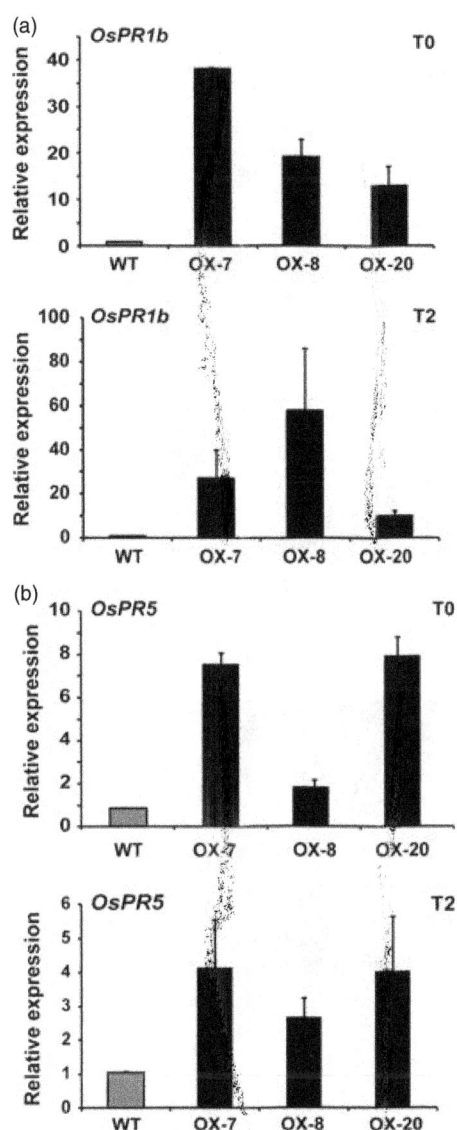

Figure 3 Enhanced expression of *PR* genes in *OsACS2*-overexpression lines. (a) Relative expression of *OsPR1b* in T0 (top) and T2 (bottom) lines. (b) Relative expression of *OsPR5* in T0 (top) and T2 (bottom) lines. Experiments were conducted once in T0 lines and three times in T2 lines. The T0 data are the average of three technical replicates; T2 data are the average of three independent biological replicates with standard error.

measuring the lesion size (an indicator of disease severity). The first isolate tested was *M. oryzae* IC17-18/1, which is moderately virulent on cv. Kitaake. The OsACS2-OX lines all showed a significant reduction in lesion number ($F = 25.7$; $P < 0.0001$), (Figure 4a,c); however, it was noted that the overexpression lines OX-8 and OX-20 showed a more significant reduction in lesion number than OX-7. All three overexpression lines also showed a significant reduction in lesion size ($F = 21.42$; $P < 0.0001$) (Figure 4b,c). A second *M. oryzae* isolate, IE1K-FN9, is highly virulent on cv. Kitaake and has the ability to overcome the *Pi-ta* resistance gene widely deployed in the US rice cultivars. Similar to the results from the previous IC17-18/1 inoculation, the OsACS2-OX lines showed significant reductions in both lesion number ($F = 23.9$; $P < 0.0001$) and lesion size ($F = 29.63$; $P < 0.0001$) (Figure 4d–f). It was noted that OX-8 showed the most significant reduction in lesion number as compared with OX-7 and OX-20;

however, all three overexpression lines were significantly more resistant as compared with cv. Kitaake. These results suggest that increased resistance conferred by overexpression of *OsACS2* is not limited to single races of *M. oryzae*.

OsACS2-overexpression lines show increased resistance to *Rhizoctonia solani*

After establishing that overexpression of *OsACS2* confers resistance to *M. oryzae* in a nonrace-specific manner, the question arose if this resistance could protect against a completely different species of fungus. The OsACS2-OX lines and wild-type control were inoculated with *R. solani* field isolate RR0140 (Wamishe *et al.*, 2007) using the mycelia ball method (Park *et al.*, 2008). The OsACS2-OX lines showed a 35%–45% reduction in lesion size compared with nontransformed cv. Kitaake ($F = 19.8$; $P < 0.0001$) (Figure 5a,b). These results demonstrate that *OsACS2*-overexpression lines are not only more resistant to the hemibiotrophic rice blast fungus, but also exhibit enhanced resistance to the necrotrophic sheath blight pathogen.

Inducible overexpression of *OsACS2* does not negatively affect seed production and agronomic traits

Regardless of the strategies and genes used to genetically improve crop cultivars, one of the most important aspects to check is the potential effect of transgene on plant growth and grain production. To gauge the effect of inducible overexpression of *OsACS2* on these agronomically important traits, twelve plants from each transgenic lines were grown in glasshouse conditions over two different seasons, and the growth rate, time until maturity and seed production were measured. There were no differences between WT cv. Kitaake and OsACS2-OX lines until 5 weeks postgermination, which was close to the end of the vegetative stage (approximately 25–30 cm). After 6 weeks, the OX-7 and OX-20 lines grew slightly slower than WT cv. Kitaake and were about 7–10 cm shorter at maturity (66 and 63 cm, respectively, compared with 74 cm for Kitaake). Interestingly, OX-8 line grew significantly slower than nontransformed and other transformed lines between 5 and 8 weeks postgermination, but underwent rapid growth after week 8 and reached a similar height to OX-7 and OX-20 at maturity (Figure 6a).

To assess the maturation rate of the OsACS2-OX lines, the time elapsed until each growth stage was recorded for each individual plant. All plants reached the booting stage at weeks 6–7, heading at weeks 7–8 and flowering between weeks 7 and 9. OX-7 and OX-20 plants reached each reproductive stage in the shortest time, flowering about 0.5 weeks earlier than the wild-type plant. Similar to the growth rate data, OX-8 reached each stage latest, 1 week behind OX-7 and OX-20 and 0.5 weeks later than Kitaake. However, in the ripening phase, OX-8 underwent rapid grain-filling and ripening as compared with all other lines, and reached maturity between 12 and 13 weeks, similar to OX-7 and cv. Kitaake. OX-20 matured in the shortest time, by 12 weeks postgermination (Figure 6b).

To analyse yield-related components, the number of panicles per plant, the number of seeds per panicle and the weight of 100 seeds were measured (Figure 6c–e). There were no significant differences for each of these parameters between WT and OsACS2-OX lines; however, OX-7 had a slightly higher number of panicles (average of 3.6 per plant, compared with 3 per plant for Kitaake, OX-8 and OX-20) as well as number of seeds per panicle (56 seeds, as compared to approximately 44 seeds in Kitaake, OX-8 and OX-20). For all lines, the weight per 100 seeds was

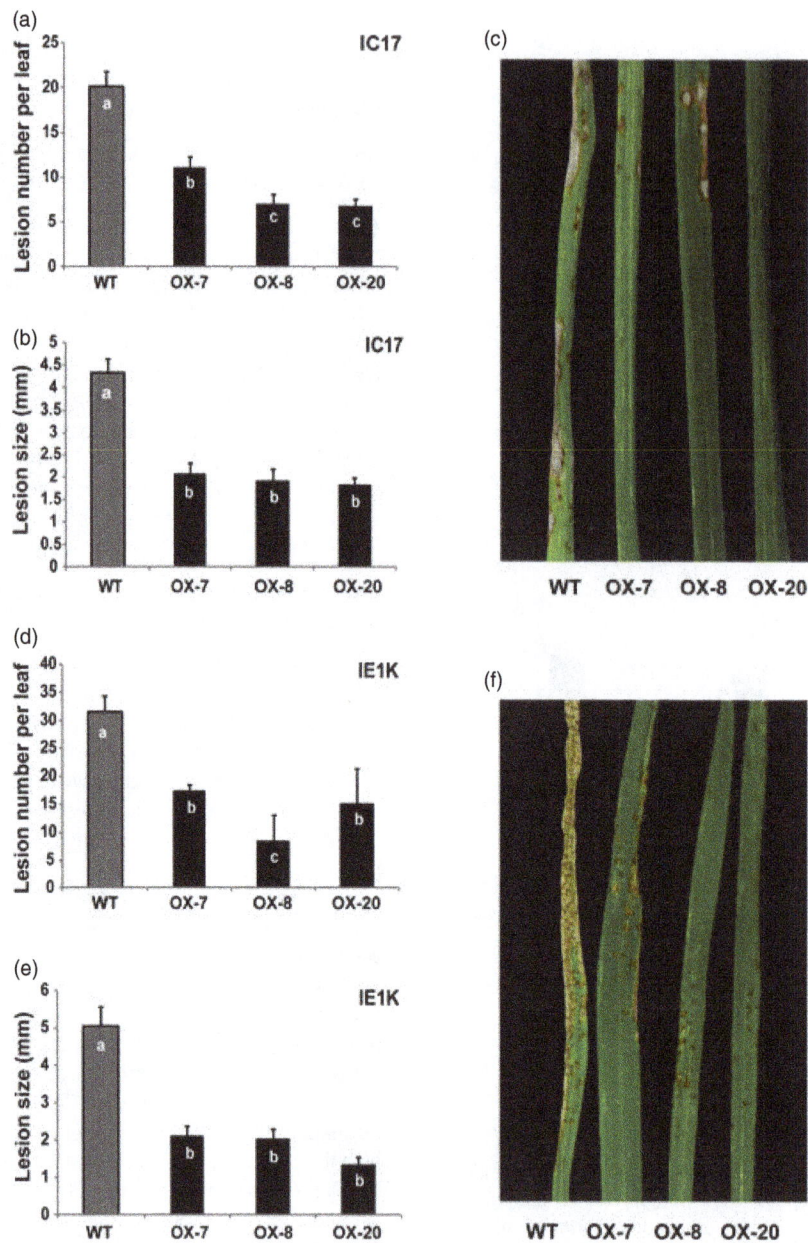

Figure 4 Increased resistance of *OsACS2*-overexpression lines to moderately virulent (IC17) and highly virulent (IE1K) isolates of *Magnaporthe oryzae*. (a) Lesion number per leaf after inoculation with *M. oryzae* IC17-18/1. (b) Lesion size after inoculation with *M. oryzae* IC17-18/1. (c) Rice blast symptoms on wild type (WT) and *OsACS2*-overexpression lines inoculated with *M. oryzae* IC17-18/1. (d) Lesion number per leaf after inoculation with *M. oryzae* IE1K-FN9. (e) Lesion size after inoculation with *M. oryzae* IE1K-FN9. (f) Rice blast symptoms on WT and *OsACS2*-overexpression lines inoculated with *M. oryzae* IE1K-FN9. About 20 plants per line were evaluated in each experiment, and the data were averaged from three independent replicates. Letters represent each significance groups, determined through one-way analysis of variance.

about 2.7 g, with little variation. From this data, we can conclude that inducible overexpression of OsACS2 has no negative effect on these agronomically important traits, and may even promote slightly more rapid maturation and yield in a glasshouse setting.

Discussion

The development of crop cultivars through transgenic methods is becoming increasingly common, due to advances in gene discovery and overall improvement in the efficiency and efficacy of stable transformation methods. The use of transgenic approaches to improve host resistance is beneficial, as it mitigates

the need for pesticides and is more efficient than conventional breeding methods (Gust *et al.*, 2010). Breeding strategies for host genetic resistance have traditionally focused on using major *R* genes to ensure a complete, efficient form of resistance. The drawbacks to this method are that the majority of *R* genes only protect against a narrow spectrum of pathogens, and are prone to breakdown due to introductions of different isolates, or mutations within pathogen populations. For example, breakdown of rice blast *R* gene *Pi-ta* occurs due to the unstable telomeric location of the corresponding *Avr Pita* gene in *M. oryzae* (Dai *et al.*, 2010; Orbach *et al.*, 2000; Zhou *et al.*, 2007). For this reason, it is important to discover other, less specialised sources of

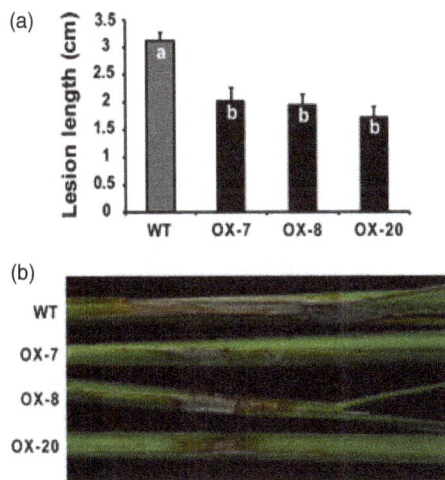

Figure 5 Increased resistance of OsACS2-overexpression lines to Rhizoctonia solani isolate RR0140. (a) Lesion length on the wild-type plant and OsACS2-OX lines. (b) Comparison of sheath blight symptoms between wild type and OsACS2-OX lines. At least eight plants per line were evaluated in each experiment, and the data represent the average of three independent replicates with standard error. Letters represent different significance groups according to one-way analysis of variance.

host genetic resistance. To date, transgenic modification or altered expression of many rice genes has been shown to increase host resistance to different fungal and bacterial pathogens. These include receptors OsWAK1 (Li et al., 2009), OsSERK1 (Hu et al., 2005) and OsBRR1 (Peng et al., 2009), salicylic acid (SA) signalling component AtNPR1 and its rice homologue OsNH1 (Chern et al., 2001, 2005), jasmonic acid (JA) biosynthetic gene OsAOS2 (Mei et al., 2006), transcription factors OsWRKY13 (Qiu et al., 2007), OsWRKY45 (Shimono et al., 2007) and OsWRKY71 (Liu et al., 2007), and PR genes such as PR5 (Datta et al., 1999), among many others.

Plant hormones such as SA, JA and ET play diverse roles in mediating defence signalling and disease resistance responses (Grant and Jones, 2009; Lopez et al., 2008; Robert-Seilaniantz et al., 2007). These hormones often invoke the activation of defence-related kinases, transcription factors and PR genes, many of which have been found to improve host resistance in transgenic rice (reviewed in Delteil et al., 2010). SA is a phenolic compound that has been shown to be involved in the activation of systemic acquired resistance in many dicotyledonous plants (Durrant and Dong, 2004; Park et al., 2007). Because rice plants contain a high basal level of SA, which is not significantly induced by pathogen infection, SA may not serve as an effective defence signal in rice (Silverman et al., 1995; Yang et al., 2004). However, endogenous SA was shown to protect rice plants from oxidative damage caused by biotic and abiotic stresses and to mediate host resistance in adult plants (Iwai et al., 2007; Yang et al., 2004). Furthermore, SA signalling components appear to be functional in rice because overexpression of the rice NPR1 homologue, OsNH1, resulted in increased PR gene expression and enhanced resistance to X. oryzae pv. oryzae (Chern et al., 2005). Possibly due to high basal level of SA in rice, which may inhibit jasmonate biosynthesis, JA levels increase only slightly after M. oryzae infection (Mei et al., 2006). However, JA was shown to be capable of mediating defence signalling and disease resistance in rice. Application of exogenous JA or methyl JA (MeJA) activates

the expression of several PR genes, along with an increased level of phytoalexins in rice (Tamogami et al., 1997). In addition, overexpression of the jasmonate biosynthetic gene allene oxide synthase 2 (OsAOS2) resulted in increased JA production and PR gene expression as well as enhanced resistance to M. oryzae (Mei et al., 2006). Recently, increasing evidence from our lab and other groups suggests that ET may play a more prominent role than SA and JA in rice disease resistance (Bailey et al., 2009; Seo et al., 2010; Singh et al., 2004). Applications of ethephon decreased the incidence of rice blast in field conditions (Singh et al., 2004) and triggered PR gene expression in rice cell cultures (Agrawal et al., 2003). More importantly, inoculation of rice by M. oryzae triggers induction of ET biosynthetic genes (Iwai et al., 2006), and knockdown of these genes results in increased susceptibility to M. oryzae (Seo et al., 2010). Therefore, it is imperative to explore the transgenic approach for improving rice disease resistance based on the genetic modification for enhanced ET biosynthesis.

In this study, we created transgenic rice exhibiting inducible overproduction of ET, by placing the ET biosynthetic gene OsACS2 under control of the PBZ1 promoter. This pathogen-inducible promoter was chosen for the purpose of fine-tuning the host response to a pathogen, as opposed to a constitutive promoter, as continuous overexpression of a transgene may potentially have a negative effect on other aspects of plant health, yield or tolerance to other environmental stresses (Brown, 2002; Kim et al., 2009; Potenza et al., 2004). OsACS2 was chosen to overexpress because this gene is significantly induced by M. oryzae and closely associated with ET-mediated rice defence response (Iwai et al., 2006). There is some variability in induction of OsACS2 in the OsACS2-OX lines, particularly in the elevated basal expression of OsACS2 in OX-8 (as seen in Figures 2a,b). This might be caused by position effects where nearby cis-elements at the transgene insertion site in the OX-8 line resulted in elevated basal expression of OsACS2. Despite this, both OsACS2 transcripts and ET levels were greatly increased in T_0 and T_2 generations after BTH- and pathogen-activation of the PBZ1 promoter as compared with basal conditions. Therefore, we can conclude that the OsACS2 transgene is inducible, functional and stable through multiple generations.

Inoculation of OsACS2-overexpression lines with a moderately virulent isolate of M. oryzae (IC17-18/1) showed about a 50% reduction in both lesion number (incidence) and lesion size (severity), along with a significant increase in ET production at 48 and 72 h postinoculation. It is important to note that disease symptoms do appear on OsACS2-overexpression lines; however, blast lesions occur later and are smaller, and the overall disease severity is lower than that of nontransformed lines, classifying this as more of a partial resistance. We also inoculated the OsACS2-overexpression lines with a highly virulent isolate, IE1K-FN9, in order to gauge the efficacy of ET-mediated resistance against a more pathogenic isolate. Infection by IE1K-FN9 resulted in many lesions and extensive damage on leaves of nontransformed cv. Kitaake, but much fewer lesions on OsACS2-overexpression lines. Interestingly, the use of an unrelated, necrotrophic fungal pathogen, R. solani, showed the same trend of a significant reduction in lesion size, but not a complete inhibition of the pathogen growth. This is consistent with our recent observation that transgenic suppression of ET biosynthesis gene via RNAi or treatment of ET biosynthetic inhibitor aminooxyacetic acid (AOA) resulted in increased susceptibility to R. solani (Yang lab, unpubl. data). Taken together, our results suggest that ET-mediated

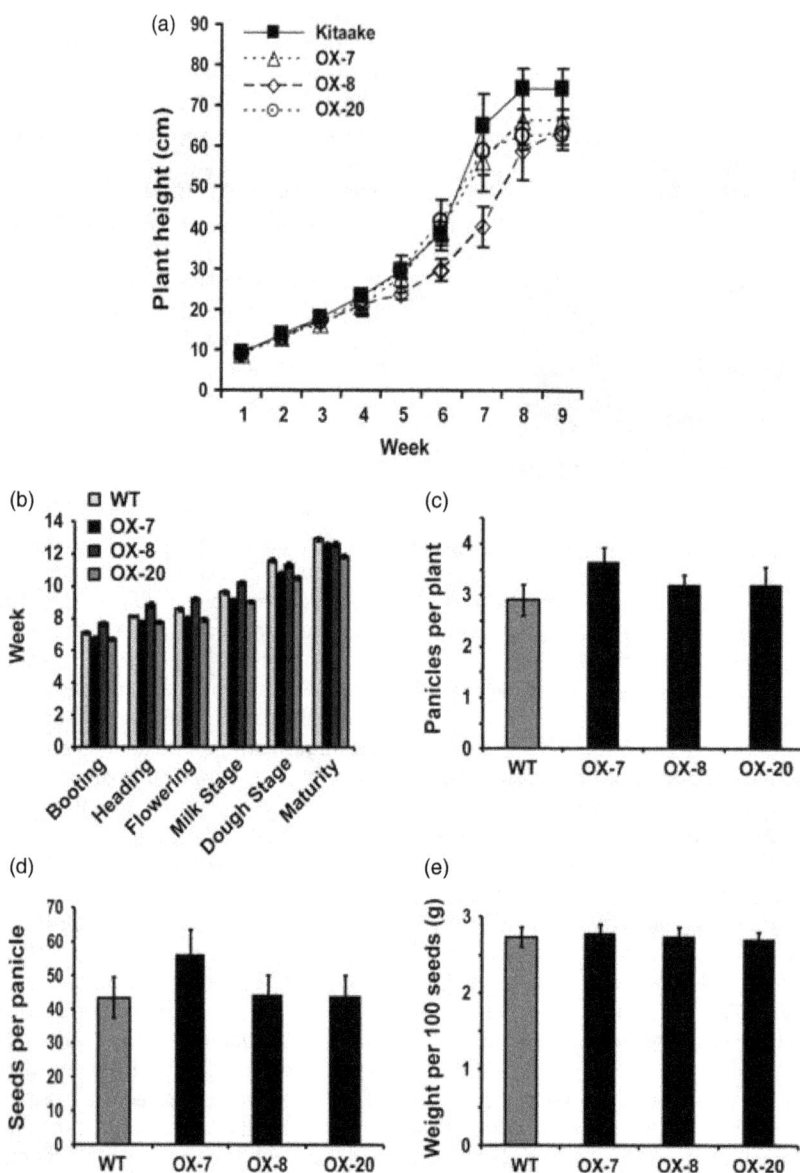

Figure 6 Evaluation of plant growth and development as well as grain production of *OsACS2*-overexpression lines. (a) Plant height during rice growth and maturation. (b) Time required to reach each growth and reproductive stage. (c) The number of panicles of each plant at maturity. (d) The number of seeds per panicle. (e) The weight per 100 seeds. At least 12 plants per line were evaluated in each experiment, and the data represent the average of two independent replicates with standard error.

disease resistance may be broad spectrum in nature, and is potentially more durable than *R* gene-mediated resistance.

The data derived from this study show that increased ET production enhances host resistance in rice against *M. oryzae* and *R. solani*. However, the mechanism of ET-mediated resistance is still not clear. A simple explanation is that it can be attributed to the enhanced expression of *PR* genes at the end of the ET signalling pathway. It has been shown that application of exogenous ET results in the activation of rice *PR* genes with GCC-box-containing promoters, including *OsPR1b* and *OsPR5* (Agrawal *et al.*, 2003). Therefore, inducible overproduction of endogenous ET may lead to increased expression of *PR* genes in the transgenic lines. Indeed, the *OsACS2*-overexpression lines were found to have increased expression of both *OsPR1b* and *OsPR5*. The higher expression of *OsPR5* in the T0 generation could tentatively be explained by the increased plant age. Aged tissues often show

many hallmarks of stress-related responses, including elevated SA levels, production of reactive oxygen species and *PR* gene expression (Kus *et al.*, 2002; Wyatt *et al.*, 1991). It is likely that in the older T0 plants, the *PBZ1* promoter is active at a low level, which increases expression of the *OsACS2* transgene.

Another possible explanation for ethylene-mediated resistance is that the increased ET production is affecting other hormonal pathways such as the abscisic acid (ABA) level and/or signalling. Under submerged conditions, ET was shown to induce ABA 8'-hydroxylase, which oxidises ABA to its inactive form, phaseic acid (Saika *et al.*, 2007). Such a reduction of active ABA may lead to enhanced rice blast resistance because ABA appears to positively regulate rice susceptibility to *M. oryzae* infection (Bailey *et al.*, 2009; Jiang *et al.*, 2010; Koga *et al.*, 2004). By contrast, suppression of an ABA-inducible MAP kinase down-regulated the ABA pathway, but up-regulated the ET pathway, leading to enhanced rice blast resistance (Bailey et al., 2009; Xiong and Yang, 2003). Therefore, it is also possible that ET indirectly promotes host resistance to rice blast by suppressing the ABA level and/or signalling.

An alternative explanation is that the increased resistance is mediated not by ET, but by cyanide. Cyanide is a by-product of the ET biosynthetic pathway and is produced in equal amounts to ET (Peiser et al., 1984). Seo et al. (2010) hypothesised that Pii-mediated resistance was due to increased cyanide production after demonstrating that exogenous application of either cyanide or the ET precursor ACC complemented ET-deficient rice lines, but ethephon failed to. However, this is contradictory to the results published by Singh et al. (2004), which demonstrated a marked reduction in blast disease incidence in several cultivars after ethephon application. In addition, results from our lab indicated that ET-insensitive rice, which is not deficient in ET or cyanide levels, has increased susceptibility to M. oryzae (Bailey et al., 2009; Yang lab, unpubl. data). Due to these contradictory results, additional studies are necessary to clarify the roles of ET biosynthesis, cyanide production, ET signalling and other factors involved in ET-mediated rice disease resistance.

In summary, we have developed transgenic rice lines with pathogen-inducible overexpression of the ET biosynthetic gene OsACS2. These lines show inducible overproduction of ET and enhanced resistance to the fungal diseases rice blast and sheath blight, without negatively impacting agronomically important traits such as yield in glasshouse conditions. Future work will focus on the field applications of OsACS2-OX lines, in terms of agronomic traits along with the efficacy against a greater diversity of pathogens. In addition, the resistance of OsACS2-overexpression lines to multiple fungal pathogens suggests that enhancement of ET-mediated resistance may be applicable in other monocot species for controlling devastating fungal diseases such as wheat blast and turfgrass grey leaf spot.

Materials and methods

Gene construct and rice transformation

The full length cDNA sequence was amplified by PCR using rice OsACS2 cDNA (AK064250) as a template with a pair of specific primers containing KpnI or SalI restriction sites (Forward primer: 5' ACG GTA CCA TGG CGT ACC AGG GCA TCG AC 3'; Reverse primer: 5' ACC GTC GAG TCT GCT GGC TTA ATC AGC TG 3'). After restriction digestion, the PCR fragment was inserted into the modified vector pCAMBIA1300P, which contains a pathogen-inducible rice PBZ1 promoter (Lee, 2002). The resulting PBZ1:: OsACS2 construct was then transformed into Agrobacterium tumefaciens strain EHA105 through electroporation. The Agrobacterium-mediated rice transformation was carried out using calli derived from mature seeds of cultivar Kitaake according to a previously described protocol (Hiei et al., 1994; Mei et al., 2006).

Plant materials and growth conditions

Seeds of both nontransformed cultivar Kitaake and self-pollinated T1 OsACS2-OX lines were germinated in water at 37 °C for 2 days before planting in MetroMix 360 soil (Sun Gro, Bellevue, WA). Rice plants were maintained in glasshouse with 12 h of light at 28 °C day and 24 °C night, respectively. Seven days postgermination, seedlings were fertilised with 0.5% ammonia sulphate and 0.1% Sprint iron solution. To select hygromycin-resistant transgenic seedlings from segregating progeny, leaf segments were placed in a 25 µg/mL hygromycin solution, and checked

2 days later for hygromycin resistance. Homozygous T2 OsACS2-OX lines were selected based on the transgene segregation, and further propagated for subsequent experimental studies.

Chemical treatments

To test the stability and inducibility of the transgene, 2-week-old rice seedlings of WT cv. Kitaake and the homozygous T2 OsACS2-OX lines were either sprayed with dH$_2$O (basal) or 0.25 mM BTH. BTH has been previously shown to effectively induce the PBZ1 promoter (Lee et al., 2001). Twenty-four hrs after the treatment, rice leaves were collected for the ET measurement as well as the RNA extraction and qRT-PCR assays.

Fungal isolates and disease assays

The two field isolates of M. oryzae used in this study were IC17-18/1 and IE1K-FN9 (Zhou et al., 2007). IC17-18/1 is moderately virulent on cv. Kitaake and its infection results in necrotic lesions without causing full leaf senescence. IEK1-FN9 is highly virulent on cv. Kitaake, by producing a higher number of lesions and causing extensive leaf damage and sometimes full leaf senescence. Both M. oryzae isolates were maintained on oatmeal agar for 7–10 days before inoculation. Two-week-old plants at a two to three leaf stage were spray-inoculated with 2.5×10^5 conidia/mL suspension plus 0.1% Tween-20 until runoff occurred. Inoculated plants were then maintained under moist conditions at room temperature for 24 h, before moving to a growth chamber (28 °C day/24 °C night, 12 h light). Disease severity was scored 7 days postinoculation by counting the number of lesions, as well as measuring the length of the three largest lesions per leaf. In contrast to small reddish lesions associated with hypersensitive response, the susceptible interaction typically results in expanding lesions with grey and necrotic centres.

Sheath blight inoculation was performed using a field isolate (RR0140) of R. solani according to the method described by Park et al. (2008). Rhizoctonia solani mycelia (maintained on potato dextrose agar at 28 °C) were inoculated into 250 mL of potato dextrose broth and incubated on a 28 °C shaker for 7 days. Excess broth was strained out, and mycelia were separated into 5-mm-diameter balls. Each mycelial ball was secured against the sheath of 6-week-old plants by aluminium foil, which was removed once disease symptom initiated at 2 days postinoculation. Disease severity was evaluated 7 days postinoculation by measuring the length of each lesion.

RNA extraction and cDNA synthesis

Rice leaf tissues were snap-frozen in liquid nitrogen, and total RNA was extracted from 100 mg of frozen tissues using the TRIzol reagent (Invitrogen, Carlsbad, CA). RNA pellets were washed with 70% ethanol and resuspended in sterile deionized water pretreated with diethylpyrocarbonate (DEPC). Each RNA sample was treated with DNase I (New England Biolabs, Ipswich, MA) to remove traces of genomic DNA. cDNA synthesis was carried out using the High Capacity cDNA Reverse Transcription Kit (Applied Biosystems, Foster City, CA) according to the manufacturer's protocol.

Quantitative real-time polymerase chain reaction

All qRT-PCR was performed using a Step One Plus Real-Time PCR system (Applied Biosystems) with DyNAmo SYBR green PCR kit (New England Biolabs) according to the manufacturer's instructions. PCR cycling conditions included a DNA denaturing stage of 94 °C for 15 min, followed by 40 cycles of 94 °C for 30 s, 60 °C for 45 s and 72 °C for 45 s. The specific primer were synthesised for detecting transcripts of *OsACS2* (forward primer: 5′ TGC GCC TTA CTA CGT CGA CTA CAT 3′; reverse primer: 5′ ACG CAC TAA CGC ACG TCT CTA CAA 3′; Accession number AK06425), *OsPR1b* (forward primer 5′ATT TAT TCG AGC GCC ACA TGA CGG 3′; reverse primer: 5′ GAC GAG TGG TCA AAC ATT GCA AGC 3′; Accession number U89895), and *OsPR5* (forward primer: 5′ TAC AAC GTC GCC ATG AGC TTC T 3′; reverse primer: 5′ TGG GCA GAA GAC GAC TTG GTA GTT 3′; Accession number X68197). Relative expression data were normalised using the rice ubiquitin 1 gene (forward primer: 5′ TGG TCA GTA ATC AGC CAG TTT G 3′; reverse primer: 5′CAA ATA CTT GAC GAA CAG AGG C 3′).

Ethylene measurement

Ethylene production in rice leaves was quantified using a gas chromatograph (Hewlett-Packard 6890, Hewlett Packard, Palo Alto, CA). Rice leaves were excised from treated or untreated plants, weighted and incubated in a sealed 4-mL glass vial for 2–3 h. Three 1-mL air samples were injected per plant sample at specific time points. ET amounts were measured in parts per million (ppm) based on an ET standard and converted to nL/g/h.

Assessment of agronomic traits

The growth rate, time until maturation, and yield were assessed using twelve plants from each OsACS2-OX line and nontransformed cv. Kitaake in glasshouse conditions (28 °C, 12 h light). Two independent biological replicates were performed in two different seasons. The growth rate was calculated by measuring each plant once per week from soil surface to meristem, and the rice growth stage was noted according to the standard published by the Rice Knowledge Bank (www.knowledgebank.irri.org). Yield was assessed by counting the number of panicles per plant as well as the number of seeds per panicle, and weighing seeds in three separate groups of 100 for each line.

Acknowledgements

This work was supported by the grants from USDA/NRI (2008-35301-19028) and NSF Plant Genome Research Programme (DBI-0922747). We would like to thank Dr. Kathleen Brown and Ms. Amy Monko for help with the gas chromatography, Dr. Germán Sandoya for help with statistical analyses, and the Rice Genome Resource Centre in Japan for providing the *OsACS2* cDNA clone.

References

Agrawal, G.K., Rawkwal, R. and Jwa, N.S. (2003) Differential induction of three pathogenesis-related genes, *PR10*, *PR1b*, and *PR5* by the ethylene generator ethephon under light and dark in rice (*Oryza sativa* L.) seedlings. *J. Plant Physiol.* **158**, 133–137.

Alexander, L. and Grierson, D. (2002) Ethylene biosynthesis and action in tomato: a model for climacteric fruit ripening. *J. Exp. Bot.* **53**, 2039–2055.

Anderson, J.P. and Singh, K.B. (2011) Interactions of *Arabidopsis* and *Medicago truncatula* with the same pathogens differ in dependence on ethylene and ethylene response factors. *Plant Signal. Behav.* **6**, 551–552.

Anderson, J.P., Lichtenzveig, J., Gleason, C., Oliver, R.P. and Singh, K.B. (2010) The B-3 ERF MtERF1-1 mediates resistance to a subset of root pathogens in *Medicago truncatula* without adversely affecting symbiosis with rhizobia. *Plant Physiol.* **154**, 861–873.

Bailey, T.A., Zhou, X., Chen, J. and Yang, Y. (2009) Role of ethylene, abscisic acid and MAP kinase pathways in rice blast resistance. In *Advances in Genetics, Genomics and Control of Rice Blast Disease* (Wang, G.-L. and Valent, B., eds), pp. 185–190. New York, NY: Springer.

Banniza, S. and Holderness, M. (2001) Pathogen biology and diversity. In *Major Fungal Diseases of Rice: Recent Advances* (Sreenivasaprasad, S. and Johnson, R., eds), pp. 201–211. Norwell, MA: Kluwer Academic Publishers.

Bonman, J. (1992) Durable resistance to rice blast disease-environmental influences. *Euphytica*, **63**, 115–123.

Broekaert, W.F., Delauré, S.L., De Bolle, M.F.C. and Cammue, B.P.A. (2006) The role of ethylene in host-pathogen interactions. *Annu. Rev. Phytopathol.* **44**, 393–416.

Brown, J.K.M. (2002) Yield penalties of disease resistance in crops. *Curr. Opin. Plant Biol.* **5**, 1–6.

Chae, H.S. and Kieber, J.J. (2005) Eto brute? Role of ACS turnover in regulating ethylene biosynthesis. *Trends Plant Sci.* **10**, 291–296.

Chern, M.S., Fitzgerald, H.A., Yadav, R.C., Canlas, P.E., Dong, X. and Ronald, P. C. (2001) Evidence for a disease-resistance pathway in rice similar to the NPR1-mediated signaling pathway in *Arabidopsis*. *Plant J.* **27**, 101–113.

Chern, M.S., Fitzgerald, H.A., Canlas, P.E., Navarre, D.A. and Ronald, P.C. (2005) Overexpression of a rice NPR1 homolog leads to constitutive activation of defense response and hypersensitivity to light. *Mol. Plant–Microbe Interact.* **18**, 511–526.

Dai, Y., Jia, Y., Correll, J., Wang, X. and Wang, Y. (2010) Diversification and evolution of the avirulence gene *AVR-pita 1* in field isolates of *Magnaporthe oryzae*. *Fungal Genet. Biol.* **47**, 974–980.

Datta, K., Velazhahan, R., Oliva, N., Ona, I., Mew, T., Khush, G.S., Muthukrishnan, S. and Datta, S.K. (1999) Over-expression of the cloned rice thaumatin-like protein (PR-5) gene in transgenic rice plants enhances environmental friendly resistance to *Rhizoctonia solani* causing sheath blight disease. *Theor. Appl. Genet.* **98**, 1138–1145.

Delteil, A., Zhang, J., Lessard, P. and Morel, J.B. (2010) Potential candidate genes for improving rice disease resistance. *Rice*, **3**, 56–71.

Durrant, W.E. and Dong, X. (2004) Systemic acquired resistance. *Annu. Rev. Phytopathol.* **42**, 185–209.

Geraats, B.P.J., Bakker, P.A.H.M., Lawrence, C.B., Achuo, E.A., Höfte, M. and van Loon, L.C. (2003) Ethylene-insensitive tobacco shows differentially altered susceptibility to different pathogens. *Phytopathology*, **93**, 813–821.

Grant, M.R. and Jones, J.D. (2009) Hormone (dis)harmony moulds plant health and disease. *Science*, **324**, 750–752.

Gust, A.A., Brunner, F. and Nürnberger, T. (2010) Biotechnological concepts for improving plant innate immunity. *Curr. Opin. Biotechnol.* **21**, 204–210.

Hiei, Y., Ohta, S., Komari, T. and Kumashiro, T. (1994) Efficient transformation of rice (*Oryza sativa* L.) mediated by *Agrobacterium* and sequence analysis of the boundaries of the T-DNA. *Plant J.* **6**, 271–282.

Hoffman, T., Schmidt, J.S., Zheng, X. and Bent, A.F. (1999) Isolation of ethylene-insensitive soybean mutants that are altered in pathogen susceptibility and gene-for-gene disease resistance. *Plant Physiol.* **119**, 935–950.

Hu, H., Xiong, L. and Yang, Y. (2005) Rice SERK1 gene positively regulates somatic embryogenesis of cultured cell and host defense response against fungal infection. *Planta*, **222**, 107–117.

Iwai, T., Miyasaka, A., Seo, S. and Ohashi, Y. (2006) Contribution of ethylene biosynthesis for resistance to blast fungus infection in young rice plants. *Plant Physiol.* **142**, 1202–1215.

Iwai, T., Seo, S., Mitsuhara, I. and Ohashi, Y. (2007) Probenazole-induced accumulation of salicylic acid confers resistance to *Magnaporthe grisea* in adult rice plants. *Plant Cell Physiol.* **48**, 915–924.

Jiang, C.J., Shimono, M., Sugano, S., Kojima, M., Yazawa, K., Yoshida, R., Inoue, H., Hayashi, N., Sakakibara, H. and Takatsuji, H. (2010) Abscisic acid interacts antagonistically with salicylic acid signaling pathway in rice-*Magnaporthe grisea* interaction. *Mol. Plant–Microbe Interact.* **23**, 791–798.

Kim, E.H., Kim, Y.S., Park, S.-H., Koo, Y.J., Choi, Y.D., Chung, Y.-Y., Lee, I.-J. and Kim, J.-K. (2009) Methyl jasmonate reduces grain yield by mediating signals to alter spikelet development in rice. *Plant Physiol.* **149**, 1751–1760.

Koga, H., Dohi, K. and Mori, M. (2004) Abscisic acid and low temperatures suppress the whole plant-specific resistance reaction of rice plants to the infection of *Magnaporthe grisea*. *Physiol. Mol. Plant Pathol.* **65**, 3–9.

Kou, Y. and Wang, S. (2010) Broad-spectrum and durability: understanding of quantitative disease resistance. *Curr. Opin. Plant Biol.* **13**, 181–185.

Kus, J.V., Zaton, K., Sankar, R. and Cameron, R.K. (2002) Age-related resistance in *Arabidopsis* is a developmentally-regulated defense response to *Pseudomonas syringae. Plant Cell,* **14**, 479–490.

Lai, X.H., Marchetti, M.A. and Petersen, H.D. (1999) Comparative slow-blasting in rice grown under upland and flooded blast nursery culture. *Plant Dis.* **83**, 681–684.

Lee, M.W. (2002) *Molecular characterization and functional analysis of a pathogen-inducible rice myb gene.* Ph.D. dissertation, Fayetteville, AR, USA: University of Arkansas.

Lee, F.N. and Rush, M.C. (1983) Rice sheath blight: a major rice disease. *Plant Dis.* **67**, 829–832.

Lee, M.W., Qi, M. and Yang, Y. (2001) A novel jasmonic acid-inducible rice myb gene associates with fungal infection and host cell death. *Mol. Plant–Microbe Interact.* **14**, 527–535.

Li, Z., Pinson, S.R.M., Marchetti, M.A., Stansel, J.W. and Park, W.D. (1995) Characterization of quantitative trait loci (QTLs) contributing to field resistance to sheath blight (*Rhizoctonia solani*). *Theor. Appl. Genet.* **91**, 382–388.

Li, H., Zhou, S.Y., Zhao, W.S., Su, S.C. and Peng, Y.L. (2009) A novel wall-associated receptor-like kinase gene, *OsWAK1*, plays important roles in rice blast disease resistance. *Plant Mol. Biol.* **69**, 337–346.

Liang, X., Shen, N.F. and Theologis, A. (1996) Li⁺-regulated 1-aminocyclopropane-1-carboxylate synthase gene expression in *Arabiodopsis thaliana. Plant J.* **10**, 1027–1036.

Liu, X., Bai, X., Wang, X. and Chu, C. (2007) *OsWRKY71*, a rice transcription factor, is involved in rice defense response. *J. Plant Physiol.* **164**, 969–979.

Liu, G., Jia, Y., Correa-Victoria, F., Prado, G.A., Yeater, K.M., McClung, A. and Correll, J.C. (2009) Mapping quantitative trait loci responsible for resistance to sheath blight in rice. *Phytopathology,* **99**, 1078–1084.

van Loon, L.C., Geraats, B.P.J. and Linthorst, H.J.M. (2006) Ethylene as a modulator of disease resistance in plants. *Trends Plant Sci.* **11**, 184–191.

Lopez, M.A., Bannenberg, G. and Castresana, C. (2008) Controlling hormone signaling is a plant and pathogen challenge for growth and survival. *Curr. Opin. Plant Biol.* **11**, 420–427.

Mei, C., Min, Q., Sheng, G. and Yang, Y. (2006) Inducible overexpression of a rice allele oxide synthase gene increases the endogenous jasmonic acid level, PR gene expression and host resistance to fungal infection. *Mol. Plant–Microbe Interact.* **19**, 1127–1137.

Moeder, W., Barry, C.S., Tauriainen, A.A., Betz, C., Tuomainen, J., Utriainen, M., Grierson, D., Sandermann, H., Langebartels, C. and Kangasjärvi, J. (2002) Ethylene synthesis regulated by biphasic induction of 1-aminocyclopropane-1-carboxylic acid synthase and 1-aminocyclopropane-1-carboxylic acid oxidase genes is required for hydrogen peroxide accumulation and cell death in ozone-exposed tomato. *Plant Physiol.* **130**, 1–9.

Orbach, M.J., Farrall, L., Sweigard, J.A., Chumley, F.G. and Valent, B. (2000) A telomeric avirulence gene determines efficacy for the rice blast resistance gene *Pi-ta. Plant Cell,* **12**, 2019–2032.

Park, S.-W., Kaimoyo, E., Kumar, D., Mosher, S. and Klessig, D.F. (2007) Methyl salicylate is a critical mobile signal for plant systemic acquired resistance. *Science,* **318**, 113–116.

Park, D.-S., Sayler, R.J., Hong, Y.-G., Nam, M.-H. and Yang, Y. (2008) A method for inoculation and evaluation of rice sheath blight disease. *Plant Dis.* **92**, 25–29.

Peiser, G.D., Wang, T.T., Hoffman, N.E., Yang, S.F., Liu, H.-W. and Walsh, C.T. (1984) Formation of cyanide from carbon 1 of 1-aminocyclopropane-1-carboxylic acid during its conversion to ethylene. *Proc. Natl. Acad. Sci. USA,* **81**, 3059–3063.

Peng, H., Zhang, Z., Li, Y., Lei, C., Zhai, Y., Sun, X., Sun, D., Sun, Y. and Lu, T. (2009) A putative leucine-rich repeat receptor kinase, OsBRR1, is involved in rice blast resistance. *Planta,* **230**, 377–385.

Penmetsa, R.V., Uribe, P., Anderson, J.P., Gish, J.C., Nan, Y.W., Xu, K., Sckisel, G., Pereira, M., Baek, J.M., Lopez-Meyer, M., Long, S.R., Harrison, M.J., Singh, K.B., Kiss, G.B. and Cook, D.R. (2008) The *Medicago truncatula* ortholog of *Arabidopsis* EIN2, Sickle, is a negative regulator of symbiotic and pathogenic microbial associations. *Plant J.* **55**, 580–595.

Pinson, S.R.M., Capdevielle, F.M. and Oard, J.H. (2005) Confirming QTLs and finding additional loci conditioning sheath blight resistance in rice using recombinant inbred lines. *Crop Sci.* **45**, 503–510.

Poland, J.A., Balint-Kurti, P.J., Wisser, R.J., Pratt, R.C. and Nelson, R.J. (2008) Shades of gray: the world of quantitative disease resistance. *Trends Plant Sci.* **14**, 21–29.

Potenza, C., Aleman, L. and Sengupta-Gopalan, C. (2004) Targeting transgene expression in research, agricultural, and environmental applications: promoters used in plant transformation. *In Vitro Cell. Dev. Biol.* **40**, 1–22.

Qiu, D., Xiao, J., Ding, X., Xiong, M., Cai, M., Cao, Y., Li, X., Xu, C. and Wang, S. (2007) OsWRKY13 mediates rice disease resistance by regulating defense related genes in salicylate and jasmonate-dependent signaling. *Mol. Plant–Microbe Interact.* **20**, 492–499.

Robert-Seilaniantz, A., Navarro, L., Bari, R. and Jones, J.D. (2007) Pathological hormone imbalances. *Curr. Opin. Plant Biol.* **10**, 372–379.

Saika, H., Okamoto, M., Miyoshi, K., Kushiro, T., Shinoda, S., Jikumaru, Y., Fujimoto, M., Arikawa, T., Takahashi, H., Ando, M., Arimura, S., Miyao, A., Hirochika, H., Kamiya, Y., Tsutsumi, N., Nambara, E. and Nakazono, M. (2007) Ethylene promotes submergence-induced expression of *OsABA8ox1*, a gene that encodes ABA 8'-hydroxylase in rice [*Oryza sativa*]. *Plant Cell Physiol.* **48**, 287–298.

Seo, S., Mitsuhara, I., Feng, J., Iwai, T., Hasegawa, M. and Ohashi, Y. (2010) Cyanide, a coproduct of plant hormone ethylene biosynthesis, contributes to the resistance of rice to blast fungus. *Plant Physiol.* **155**, 502–514.

Shimono, M., Sugano, S., Nakayama, A., Jiang, C.J., Ono, K., Toki, S. and Takatsuji, H. (2007) Rice WRKY45 plays a crucial role in benzothiadiazole-inducible blast resistance. *Plant Cell,* **19**, 2064–2076.

Silverman, P., Seskar, M., Kanter, D., Schweizer, P., Métraux, J.-P. and Raskin, I. (1995) Salicylic acid in rice: biosynthesis, conjugation, and possible role. *Plant Physiol.* **108**, 633–639.

Singh, M.P., Lee, F.N., Counce, P.A. and Gibbons, J.H. (2004) Mediation of partial resistance to rice blast through anaerobic induction of ethylene. *Phytopathology,* **94**, 819–825.

Skamnioti, P. and Gurr, S.J. (2009) Against the grain: safeguarding rice from blast disease. *Trends Biotechnol.* **27**, 141–150.

Steffens, B. and Sauter, M. (2009) Epidermal cell death in rice is confined to cells with a distinct molecular identity and is mediated by ethylene and H_2O_2 through an autoamplified signal pathway. *Plant Cell,* **21**, 184–196.

Tamogami, S., Rakwal, R. and Kodama, O. (1997) Phytoalexin production elicited by exogenously applied jasmonic acid in rice leaves (*Oryza sativa* L.) is under the control of cytokinins and ascorbic acid. *FEBS Lett.* **412**, 61–64.

Wamishe, Y., Jia, Y., Singh, P. and Cartwright, R.D. (2007) Identification of field isolates of *Rhizoctonia solani* to detect quantitative resistance in rice under greenhouse conditions. *Front. Agric. China,* **1**, 361–367.

Wisser, R.J., Sun, Q., Hulbert, S.H., Kresovich, S. and Nelson, R.J. (2005) Identification and characterization of regions of the rice genome associated with broad-spectrum, quantitative disease resistance. *Genetics,* **169**, 2277–2293.

Wyatt, S., Pan, S. and Kuc, J. (1991) β-1,3-glucanase, chitinase, and peroxidase activities in tobacco tissues resistance and susceptible to blue mould as related to flowering, age, and sucker development. *Physiol. Mol. Plant Pathol.* **39**, 433–440.

Xiong, L. and Yang, Y. (2003) Disease resistance and abiotic stress tolerance in rice are inversely modulated by an abscisic acid-inducible mitogen-activated protein kinase. *Plant Cell,* **15**, 745–759.

Yang, Y., Qi, M. and Mei, C. (2004) Endogenous salicylic acid protects rice plants from oxidative damage caused by aging as well as biotic and abiotic stress. *Plant J.* **40**, 909–919.

Young, T.E., Meeley, R.B. and Gallie, D.R. (2004) ACC synthase expression regulates leaf performance and drought tolerance in maize. *Plant J.* **40**, 813–825.

Zhou, E., Jia, Y., Correll, J. and Lee, F.N. (2007) Instability of the *Magnaporthe oryzae* avirulence gene *AVR-Pita* alters virulence. *Fungal Genet. Biol.* **44**, 1024–1034.

8

High-level hemicellulosic arabinose predominately affects lignocellulose crystallinity for genetically enhancing both plant lodging resistance and biomass enzymatic digestibility in rice mutants

Fengcheng Li[1,2,3,†], Mingliang Zhang[1,2,3,†], Kai Guo[1,2,4], Zhen Hu[1,2,3], Ran Zhang[1,2,3], Yongqing Feng[1,2,3], Xiaoyan Yi[1,2,3], Weihua Zou[1,2,3], Lingqiang Wang[1,2,3], Changyin Wu[1,4], Jinshan Tian[5], Tiegang Lu[6], Guosheng Xie[1,2,3]* and Liangcai Peng[1,2,3,*]

[1]National Key Laboratory of Crop Genetic Improvement and National Centre of Plant Gene Research, Huazhong Agricultural University, Wuhan, China

[2]Biomass and Bioenergy Research Centre, Huazhong Agricultural University, Wuhan, China

[3]College of Plant Science and Technology, Huazhong Agricultural University, Wuhan, China

[4]College of Life Science and Technology, Huazhong Agricultural University, Wuhan, China

[5]Yichang Academy of Agricultural Science, Yichang, China

[6]Biotechnology Research Institute, Chinese Academy of Agricultural Sciences/National Key Facility for Gene Resources and Genetic Improvement, Beijing, China

*Correspondence

emails lpeng@mail.hzau.edu.cn; xiegsh@mail.hzau.edu.cn
[†]These two authors contributed equally to this work.

Keywords: rice, biomass digestibility, lodging resistance, cell wall, genetic modification, GH9B and XAT.

Summary

Rice is a major food crop with enormous biomass residue for biofuels. As plant cell wall recalcitrance basically decides a costly biomass process, genetic modification of plant cell walls has been regarded as a promising solution. However, due to structural complexity and functional diversity of plant cell walls, it becomes essential to identify the key factors of cell wall modifications that could not much alter plant growth, but cause an enhancement in biomass enzymatic digestibility. To address this issue, we performed systems biology analyses of a total of 36 distinct cell wall mutants of rice. As a result, cellulose crystallinity (CrI) was examined to be the key factor that negatively determines either the biomass enzymatic saccharification upon various chemical pretreatments or the plant lodging resistance, an integrated agronomic trait in plant growth and grain production. Notably, hemicellulosic arabinose (Ara) was detected to be the major factor that negatively affects cellulose CrI probably through its interlinking with β-1,4-glucans. In addition, lignin and G monomer also exhibited the positive impact on biomass digestion and lodging resistance. Further characterization of two elite mutants, Osfc17 and Osfc30, showing normal plant growth and high biomass enzymatic digestion in situ and in vitro, revealed the multiple GH9B candidate genes for reducing cellulose CrI and XAT genes for increasing hemicellulosic Ara level. Hence, the results have suggested the potential cell wall modifications for enhancing both biomass enzymatic digestibility and plant lodging resistance by synchronically overexpressing GH9B and XAT genes in rice.

Introduction

Crop residues are considered to be a major biomass resource for biofuel production (Service, 2007). As one staple food crop worldwide, rice produces about 800 million metric tons of straws annually (Domínguez-Escribá and Porcar, 2009). Principally, biomass conversion into bioethanol involves in three major steps: physical and chemical pretreatments for cell wall disassociation, enzymatic digestion for soluble sugar release and yeast fermentation for ethanol production. However, lignocellulosic recalcitrance is a great hindrance for biomass conversion because plant cell walls have evolved complex structural and chemical mechanisms for resisting the physical and biochemical digestions in nature. Genetic modification of plant cell walls has been proposed as a promising solution to lignocellulosic recalcitrance (Xie and Peng, 2011). As genetic modifications of cell walls are mostly associated with defects in plant growth and development

(Abramson et al., 2010; Casler et al., 2002; Li et al., 2008), it becomes critical to find out the key factors of plant cell wall modifications that could not much affect plant growth, but lead to an enhancement in biomass digestibility (Xie and Peng, 2011).

Plant cell walls are primarily composed of cellulose, hemicelluloses, lignin and pectic polysaccharides with minor structural proteins. Cellulose is an unbranched β-1,4-linked glucans and assembled into crystalline microfibrils (Somerville, 2006). The hydrogen bonds formed between β-1,4-glucan chains significantly determine cellulose crystallinity (Bansal et al., 2010; Park et al., 2010). Cellulose crystallinity can be defined as measuring the cellulose crystalline index (CrI) that reflects the relative amount of crystalline cellulose (Park et al., 2010; Zhang et al., 2013). It has been reported that cellulose synthesis is catalyzed by a superfamily of CESA enzymes in plants (Arioli et al., 1998; Pear et al., 1996). In rice, OsCESA1, 3 and 8 are involved in primary wall cellulose biosynthesis, whereas OsCESA4, 7 and 9 are reported for secondary cell wall formation (Tanaka *et al.*, 2003; Wang *et al.*, 2010). Recently, *OsGH9B1*, *B3* and *B16* genes have been suggested with a role in cellulose crystallinity modification in rice (Xie *et al.*, 2013). In addition, *COBRA-like 1* gene may have function in cellulose assembly in rice (Liu *et al.*, 2013).

Hemicelluloses are a class of heterogeneous polysaccharides with various monosaccharides, and arabinoxylan is a major hemicellulose in the mature tissues of grass plants (Scheller and Ulvskov, 2010). Several glycosyltransferase (GT) gene families, such as *GT43* and *GT61*, have been reported to involve in both main chain biosynthesis and side chain substitution of hemicelluloses in grasses (Anders *et al.*, 2012; Chiniquy *et al.*, 2013). For instance, a homologue gene *TaXAT2* knock-down mutants strongly decrease α-(1, 3)-linked arabinosyl substitution of xylan in wheat (Anders *et al.*, 2012). *OSIRX9* and *OSIRX14*, homologues of the *Arabidopsis IRX9* and *IRX14* genes, have been identified in building the xylan backbone (Chiniquy *et al.*, 2013). Lignin is a phenolic polymer composed mainly of *p*-coumaryl alcohol (H), coniferyl alcohol (G) and sinapyl alcohol (S). The three monomers are cross-linked by ether-, ester- and C-C bonds to form a stable and water-proofing lignin complex (Li *et al.*, 2014b; Ralph *et al.*, 2004; Sun *et al.*, 2013). More than 90 genes derived from 10 super-families are involved in lignin biosynthesis and polymerization (Raes *et al.*, 2003; Xu *et al.*, 2009); however, only a few member genes have been functionally identified in rice.

Principally, lignocellulosic recalcitrance is determined by cell wall compositions, wall polymer features and wall network styles. Cellulose CrI has been reported to be the key factor negatively affecting biomass enzymatic digestibility in all plants examined, whereas hemicelluloses can negatively affect lignocellulose CrI for high biomass digestibility in *Miscanthus* (Li *et al.*, 2013; Xu *et al.*, 2012). Notably, arabinose (Ara) substitution degree of xylans is the positive factor on biomass enzymatic saccharification upon various chemical pretreatments in the grass plants examined (Li *et al.*, 2013). By comparison, lignin could play dual roles in biomass enzymatic digestions, due to its structural diversity and distinctive heterogeneity in different plant species (Chen and Dixon, 2007; Grabber, 2005; Jung *et al.*, 1994, 2012; Penning *et al.*, 2014). Despite that the cell wall factors on biomass saccharification have been examined in plants, little is known about the factors associated with both high biomass digestibility and normal plant growth.

In rice, lodging is a major and integrated agronomic trait in plant growth and grain production, because it causes poor grain filling and yield loss and reduces grain quality and mechanical harvesting efficiency (Berry *et al.*, 2004). Rice lodging index (LI) arises from the bending or breaking of the lower culm internodes (Sirajul Islam *et al.*, 2007) and is highly associated with plant height, fresh weight, stem diameter and others (Crook and Ennos, 1994; Sirajul Islam *et al.*, 2007). Although plant cell wall composition and features can greatly affect plant mechanic strength (Tanaka *et al.*, 2003; Zhang and Zhou, 2011), little is known about their impacts on plant LI in rice and other plants (Halpin *et al.*, 1998; Ma, 2009).

Plants constitute numerous different cell types with diverse cell wall components, and thus, it is difficult in technique to identify the key factors of plant cell walls for high biomass enzymatic digestibility and normal plant growth using classic approaches such as selection of one-gene transgenic plant or one genetic mutant. However, systems biology analysis has been considered as a powerful approach for the multiple traits and factors using large population of samples (Atias *et al.*, 2009; Farrokhi *et al.*, 2006; Guo *et al.*, 2014). Hence, in this study, we performed comparative and correlative analyses among cell wall composition and features, biomass enzymatic saccharification and lodging resistance using a total of 36 distinct cell wall mutants of rice, leading to identification of the key factors on cell wall modifications. Characterization of two elite mutants reveals several important candidate genes for genetic enhancing biomass digestibility and lodging resistance in rice.

Results

Selection of large population of rice cell wall mutants

In this study, we initially performed large-scale screening of cell wall mutants using the mutagenesis pools of genome-wide T-DNA insertions and chemical EMS inductions with more than 10 000 individual rice lines (Wu *et al.*, 2003; Xie and Peng, 2011). With several generations of multiplications, a total of 36 rice homozygous mutants were selected as experimental materials in this study (Figure 1). Compared with wild type variety (Nipponbare, NPB), the rice mutants displayed large variations in the three wall polymer levels in the mature straws (Figure 1a). For instance, cellulose levels varied from 14.05% to 31.62%, hemicelluloses levels from 8% to 23.37% and lignin levels from 10.63% to 18.5% on a dry matter basis (Figure 1a).

Biomass enzymatic digestibility (or saccharification) has been defined by calculating either the hexoses yield (% cellulose)

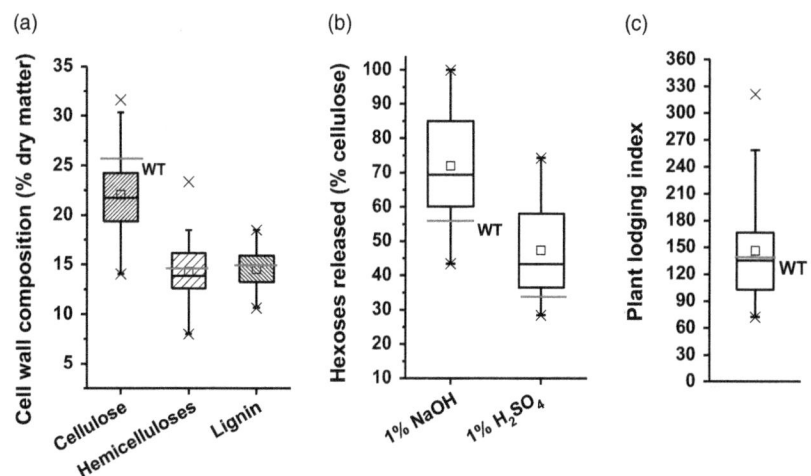

Figure 1 Variations of total 36 genetic mutants of rice (*n* = 37). (a) Diverse three major wall polymer levels. (b) Varied hexoses yields released from the mixed-cellulases hydrolysis after 1% NaOH and 1% H_2SO_4 pretreatments. (c) Varied plant lodging indexes. Blue line indicated as wild type.

released from hydrolysis by a crude cellulase mixture of lignocellulose after pretreatment or total sugars yield (% dry matter) released from both enzymatic hydrolysis and pretreatment (Li et al., 2013; Xu et al., 2012). Due to their diverse cell wall compositions, the 36 rice mutants also exhibited varied biomass enzymatic digestibility upon 1% NaOH and 1% H_2SO_4 pretreatments (Figure 1b). In particular, the hexoses yields varied from 43% to 100% (% cellulose) by 1% NaOH pretreatment and from 28% to 74% by 1% H_2SO_4 pretreatment.

As plant cell wall greatly affects plant mechanic strength (Ma, 2009; Tanaka et al., 2003; Zhang and Zhou, 2011), we measured plant LI in the rice mutants. As determined in cell wall compositions and biomass digestibility, the mutants also exhibited large variations in plant LI (Figure 1c). Compared with the wild type, more than 50% mutants displayed a higher lodging resistance due to their relative lower LI values. Moreover, we performed a correlation analysis between LI and morphological traits including plant height, fresh weight and breaking force (Table S1). As a result, the plant LI displayed a positive correlation with plant height and fresh weight ($P < 0.01$), but was not significantly correlated with the plant breaking force.

To examine the genetic stability of the homozygous mutants used in this study, we also performed a correlation analysis among the cell wall composition, hexoses yields, breaking forces

and LI using a total of 30 mutants samples harvested between 2009 and 2010 seasons (Table S2). Notably, all those parameters measured in two season's samples showed a positive correlation at $P < 0.05$ and 0.01.

Effects of wall polymer levels on hexoses yields and LI

Pairwise correlation has been extensively applied to investigate biological traits relationships or associations using large populations of samples (Li et al., 2013; Zhang et al., 2013). In this study, a correlation analysis was conducted to find out effects of plant cell wall composition on biomass digestibility and LI in rice mutants (Figure 2). Significantly, cellulose level negatively affected hexoses yields under 1% NaOH or 1% H_2SO_4 pretreatment whereas both hemicelluloses and lignin contents displayed a positive impact at $P < 0.05$ and 0.01 ($n = 37$), different from observations in wheat showing hemicelluloses and lignin as negative factors (Wu et al., 2013). By contrast, cellulose level positively affected LI, but hemicelluloses and lignin levels showed the negative impacts at significance levels ($P < 0.01$). Because LI adversely indicates degree of plant lodging resistance, both hemicelluloses and lignin were examined to be the positive factors on plant lodging resistance in rice. Hence, three major wall polymer levels could distinctively affect biomass enzymatic digestibility and plant lodging resistance in rice.

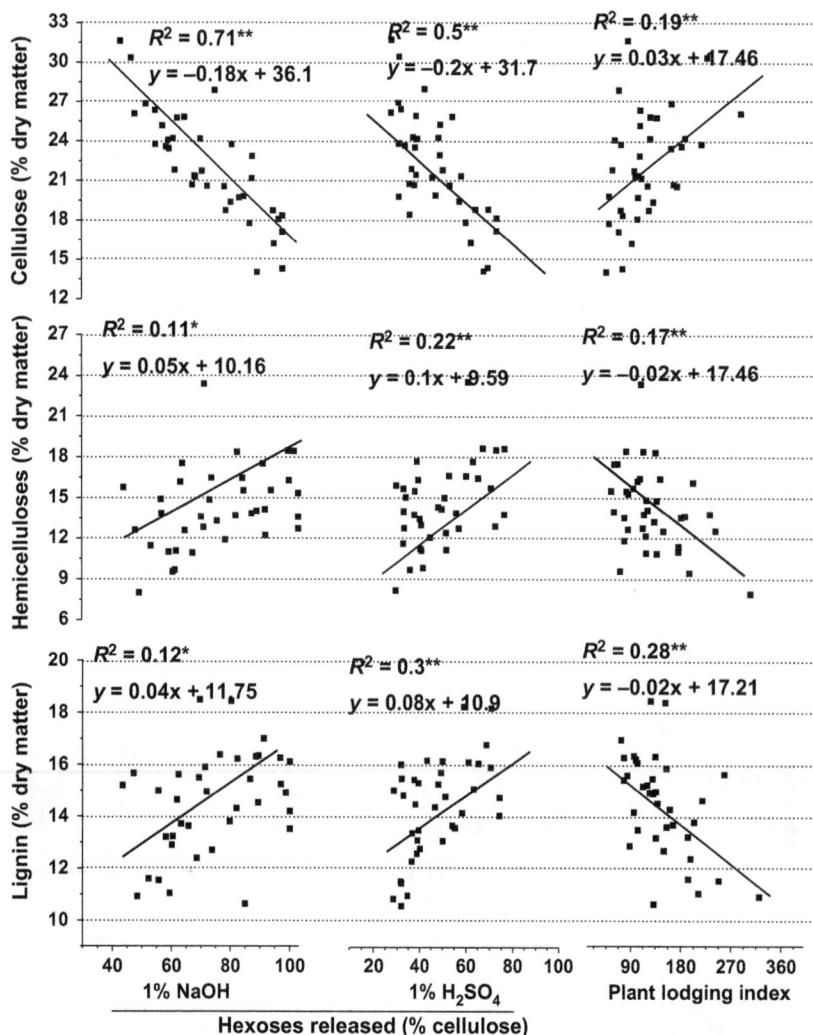

Figure 2 Effects of the wall polymer levels on hexoses yields and lodging indexes in rice mutants and wild type ($n = 37$). * and ** indicated significant correlations at $P < 0.05$ and 0.01, respectively.

Effects of wall polymer features on biomass saccharification and lodging resistance

It has been characterized that three major wall polymer features, rather than wall polymer levels, could predominately affect biomass enzymatic digestibility in the grass plants examined (Li et al., 2014a; Wu et al., 2013). However, little is yet known about the wall polymer feature impacts on plant agronomic traits. In this work, three major wall polymer features were detected including cellulose CrI, monosaccharide composition of hemicelluloses and three monomer constitution of lignin in a total of 36 rice mutants and wild type (Table S3). As expected, the cellulose CrI showed a negative impact on hexoses yields under 1% NaOH and 1% H_2SO_4 pretreatments at $P < 0.01$ (Figure 3a), consistent with the findings in other plants (Li et al., 2014a; Wu et al., 2013; Zhang et al., 2013). However, the cellulose CrI exhibited a positive impact on the LI at $P < 0.05$ (Figure 3a). As the LI was a negative parameter on lodging resistance, to our knowledge, this was the first time to report that the cellulose CrI was the negative factor on both biomass digestibility and lodging resistance in plants.

With respect to the hemicelluloses feature, Xyl and Ara were determined to be two major monosaccharides covering 80% and 12% of total (Table S3), similar to other grass plants (Li et al., 2014a; Wu et al., 2013; Zhang et al., 2013). However, only Ara level of hemicelluloses, other than Xyl and Xyl/Ara, showed a significant correlation either positively with hexoses yields upon 1% NaOH and 1% H_2SO_4 pretreatments or negatively with LI at $P < 0.01$ (Figure 3b). Hence, the Ara level was the positive factors on both biomass digestibility and lodging resistance in rice, in contrast to the cellulose CrI being the negative factors. In addition, despite that the Xyl/Ara showed a negative impact on hexoses yields, similar to the findings in other grasses examined (Li et al., 2013, 2014a; Wu et al., 2013), it did not exhibit correlation with LI at significant level in rice mutants. In terms of three monomers of lignin in rice mutants, the G monomer exhibited a significant impact on both hexoses yields from 1% NaOH or 1% H_2SO_4 pretreatment and LI in rice mutants at $P < 0.05$ (Figure 3c). As a comparison, the H and S monomers showed a positive correlation with the hexoses yield from 1% H_2SO_4 pretreatment, other than from 1% NaOH pretreatment. Therefore, the Ara and G monomer levels could predominately determine hemicelluloses and lignin's positive impacts on biomass enzymatic digestibility and lodging resistance in rice mutants, respectively.

Association among three wall polymers

Using the 36 rice mutants, we further performed a correlative analysis among the three major wall polymers (Figure 4). Both cellulose level and CrI were negatively correlated with hemicelluloses levels at $P < 0.01$, but did not show any significant correlation with lignin levels (Figure 4a,b), indicating that cellulose should mainly have an interaction with hemicelluloses other than lignin in rice mutants as reported in Miscanthus (Xu et al., 2012). Notably, the cellulose CrI exhibited the significant correlation with hemicellulosic Ara, but remained no correlation with G monomer and other two monomers (S and H) of lignin (Figures 4c and S1a). Despite that cellulose CrI is basically affected by cellulose level (Figure S1b), the results revealed that the Ara level, other than G monomer, could negatively affect cellulose CrI via its interaction with β-1,4-glucans for positively enhancing lignocellulosic enzymatic digestibility and lodging resistance in rice mutants.

Furthermore, the hemicelluloses and Ara levels, respectively, exhibited a positive correlation with lignin and G monomer levels at $P < 0.01$, indicating that both polymers should have an association (Figure 4d). In addition, the H monomer was also positively correlated with Ara at $P < 0.05$ (Figure S1c), but it remains lower coefficient values than that of G monomer. Therefore, the data suggested that the hemicellulosic Ara may mainly interlink with G monomer, other than H and S monomers of lignin in rice mutants.

Characterization of two standard rice mutants

Among the 36 rice mutants, we characterized two standard rice mutants termed as Osfc17 and Osfc30 (Figure 5). In general, the two mutants, like wild type (NPB), exhibited normal agronomic traits in the field (Figure 5a,b). However, the Osfc17 mutant could even show the higher dry spike (total grain yield) by 17% than that of the wild type at $P < 0.05$, whereas the Osfc30 mutant had much lower LI by 27% at $P < 0.01$ (Table S4). Notably, both mutants showed much higher hexoses yields (% cellulose) or total sugar yields (% dry matter) up to 1.3-fold to twofold than that of the wild type after pretreatments with H_2SO_4 and NaOH at three concentrations (Figure 5c and Table S5). By comparison, the Osfc30 mutant could even exhibit higher hexoses and total sugars yields up to onefold than that of the Osfc17 mutant. Hence, both mutants might be directly applied as desirable bioenergy crops for biofuel purpose.

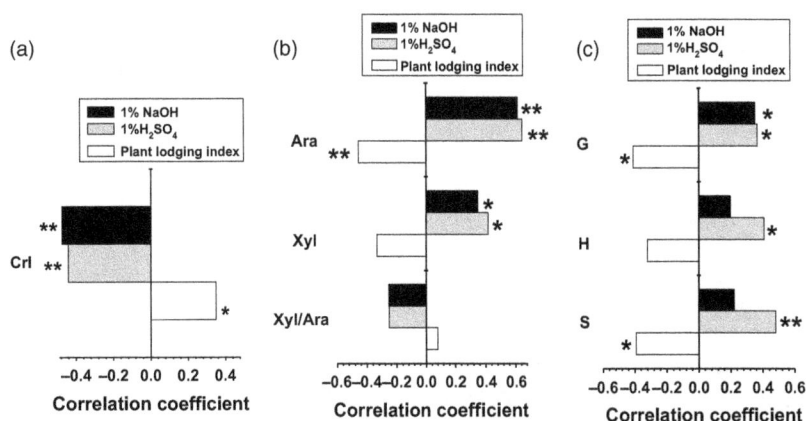

Figure 3 Effects of the wall polymer features on hexoses yields and lodging indexes in rice mutants and wild type ($n = 37$). (a) Cellulose crystallinity (CrI). (b) Hemicellulosic Ara and Xyl. (c) Three monomers (G, S, H) of lignin. * and ** indicated significant correlations at $P < 0.05$ and 0.01, respectively.

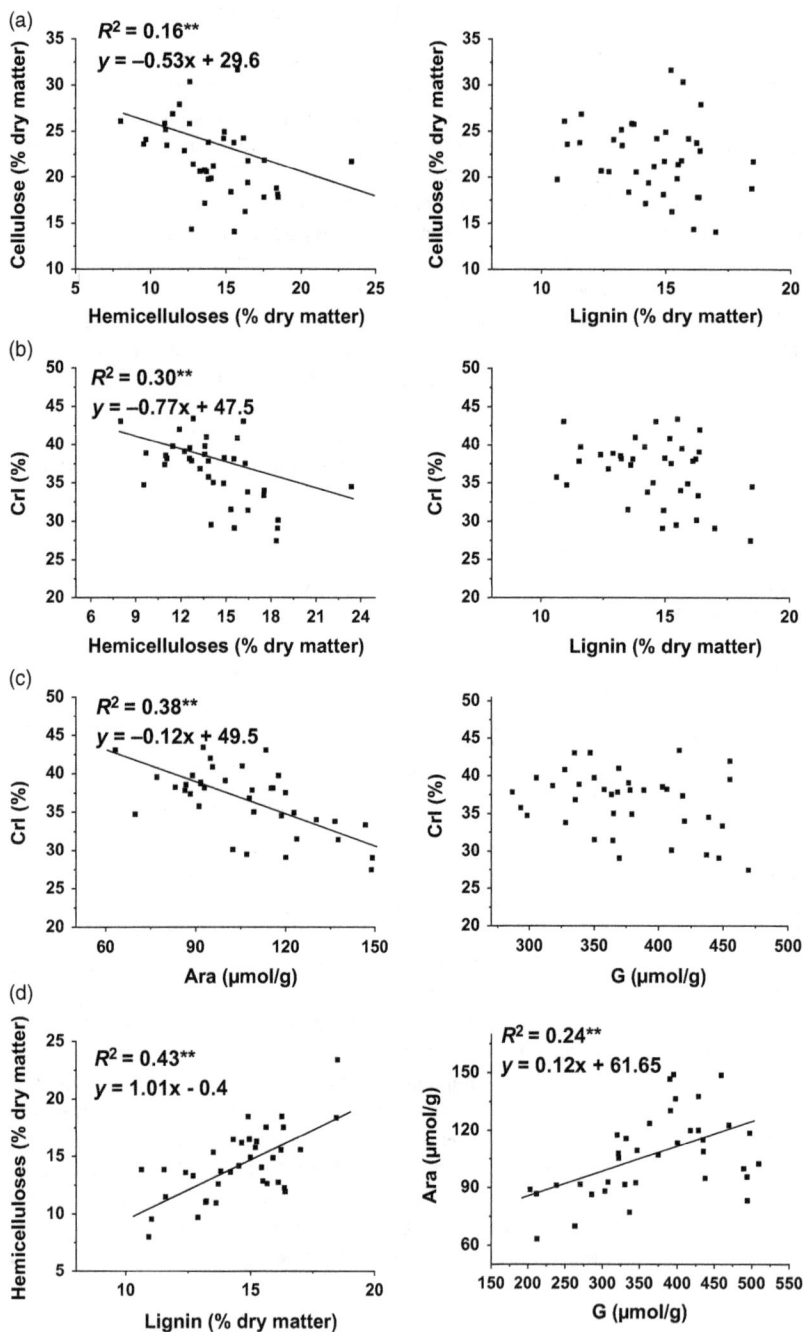

Figure 4 Correlations among the wall polymer levels and features in rice mutants and wild type (n = 37). (a) Correlations of cellulose with hemicelluloses and lignin. (b) Correlations of CrI with hemicelluloses and lignin. (c) Correlations of CrI with hemicellulosic Ara and G monomer. (d) Correlations between hemicelluloses and lignin. ** indicated significant correlations at $P < 0.01$.

Furthermore, the two mutants were determined with distinct alternations in cell wall compositions and features (Table 1). Compared with the wild type, the *Osfc17* mutant exhibited increased hemicelluloses and lignin levels by 57% and 23% at $P < 0.01$, whereas the *Osfc30* mutant showed a major decrease in cellulose level by 29%. In terms of the wall polymer features, both mutants exhibited decreased cellulose CrI by 10% and 21% than that of the wild type (Table 1), confirming that the cellulose CrI is the negative factor on biomass enzymatic digestibility. In addition, the *Osfc30* mutant showed much lower cellulose CrI value than that of the *Osfc17* mutant, consistent with their difference in biomass enzymatic digestibility (Table S5). Meanwhile, compared with the wild type, the *Osfc17* and *Osfc30* mutants exhibited increased Ara levels by 42% and 23%, but did

not show much different G monomer levels (Table 1), which support the previous findings about the negative impact of Ara (other than G monomer) on cellulose CrI for high biomass digestibility (Figure 4c). As the *Osfc17* mutant showed much less decreased cellulose level (−19%, relative to the wild type) than that of the *Osfc30* mutant (−29%), its much more increased Ara level (42%, relative to 23% in *Osfc30*) may mainly attribute for the reduced cellulose CrI (Table 1). Hence, the reduced cellulose level and increased Ara level in the *Osfc30* mutant should both attribute for the much reduced cellulose CrI. Taken all together, it demonstrated that the cell wall modifications could not only maintain plant normal growth and grain yield, but also enhance plant lodging resistance and biomass enzymatic digestibility in rice mutants.

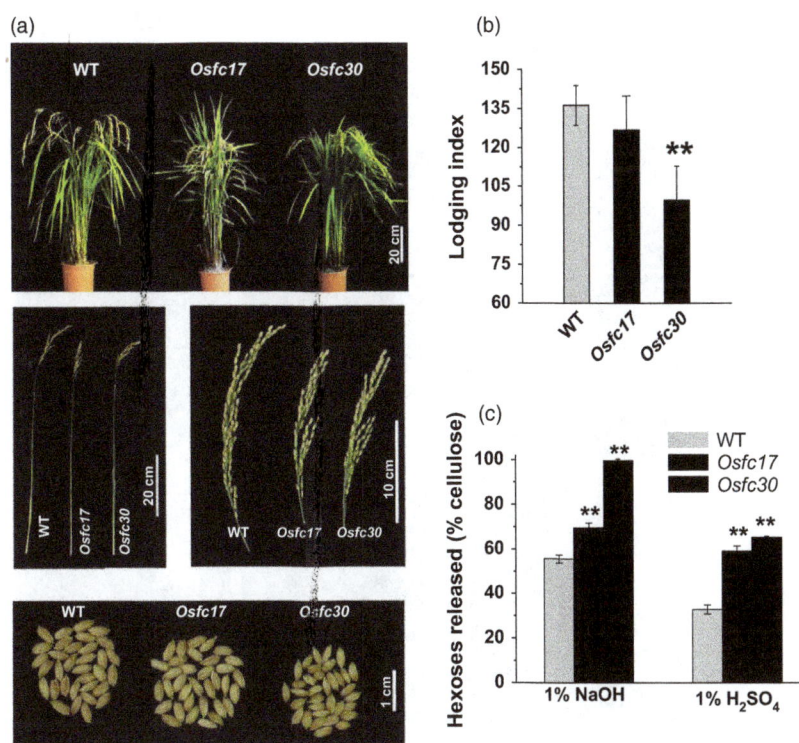

Figure 5 Comparisons between two elite mutants (*Osfc17*, *Osfc30*) and wild type (NPB). (a) Plant growth and grain phenotypes. (b) Plant lodging index. (c) Hexoses yields released from enzymatic hydrolysis after 1% NaOH and 1% H_2SO_4 pretreatments. ** indicated significant difference between the mutants and wild type by *t*-test at $P < 0.01$.

Table 1 Three major wall polymer levels and features in the two mutants and wild type

Sample	Wall polymer level (% dry matter)			Wall polymer feature		
	Cellulose	Hemicelluloses	Lignin	CrI[†] (%)	Ara (µmol/g)	G (µmol/g)
WT	26.34 ± 1.64	14.9 ± 1.46	14.99 ± 0.13	38.23	83.17 ± 1.7	494.42 ± 4.64
Osfc17	21.29 ± 0.61*	23.37 ± 0.63**	18.5 ± 0.14**	34.52	118.50 ± 3.18**	497.27 ± 14.01
	(−19%)[‡]	(57%)	(23%)	(−10%)	(42%)	(1%)
Osfc30	18.75 ± 0.36**	18.47 ± 0.47**	16.25 ± 0.1*	30.13	102.37 ± 1.87**	509.15 ± 11.32
	(−29%)	(24%)	(8%)	(−21%)	(23%)	(3%)

[†]CrI method was detected at ±0.05–0.15 at $P < 0.01$ using five representative samples.

* and ** indicated significant difference between the mutants and wild type by *t*–test at $P < 0.05$ and 0.01, respectively.

[‡]Percentage of the increased and decreased level between the mutants and wild type.

Observations of cell tissues *in situ* and biomass residues *in vitro*

To confirm the biomass enzymatic digestibility of the two standard mutants (*Osfc17*, *Osfc30*), we observed their stem cell tissues *in situ* and biomass residues *in vitro* under scanning electron microscopy. Without any pretreatment and enzymatic digestion, the young stems of two mutants and wild type did not show any visible difference *in situ* (Figure 6). However, both mutants exhibited much more destructed and lost cells than that of the wild type after 1% NaOH or 1% H_2SO_4 pretreatment and sequential enzymatic hydrolysis. Hence, the effective destruction *in situ* of stem tissues in the mutants could attribute for high biomass enzymatic digestibility *in vitro*. Furthermore, the two mutants were observed with much rougher surfaces of biomass residues *in vitro* compared with the wild type under 1% NaOH or 1% H_2SO_4 pretreatments and sequential enzymatic hydrolysis (Figure S2), consistent with the findings in other grass plants (Li

et al., 2013, 2014a; Xu *et al.*, 2012). Therefore, both observations *in situ* and *in vitro* confirmed a high biomass enzymatic digestibility in the two mutants.

Detection of the major gene candidates for cell wall modifications

It has been reported that more than one thousand genes should be associated with plant cell wall formation and modification (Ralph *et al.*, 2004; Scheller and Ulvskov, 2010; Somerville, 2006; Xie *et al.*, 2013). In this study, we detected the transcription alterations of the major genes that lead to reducing cellulose CrI and increasing hemicellulosic Ara levels in the *Osfc17* and *Osfc30* mutants (Figure 7). Relative to the wild type, the *Osfc17* and *Osfc30* mutants exhibited much higher transcription levels of *GH9B1*, *3*, *16* genes by onefold to sevenfold, with lower *CESAs* gene expressions by 0.5- to twofold (Figure 7a). As CESA enzymes are involved in cellulose biosynthesis and GH9Bs have enzymatic activity for decreasing cellulose CrI (Xie *et al.*, 2013),

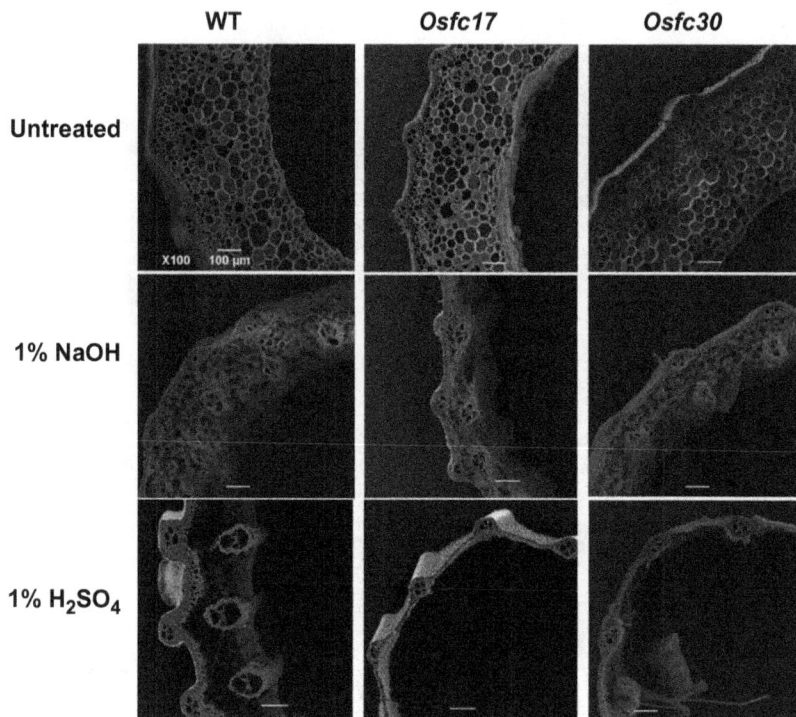

Figure 6 Observations *in situ* of stem tissues in the two mutants (*Osfc17*, *Osfc30*) and wild type (NPB) after pretreated with 1% NaOH and 1% H_2SO_4 and sequentially digested with mixed-cellulase enzymes using scanning electron microscopy.

the data were consistent with the reduced cellulose level and CrI in the mutants (Table 1). With regard to the increased hemicellulose and Ara levels, the two mutants were detected with much higher transcription levels of *OsXAT2, 3* and *OsIRX9, 14* genes up to 0.4- to 2.1-fold compared with the wild type (Figure 7b). However, the *OsXAT2, 3* transcription levels were higher than that of *OsIRX9, 14*. As the OSIRX9 and OSIRX14 are involved in backbone chain synthesis of hemicelluloses (Chiniquy *et al.*, 2013), and OSXAT2 and OSXAT3 are specific for the Ara side chain elongation (Anders *et al.*, 2012), the increased *OsXAT2* and *OsXAT3* expression levels were in support of the findings about much higher Ara levels in the two mutants. Hence, three *GH9B* and two *OsXAT* genes could be considered for genetic enhancing biomass enzymatic digestibility and plant lodging resistance in rice.

In addition, as total 10 gene families are involved in the lignin biosynthesis in monocot and dicot plants (Raes *et al.*, 2003; Xu *et al.*, 2009), we also detected transcription levels of the 19 representative genes based on microarray data and coexpression analysis in rice (Guo *et al.*, 2014). Compared with the wild type, the *Osfc17* and *Osfc30* mutants showed much higher expression levels in the 14 genes, but remained lower transcription levels in four genes (Figure S3).

Discussion

Large-scale screening of rice mutants for wall factor identification

Selection of genetic mutants and transgenic plants with cell wall modifications has been considered for enhancing biomass enzy-

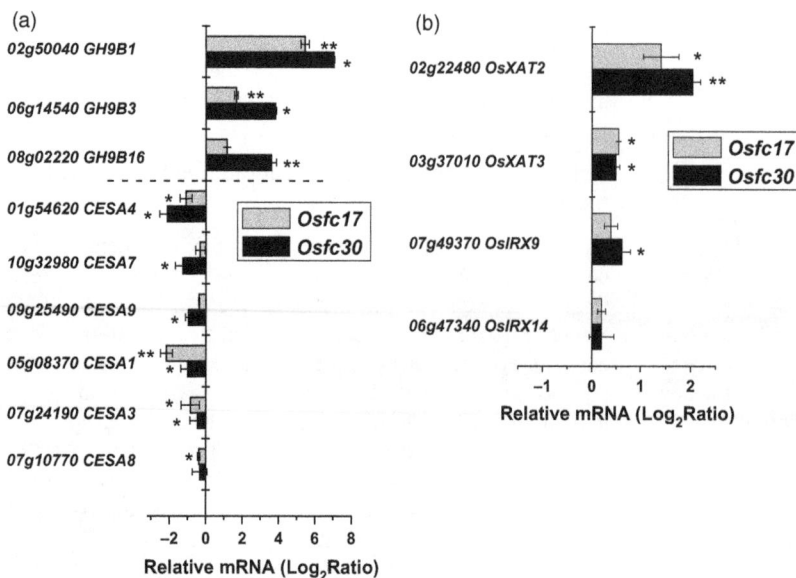

Figure 7 Gene expression ratios of the two mutants (*Osfc17*, *Osfc30*) relative to the wild type (NPB) by qRT-PCR analysis. (a) Genes involved in cellulose biosynthesis (*CESAs*) and cellulose crystallinity modifications (*GH9Bs*). (b) Genes involved in Xylan backbone synthesis (*OsIRXs*) and branched Ara elongation (*OsXATs*). * and ** indicated significant different transcription levels between the two mutants and wild type by *t*-test at *P* < 0.05 and 0.01, respectively.

matic digestibility (Pauly and Keegstra, 2010; Xie and Peng, 2011). Due to the complicated structures and diverse functions of plant cell walls, however, the selected mutants and transgenic plants have exhibited the defects to some degrees in plant growth and development (Abramson et al., 2010; Li et al., 2008). Hence, it becomes essential to find out the key factors of plant cell wall modifications for normal plant growth and high biomass enzymatic saccharification using systems biology approach. Principally, the systems biological approach is powerful for analysis of the multiple traits and factors, but it requires large population of samples (Atias et al., 2009; Farrokhi et al., 2006; Guo et al., 2014). As rice is a cultivar, large-scale screening of rice cell wall mutants is optimal for collecting large populations of samples. In this study, a total of 36 cell wall mutants have exhibited diverse biomass enzymatic digestibility and varied agronomic traits, in particular on the plant LI that is a major and integrated agronomic trait in plant growth and grain production. Notably, the mutant samples harvested from two field seasons exhibit a significantly positive correlation in various agronomic traits and biological parameters measured, indicating that the rice mutants are homozygous and genetic stable for any experiments in this work. Using those mutants, therefore, we could perform a correlative and comparative analysis, leading to finding out the key factors on cell wall modifications in rice. Notably, we have identified the several rice mutants (for instance, Osfc30) that exhibit both lodging resistance and biomass enzymatic digestibility higher than that of the wild type, suggesting that the mutants could be directly used as bioenergy crops.

Hypothetical model of the key factors that direct for cell wall modifications

Based on the systems biology analysis of the 36 cell wall mutants, to our knowledge, this study at the first time has reported the distinct effects of three major wall polymer features on both biomass enzymatic digestibility and plant lodging resistance. Hence, a hypothetical model could be proposed to elucidate the key factors on cell wall modifications (Figure 8). First, the cellulose crystallinity (CrI) is the key factor that negatively determines biomass enzymatic hydrolysis and plant lodging resistance (Figure 3a). Second, the hemicellulosic Ara is the major factor negatively affecting cellulose CrI probably through its interlinking with β-1,4-glucans via hydrogen bonds (Figure 4b,c). Third, the lignin and G monomer may positively affect biomass digestibility and lodging resistance by interaction with hemicellulosic Ara (Figures 3b,c and 4d). In addition, the G monomer and lignin may also have a direct and minor impact on biomass digestion and lodging resistance, but the mechanism remains unclear.

Despite the cellulose CrI has been demonstrated to be the key negative factor on biomass enzymatic saccharification in all grass plants examined (Abramson et al., 2010; Li et al., 2014a; Wu et al., 2013; Zhang et al., 2013), it remains under test in terms of its negative impact on lodging resistance in all other crops, because the LI is the integrated parameter associated with the multiple agronomic traits that are distinctive in different crops (Ma, 2009; Sirajul Islam et al., 2007). As hemicellulosic Ara is the major branched sugar in grass plants, other branched monosaccharides in hemicelluloses may also negatively affect cellulose CrI for enhancing biomass digestibility and lodging resistance. In addition, lignin is important for lodging resistance in plants (Ma, 2009), but to our surprise, it has also exhibited significantly positive impact on biomass enzymatic hydrolysis in rice mutants.

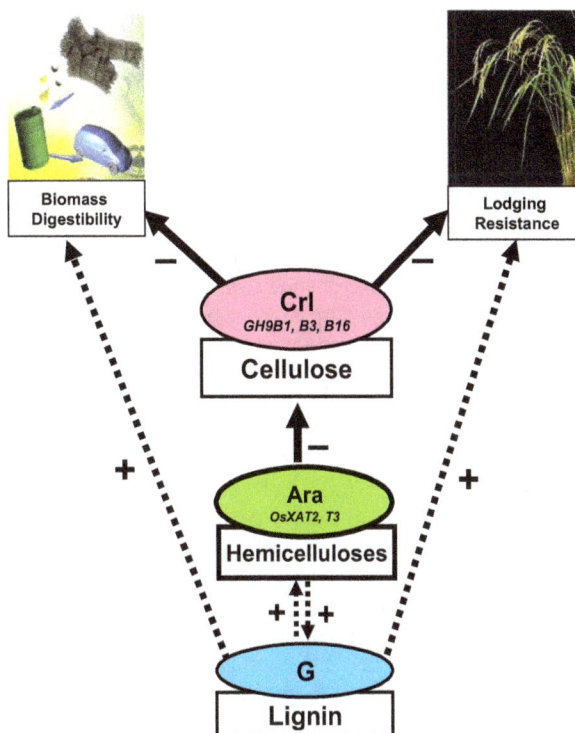

Figure 8 Hypothetical model on the key factors of cell wall modifications for genetic enhancing both biomass enzymatic digestibility and plant lodging resistance in rice. Cellulose crystallinity (CrI) is the key factor negatively determining biomass enzymatic saccharification and plant lodging resistance, and GH9Bs enzymes have activities for reducing cellulose crystallinity; hemicellulosic Ara is the main factor negatively affecting cellulose CrI through its direct interaction with β-1,4-glucans, and OSXATs enzymes have activities for Ara elongations; G monomer and lignin may directly affect biomass digestibility and lodging resistance and/or have an interaction with hemicelluloses and Ara as an indirect impact.

Thus, we assume that the increased G and other two monomers (H, S) may not well form the interlinking networks via various chemical bonds, leading to an easy removal by chemical pretreatments. In other case, the specific selection of cell wall mutants for improved biomass digestibility and agronomic traits may result in the deposition of lignin with altered cell wall ultrastructure or anatomical distributions in the rice tissues. Therefore, the modified lignin may be required for structural integrity of cell walls, but apparently not enough to impede cell wall polysaccharides digestibility after acid or base pretreatment. A similar result was also reported in Lucene (Medicago sativa L.) with improved forage quality (Jung et al., 1994). Obviously, the detailed mechanism of lignin and monomers in the cell wall digestibility deserves further investigation.

Potential genetic modification of cell walls for enhancing both biomass digestibility and plant lodging resistance

Genetic modification of plant cell walls has been considered as a promising solution for reducing biomass recalcitrance and maintaining plant normal growth (Xie and Peng, 2011). Based on the proposed hypothetical model, we have further identified the major genes that could be applied for genetic modifications of plant cell walls towards enhancing both biomass enzymatic digestibility and plant lodging resistance in rice (Figure 8). With respect to the cellulose crystallinity that is the key and negative factor, genetic reducing cellulose CrI becomes critical by synchronically overexpressing both three *GH9B* and two *OsXAT* candidate genes (Figure 7a,b), which have been demonstrated in the two elite mutants (*Osfc17* and *Osfc30*) (Figure 5). It has been assumed that GH9B enzymes may have activities to produce the noncrystalline cellulose in the surface of microfibers (Xie *et al.*, 2013), and thus, more hemicellulosic Ara catalyzed by XAT may be required to fill in the noncrystalline cellulose regions to maintain cell wall strength and integrity, which supports that the Ara may have interaction with noncrystalline cellulose via hydrogen bonds (Li *et al.*, 2013). Furthermore, our preliminary data have showed that over-expressing single *OsGH9B* gene could result in plant growth defects in the transgenic rice lines (data not shown). Hence, overproduction of both GH9B and XAT enzymes should be the potential approach for largely enhancing both biomass enzymatic digestibility and lodging resistance in transgenic rice. In addition, as described above, the enhanced G monomer and lignin may also aid to maintain cell wall strength and integrity in the transgenic plants or mutants.

However, there may be an alternative approach by genetic manipulation of single genes that regulate multiple genes (such as *GH9B*, *XAT* and others), as observed in *Osfc17* and *Osfc30* mutants. Our preliminary data have showed that *Osfc17* may encode the dynamic-related protein and *Osfc30* may be the upstream transcription regulator (data not shown). Hence, characterization of *Osfc17* and *Osfc30* mutants could offer other desire single genes for genetic modification of plant cell walls in rice and beyond.

A rapid evaluation of biomass digestibility using young stem tissues

Fourier transform infrared attenuated total reflectance spectroscopy has been used to predict cell wall composition and biomass digestibility in various plants such as *Switchgrass*, *Bluestem* grass, prairie biomasses, hardwood and corn stover (Sills and Gossett, 2012). A near infrared spectroscopic approach was also applied to evaluate biomass enzymatic digestibility in *Miscanthus* (Huang *et al.*, 2012). However, this machine-based method requires the equation formula based on chemical analysis of large populations of biomass samples. In this study, we could pre-evaluate biomass enzymatic saccharification by observing *in situ* the enzymatic digestion of young stem tissues at heading stage of rice (Figure 6). To confirm the desire mutants, we have further performed time-course observations of the stem tissue digestions between wild type and mutants. Therefore, this method can be used to quickly identify the genetic mutants and transgenic plants with high biomass enzymatic digestibility.

Conclusions

Using systems biology analysis of a total of 36 cell wall mutants, cellulose crystallinity (CrI) has been examined to be the key factor negatively determining biomass enzymatic saccharification and plant lodging resistance, whereas hemicellulosic Ara is the major factor that negatively affects cellulose CrI. Characterization of two elite mutants (*Osfc17* and *Osfc30*) further suggests the potential cell wall modifications for enhancing both biomass enzymatic digestibility and plant lodging resistance by synchronically expressing *GH9B* and two *OsXAT* candidate genes in rice.

Experimental procedures

Plant materials

Selections of rice T-DNA insertion and EMS induction mutagenesis pools were performed in 2008 and 2009 as described by Xie *et al.* (2013) and Wu *et al.* (2013). The samples of a total of 36 homozygous rice mutants were collected from the Huazhong Agricultural University experimental fields in 2009 and 2010. The collected mature straws were dried at 60 °C to the constant weight, grounded into powders through 40 mesh (0.425 mm × 0.425 mm) and stored in the dry container until use.

Measurement of LI in rice

Lodging index was detected using rice stem tissue in 30 days after heading. Length of the fourth stem internode was measured, and its breaking force was detected using a Prostrate Tester (DIK 7401; Daiki, Osaka, Japan), with the distance between fulcra of the tester at 5 cm. Fresh weight (W) of the upper portion of the plant was measured including panicle, four internodes and leaves. Bending moment (BM) was calculated using the following formula: BM = length of the fourth internode × W, and thus, LI was calculated as follows: LI = BM/breaking force × 100%. Measurements of lodging indexes of all samples were duplicated with six independent biological experiments.

Dry spike measurement

The spike weights of rice mutants were calibrated after these samples were dried in the oven at 60 °C to a constant weight.

Plant cell wall fractionation

The plant cell wall fractionation procedure was applied to extract cellulose and hemicelluloses, as described by Peng *et al.* (2000) with minor modification by Wu *et al.* (2013). The soluble sugar, lipids and starch of the samples were successively removed by potassium phosphate buffer (pH 7.0), chloroform-methanol (1 : 1, v/v) and DMSO-water (9 : 1, v/v). The remaining pellets as total crude cell wells were suspended in 0.5% (w/v) ammonium oxalate and heated for 1 h in a boiling water bath, and the supernatants were combined as total pectin. The remaining pellets were suspended in 4 M KOH containing 1.0 mg/mL sodium borohydride for 1 h at 25 °C, and the combined supernatants were neutralized, dialyzed and lyophilized as KOH-extractable hemicelluloses. The remaining pellets were sequentially extracted with TFA as non-KOH-extractable hemicelluloses. The pellets were further extracted with acetic–nitric acids–water (8 : 1 : 2) for 1 h at 100 °C, and the remaining materials were regarded as crystalline cellulose. All experiments were carried out in biological triplicate.

Colorimetric assay of hexoses and pentoses

UV–VIS Spectrometer (V-1100D; Shanghai MAPADA Instruments Co., Ltd. Shanghai, China) was applied for total hexoses and pentoses assays. Hexoses were detected by anthrone/H_2SO_4 method, and pentoses were detected by orcinol/HCl method (Dische, 1962). Regarding the high pentoses level can affect the absorbance reading at 620 nm for hexoses content by the anthrone/H_2SO_4 method, the deduction from pentoses reading at 660 nm was carried out for final hexoses calculation. A series of xylose concentrations were analysed for plotting the standard curve referred for the deduction, which was verified by

GC-MS analysis. Cellulose level was measured by the anthrone/H_2SO_4 method, and hemicelluloses level was calculated according to total hexoses and pentoses detected. Total sugar yields of biomass samples released from pretreatment and enzymatic hydrolysis were subject to the sum total of hexoses and pentoses. All experiments were carried out in biological triplicate.

Detection of cellulose crystallinity

X-ray diffraction method described by Zhang et al. (2013) was applied for detection of cellulose crystallinity index (CrI) using Rigaku-D/MAX instrument (Uitima III, Tokyo, Japan). The well-mixed powders of biomass samples were detected under plateau conditions. Ni-filtered Cu Kα radiation ($\lambda = 0.154056$ nm) generated at voltage of 40 kV and current of 18 mA and scanned at speed of 0.0197°/s from 10° to 45°. The crystallinity index (CrI) was calculated using the intensity of the 200 peak (I_{200}, $\theta = 22.5°$) and the intensity at the minimum between the 200 and 110 peaks (I_{am}, $\theta = 18.5°$) as the follow: $CrI = 100 \times (I_{200} - I_{am})/I_{200}$. I_{200} represents both crystalline and amorphous materials while I_{am} represents amorphous material. Standard error of the CrI method was detected at ±0.05–0.15 using five representative samples in triplicate.

Hemicelluloses monosaccharide determination by GC-MS

The sample preparations and GC-MS analysis were conducted as previously described by Li et al. (2013).

Total lignin assay

Total lignin determinations were performed by two-step acid hydrolysis method according to Laboratory Analytical Procedure of the National Renewable Energy Laboratory. The acid-insoluble lignin (AIL) was calculated gravimetrically as acid-insoluble residues after subtraction for ash, and the acid-soluble lignin (ASL) was detected by UV spectroscopy.

Acid-insoluble lignin assay

0.5 g sample as W_1 was extracted with benzene-ethanol (2 : 1, v/v) in a Soxhlet for 4 h, and air-dried in hood overnight. The sample was hydrolyzed with 10 mL 72% H_2SO_4 (v/v) in shaker at 30 °C for 1.5 h. After hydrolysis, the acid was diluted to 2.88% and placed in the autoclave for 1 h at 121 °C (15 psi). The autoclaved hydrolysis was vacuum-filtered through the previously weighed filtering crucible. The filtrate was captured in a filtering flask for ASL. The lignin was washed free of acid with hot distilled water and the crucible, and the acid-insoluble residue was dried in an oven at 80 °C. The weight of the crucible and dry residue was recorded to the nearest 0.1 mg (W_2). The dried residue was burn into ash in a muffle furnace at 200 °C for 30 min and 575 °C for 4 h. The crucibles and ash were weighed to the nearest 0.1 mg as W_3. AIL was calculated according to the equation: $AIL (\%) = (W_2 - W_3) \times 100/W_1\%$.

Acid-soluble lignin assay

The hydrolysis liquor obtained previously was transferred into 250-mL volumetric flask and brought up to 250 mL with 2.88% sulphuric acid. The absorbance was read at 205 nm on a UV–Vis spectroscopy (Du800; Beckman Coulter Inc. Brea, California, USA), and 2.88% sulphuric acid was used as blank. The calculation of the ASL was based on the equation: ASL $(\%) = (A \times D \times V/1000 \times K \times W_1) \times 100\%$. A (absorption value), D (Dilution ratio of the sample), K (absorptivity constant) = 110 L/g/cm. Total lignin (%) = ASL% + AIL%. All experiments were carried out in biological triplicate.

Lignin monomer detection by HPLC

Lignin monomers were determined by HPLC as previously described by Xu et al. (2012) and Wu et al. (2013).

Determination of biomass digestibility

Chemical pretreatments and the following residues enzymatic hydrolysis were performed as previously described by Huang et al. (2012).

Scanning electron microscopic (SEM) observation

We prepared a transverse section of the 2nd stem internode tissues at heading stages. Stem transverse sections were pretreated with 1% NaOH or 1% H_2SO_4 as described above; the stem transverse sections were washed with distilled water until pH 7.0 and hydrolyzed with the mixed-cellulase for 2 h at 50 °C. The mixed-cellulases enzyme containing β-glucanase ($\geq 6 \times 10^4$ U), cellulase (≥ 600 U) and xylanase ($\geq 1.0 \times 10^5$ U) was purchased from Imperial Jade Bio-technology Co., Ltd. (Ningxia, China). After enzymatic hydrolysis, the sample surfaces were sputter-coated with gold and observed the tissue degradation intensity under scanning electron microscope (SEM JSM-6390/LV; Hitachi, Tokyo, Japan). The well-mixed biomass residues from pretreatment and sequential enzymatic hydrolysis were observed as previously described by Wu et al. (2013). Each sample was observed for 5–10 times, and the representative image was used in this study.

qRT-PCR analysis

Total RNA was isolated using RNAprep pure Plant Kit (DP432; TIANGEN BIOTECH, Beijing, China), and 5 μg total RNA was reverse transcribed with an oligo(dT)18 primer in a 50 μL reaction using an M-MLV Reverse Transcriptase (Promega, Madison, Wisconsin, USA) according to the manufacturer's instructions. The qRT-PCR was performed in a 20 μL reaction system: cDNA template 2.0 μL, 2× SYBR Green1 Mix 10 μL, primer-F 0.5 μL, primer-R 0.5 μL, MilliQ 7.0 μL with SYBR Green qPCR kit (ZOMANBIO, Beijing, China) on Two Color Real-time PCR Detection System (MyiQ2; Bio-Rad, Hercules, California, USA) using the following program: 2 min at 95 °C followed by 40 cycles of 15 s at 95 °C, 15 s at 60 °C, 25 s at 72 °C. Ubiquitin gene (AK059011) was used as an internal standard in the qRT-PCR. The gene expression unit was subjective to the percentage of the target gene expression value relative to the internal standard (Ubiquitin gene). All quantitative PCR experiments were performed in biological triplicate. All the gene-specific primers used were listed in Table S6.

Statistical calculation of correlation coefficients

Correlation coefficients were calculated by performing Spearman rank correlation analysis for all pairs of measured traits across the whole population. This analysis used average values calculated from all original determinations for a given traits pair.

Acknowledgements

We specially thank Prof. Qifa Zhang for kindly providing the rice T-DNA mutant pools. This work was supported in part by grants from the 111 Project of MOE (B08032), the Transgenic Plant and Animal Project of MOA (2009ZX08009-119B), the 973 Pre-project of MOST (2010CB134401) and HZAU Changjiang Scholar Promoting Project (52204-07022).

References

Abramson, M., Shoseyov, O. and Shani, Z. (2010) Plant cell wall reconstruction toward improved lignocellulosic production and processability. *Plant Sci.* **178**, 61–72.

Anders, N., Wilkinson, M.D., Lovegrove, A., Freeman, J., Tryfona, T., Pellny, T.K., Weimar, T., Mortimer, J.C., Stott, K., Baker, J.M., Defoin-Platel, M., Shewry, P.R., Dupree, P. and Mitchell, R.A. (2012) Glycosyl transferases in family 61 mediate arabinofuranosyl transfer onto xylan in grasses. *Proc. Natl Acad. Sci. USA*, **109**, 989–993.

Arioli, T., Peng, L., Betzner, A.S., Burn, J., Wittke, W., Herth, W., Camilleri, C., Hofte, H., Plazinski, J., Birch, R., Cork, A., Glover, J., Redmond, J. and Williamson, R.E. (1998) Molecular analysis of cellulose biosynthesis in *Arabidopsis*. *Science*, **279**, 717–720.

Atias, O., Chor, B. and Chamovitz, D.A. (2009) Large-scale analysis of *Arabidopsis* transcription reveals a basal co-regulation network. *BMC Syst. Biol.* **3**, 86.

Bansal, P., Hall, M., Realff, M.J., Lee, J.H. and Bommarius, A.S. (2010) Multivariate statistical analysis of X-ray data from cellulose: a new method to determine degree of crystallinity and predict hydrolysis rates. *Bioresour. Technol.* **101**, 4461–4471.

Berry, P.M., Sterling, M., Spink, J.H., Baker, C.J., Sylvester-Bradley, R., Mooney, S.J., Tams, A.R. and Ennos, A.R. (2004) Understanding and reducing lodging in cereals. *Adv. Agron.* **84**, 217–271.

Casler, M.D., Buxton, D.R. and Vogel, K.P. (2002) Genetic modification of lignin concentration affects fitness of perennial herbaceous plants. *Theor. Appl. Genet.* **104**, 127–131.

Chen, F. and Dixon, R.A. (2007) Lignin modification improves fermentable sugar yields for biofuel production. *Nat. Biotechnol.* **25**, 759–761.

Chiniquy, D., Varanasi, P., Oh, T., Harholt, J., Katnelson, J., Singh, S., Auer, M., Simmons, B., Adams, P.D., Scheller, H.V. and Ronald, P.C. (2013) Three novel rice genes closely related to the *Arabidopsis IRX9*, *IRX9L*, and *IRX14* genes and their roles in xylan biosynthesis. *Front. Plant Sci.* **4**, 1–13.

Crook, M.J. and Ennos, A.R. (1994) Stem and root characteristics associated with lodging resistance in four winter wheat cultivars. *J. Agric. Sci.* **123**, 167–174.

Dische, Z. (1962) Color reactions of carbohydrates. In *Methods in Carbohydrate Chemistry*, Vol. **1**(Whistler, R.L. and Wolfrom, M.L., eds), pp. 477–512. New York: Academic Press.

Domínguez-Escribá, L. and Porcar, M. (2009) Rice straw management: the big waste. *Biofuels, Bioprod. Bioref.* **4**, 154–159.

Farrokhi, N., Burton, R.A., Brownfield, L., Hrmova, M., Wilson, S.M., Bacic, A. and Fincher, G.B. (2006) Plant cell wall biosynthesis: genetic, biochemical and functional genomics approaches to the identification of key genes. *Plant Biotechnol. J.* **4**, 145–167.

Grabber, J.H. (2005) How do lignin composition, structure, and cross-linking affect degradability? A review of cell wall model studies. *Crop Sci.* **45**, 820–831.

Guo, K., Zou, W., Feng, Y., Zhang, M., Zhang, J., Tu, F., Xie, G., Wang, L., Wang, Y., Klie, S., Persson, S. and Peng, L. (2014) An integrated genomic and metabolomic framework for cell wall biology in rice. *BMC Genomics*. **15**, 596.

Halpin, C., Holt, K., Chojecki, J., Oliver, D., Chabbert, B., Monties, B., Edwards, K., Barakate, A. and Foxon, G.A. (1998) *Brown-midrib maize* (*bm1*)-a mutation affecting the cinnamyl alcohol dehydrogenase gene. *Plant J.* **14**, 545–553.

Huang, J., Xia, T., Li, A., Yu, B., Li, Q., Tu, Y., Zhang, W., Yi, Z. and Peng, L. (2012) A rapid and consistent near infrared spectroscopic assay for biomass enzymatic digestibility upon various physical and chemical pretreatments in *Miscanthus*. *Bioresour. Technol.* **121**, 274–281.

Jung, H.G., Smith, R.R. and Endres, C.S. (1994) Cell wall composition and degradability of stem tissue from *lucerne* divergently selected for lignin and *in vitro* dry-matter disappearance. *Grass Forage Sci.* **49**, 295–304.

Jung, J.H., Fouad, W.M., Vermerris, W., Gallo, M. and Altpeter, F. (2012) RNAi suppression of lignin biosynthesis in sugarcane reduces recalcitrance for biofuel production from lignocellulosic biomass. *Plant Biotechnol. J.* **10**, 1067–1076.

Li, X., Weng, J.K. and Chapple, C. (2008) Improvement of biomass through lignin modification. *Plant J.* **54**, 569–581.

Li, F., Ren, S., Zhang, W., Xu, Z., Xie, G., Chen, Y., Tu, Y., Li, Q., Zhou, S., Li, Y., Tu, F., Liu, L., Wang, Y., Jiang, J., Qin, J., Li, S., Jing, H.C., Zhou, F., Gutterson, N. and Peng, L. (2013) Arabinose substitution degree in xylan positively affects lignocellulose enzymatic digestibility after various NaOH/H_2SO_4 pretreatments in *Miscanthus*. *Bioresour. Technol.* **130**, 629–637.

Li, M., Feng, S., Wu, L., Li, Y., Fan, C., Zhang, R., Zou, W., Tu, Y., Jing, H., Li, S. and Peng, L. (2014a) Sugar-rich sweet sorghum is distinctively affected by wall polymer features for biomass digestibility and ethanol fermentation in bagasse. *Bioresour. Technol.* **167**, 14–23.

Li, Z., Zhao, C., Zha, Y., Wan, C., Si, S., Liu, F., Zhang, R., Li, F., Y, B., Yi, Z., Xu, N., Peng, L. and Li, Q. (2014b) The minor wall-networks between monolignols and interlinked-phenolics predominantly affect biomass enzymatic digestibility in *Miscanthus*. *PLoS ONE*, **9**, e105115.

Liu, L., Shang-Guan, K., Zhang, B., Liu, X., Yan, M., Zhang, L., Shi, Y., Zhang, M., Qian, Q., Li, J. and Zhou, Y. (2013) Brittle Culm1, a COBRA-Like protein, functions in cellulose assembly through binding cellulose microfibrils. *PLoS Genet.* **9**, e1003704.

Ma, Q.H. (2009) The expression of caffeic acid 3-O-methyltransferase in two wheat genotypes differing in lodging resistance. *J. Exp. Bot.* **60**, 2763–2771.

Park, S., Baker, J.O., Himmel, M.E., Parilla, P.A. and Johnson, D.K. (2010) Cellulose crystallinity index: measurement techniques and their impact on interpreting cellulase performance. *Biotechnol. Biofuels*, **3**, 10.

Pauly, M. and Keegstra, K. (2010) Plant cell wall polymers as precursors for biofuels. *Curr. Opin. Plant Biol.* **13**, 305–312.

Pear, J.R., Kawagoe, Y., Schreckengost, W.E., Delmer, D.P. and Stalker, D.M. (1996) Higher plants contain homologs of the bacterial *celA* genes encoding the catalytic subunit of cellulose synthase. *Proc. Natl Acad. Sci. USA*, **93**, 12637–12642.

Peng, L., Hocart, C.H., Redmond, J.W. and Williamson, R.E. (2000) Fractionation of carbohydrates in *Arabidopsis* root cell walls shows that three radial swelling loci are specifically involved in cellulose production. *Planta*, **211**, 406–414.

Penning, B.W., Sykes, R.W., Babcock, N.C., Dugard, C.K., Held, M.A., Klimek, J.F., Shreve, J.T., Fowler, M., Ziebell, A., Davis, M.F., Decker, S.R., Turner, G.B., Mosier, N.S., Springer, N.M., Thimmapuram, J., Weil, C.F., McCann, M.C. and Carpita, N.C. (2014) Genetic determinants for enzymatic digestion of lignocellulosic biomass are independent of those for lignin abundance in a maize recombinant inbred population. *Plant Physiol.* **165**, 1475–1487.

Raes, J., Rohde, A., Christensen, J.H., Van de Peer, Y. and Boerjan, W. (2003) Genome-wide characterization of the lignification toolbox in *Arabidopsis*. *Plant Physiol.* **133**, 1051–1071.

Ralph, J., Lundquist, K., Brunow, G., Lu, F., Kim, H., Schatz, P.F., Marita, J.M., Hatfield, R.D., Ralph, S.A., Christensen, J.H. and Boerjan, W. (2004) Lignins: natural polymers from oxidative coupling of 4-hydroxyphenyl-propanoids. *Phytochem. Rev.* **3**, 29–60.

Scheller, H.V. and Ulvskov, P. (2010) Hemicelluloses. *Annu. Rev. Plant Biol.* **61**, 263–289.

Service, R.F. (2007) Biofuel researchers prepare to reap a new harvest. *Science*, **315**, 1488–1491.

Sills, D.L. and Gossett, J.M. (2012) Using FTIR spectroscopy to model alkaline pretreatment and enzymatic saccharification of six lignocellulosic biomasses. *Biotechnol. Bioeng.* **109**, 894–903.

Sirajul Islam, M., Peng, S., Visperas, R.M., Ereful, N., Sultan Uddin Bhuiya, M. and Julfiquar, A.W. (2007) Lodging-related morphological traits of hybrid rice in a tropical irrigated ecosystem. *Field Crop Res.* **101**, 240–248.

Somerville, C. (2006) Cellulose synthesis in higher plants. *Annu. Rev. Cell Dev. Biol.* **22**, 53–78.

Sun, H., Li, Y., Feng, S., Zou, W., Guo, K., Fan, C., Si, S. and Peng, L. (2013) Analysis of five rice 4-coumarate: coenzyme A ligase enzyme activity and stress response for potential roles in lignin and flavonoid biosynthesis in rice. *Biochem. Biophys. Res. Commun.* **430**, 1151–1156.

Tanaka, K., Murata, K., Yamazaki, M., Onosato, K., Miyao, A. and Hirochika, H. (2003) Three distinct rice cellulose synthase catalytic subunit genes required for cellulose synthesis in the secondary wall. *Plant Physiol.* **133**, 73–83.

Wang, L., Guo, K., Li, Y., Tu, Y., Hu, Z., Wang, B., Cui, X. and Peng, L. (2010) Expression profiling and integrative analysis of the CESA/CSL superfamily in rice. *BMC Plant Biol.* **10**, 282.

Wu, C., Li, X., Yuan, W., Chen, G., Kilian, A., Li, J., Xu, C., Zhou, D.X., Wang, S. and Zhang, Q. (2003) Development of enhancer trap lines for functional analysis of the rice genome. *Plant J.* **35**, 378–427.

Wu, Z., Zhang, M., Wang, L., Tu, Y., Zhang, J., Xie, G., Zou, W., Li, F., Guo, K., Li, Q., Gao, C. and Peng, L. (2013) Biomass digestibility is predominantly affected by three factors of wall polymer features distinctive in wheat accessions and rice mutants. *Biotechnol. Biofuels*, **6**, 183.

Xie, G. and Peng, L. (2011) Genetic engineering of energy crops: a strategy for biofuel production in China. *J. Integr. Plant Biol.* **53**, 143–150.

Xie, G., Yang, B., Xu, Z., Li, F., Guo, K., Zhang, M., Wang, L., Zou, W., Wang, Y. and Peng, L. (2013) Global identification of multiple OsGH9 family members and their involvement in cellulose crystallinity modification in rice. *PLoS ONE*, **8**, e50171.

Xu, Z., Zhang, D., Hu, J., Zhou, X., Ye, X., Reichel, K.L., Stewart, N.R., Syrenne, R.D., Yang, X., Gao, P., Shi, W., Doeppke, C., Sykes, R.W., Burris, J.N., Bozell, J.J., Cheng, M.Z., Hayes, D.G., Labbe, N., Davis, M., Stewart, C.N. and Yuan, J.S. (2009) Comparative genome analysis of lignin biosynthesis gene families across the plant kingdom. *BMC Bioinformatics*, **10**, S3.

Xu, N., Zhang, W., Ren, S., Liu, F., Zhao, C., Liao, H., Xu, Z., Huang, J., Li, Q., Tu, Y., Yu, B., Wang, Y., Jiang, J., Qin, J. and Peng, L. (2012) Hemicelluloses negatively affect lignocellulose crystallinity for high biomass digestibility under NaOH and H_2SO_4 pretreatments in *Miscanthus*. *Biotechnol. Biofuels*, **5**, 58.

Zhang, B. and Zhou, Y. (2011) Rice brittleness mutants: a way to open the 'black box' of monocot cell wall biosynthesis. *J. Integr. Plant Biol.* **53**, 136–142.

Zhang, W., Yi, Z., Huang, J., Li, F., Hao, B., Li, M., Hong, S., Lv, Y., Sun, W., Ragauskas, A., Hu, F., Peng, J. and Peng, L. (2013) Three lignocellulose features that distinctively affect biomass enzymatic digestibility under NaOH and H_2SO_4 pretreatments in *Miscanthus*. *Bioresour. Technol.* **130**, 30–37.

9

The *superwoman1-cleistogamy2* mutant is a novel resource for gene containment in rice

Fabien Lombardo[1,a], Makoto Kuroki[2,3,a], Shan-Guo Yao[4,†,a], Hiroyuki Shimizu[2], Tomohito Ikegaya[2], Mayumi Kimizu[4], Shinnosuke Ohmori[4], Takashi Akiyama[1], Takami Hayashi[2,5], Tomoya Yamaguchi[5,§], Setsuo Koike[5], Osamu Yatou[4] and Hitoshi Yoshida[1,4,*]

[1]Division of Applied Genetics, Institute of Agrobiological Sciences, National Agriculture and Food Research Organization (NARO), Ibaraki, Japan
[2]Division of Crop Breeding Research, Hokkaido Agricultural Research Center, NARO, Hokkaido, Japan
[3]Division of Rice Research, Institute of Crop Science, NARO, Ibaraki, Japan
[4]Division of Crop Development, Central Region Agricultural Research Center, NARO, Niigata, Japan
[5]Division of Agro-Production Technologies and Management Research, Tohoku Agricultural Research Center, NARO, Iwate, Japan

*Correspondence
email yocida@affrc.go.jp

[†]Present address: Center for Genome Biology, Institute of Genetics and Developmental Biology, Chinese Academy of Sciences, Beijing 100101, China.
[§]Present address: Agriculture, Forestry and Fisheries Research Council, Ministry of Agriculture, Forestry and Fisheries of Japan, Tokyo 100-8950, Japan.
[a]These authors contributed equally to this work.

Keywords: cleistogamy, MADS-box gene, breeding, GMO, gene flow, lodicule.

Summary

Outcrossing between cultivated plants and their related wild species may result in the loss of favourable agricultural traits in the progeny or escape of transgenes in the environment. Outcrossing can be physically prevented by using cleistogamous (i.e. closed-flower) plants. In rice, flower opening is dependent on the mechanical action of fleshy organs called lodicules, which are generally regarded as the grass petal equivalents. Lodicule identity and development are specified by the action of protein complexes involving the SPW1 and OsMADS2 transcription factors. In the *superwoman1-cleistogamy1* (*spw1-cls1*) mutant, SPW1 is impaired for heterodimerization with OsMADS2 and consequently *spw1-cls1* shows thin, ineffective lodicules. However, low temperatures help stabilise the mutated SPW1/OsMADS2 heterodimer and lodicule development is restored when *spw1-cls1* is grown in a cold environment, resulting in the loss of the cleistogamous phenotype. To identify a novel, temperature-stable cleistogamous allele of *SPW1*, targeted and random mutations were introduced into the *SPW1* sequence and their effects over SPW1/OsMADS2 dimer formation were assessed in yeast two-hybrid experiments. In parallel, a novel cleistogamous allele of *SPW*1 called *spw1-cls2* was isolated from a forward genetic screen. In *spw1-cls2*, a mutation leading to a change of an amino acid involved in DNA binding by the transcription factor was identified. Fertility of *spw1-cls2* is somewhat decreased under low temperatures but unlike for *spw1-cls1*, the cleistogamous phenotype is maintained, making the line a safer and valuable genetic resource for gene containment.

Introduction

In rice (*Oryza sativa* L.) stigmas and pollen mature before flower opening, thereby promoting self-pollination. Outcrossing rates in cultivated rice are estimated to be less than one per cent (Messeguer et al., 2001); however, figures well above twenty per cent have been reported for crosses between cultivated and wild rice species when grown in specific conditions (Marathi and Jena, 2014; Phan et al., 2012). Cross-pollination between cultivated species and weedy wild relatives is problematic to rice producers as it may hinder the effectiveness of weed management and/or lead to the loss of favourable traits in a given line (Gealy et al., 2015). For example, Clearfield® rice is a commercial, nontransgenic variety of rice which is resistant to a herbicide called imazethapyr. The herbicide has been used efficiently to weed out red rice, an invasive wild relative, from fields cultivated with Clearfield® rice. However, outcrosses between Clearfield® rice and weedy red rice in some areas have resulted in an imazethapyr-resistant hybrid rice, making the application of the herbicide inefficient (Sudianto et al., 2013). The eventuality of such 'gene flow' from cultivated species to their wild relatives, and most particularly in case of genetically modified (GM) crops, is a cause of concern and has called for the development of so-called gene containment methods (Daniell, 2002; Gressel, 2014). Various strategies have been designed, such as transgene removal from pollen by molecular excision, transgene splitting using intein flanking sequences, male sterility or generation of cleistogamous (i.e. self-pollinating, closed-flower) plants to prevent pollen dispersal (Moon et al., 2011; Ohmori et al., 2012; Shinoyama et al., 2011; Toppino et al., 2011; Wang et al., 2014). In rice, the cleistogamy trait can be introduced by modification of a single gene. In contrast to most other strategies, engineering cleistogamy in rice is relatively straightforward and does not necessarily require the introduction of several transgenes, as described hereafter.

Rice flowers bear specialised perianth structures, the lemma and palea, which tightly enclose the inner sexual organs. At anthesis, the flower opens under the mechanical pressure exerted by the lodicules, two fleshy organs which swell and push the lemma and palea open to allow anther exertion and pollen dispersal. Cleistogamy can be induced by altering lodicule morphology so that the organs are unable to exert sufficient outward pressure to trigger flower opening, such as in the *spw1-cls* mutant (Yoshida et al., 2007). The pollen then remains entrapped within the closed flower and no outcrossing can occur (Ohmori et al., 2012).

At the molecular level, rice lodicule identity is mainly specified by the action of two transcription factors, SUPERWOMAN1 (SPW1) and OsMADS2. Both proteins are encoded by MIKC-type MADS-box genes and consist of a highly conserved N-terminal

DNA-binding domain (MADS domain) followed by two domains involved in protein–protein interactions (Intervening and, predominantly, Keratin-like domain) and a C-terminal domain supporting different functions, notably transcriptional control (Kaufmann et al., 2005). Upon heterodimerization, SPW1 and OsMADS2 are thought to bind to target DNA motifs called CArG boxes and regulate gene expression (Shore and Sharrocks, 1995; Yao et al., 2008), although it is likely that the heterodimer functions within larger protein complexes (Theissen and Saedler, 2001). In addition to specifying lodicule identity, SPW1 and OsMADS2 are also driving stamen development in the third floral whorl (Yoshida and Nagato, 2011). No osmads2 mutant has been isolated so far, in contrast several spw1 mutants have been described in the literature. In an spw1 loss-of-function mutant, such as in the spw1-1 mutant, lodicules are homeotically transformed into glume-like organs and male organs (i.e. stamen) into female (i.e. carpel-like) organs, hence the superwoman1 gene designation (Nagasawa et al., 2003). Over-expression of the SPW1 gene results in a phenotype opposite to that of spw1, that is a female to male (i.e. carpel to stamen-like) organ transformation, a phenotype also referred to as 'superman' (Lee et al., 2003). In the weak superwoman1-cleistogamy (spw1-cls) allele, the isoleucine in position 45 of the SPW1 protein is replaced by a threonine (I45T; Yoshida et al., 2007). The amino acid I45 is part of a β-strand within the dimerisation interface consisting of predominantly hydrophobic amino acids (Figure 1; Shore and Sharrocks, 1995). The I45T change lowers the hydrophobicity of the region and disturbs SPW1/OsMADS2 dimerisation. Consequently, lodicules are elongated in spw1-cls and flowers are unable to open. Although stamen formation is also dependent on the SPW1/OsMADS2 dimer activity, the stamens of spw1-cls show no visible defects and the plants are fertile. Most likely, the remaining biological activity of the SPW1^{I45T}/OsMADS2 heterodimer is sufficient to support proper stamen development in spw1-cls (Yoshida, 2012; Yoshida et al., 2007). Furthermore, SPW1 can also dimerise with OsMADS4, which is encoded by a paralog of OsMADS2 mainly expressed in the third whorl, and the SPW1^{I45T}/OsMADS4 dimer is also expected to contribute to stamen development in spw1-cls (Yao et al., 2008).

The single-locus, non-transgenic nature of the cleistogamy trait in spw1-cls makes the line attractive for gene containment purposes; however, extensive studies have revealed that when plants are grown in a cold environment the SPW1^{I45T}/OsMADS2 dimer regains enough stability to drive substantial lodicule development and consequently the cleistogamous phenotype of spw1-cls is lost (Yoshida et al., 2007; SO and HY, unpublished data).

In the present work, it was set out to identify mutants that show substantial lodicule elongation, ensuring a stable cleistogamy even in a cold environment, but also normal stamen development to maintain satisfactory fertility rates. The possibility of fine-tuning the biological activity of the SPW1/OsMADS2 heterodimer was investigated in a reverse genetic approach. Dimer formation between OsMADS2 and directly and randomly mutated versions of the SPW1 protein was assessed in yeast two-hybrid experiments. Subsequently, transgenic lines were generated using selected mutant constructs and their phenotypes were confirmed in planta.

Concomitantly, a chemically mutagenised population of rice was screened in a forward genetic approach to allow for the identification of cold-stable cleistogamous lines not restricted to heterodimerization mutants. The isolation of a novel cleistogamous allele of SPW1, called spw1-cls2, is described. The mutation in spw1-cls2 leads to a change of an amino acid involved in DNA binding and, unlike for spw1-cls, cleistogamy is maintained when plants are grown in a cold environment. The present work shows that the spw1-cls2 line can be used advantageously in gene containment strategies.

Results

Identification of mutated versions of the SPW1 protein impaired for heterodimerization with OsMADS2.

With the recent techniques collectively referred to as genome editing, directed mutagenesis in a plant genome is becoming possible (Osakabe and Osakabe, 2015; Schaeffer and Nakata, 2015). In this context, it was set out to confirm in an initial study that lodicule development could be manipulated by introducing various mutations in the SPW1 sequence without compromising plant fertility. To identify mutations causing intermediate destabilisations of the SPW1/OsMADS2 heterodimer, the SPW1 protein was modified at specific amino acid positions and dimerisation with OsMADS2 was estimated for each generated mutant using a yeast two-hybrid colorimetric filter assay. In spw1-cls, the mutated amino acid (I45) belongs to a hydrophobic β-strand involved in dimerisation with OsMADS2 and thus mutations at this position were initially favoured in the experimental design. A total of 11 cDNAs encoding for amino acids with gradually decreasing hydrophobicity indices were designed, including amino acids with charged (I45D, I45R), aromatic (I45Y) and unique (I45G) side chains. The binding affinities between OsMADS2 and each of the mutated SPW1 proteins were estimated from the resulting colouring intensities in the filter assay. It was hoped to identify mutations slightly more severe than in spw1-cls, which would correspond to coloration intensities somewhat weaker than the one obtained for the SPW1^{I45T} control. In our experimental conditions, filter spots corresponding

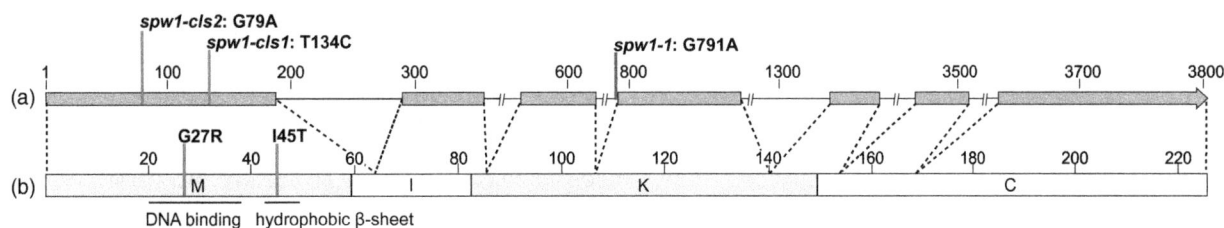

Figure 1 Representation of the SPW1 gene (a) and its product (b). a: Exons are represented with grey boxes and introns with black solid lines. Domains of the SPW1 protein are abbreviated as follows: M for MADS, I for intervening, K for keratin-like and C for C-terminal. b: Positions of the highly conserved region critical for DNA binding and the β-sheet involved in dimerisation mutated in spw1-cls2 and spw1-cls1, respectively, are underlined.

to the SPW1^{WT}, SPW1^{I45T} and empty vector controls showed a strong, fair or no coloration, respectively (Table 1). Substitution with amino acids with hydrophilic (S, Q, R, N, D) or neutral (G) side chains all resulted in a faint colouring, indicative of a very low binding affinity between the two proteins. For amino acids with hydrophobic side chains excepted cysteine (i.e. F, M, Y, A), colouring intensities were comparable or stronger than the one of the SPW1^{I45T} control. Substitution with cysteine (C) also resulted in a faint colouring. These results indicate that up to now the investigated mutations had either too weak or too strong effects over SPW1/OsMADS2 dimerisation to be considered candidates for a prospective genome editing.

To test for mutations outside the 45th position of the SPW1 protein, a random mutagenesis strategy was followed and a yeast two-hybrid library of *SPW1* mutant cDNAs was created using error-prone PCR amplification. About 10 000 independent clones were screened in a colorimetric assay. This time, several clones showing intermediate colouring intensities were isolated and, after identification of the mutated loci by sequencing, nine clones were eventually selected for further analysis. Among these, five clones carried two mutations affecting different domains of the SPW1 protein: one clone was mutated for the MADS domain (E34G/I46V), two clones for both the MADS and the K domains (E34G/E82G; T51A/L79P), and two other clones for both the K and the C-terminal domains (N98D/A131V; G110R/T144I). The remaining four clones each carried a single mutation affecting the MADS domain (L35F; F57S), the I domain (I67T), or the K domain (W80R).

To provide a more precise assessment of the severity of the isolated mutations, binding affinities between the corresponding mutated SPW1 proteins and OsMADS2 were evaluated comparatively to that of SPW1^{I45T} in a liquid β-galactosidase assay. β-Galactosidase activities ranged from about 35% to 90% to that of the SPW1^{I45T}/OsMADS2 dimer, indicating that the screen was successful in identifying mutations with intermediate destabilising effects (Figure 2a).

Table 1 Colouring intensities obtained for various mutants of *spw1* in a colorimetric assay

	Amino acid in position 45	Colouring intensity	Hydrophobicity index
Controls	Isoleucine (I; WT)	+++++	99
	Threonine (T; *cls1*)	++++	13
	Null (empty vector)	–	n.a.
Mutants via directed mutagenesis	Phenylalanine (F)	++++	100
	Methionine (M)	+++++	74
	Tyrosine (Y)	++++	63
	Cysteine (C)	+	49
	Alanine (A)	++++	41
	Glycine (G)	+	0
	Serine (S)	+	–5
	Glutamine (Q)	+	–10
	Arginine (R)	+	–14
	Asparagine (N)	+	–28
	Aspartic acid (D)	+	–55

Colouring intensities reflect the binding affinities between OsMADS2 and each mutated SPW1 protein in a yeast two-hybrid filter assay. Hydrophobicity indices are given for pH = 7 (Monera *et al.*, 1995).

To evaluate to what extent the newly identified mutations would actually affect floral development, five mutants spanning the obtained range of β-galactosidase activities (E34G/E82G; I67T; W80R; N98D/A131V and G110R/T44I) were selected for testing *in planta*. Each of the selected mutation was introduced into the SPW1 genomic clone (gSPW1) and the resulting constructs were used to complement the spw1-1 null mutant, along with wild type (WT) and spw1-cls controls. A minimum of 24 transgenic lines were generated and scored for floral organ development for each of the seven constructs.

Control lines transformed with the WT allele displayed four distinct phenotypes: about two-thirds of the lines showed the expected WT phenotype, and about 17% showed either a cleistogamous (*cls*-like) phenotype or a loss-of-function phenotype, indicative of a partial and lack of complementation, respectively; the remaining 17% of the lines showed a 'superman' phenotype, indicative of overexpression of the transgene (Figure 2b, d–g). Similarly, complementation of the loss-of-function spw1-1 mutant with the gSPW1^{I45T} construct resulted in most lines showing the expected spw1-cls phenotype but also in around ten per cent of lines in which complementation had failed, resulting in a loss-of-function phenotype. While complementation had failed in several lines, most likely due to artefactual variations of the level of expression of the transgene based on its insertion locus, the majority of transgenic lines showed the expected complementation phenotypes in both control experiments. Based on this observation, it was assumed that the most frequent phenotype obtained for a given mutant was representative of the severity of its mutation. Lines carrying mutations that resulted in the lowest β-galactosidase activities in the yeast two-hybrid assay (e.g. G110R/T144I) showed for the most part a loss-of-function phenotype and were completely sterile (Figure 2b, j). Lines carrying mutations that resulted in more intermediate β-galactosidase values (e.g. W80R) were for the most part producing seeds and showing a *cls*-like phenotype (Figure 2b, i). Thus, mutations resulting in around 50 per cent of β-galactosidase activity in the yeast two-hybrid assay are good candidates for genetic engineering. Altogether, these results show that progressive disruption of the heterodimer leads to an increasing loss of lodicule identity, validating our prerequisite for potential fine-tuning of the SPW1/OsMADS2 heterodimer activity.

A novel cleistogamous allele of *SPW1* isolated in a forward genetic screen.

A cleistogamous line, provisionally named *cleistogamy2* (*cls2*), was isolated from a chemically mutagenised rice population (cv Kita-aoba). In contrast to the WT, stamen exertion could not be seen in *cls2* (Figure 3). Microscopic observations of *cls2* flowers revealed that the lodicules were thin and elongated, resembling the lodicules of the spw1-cls mutant (Figure 4c, e). Furthermore, the surface of the lodicules in *cls2* was populated by both round cells and elongated cells, a pattern also observed in spw1-cls, suggesting a partial loss of lodicule identity (Figure 4d, f). Growth chamber experiments revealed that, unlike for spw1-cls, the cleistogamous phenotype of *cls2* was maintained in a cold environment: for the spw1-cls mutant, open flowers could be seen from temperature averages below 26 °C and about half of spw1-cls flowers were open for averages of 23 °C (Figure 5). In contrast, flowers of the *cls2* mutant remained closed for temperature averages as low as 17 °C.

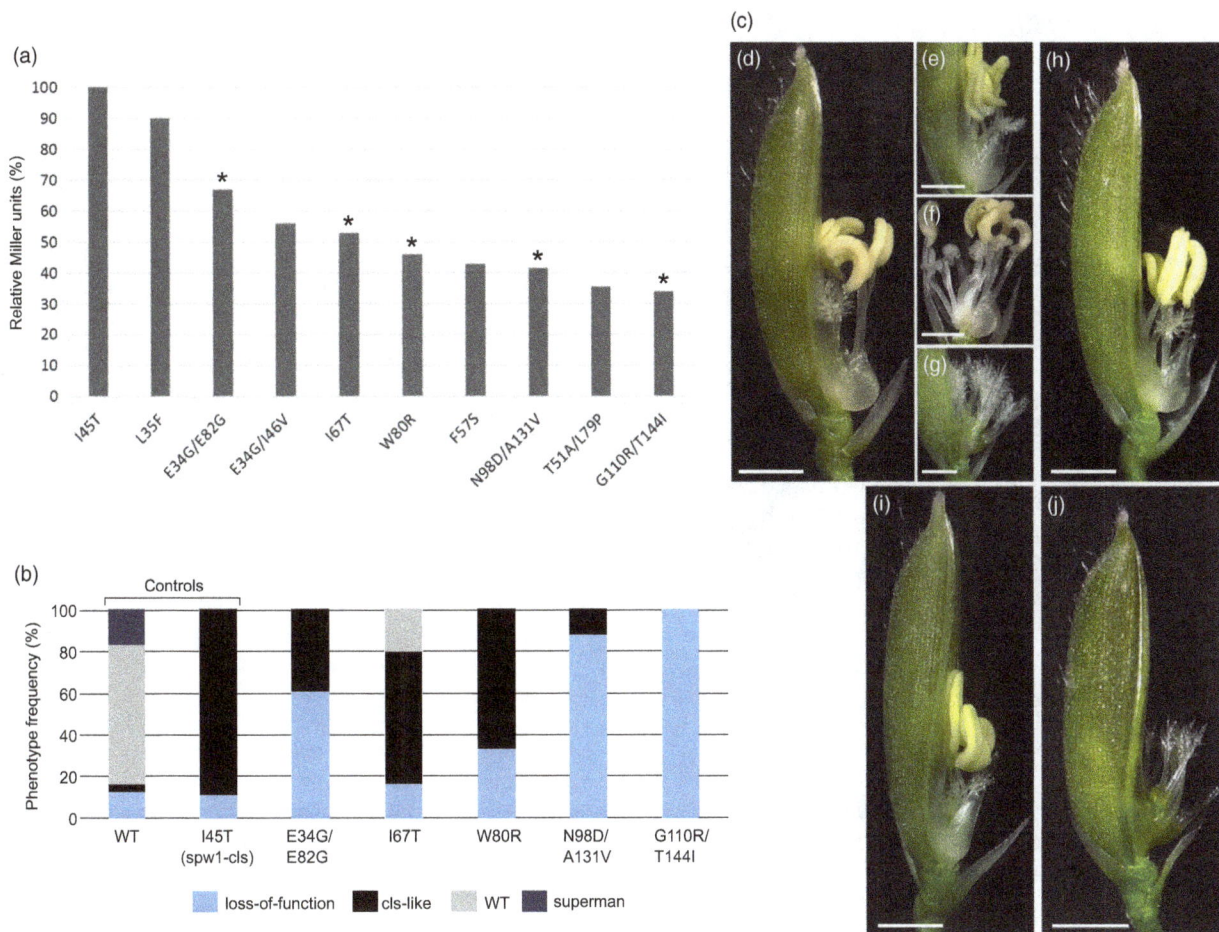

Figure 2 Yeast two-hybrid and complementation analysis of selected *spw1* mutants. a: β-Galactosidase activity of selected mutants relative to that of *spw1-cls* (I45T) in a yeast hybrid assay. Mutations selected for further analysis displayed in b are indicated by asterisks. b: Phenotype frequency in *spw1-1* lines complemented with selected mutated genomic constructs. c: Characteristic floral phenotypes observed in complementation lines. From (d) to (g): lines complemented with the WT allele showing a WT (d), cls-like (e), superman (f) or a loss-of-function (g) phenotype. Complementation with an I45T (h), W80R (i) or a G110/T144I (j) construct. For all pictures the lemma was removed to allow observation of the inner organs. Bars = 2 mm.

The *SPW1* and *OsMADS2* genes being strong candidates for the causal gene of the elongated lodicule phenotype, both genomic sequences were checked for mutations in *cls2*. A guanine to adenine transition was found in *SPW1*, leading to a glycine to arginine change in the MADS domain of the protein (G27R; Figure 1). Allelism tests confirmed that *SPW1* was the causal gene of the cleistogamous phenotype in *cls2* (Supplemental Figure 1). The *cls2* and *spw1-cls* alleles were therefore renamed to *spw1-cls2* and *spw1-cls1*, respectively.

To evaluate field performance and agronomic traits of the *spw1-cls2* line, plants were grown in paddy fields in three locations of Japan: Tsukubamirai, Joetsu and Sapporo, in order of increasingly colder climates. Preliminary data indicate that the fertility of *spw1-cls2* is negatively affected by low temperatures, with about 90% (Tsukubamirai), 75% (Joetsu) and 65% (Sapporo) of the wild type fertility. Further study in Tsukubamirai showed that *spw1-cls2* was comparable to the WT for heading date, culm number and length, panicle length and spikelet number (Table 2). In all three locations only rare flower openings were observed, with rates below one per cent at the coldest location of Sapporo, indicating that the cleistogamous phenotype of *spw1-cls2* is remarkably stable in field conditions.

The *spw1-cls2* mutant is impaired for CArG box recognition by the SPW1/OsMADS2 heterodimer.

The glycine in position 27 is a conserved residue of the MADS domain (Jeon *et al.*, 2000; Silva *et al.*, 2016) and has been shown to be involved in binding to DNA target elements called CArG boxes (Schwarz-Sommer *et al.*, 1992). In the *spw1-cls2* mutant, the G27R change in the SPW1 protein is therefore likely to interfere with its DNA-binding function.

CArG boxes are variations of a core $CC(A/T)_6GG$ motif, and different CArG boxes are specifically bound by different MADS-box protein complexes (Shore and Sharrocks, 1995). CArG loci bound by SPW1 transcriptional complexes have yet to be identified; however, data about *SPW1* and *OsMADS2* orthologs are available. In *Antirrhinum majus, DEFICIENS* (*DEF*) and *GLOBOSA* (GLO) are the respective orthologs of *SPW1* and *OsMADS2*. The DEF/GLO heterodimer has been shown to bind efficiently a CArG box located in the *DEF* promoter region (Schwarz-Sommer *et al.*, 1992; Zachgo *et al.*, 1995). Using a probe corresponding to the *DEF* promoter CArG box sequence, heterodimers of OsMADS2 and SPW1[I45T], SPW1[G27R] or SPW1[WT] were compared for their binding capacities in an electrophoretic

Figure 3 Details of a wild type inflorescence after anthesis compared with a *cls2* mutant inflorescence. Bar = 1 cm.

mobility shift assay (EMSA). A pronounced band shift was observed in the lane loaded with the WT heterodimer, confirming that the *DEF*-CArG probe is significantly bound by SPW1^{WT}/OsMADS2 (Figure 6). In comparison, the band shifted by SPW1^{I45T}/OsMADS2 was faint and was even fainter in the SPW1^{G27R}/OsMADS2 lane, indicating a weak binding between the probe and both of the mutated dimers. This result supports the idea that a defective binding of SPW1 transcriptional complexes to target CArG box loci is responsible for the mutant phenotypes in *spw1-cls1* and *spw1-cls2*, with *spw1-cls2* being more severely affected.

It was previously demonstrated that the phenotype of *spw1-cls1* is caused by a destabilisation of the SPW1^{I45T}/OsMADS2 dimer (Yoshida *et al.*, 2007). The glycine in position 27 of SPW1 sequence does not belong to a region involved in protein dimerisation, and so is not expected to disturb dimer formation (Shore and Sharrocks, 1995; Silva *et al.*, 2016). To assess the heterodimerization ability of SPW1/OsMADS2 in *spw1-cls2*, the binding affinity between SPW1^{G27R} and OsMADS2 was investigated in a yeast two-hybrid assay. There were no significant differences in β-galactosidase activities between the SPW1^{WT}/OsMADS2 and the SPW1^{G27R}/OsMADS2 dimer samples, indicating that, unlike for *spw1-cls1*, heterodimer formation is not affected in *spw1-cls2* (Figure 7).

Taken together, the results of the yeast two-hybrid and EMSA experiments strongly suggest that the *spw1-cls2* phenotype is caused by a defective binding of transcriptional complexes involving SPW1^{G27R} to target CArG box elements.

Discussion

Isolation of an *spw1* cold-stable cleistogamous line

The ability to prevent cultivated plants from outcrossing with plants of the environment is of particular importance in modern agriculture, all the more so in the light of the environmental

concerns expressed by the scientific community and the general public regarding the development and commercialisation of GM crops (Bennett *et al.*, 2013). Cultivated rice is mostly self-pollinating; however, outcrossing with wild relatives is a matter of concern, particularly when plants are grown in vicinity (Marathi and Jena, 2014; Phan *et al.*, 2012). Among the various strategies that have been developed to prevent genetic material exchange between cultivated crops and plants of the environment, often collectively referred to as gene containment, cleistogamy has the advantage of strictly preventing cross-pollination (as opposed to male sterility for example in which flowers are potential pollen recipients) while maintaining plants fertile. In a previous work, it was shown that a single amino acid change in the sequence of the rice transcription factor SPW1 was responsible for the cleistogamous phenotype of the *spw1-cls1* mutant (Yoshida *et al.*, 2007). The one-base genetic change in *spw1-cls1* makes the cleistogamous trait of the mutant straightforward to manage from a breeding perspective. Furthermore, *spw1-cls1* being a non-transgenic line, its commercialisation would be dispensed with the lengthy and costly regulatory approval processes usually required the in the case of GM material (Mullins, 2014). However, a major drawback of the *spw1-cls1* line is that a significant number of flowers open when the plants are grown under cool weather (Yoshida *et al.*, 2007; Ohmori *et al.*, unpublished data). It is thought that low temperatures allow the SPW1/OsMADS2 heterodimer stabilising, resulting in the recovery of the activity of related transcriptional complexes (Yoshida, 2012). Unlike *spw1-cls1*, *spw1-cls2* is not impaired for dimerisation with OsMADS2 (Figure 7). The presence of a mutation in the MADS domain of the protein and the observed lower binding affinity to a canonical CArG box probe (Figure 6) strongly suggest that the phenotype in *spw1-cls2* is caused by a defective binding of SPW1 complexes to physiological DNA targets. Growth chamber experiments revealed that the flowers of *spw1-cls2* remained closed despite temperature averages as low as 17 °C, contrasting sharply with the phenotype of *spw1-cls1* (Figure 5). Technical limitations imposed by the growth chamber system did not allow testing for the effects of large amplitude and/or irregular temperature patterns that can typically be observed in fields. Nevertheless, the stability of the cleistogamous trait towards cold in *spw1-cls2* was confirmed in standard cultivation conditions. Plants grown in fields around the northern city of Sapporo in Japan, which experiences very cool summers, showed only rare (less than one per cent) open flowers.

Phenotypic fine-tuning by structural modifications of MADS-box proteins

Recent structural studies have highlighted the high functional flexibility of the keratin-like domain of floral MADS-box proteins (Puranik *et al.*, 2014; Silva *et al.*, 2016). The critical role of specific amino acids of the keratin-like domain in the heterodimerization of APETALA3 (AP3) and PISTILLATA (PI), the respective *Arabidopsis thaliana* orthologs of SPW1 and OsMADS2, had already been demonstrated in two earlier studies (Yang *et al.*, 2003a,b). Numerous evidence indicate that relatively minor amino acid changes affecting multimerization can have significant physiological repercussions, a feature hypothesised to subtend the central role of MADS-box genes in floral evolution (van Dijk *et al.*, 2010). It can be argued that the MADS domain is more evolutionary static while domains involved in protein–protein interactions allow for diversification and specialisation of the regulatory response (Silva *et al.*, 2016). Unsurprisingly,

Figure 4 Microscopic observations of lodicules of the *spw1-cls* and *cls2* mutants. Light and electronic microscopic observations of WT (a, b), *spw1-cls* (c, d) and *cls2* (e, f) lines. Lemmas were removed for electronic microscopy pictures and both lemma and palea were removed for light microscopy pictures. Bars= 500µm unless indicated otherwise.

modifications of conserved amino acids of the MADS domain often result in severe phenotypes as the transcription factor DNA-binding function is challenged, as for examples in the *lhs1*, *mfo1-1* or *soc1* floral mutants (Jeon *et al.*, 2000; Lee *et al.*, 2008; Ohmori *et al.*, 2009). The glycine in position 27 is conserved among MADS-box proteins and is involved in DNA contact (Pellegrini *et al.*, 1995). Several floral mutants altered for G27, like *spw1-cls2*, have been described in the literature: the *apetala1-2* and *cauliflower-3* mutants of *Arabidopsis thaliana* as well as the *def-nicotianoides* mutant of *Antirrhinum majus* all show a glycine to aspartic acid (G27D) change. Flower development in all these three mutants is severely altered, indicating a loss of function of the mutated proteins (Kempin *et al.*, 1995; Mandel *et al.*, 1992; Schwarz-Sommer *et al.*, 1992). The isolation

of *spw1-cls2* suggests, however, that mutating any MADS-box gene as to generate a G27R change would provide with weak alleles, which may be advantageous for phenotypic analysis or crop improvement.

The increasing number of open flowers in *spw1-cls1* as temperature decreases suggested that there was, at least to some extent, a linear correlation between the stability of SPW1/OsMADS2 transcriptional complexes and lodicule development. In the present study, it was confirmed that destabilisation of the SPW1/OsMADS2 heterodimer (Figure 2a) is associated with hindrance of lodicule development. The range of phenotypes seen in the complementation lines further supports the idea that SPW1/OsMADS2 complexes act quantitatively on target organ development (Figure 2b, c). In addition to the interactions

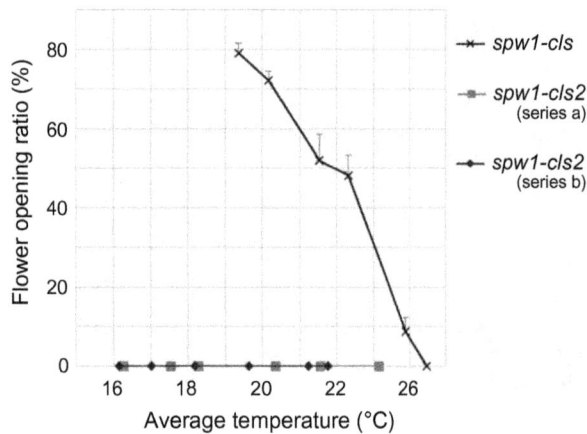

Figure 5 Effect of temperature on flower opening in the *spw1-cls* and *cls2* mutant lines. Plants were transferred to a temperature-controlled growth chamber between four to seven weeks before heading. Each series corresponds to a single experiment.

Table 2 Agronomic characteristics of the *spw1-cls2* line compared with the wild type

	Time to heading (days)	Culm length (cm)	Panicle length (cm)	No. of panicle per plant	No. of spikelet per panicle
spw1-cls2	77	63.9 ± 2.6	18.6 ± 1.0	14.2 ± 1.7	182.0 ± 24.1
Wild type	76	61.5 ± 3.6	18.0 ± 1.0	12.6 ± 1.2	189.7 ± 29.9

Data represent mean values ± standard deviation from 10 plants grown in paddy fields located in Tsukubamirai. Time to heading corresponds to the number of days from seedling to heading.

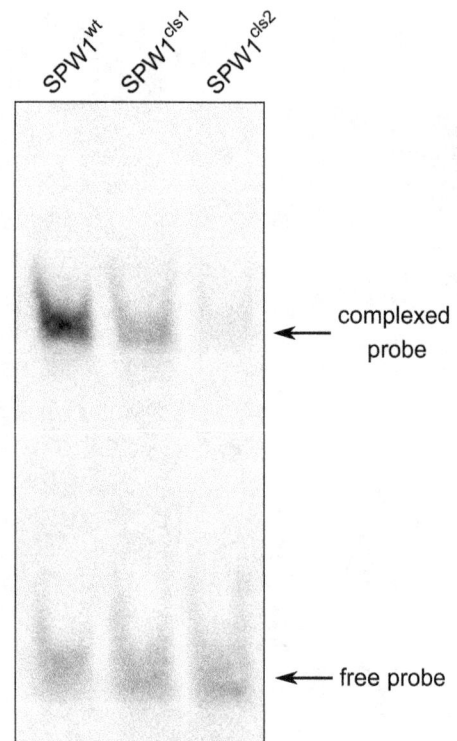

Figure 6 Electrophoretic mobility shift assay between a *DEF* promoter CArG probe and heterodimers of OsMADS2 and SPW1^{WT}, SPW1^{cls1} (SPW1^{I45T}) or SPW1^{cls2} (SPW1^{G27R}).

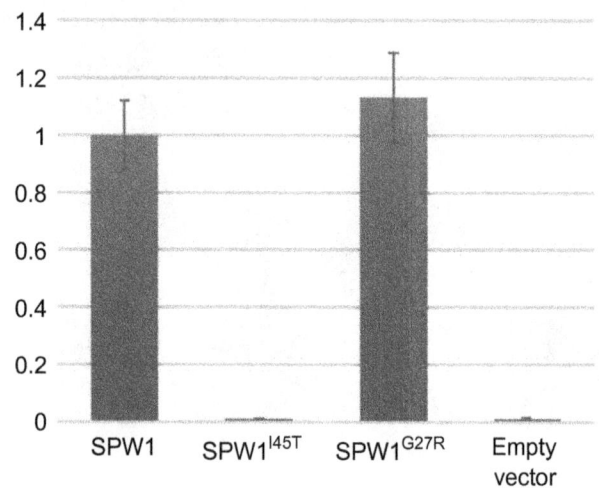

Figure 7 Relative β-galactosidase activities of SPW1 protein variants and OsMADS2 in a yeast two-hybrid liquid assay. β-Galactosidase activities generated in a liquid assay by heterodimers of OsMADS2 and SPW1^{I45T} (*spw1-cls1*), SPW1^{G27R} (*spw1-cls2*) relative to that of an OsMADS2/SPW1 (WT) heterodimer was measured at 28 °C.

between OMADS2 and OsMADS4, SPW1 has been shown to also interact with OsMADS3, OsMADS6, OsMADS7, OsMADS8, OsMADS14, OsMADS17 and OsMADS58 within different complexes (Lee *et al.*, 2003; Seok *et al.*, 2010; Yun *et al.*, 2013). It is possible that one or more mutants from the random mutagenesis (Figure 2) are also altered for the formation of complexes involving the MADS proteins listed above. As these MADS proteins have been shown to affect floral development to various extents, this could explain why the correlation between the decrease in β-galactosidase activity (Figure 2a) and the phenotype severity (Figure 2b) of the transgenic plants is not perfect. Altogether, these data indicate a direct correlation between lodicule development and the binding affinity of SPW1 and OsMADS2, suggesting that it is possible to 'fine-tune' the transcription factor complex activity to obtain a desired phenotype. However, despite the availability of a wealth of information on the structural biology of MADS-box proteins, predicting the steric effects of a given mutation and the associated downstream physiological changes remains extremely challenging. In this work, it was confirmed that assessing the binding affinity between SPW1 and OsMADS2 via yeast two-hybrid experiments could provide for a quick and fairly reliable indication of the severity of a given mutation towards floral development. Most likely, following a similar strategy to the one outlined in this work, which is originally based on (Yang *et al.*, 2003a), would allow identifying weak alleles in MADS-box genes other than *SPW1*.

Fertility rates are expected to decrease when the SPW1/OsMADS2 dimer activity falls below a certain threshold as stamen development is also dependent on the heterodimer function. Data from fields located in Tsukubamirai showed around a ten per cent reduction in fertility of *spw1-cls2* compared with that of the wild type. An unexpected result, however, is the further decreased fertility observed in fields located in the colder areas of Joetsu and Sapporo. In the case of *spw1-cls1*, low temperatures allow restoring the SPW1/OsMADS2 function and consequently lodicule and stamen development. Although mutations in *spw1-cls1* and *spw1-cls2* affect the functions of the respective transcription factors via different mechanisms, fertility was not anticipated to be negatively affected by low temperatures in *spw1-cls2*. It is possible that the decrease in the SPW1/OsMADS2 activity leads to a greater susceptibility to cold stress, however, by which mechanisms this would occur is left to speculation. One solution to maintain high fertility rates might be to mutate the MADS domain of OsMADS2 instead of that of SPW1 as this would maintain SPW1/OsMADS4 activity in the third whorl, which is also supporting stamen development (Yao *et al.*, 2008; Yoshida, 2012).

Concluding remarks

Despite a lower fertility, the *spw1-cls2* can be advantageously used for specific purposes. Coupled with isolation cultivation practices, *spw1-cls2* would provide minimal risks of transgene escape and, to the best of our knowledge, has no equivalent to date in rice. The use of *spw1-cls2* is not limited to transgene containment and rice producers who are particularly inclined to preserving the purity of a cultivar would also benefit from its adoption. As such, the cleistogamous line isolated in this study constitutes a valuable resource for rice cultivation.

Experimental procedures

Yeast two-hybrid assays

The coding sequence of *OsMADS2* as well as the wild type and mutated coding sequences of *SPW1* comprising the M, I and K domains were subcloned into the two-hybrid vectors pAS2-1 and pACT2, respectively (Clonetech). All yeast transformations were performed using Mav203 as the host strain as described previously (Yoshida *et al.*, 2007). For colorimetric filter assay, yeast cells were incubated at 18 °C to allow for visual detection of weak interactions (Yoshida *et al.*, 2007) and *lacZ* reporter gene activity in yeast cells was monitored visually using a 5-bromo-4-chloro-3-indolyl-β-d-galactopyranoside (X-Gal). For quantification of β-galactosidase activity, yeast cells were grown in liquid cultures to log phase at 25 °C unless otherwise stated and ONPG was used as substrate as described in (Yoshida *et al.*, 2007).

Screening of mutagenised cDNAs of SPW1

Mutated *SPW1* cDNAs were created using error-prone PCR amplification (Tarun *et al.*, 1998) and cloned into the pACT2 vector to generate a library of about 10 000 independent clones, following Clonetech protocol. Clones were screened in a coloration filter assay as described above.

Plant transformation

Mutations selected from the yeast two-hybrid assay were introduced into a genomic DNA fragment of *SPW1* and used for complementation of the *spw1-1* mutant. Mutated and WT control DNA fragments were cloned into the pZH2B vector (Kuroda *et al.*, 2010). Resulting vectors were introduced into scutellum-derived calli of *spw1-1* by *Agrobacterium*-mediated transformation under selection with hygromycin, as described in

(Oikawa *et al.*, 2004). The genetic background of *spw1-1* is the cultivar 'Kinmaze'.

Identification of the *spw1-cls2* mutant

To identify cleistogamous mutants of rice, an M_2 population of *Oryza sativa* L. ssp. *japonica* cv. Kita-aoba mutagenised with ethyl methanesulfonate was screened. The plants were grown in a paddy field of the NARO Hokkaido Agricultural Research Center (Sapporo, Japan, formerly the National Agricultural Research Center for Hokkaido Region) under standard conditions. Cleistogamous plants were selected following the method described by Yoshida *et al.* (2007) and subsequent examination revealed one promising line in which cleistogamy was inherited as a single recessive trait.

Temperature gradient growth chamber experiments

Plants of *spw1-cls1* and *spw1-cls2* were grown in a natural lit chamber at the NARO Tohoku Agricultural Research Center (Morioka, Japan) with a 26 °C/20 °C day/night temperature regime. A minimum of 15 seeds were sown in three-litre circular plastic pots with 0.9 g each of N, P_2O_5 and K_2O. At panicle formation stage, pots were transferred to a temperature gradient chamber. Six pots of each line were placed in different positions along the temperature gradient. Open flowers were scored and marked every two to three days and the total number of flowers was counted after full maturation.

Agronomic traits of *spw1-cls2*

To evaluate the agronomic characteristics of *spw1-cls2*, including flower opening rate and fertility, plants were cultivated under standard conditions in paddy fields of the NARO Institute of Crop Science (Tsukubamirai, Ibaraki, JAPAN), NARO Agricultural Research Center (Joetsu, Niigata, JAPAN) and NARO Hokkaido Agricultural Research Center (Sapporo, Hokkaido, JAPAN) during summers of 2011 and 2012.

Electrophoretic mobility shift assay

For each protein, the corresponding full-length CDS was cloned into a pF3A-WG expression vector (Promega). Each one of the three (SPW1, spw1-cls and spw1-cls2) proteins was co-synthesised with OsMADS2 using a TNT SP6 High-Yield wheat germ protein expression system (Promega) according to the manufacturer's instructions. A synthetic probe corresponding to a *DEFICIENS* CArG box (5'-GGCAACTCTTTCCTTTTAGGTCGCA TATGG-3') was labelled with digoxigenin using a DIG gel shift kit reagents (Roche) and 100 fmol of the probe was treated with 0.03 units of proteinase K (Sigma P2308) for 30 min at 37 °C followed by a 60-min deactivation at 60 °C to remove DIG transferase contamination from the probe sample. A total of about 1 μg of protein was put in presence of 20 fmol of labelled probe for 20 min at 25 °C in a 20 μL binding solution containing 2.5% glycerol, 5 mM $MgCl_2$, 50 ng Poly (dI·dC) and 0.05% NP-40 and 1X binding buffer (Thermo Scientific 20148A). Samples were loaded on a 6% polyacrylamide gel and detected after blotting in a chemiluminescent reaction using a LAS-3000 imager (Fujifilm).

Acknowledgements

We thank K. Tsukada, K. Yukawa and N. Ichimura for their excellent assistance in various experiments; M. Kuroda for generously providing with the pZH2B vector; Members in our laboratories for helpful discussion and encouragement; T. Genba, S. Yuminamochi and K. Koide for their help in rice cultivation. This work was partly supported by a grant from the Ministry of Agriculture, Forestry, and Fishery of Japan (Research project for Genomics for Agricultural Innovation GRA-203-1-1) and by JSPS KAKENHI Grant Number 23580012 and 26292008.

References

Bennett, A.B., Chi-Ham, C., Barrows, G., Sexton, S. and Zilberman, D. (2013) Agricultural biotechnology: economics, environment, ethics, and the future. *Annu. Rev. Environ. Resour.* **38**, 249–279.

Daniell, H. (2002) Molecular strategies for gene containment in transgenic crops. *Nat. Biotechnol.* **20**, 581–586.

van Dijk, A.D.J., Morabito, G., Fiers, M., van Ham, R.C.H.J., Angenent, G.C. and Immink, R.G.H. (2010) Sequence motifs in MADS transcription factors responsible for specificity and diversification of protein-protein interaction. *PLoS Comput. Biol.* **6**, e1001017.

Gealy, D., Burgos, N.R., Yeater, K.M. and Jackson, A.K. (2015) Outcrossing potential between U.S. blackhull red rice and indica rice cultivars. *Weed Sci.* **63**, 647–657.

Gressel, J. (2014) Dealing with transgene flow of crop protection traits from crops to their relatives. *Pest Manag. Sci.* **71**, 658–667.

Jeon, J.S., Jang, S., Lee, S., Nam, J., Kim, C., Lee, S.H., Chung, Y.Y. *et al.* (2000) *leafy hull sterile1* is a homeotic mutation in a rice MADS box gene affecting rice flower development. *Plant Cell*, **12**, 871–884.

Kaufmann, K., Melzer, R. and Theissen, G. (2005) MIKC-type MADS-domain proteins: structural modularity, protein interactions and network evolution in land plants. *Gene*, **347**, 183–198.

Kempin, S., Savidge, B. and Yanofsky, M. (1995) Molecular basis of the cauliflower phenotype in *Arabidopsis. Science*, **267**, 522–525.

Kuroda, M., Kimizu, M. and Mikami, C. (2010) A simple set of plasmids for the production of transgenic plants. *Biosci. Biotechnol. Biochem.* **74**, 2348–2351.

Lee, S., Jeon, J.-S., An, K., Moon, Y.-H., Lee, S., Chung, Y.-Y. and An, G. (2003) Alteration of floral organ identity in rice through ectopic expression of *OsMADS16. Planta*, **217**, 904–911.

Lee, J., Oh, M., Park, H. and Lee, I. (2008) SOC1 translocated to the nucleus by interaction with AGL24 directly regulates *LEAFY. Plant J.* **55**, 832–843.

Mandel, M.A., Gustafson-Brown, C., Savidge, B. and Yanofsky, M.F. (1992) Molecular characterization of the *Arabidopsis* floral homeotic gene *APETALA1. Nature*, **360**, 273–277.

Marathi, B. and Jena, K.K. (2014) Floral traits to enhance outcrossing for higher hybrid seed production in rice: present status and future prospects. *Euphytica*, **201**, 1–14.

Messeguer, J., Fogher, C., Guiderdoni, E., Marfà, V., Català, M.M., Baldi, G. and Melé, E. (2001) Field assessments of gene flow from transgenic to cultivated rice (Oryza sativa L.) using a herbicide resistance gene as tracer marker. *Theor. Appl. Genet.* **103**, 1151–1159.

Monera, O.D., Sereda, T.J., Zhou, N.E., Kay, C.M. and Hodges, R.S. (1995) Relationship of sidechain hydrophobicity and alpha-helical propensity on the stability of the single-stranded amphipathic alpha-helix. *J. Pept. Sci.* **1**, 319–329.

Moon, H.S., Abercrombie, L.L., Eda, S., Blanvillain, R., Thomson, J.G., Ow, D.W. and Stewart, C.N. (2011) Transgene excision in pollen using a codon optimized serine resolvase CinH-RS2 site-specific recombination system. *Plant Mol. Biol.* **75**, 621–631.

Mullins, E. (2014) Engineering for disease resistance: persistent obstacles clouding tangible opportunities. *Pest Manag. Sci.* **71**, 645–651.

Nagasawa, N., Miyoshi, M., Sano, Y., Satoh, H., Hirano, H., Sakai, H. and Nagato, Y. (2003) *SUPERWOMAN1* and *DROOPING LEAF* genes control floral organ identity in rice. *Development*, **130**, 705–718.

Ohmori, S., Kimizu, M., Sugita, M., Miyao, A., Hirochika, H., Uchida, E., Nagato, Y. *et al.* (2009) *MOSAIC FLORAL ORGANS1*, an *AGL6*-like MADS box gene, regulates floral organ identity and meristem fate in rice. *Plant Cell*, **21**, 3008–3025.

Ohmori, S., Tabuchi, H., Yatou, O. and Yoshida, H. (2012) Agronomic traits and gene containment capability of cleistogamous rice lines with the *superwoman1-cleistogamy* mutation. *Breed. Sci.* **62**, 124–132.

Oikawa, T., Koshioka, M., Kojima, K., Yoshida, H. and Kawata, M. (2004) A role of OsGA20ox1, encoding an isoform of gibberellin 20-oxidase, for regulation of plant stature in rice. *Plant Mol. Biol.* **55**, 687–700.

Osakabe, Y. and Osakabe, K. (2015) Genome editing with engineered nucleases in plants. *Plant Cell Physiol.* **56**, 389–400.

Pellegrini, L., Tan, S. and Richmond, T.J. (1995) Structure of serum response factor core bound to DNA. *Nature*, **376**, 490–498.

Phan, P.D.T., Kageyama, H., Ishikawa, R. and Ishii, T. (2012) Estimation of the outcrossing rate for annual Asian wild rice under field conditions. *Breed. Sci.* **62**, 256–262.

Puranik, S., Acajjaoui, S., Conn, S., Costa, L., Conn, V., Vial, A., Marcellin, R. *et al.* (2014) Structural basis for the oligomerization of the MADS domain transcription factor *SEPALLATA3* in *Arabidopsis. Plant Cell*, **26**, 3603–3615.

Schaeffer, S.M. and Nakata, P.A. (2015) CRISPR/Cas9-mediated genome editing and gene replacement in plants: transitioning from lab to field. *Plant Sci.* **240**, 130–142.

Schwarz-Sommer, Z., Hue, I., Huijser, P., Flor, P.J., Hansen, R., Tetens, F., Lönnig, W.E. *et al.* (1992) Characterization of the *Antirrhinum* floral homeotic MADS-box gene deficiens: evidence for DNA binding and autoregulation of its persistent expression throughout flower development. *EMBO J.* **11**, 251–263.

Seok, H.-Y., Park, H.-Y., Park, J.-I., Lee, Y.-M., Lee, S.-Y., An, G. and Moon, Y.-H. (2010) Rice ternary MADS protein complexes containing class B MADS heterodimer. *Biochem. Biophys. Res. Commun.* **401**, 598–604.

Shinoyama, H., Sano, T., Saito, M., Ezura, H., Aida, R., Nomura, Y. and Kamada, H. (2011) Induction of male sterility in transgenic chrysanthemums (*Chrysanthemum morifolium* Ramat.) by expression of a mutated ethylene receptor gene, *Cm-ETR1/H69A*, and the stability of this sterility at varying growth temperatures. *Mol. Breed.* **29**, 285–295.

Shore, P. and Sharrocks, A.D. (1995) The MADS-box family of transcription factors. *Eur. J. Biochem.* **229**, 1–13.

Silva, C.S., Puranik, S., Round, A., Brennich, M., Jourdain, A., Parcy, F., Hugouvieux, V. *et al.* (2016) Evolution of the plant reproduction master regulators *LFY* and the MADS transcription factors: the role of protein structure in the evolutionary development of the flower. *Front. Plant Sci.* **6**, 1–18.

Sudianto, E., Beng-Kah, S., Ting-Xiang, N., Saldain, N.E., Scott, R.C. and Burgos, N.R. (2013) Clearfield® rice: its development, success, and key challenges on a global perspective. *Crop Prot.* **49**, 40–51.

Tarun, A.S., Lee, J.S. and Theologis, A. (1998) Random mutagenesis of 1-aminocyclopropane-1-carboxylate synthase: a key enzyme in ethylene biosynthesis. *Proc. Natl Acad. Sci.* **95**, 9796–9801.

Theissen, G. and Saedler, H. (2001) Plant biology: floral quartets. *Nature*, **409**, 469–471.

Toppino, L., Kooiker, M., Lindner, M., Dreni, L., Rotino, G.L. and Kater, M.M. (2011) Reversible male sterility in eggplant (*Solanum melongena* L.) by artificial microRNA-mediated silencing of general transcription factor genes. *Plant Biotechnol. J.* **9**, 684–692.

Wang, X.-J., Jin, X., Dun, B.-Q., Kong, N., Jia, S.-R., Tang, Q.-L. and Wang, Z.-X. (2014) Gene-splitting technology: a novel approach for the containment of transgene flow in *Nicotiana tabacum. PLoS One*, **9**, e99651.

Yang, Y., Fanning, L. and Jack, T. (2003a) The K domain mediates heterodimerization of the *Arabidopsis* floral organ identity proteins, APETALA3 and PISTILLATA. *Plant J.* **33**, 47–59.

Yang, Y., Xiang, H. and Jack, T. (2003b) *pistillata-5*, an *Arabidopsis* B class mutant with strong defects in petal but not in stamen development. *Plant J.* **33**, 177–188.

Yao, S.G., Ohmori, S., Kimizu, M. and Yoshida, H. (2008) Unequal genetic redundancy of rice *PISTILLATA* orthologs, *OsMADS2* and *OsMADS4*, in lodicule and stamen development. *Plant Cell Physiol.* **49**, 853–857.

Yoshida, H. (2012) Is the lodicule a petal: molecular evidence? *Plant Sci.* **184**, 121–128.

Yoshida, H. and Nagato, Y. (2011) Flower development in rice. *J. Exp. Bot.* **62**, 4719–4730.

Yoshida, H., Itoh, J., Ohmori, S., Miyoshi, K., Horigome, A., Uchida, E., Kimizu, M. *et al.* (2007) *Superwoman1-cleistogamy*, a hopeful allele for gene containment in GM rice. *Plant Biotechnol. J.* **5**, 835–846.

Yun, D., Liang, W., Dreni, L., Yin, C., Zhou, Z., Kater, M.M. and Zhang, D. (2013) *OsMADS16* genetically interacts with *OsMADS3* and *OsMADS58* in specifying floral patterning in rice. *Mol. Plant* **6**, 743–756.

Zachgo, S., Silva, E.D.A., Motte, P., Tröbner, W., Saedler, H. and Schwarz-Sommer, Z. (1995) Functional analysis of the *Antirrhinum* floral homeotic *DEFICIENS* gene in vivo and in vitro by using a temperature-sensitive mutant. *Development*, **121**, 2861–2875.

10

Creation of fragrant rice by targeted knockout of the *OsBADH2* gene using TALEN technology

Qiwei Shan[1], Yi Zhang[1], Kunling Chen[1], Kang Zhang[1,2] and Caixia Gao[1,*]

[1]*State Key Laboratory of Plant Cell and Chromosome Engineering, Institute of Genetics and Developmental Biology, Chinese Academy of Sciences, Beijing, China*
[2]*Beijing Genovo Biotechnology Co. Ltd, Beijing, China*

Correspondence

email cxgao@genetics.ac.cn

Keywords: genome editing, transcription activator-like effector nucleases, betaine aldehyde dehydrogenase, 2AP, multiplex gene knockout.

Summary

Fragrant rice is favoured worldwide because of its agreeable scent. The presence of a defective *badh2* allele encoding betaine aldehyde dehydrogenase (BADH2) results in the synthesis of 2-acetyl-1-pyrroline (2AP), which is a major fragrance compound. Here, transcription activator-like effector nucleases (TALENS) were engineered to target and disrupt the *OsBADH2* gene. Six heterozygous mutants (30%) were recovered from 20 transgenic hygromycin-resistant lines. Sanger sequencing confirmed that these lines had various indel mutations at the TALEN target site. All six transmitted the *BADH2* mutations to the T1 generation; and four T1 mutant lines tested also efficiently transmitted the mutations to the T2 generation. Mutant plants carrying only the desired DNA sequence change but not the TALEN transgene were obtained by segregation in the T1 and T2 generations. The 2AP content of rice grains of the T1 lines with homozygous mutations increased from 0 to 0.35–0.75 mg/kg, which was similar to the content of a positive control variety harbouring the *badh2-E7* mutation. We also simultaneously introduced three different pairs of TALENs targeting three separate rice genes into rice cells by bombardment and obtained lines with mutations in one, two and all three genes. These results indicate that targeted mutagenesis using TALENs is a useful approach to creating important agronomic traits.

Introduction

Fragrant rice (*Oryza sativa*) is gaining popularity worldwide for the characteristic fragrance of its grains. Its market price, particularly that of the Indian Basmati and the Thai Jasmine types, is much higher than that of the conventional nonfragrant rice (Bhattacharjee *et al.*, 2002). More than one hundred volatile compounds were detected in the flavour of cooked fragrant rice. Of these, 2-acetyl-1-pyrroline (2AP) is the most abundant and is considered responsible for the fragrance. No apparent 2AP (at least about two orders of magnitude less) is detected in nonfragrant rice (Jezussek *et al.*, 2002; Lorieux *et al.*, 1996).

Genetic analysis of the fragrant trait has shown that fragrance is controlled by a single recessive gene (initially referred to as *fgr*) on chromosome 8. Subsequently, positional cloning suggested that *fgr* encodes betaine aldehyde dehydrogenase 2 (BADH2) (Bradbury *et al.*, 2005). The corresponding full-length gene *OsBADH2* comprises 15 exons and 14 introns, encoding 503 amino acids. Sequence alignment of this gene between fragrant and nonfragrant rice varieties identified multiple types of mutation in fragrant rice, such as a 7-bp deletion in exon 2 (designated *badh2-E2*), an 8-bp deletion together with a 3-bp SNP in exon 7 (designated *badh2-E7*) and an 803-bp deletion between exons 4 and 5 (designated *badh2-E4/5*) (Bradbury *et al.*, 2005; Kovach *et al.*, 2009; Shao *et al.*, 2011; Shi *et al.*, 2008). Further confirmation that mutations in *OsBADH2* are responsible for the fragrant phenotype came from transgene complementation (Chen *et al.*, 2008) and RNA-induced down-regulation experiments (Chen *et al.*, 2012; Niu *et al.*, 2008). It is not yet clear how *OsBADH2* influences 2AP biosynthesis. One likely hypothesis is

that BADH2 inhibits 2AP synthesis by diverting GABald, an upstream precursor of 2AP, to GABA; when BADH2 is dysfunctional, GABald accumulates and is converted to 2AP (Figure 1a) (Bradbury *et al.*, 2008; Chen *et al.*, 2008).

Currently, the main rice breeding technique is conventional hybrid breeding. In conventional breeding of fragrant rice, it takes several generations to transfer a natural *badh2* mutant gene into elite rice varieties by crossing and backcrossing. The process is laborious, time-consuming and expensive. Transgenic breeding using RNAi-based down-regulation of *OsBADH2* expression is an alternative approach (Chen *et al.*, 2012; Niu *et al.*, 2008). However, RNAi-mediated inhibition of *OsBADH2* expression is often incomplete (Chen *et al.*, 2012); transgene expression varies in different lines, so it is necessary to screen a large number of transgenic plants to identify candidate lines in which the transgene is stable expressed over generations; besides, rice lines derived by this method are regarded as transgenic and are subject to costly regulatory processes.

Genome editing technologies using sequence-specific nucleases (SSNs), including meganucleases, ZFNs, TALENs and the CRISPR/Cas9 system, have been developed to create targeted DNA double-strand breaks (DSBs) in various model and crop plant species (Voytas and Gao, 2014). The DSBs are mainly repaired by error-prone nonhomologous end joining (NHEJ) or by high-fidelity homologous recombination (HR). NHEJ often causes small insertions or deletions (indels) at the sites of breaks, frequently generating knockout mutations. Here, we report the creation of fragrant rice from a nonfragrant variety via targeted knockout of *OsBADH2* using the TALEN method. We previously described one pair of TALENs engineered to cleave *OsBADH2* (T-OsBADH2b)

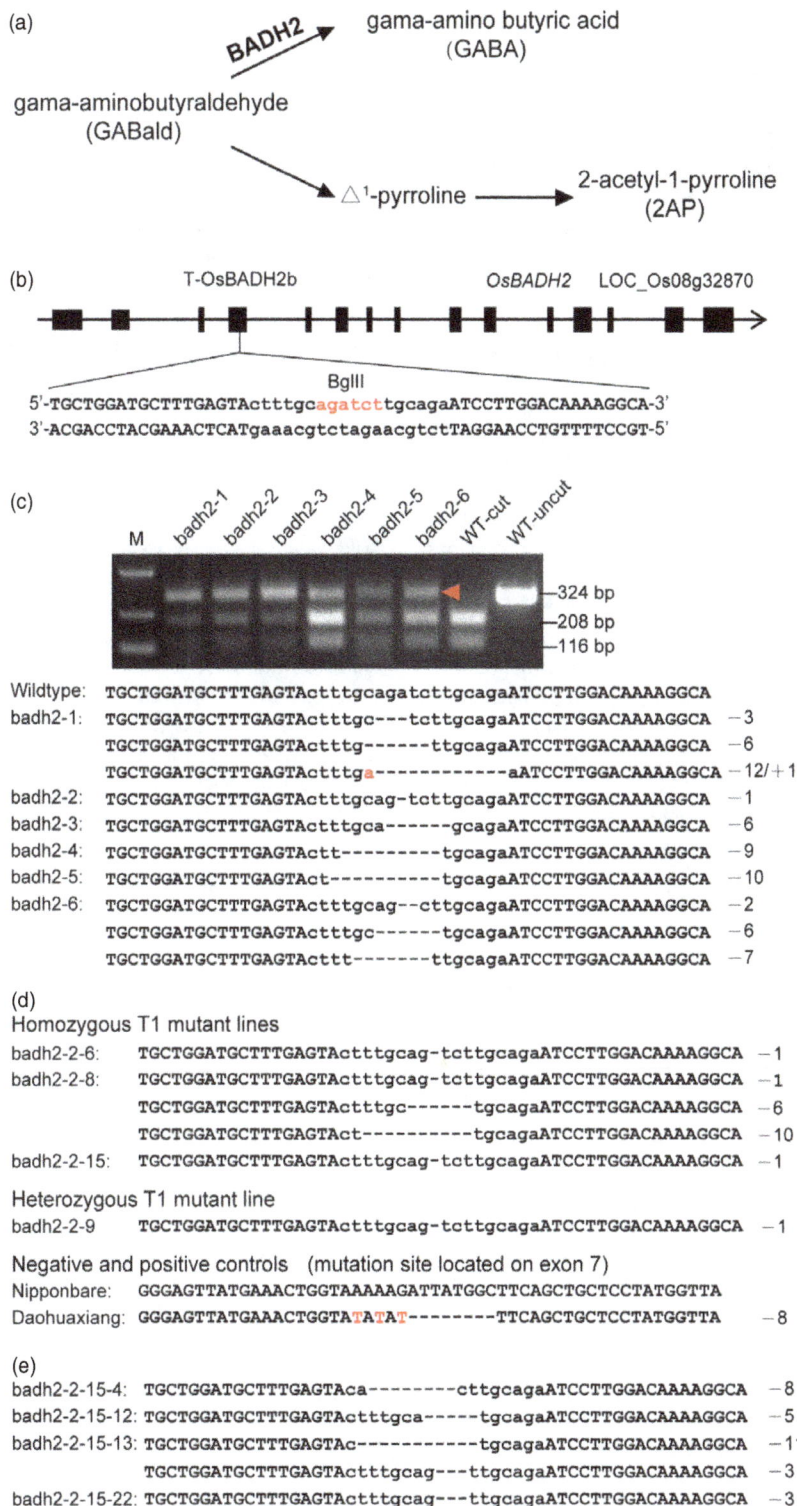

Figure 1 Transcription activator-like effector nuclease (TALEN)-induced mutations in the *OsBADH2* gene. (a) The 2AP pathway in rice. *OsBADH2* is responsible for the conversion of GABald into GABA; when BADH2 is inactive, GABald accumulates and 2AP is formed. (b) Schematic of *OsBADH2* gene structure and TALEN binding sites. *OsBADH2* contains 15 exons, indicated by black rectangles. The target site of T-OsBADH2b is shown beneath, with the TALEN binding sites in upper case and the spacer in lower case; the restriction enzyme site BglII is highlighted in red. (c) Detection of NHEJ-introduced indel mutations in six heterozygous T0 transgenic plants by PCR/RE assay and DNA sequencing. Mutations in independent plants were identified with *OsBADH2*-specific primers, and the PCR amplicons were digested with BglII. The red arrowhead indicates bands with mutations. The mutation sequences of six T0 plants are aligned under the gel. Deletions and insertions are indicated by dashes and red letters, respectively, and the numbers on the right side give the indel sizes. M, DNA molecular weight marker. (d) DNA sequences of three homozygous T1 lines, one heterozygous T1 line, a positive control (cv. Daohuaxiang, *badh2-E7*) and a negative control (cv. Nipponbare) used for measuring 2AP content. (e) New indel mutation sequences detected in four T2 seedlings of the *badh2-2* plant.

(Shan *et al.*, 2013). We have now used the same TALEN pair to create several heritable homozygous mutant rice lines. In these lines, there was substantial 2AP content, whereas no 2AP was detected in the wild-type control plant. We also attempted to knockout three genes (*OsBADH2*, *OsCKX2* and *OsDEP1*) simultaneously by particle bombardment and obtained a series of transgenic rice plants with single, double and triple knockouts, thus demonstrating the ability of TALENs to create multiplex gene knockouts.

Results

TALEN design and recovery of rice plants with *BADH2* mutations

The construction and validation of TALEN pairs has been described previously (Shan *et al.*, 2013). T-OsBADH2b targeting to the fourth exon of *OsBADH2* was used to create stable transgenic plants using *Agrobacterium*-mediated transformation. The TALEN pair recognizes two 17-bp sequences of contiguous DNA separated by an 18-bp spacer DNA that contains a BglII restriction site in its centre for identifying mutations by PCR restriction enzyme (PCR/RE) digestion assays (Figure 1b). A total of six T0 heterozygous mutant *BADH2* rice plants (badh2-1 to badh2-6) were recovered from 20 transgenic hygromycin-resistant plants (30%) from one *Agrobacterium*-mediated transformation experiment in rice variety Nipponbare (Figure 1c). Sequencing of the mutations showed that most were small deletions of 1–10 bp in the spacer region (Figure 1c and Table 1). Mutants badh2-3 and badh2-4 had 6-bp and 9-bp deletions, respectively, which did not result in frame shifts mutation. Plants badh2-2 and badh2-5 with 1-bp and 10-bp deletions, respectively, caused frameshifts and could have inactivated the gene. Hence, homozygous progeny of badh2-2 and badh2-5 were chosen for phenotypic analysis. In addition, we detected multiple changes at a single target site in plants badh2-1 and badh2-6, probably because these mutations occurred in different somatic cells. To demonstrate the wide application, the same TALEN pair was also transformed into an elite rice cultivar LPK, and 10 heterozygous mutant *BADH2* rice plants (19.2%) were regenerated from 52 transgenic T0 plants (Figure S1).

Transmission of TALEN-induced mutations to T1 and T2 generations

To see whether the mutations induced by T-OsBADH2b were transmitted to the next generation, all 6 T0 mutant plants were self-pollinated and individually genotyped, and a total of 171 T1 plants from the 6 T0 mutants were genotyped by PCR/RE to test for transmission of the mutations (Table 1). We found that all six mutations were transmitted to the T1 generation, the proportion ranging from 28.6% to 90.6%. Most of the T1 plants carrying the mutations were heterozygotes, except for the offspring of badh2-2. Among 32 T1 progeny of badh2-2, we identified four homozygous and 25 heterozygous mutant plants. No homozygous mutants were detected in the T1 generation of the other five T0 plants, probably because of the small number of T1 seeds (badh2-3, badh2-4 and badh2-5) or the mutations occurring in somatic cells that could not participate in the production of gametes (badh2-1, badh2-6). The progeny of three homozygous (badh2-2-6, badh2-2-8 and badh2-2-15) and one heterozygous (badh2-2-9) T1 mutant plants were investigated further (Figure 1d). All three homozygous mutations were faithfully transmitted, and the mutation in the heterozygous plant was transmitted in a Mendelian fashion (homozygote: heterozygote: wild type = 1 : 2 : 1) to the T2 generation (Table 2). Interestingly, some new indel mutations were detected in the T1 and T2 progeny of badh2-2. For example, whereas the mutation in badh2-2 was a 1-bp deletion, we identified additional 6-bp and 10-bp deletions in one of its offspring (badh2-2-8) (Table 2), and new indel mutation types were also identified in the progeny of line badh2-2-15. A detailed list of the new mutations detected and their sequences is given in Figure 1d and e. This suggests that the 1-bp deletion in the spacer region of badh2-2 did not prevent T-OsBADH2b from recognizing and further cleaving the target.

Generation of TALEN-free mutant rice lines

To obtain rice lines harbouring the desired *OsBADH2* mutations but not the selective marker or the TALEN construct, the PCR-based assay was used: primer pair F1/R1 to amplify the maize *Ubiquitin 1* promoter and N-terminus of TALEN-L, F2/R2 to amplify the C-terminus of TALEN-R and the *Nos* terminator and F3/R3 to amplify the hygromycin resistance gene (Figure 2a). A primer set amplifying the endogenous *BADH2* gene was used as an internal control in all three PCRs. Although all four T1 plants (badh2-2-6, badh2-2-8, badh2-2-9 and badh2-2-15) used for analysing mutation transmission from T1 to T2 carried TALEN constructs, the PCR assay failed to detect any T-DNA construct genes in 16 of 113 (14.2%) T1 plants derived from the 6 T0 lines and 7 of 96 (7.3%) T2 plants derived from the four T1 lines that contained the desired genetic modifications (Figure 2b, Tables 1

Table 1 TALEN-induced mutations in *OsBADH2* and their transmission to the T1 generation

T0 plant ID	T0 genotype	T0 mutation type (bp)	Mutation segregation in T1				Transmission ratio (%)*	TALEN-free (%)[†]
			Total	Homo	Hetero	WT		
badh2-1	Bb	−3, −6, −12/+1	10	0	5	5	50	0
badh2-2	Bb	−1	32	4	25	3	90.6	20.7 (6/29)
badh2-3	Bb	−6	5	0	2	3	40	0
badh2-4	Bb	−9	12	0	6	6	50	0
badh2-5	Bb	−10	14	0	4	10	28.6	0
badh2-6	Bb	−2, −6, −7	98	0	67	31	68.4	14.9 (10/67)

−n, nucleotide deletion of the indicated number; −n/+n, simultaneous nucleotide deletion and insertion of the indicated number at the same site.

*Transmission ratio was calculated based on the number of plants carrying the mutations over the total number of plants tested.

[†]TALEN-free ratio was calculated based on the number of mutant plants not harbouring the T-DNA construct over the total number of plants tested.

Table 2 Genetic analysis of mutations in *OsBADH2* and their transmission to the T2 generation

T1 plant ID	T1 genotype	T1 mutation type (bp)	Mutation segregation in T2				Transmission ratio (%)*	TALEN-free (%)†
			Total	Homo	Hetero	WT		
badh2-2-6	bb	−1	24	24	0	0	100	0
badh2-2-8	bb	−1, −6, −10	24	24	0	0	100	0
badh2-2-15	bb	−1	24	24	0	0	100	29.2 (7/24)
badh2-2-9	Bb	−1	24	5	13	6	75‡	0

‡Indicated that the segregation of the heterozygous lines (badh2-2-9) confirms to the Mendelian ratio (1 : 2 : 1) according to the χ^2 test ($P > 0.5$).

*Transmission ratio was calculated based on the number of plants carrying the mutations over the total number of plants tested.

†TALEN-free ratio was calculated based on the number of mutant plants not harbouring the T-DNA construct over the total number of plants tested.

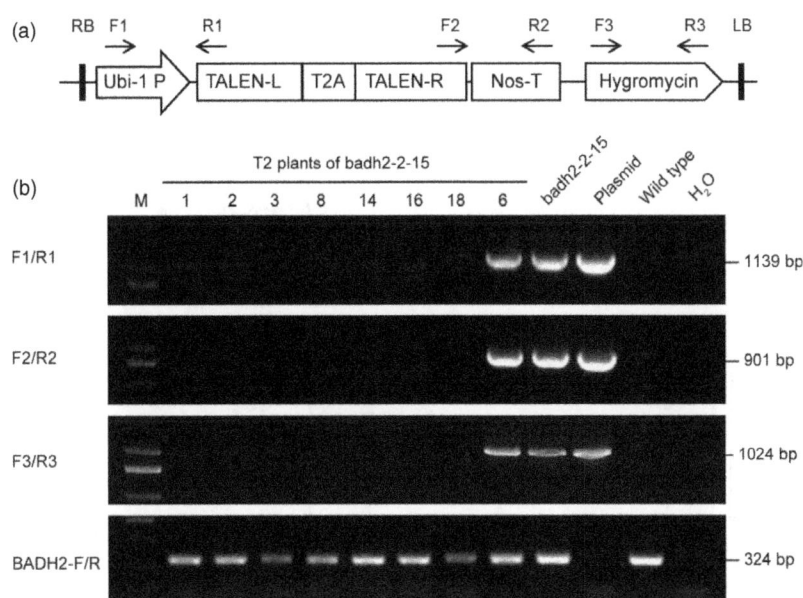

Figure 2 Segregation of the TALEN transgene in *badh2* mutants. (a) Schematic of the TALEN T-DNA construct showing the position of the three pairs of PCR primers used to survey different region of the TALEN transgene in the progeny of *badh2* mutants. F1/R1, for the *Ubiquitin 1* promoter and N-terminus of TALEN-L; F2/R2, for the C-terminus of TALEN-R and the *Nos* terminator; F3/R3, for the hygromycin resistance gene. (b) Gel images of the PCR products obtained with the three pairs of PCR primers and primers for the *OsBADH2* gene as an internal control in each reaction. M, DNA molecular weight marker. Numbers above the gel image refer to representative individual T2 plants of badh2-2-15. badh2-2-15, T1 plants; Plasmid, T-DNA expression plasmid containing T-OsBADH2b; Wild type, DNA from a nontransgenic wild-type rice plant; H_2O, negative control without any DNA. Numbers to the right of each gel indicate the sizes of amplicons.

and 2), suggesting that the T-DNA construct had been eliminated from these plants by segregation as a single genetic locus. This indicates that TALEN-free rice plant with the desired mutations can be relatively easily obtained by segregation among the progeny.

2AP content of mutant rice grains

Rice grains (T2 generation) gathered from three homozygous T1 lines (badh2-2-6, badh2-2-8 and badh2-2-15) and 1 heterozygous T1 line (badh2-2-9) were assayed for 2AP content by gas chromatography–mass spectrometry (GC-MS). It should be noted that the T2 grains of three homozygous T1 lines were all homozygote and the T2 grain of heterozygous T1 line was a mixture of homozygote, heterozygote and wild type. Grains of fragrant rice cultivar Daohuaxiang with the *badh2-E7* genotype were used as a positive control, and the nonfragrant rice cultivar Nipponbare served as a negative control (Figure 1d). 2, 4, 6- trimethyl pyridine (TMP) was used as internal standard

because of its similar molecular weight and chemical characteristics to 2AP in GC-MS. As shown in Figure 3a–f, a 2AP peak was detected from all the homozygous, heterozygous and positive control lines, but none from the negative control line. After normalizing 2AP content based on the TMP yields, the grains from 3 homozygous lines had as high or slightly higher 2AP levels than the positive control Daohuaxiang (0.5–0.75 mg/kg versus 0.5 mg/kg); in the grains from the heterozygous line, the 2AP level was lower than in the positive control (0.35 versus 0.5 mg/kg). However, 2AP level was clearly produced in all the homozygous and heterozygous mutant lines created by TALENs (Figure 3g).

Multiplex gene knockout by bombardment with three pairs of TALENs

The development of multiplex genome engineering, in which several genes are altered simultaneously, will facilitate the generation of multiple agronomically important traits and enable the manipulation of complex traits. In diploid rice, a trait is often

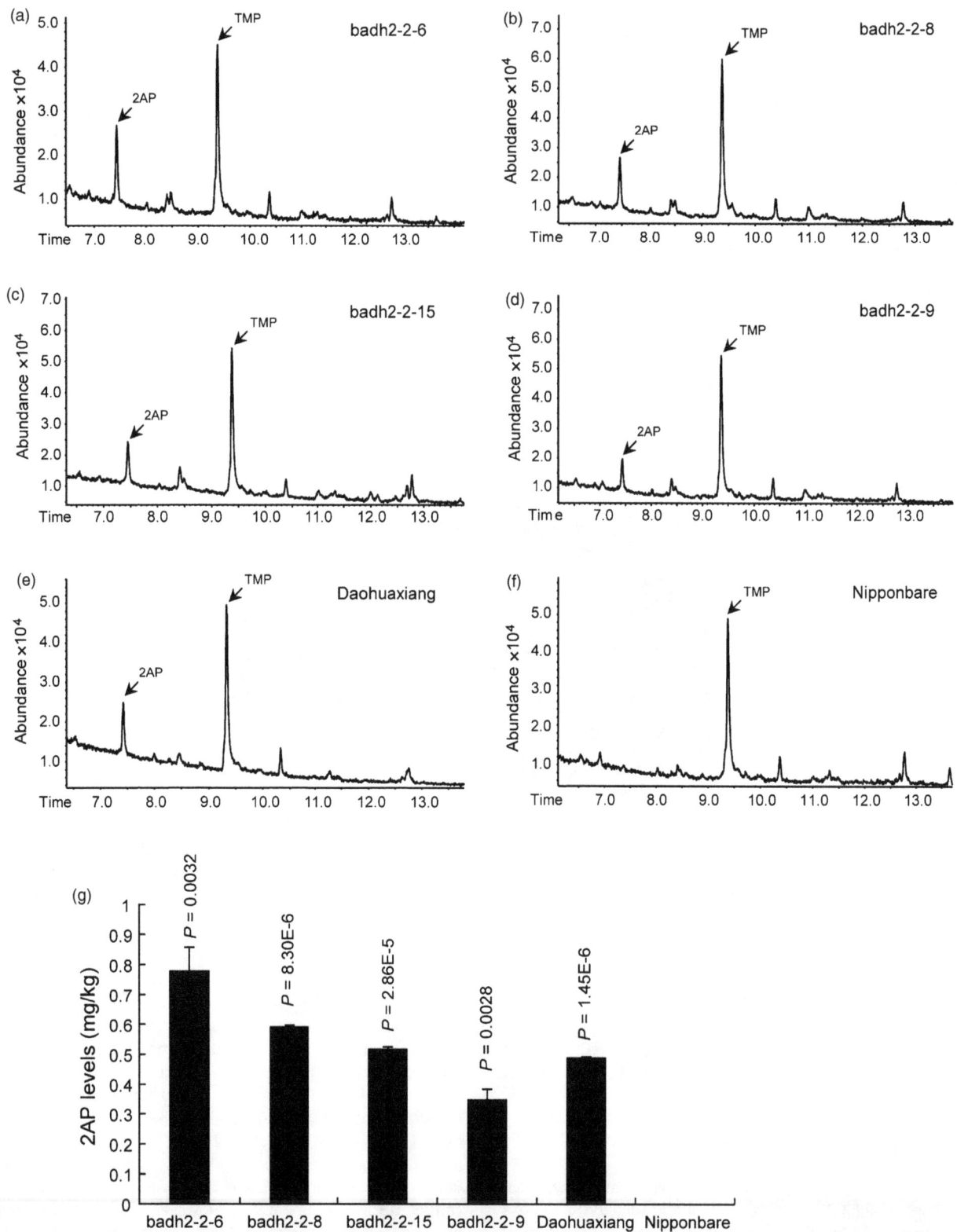

Figure 3 2AP contents of *badh2* mutant grains as measured by GC-MS. (a-f) Total ion chromatograms (TIC) of 2AP and TMP (as internal standard) in the TALEN-induced *badh2* mutant lines and control lines: (a) Grains of the homozygous T1 line, badh2-2-6. (b) Grains of the homozygous T1 line, badh2-2-8. (c) Grains of the homozygous T1 line, badh2-2-15. (d) Grains of the heterozygous T1 line, badh2-2-9. (e) Grains of the positive control, cv. Daohuaxiang. (f) Grains of the negative control, cv. Nipponbare. (g) 2AP levels of the *badh2* and control lines. Values are means ± SD of three replications. A Student's *t*-test was applied to generate *P*-values.

controlled by several genes, and a single gene mutation may not produce phenotypic change. To examine whether TALENs can disrupt multiple target genes, three TALEN pairs targeting *OsBADH2*, *OsCKX2* and *OsDEP1*, respectively, were introduced into rice calli by bombardment (Figures 1b and 4a). Mutation of *OsCKX2* and *OsDEP1* had been reported to increase the yield of rice grain (Ashikari *et al.*, 2005; Huang *et al.*, 2009). A total of 207 transgenic rice lines were regenerated from hygromycin-tolerant calli after 3-month selection, and PCR/RE assays were used to detect indel mutations. Among these transgenic lines were 20 (9.7%) *OsBADH2* mutants, 53 (25.6%) *OsCKX2* mutants and 19 (9.2%) *OsDEP1* mutants. Most were heterozygotes (Table 3). We examined all the mutants and found 4 plants (4/207, 1.9%) (C10, E5, F4, F9) containing mutations of all three target genes (Figure 4b). Sequencing of the PCR products spanning the TALEN target sites revealed that all these triply mutated rice plants contained site-specific indel mutations in the gene coding regions, and plants F4 and F9 were homozygous for *OsCKX2* and *OsDEP1* (Figure 4c). In addition to the triple-gene mutation, we also found a series of single and double mutants at various frequencies (Table 4, Figure S2). Previous studies have reported a low frequency of chromosome translocations between pairs of targeted DSBs on different mammalian chromosomes (Brunet *et al.*, 2009; Piganeau *et al.*, 2013). However, we did not detect any chromosome translocation among all the double and triple mutants (data not shown). These results demonstrate that simultaneous multiple gene mutations can be generated in rice by co-transformation of multipairs of TALENs.

Discussion

Genome editing using SSNs provides good opportunities for crop improvement. To date, SSNs have been used to create gene knockout plants in a variety of important crops, such as rice, maize, wheat, barley and soya bean (Haun *et al.*, 2014; Li *et al.*, 2012; Liang *et al.*, 2014; Shukla *et al.*, 2009; Wang *et al.*, 2014; Wendt *et al.*, 2013). However, examples of real improvement of agronomic traits or the creation of excellent novel genotypes using SSNs are very limited. One of the first examples was the use of ZFNs to target the maize *IPK1* gene, which encodes an enzyme that catalyses the final step in phytate biosynthesis (Shukla *et al.*, 2009). Reducing the level of phytate in seeds is of value because phytate is an antinutritional component and also contributes to environment pollution. In rice, TALENs were used to create a mutation in the pathogen TAL effector binding site in the promoter of *OsSWEET14*, which contributes to pathogen survival and virulence, thereby eliminating the transcription of this gene and reducing the pathogen's virulence (Li *et al.*, 2012). Another good example occurred in bread wheat. By knocking out all six alleles encoding the MILDEW-RESISTANCE LOCUS (MLO) protein with one pair of TALENs, the authors generated a mutant line with broad-spectrum resistance to powdery mildew, a devastating fungal disease (Wang *et al.*, 2014). Haun *et al.* (2014)

recently created soya bean lines that are low in polyunsaturated fats by introducing mutations in two fatty acid desaturase 2 genes (FAD2-1A and FAD2-1B). Here, we have provided another example of the use of SSN to disrupt biochemical pathways and create plants that accumulate valuable biosynthetic intermediates. We used TALEN technology to knockout *OsBADH2* and obtain homozygous mutants with significantly increased content of the fragrant chemical 2AP; the level produced is similar to or even higher than in the fragrant rice cultivar Daohuaxiang. This method provides rice breeders with a new way to breed fragrant rice. Precise molecular breeding using SSNs is superior to conventional or transgene-based breeding methods such as RNAi or genetic engineering. For example, (1) it can modify a target gene accurately; (2) there is no need for laborious crossing and backcrossing, so it is time-saving and convenient; and (3) 'clean' plants (with the selective marker and SSN transgene segregated away from the genome) can be obtained.

The mutations in the six T0 plants induced by TALENs were transmitted to the T1 generation, but not in a Mendelian ratio. For lines badh2-2, 2-3, 2-4 and 2-5, this may have been because of the small numbers of T1 seeds. In lines badh2-1 or 2-6, we detected multiple indel mutations at a single target site, and such chimeric mutations in the T0 plants may have resulted from delayed cleavage in the primary embryogenic cell. These mutations occurred in somatic cells that did not participate in the production of gametes. A similar phenomenon has been reported in rice, wheat, corn and *Arabidopsis* (Feng *et al.*, 2013, 2014; Liang *et al.*, 2014; Wang *et al.*, 2014).

It is noteworthy that sequencing revealed some new indel mutations among the T1 and T2 offspring of badh2-2. As only one nucleotide were deleted in the middle of the 18-bp spacer in T0 badh2-2, the deletion did not destroy either the left or the right TALEN binding sequence; hence, the FokI nuclease may still have been able to form active dimers and cleave the 17-bp spacer. The new indel could then have resulted from continued TALEN cleavage of the target. We identified several mutant lines that no longer contained the TALEN construct; these were generated in the progeny by segregation. No new indel mutations were detected in such progeny lines, for example badh2-2-15-1, 2-2-15-2, 2-2-15-3 and 2-2-15-8 (Figure 2). This suggests that it is important to segregate the TALEN transgene from the plant genome to stabilize the induced mutations as well as to satisfy biosafety concerns. Note that 'clean' fragrant lines of this kind differ from the wild type by only one nucleotide base pair.

Interestingly, we observed differences in the accumulation of 2AP in the grains of different homozygous T1 mutant lines derived from the same T0 plant, badh2-2: the 2AP content of badh2-2-6 was significantly higher than that of badh2-2-8 and badh2-2-15 (Figure 3g). The reason for this is currently unclear; however, it underlines the importance of selecting the mutant lines with the highest 2AP content for breeding programme. At the same time, the grains from the heterozygous line badh2-2-9 contained about half the 2AP content of homozygous badh2-2-6.

Table 3 Multiple gene knockouts in rice using TALENs, showing the frequencies of mutations in each of the three genes targeted

No. of tested plants	Mutations in *OsBADH2* (%)			Mutations in *OsCKX2* (%)			Mutations in *OsDEP1* (%)		
	Total	Homo	Hetero	Total	Homo	Hetero	Total	Homo	Hetero
207	20 (9.7)	2 (1.0)	18 (8.7)	53 (25.6)	19 (9.2)	34 (16.4)	19 (9.2)	6 (2.9)	13 (6.3)

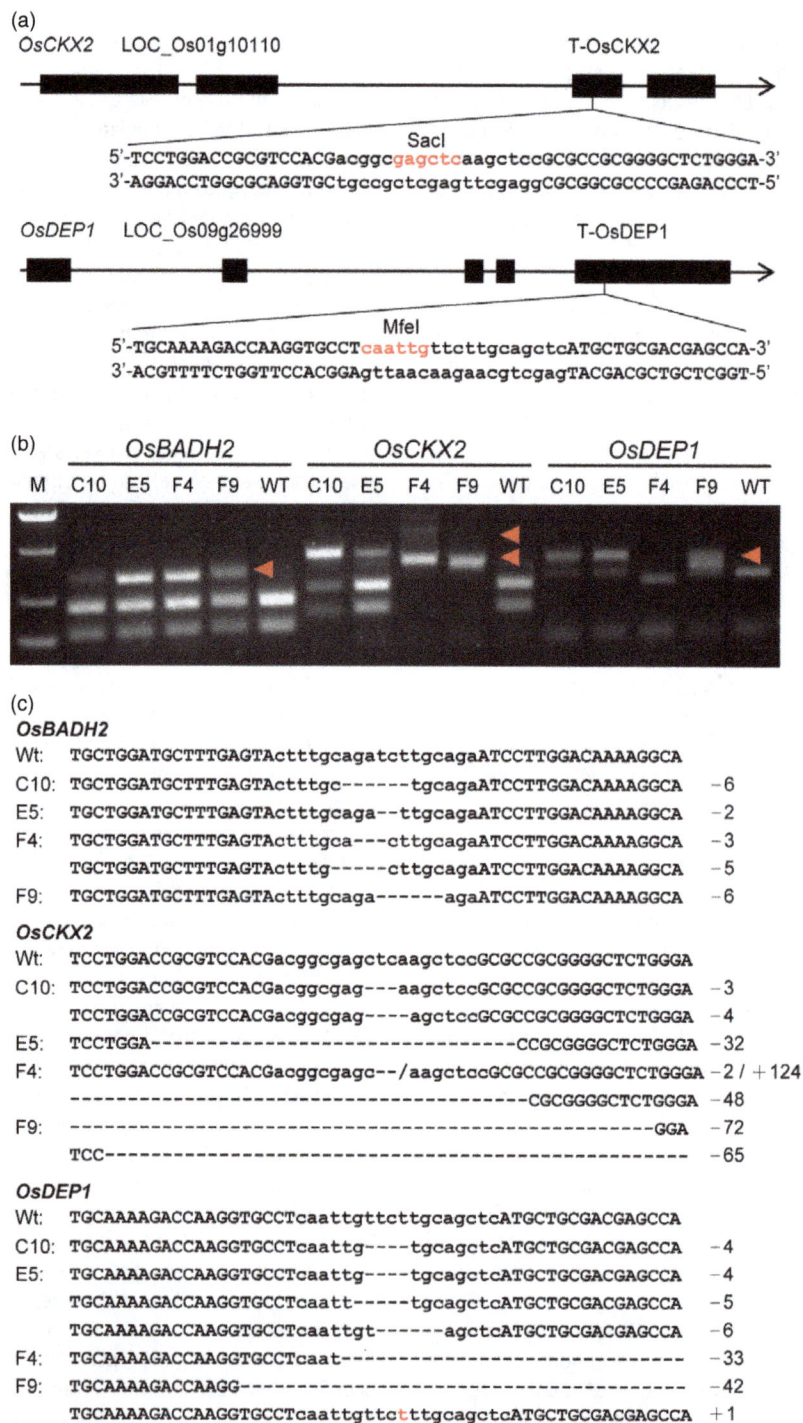

Figure 4 Multiple gene knockouts by co-transformation of three TALEN pairs. (a) Schematic of the *OsCKX2* and *OsDEP1* genes and the corresponding TALEN binding sites. *OsCKX2* contains 4 exons, and *OsDEP1*, 5 exons, indicated by black rectangles. The target sites of T-OsCKX2 and T-OsDEP1 are shown beneath, with the TALEN binding sites in upper case and spacer in lower case; the SacI and MfeI sites are highlighted in red. (b) Detection by PCR/RE assay of NHEJ-introduced indel mutations in four T0 plants mutated in all three genes (C10, E5, F4, F9). Mutations in independent plants were identified with gene-specific primers, and the PCR amplicons were digested with BglII, SacI and MfeI, respectively. Red arrowheads indicate mutant bands. (c) Sequencing of the TALEN-induced mutant alleles in each of the triple-gene mutated rice plants. Deletions and insertions are indicated by dashes and the '/' symbol, respectively, and the numbers on the right side show the sizes of the indels.

The grains of badh2-2-9 were in fact a mixture of homozygous mutant, heterozygote and wild type, and the reduced 2AP content was due to dilution of the homozygous mutant by the wild type and heterozygote.

TALENs are reported to have higher target specificity than ZFNs or the CRISPR/Cas9 system, as their binding sequences are longer (typically each TALEN monomer recognizes 15–20 bp) (Carroll, 2014). Although, almost all the SSNs available today can

Table 4 Multiple gene knockouts in rice using TALENs, showing the frequencies of mutations for all the combinations of single, double and triple genes targeted

No. of tested plants	Single (%)			Double (%)			Triple (%)
	BADH2	CKX2	DEP1	BADH2/CKX2	BADH2/DEP1	CKX2/DEP1	BADH2/CKX2/DEP1
207	6 (2.9)	30 (14.5)	2 (1.0)	8 (3.9)	2 (1.0)	11 (5.3)	4 (1.9)

accommodate one to several mutations within their target site, TALENs can only accommodate a relatively small number of position-dependent mismatches (Juillerat et al., 2014). A study in human cells demonstrated that off-target effects were extremely rare even if there was only one nucleotide mismatch (Mussolino et al., 2011). We identified several potential off-target sites for T-OsBADH2b using the PROGNOS program (Fine et al., 2013) with less stringent criteria that allowed up to a 6-bp mismatch and 10- to 30-bp spacers (Table S1). We then examined three of the most likely off-target sites with 9-bp (OffT-1), 11-bp (OffT-5) and 10-bp (OffT-17) mismatches in their recognition sequences in the six T0 plants and found none. This suggests that this T-OsBADH2b TALEN pair induces mutations site specifically. It also suggests that it is worth using online tools to predict off-target TALEN sites.

Multiplex gene knockout is important for analysing polyploid organisms (such as bread wheat and Brassica napus) and diploid organisms with duplicated genomes (such as soya bean). In such cases, four or six alleles may need to be targeted simultaneously in order to introduce a gene function or phenotype change. One strategy for multiplex targeted mutagenesis using sequence-specific nucleases is to design a pair of SSNs targeting the conserved region in different genomes or duplicated genes and to select for organisms with all the targeted copies mutated (Li et al., 2013a). Using this strategy, our group successfully disrupted the wheat MLO gene using a pair of TALENs and obtained a mlo-aabbdd mutant with high levels of resistance to powdery mildew (Wang et al., 2014). Alternatively, if the corresponding genes in the different genomes have low homology, or one wishes to target unrelated genes, two or more SSN pairs targeting different gene-specific regions can be designed and introduced together into cells, followed by selection for lines with mutations in all the relevant genes. This strategy has been applied in human cells and model animals such as mice, rat, zebrafish, Bombyx mori and Xenopus laevis (Cong et al., 2013; Jao et al., 2013; Li et al., 2013b, 2014; Ma et al., 2014; Ota et al., 2014; Sakane et al., 2013; Sakuma et al., 2014; Wang et al., 2013). Only two of these examples used the TALEN technique (Li et al., 2014; Sakane et al., 2013), while the others employed the CRISPR/Cas9 system. We have described above the first successful example of multiplex gene knockout in crop plants; it involved co-transforming three pairs of TALENs targeting, respectively, the genes OsBADH2, OsCKX2 and OsDEP1 related to rice quality and yield. Although the CRISPR/Cas9 system is straightforward and effective for multiplex genome editing because an sgRNA of only 100-nt is required to guide Cas9 nuclease to the target sites, our results suggest that TALEN technology can also be used efficiently for multiple gene disruption in crop plants.

Experimental procedures

Plasmid construction

TALEN-coding plasmids were constructed as previously described (Shan et al., 2013). Three pairs of TALENs were employed in this work: pGW3-T-OsBADH2b, pGW3-T-OsCKX2 and pGW3-T-OsDEP1 targeting OsBADH2, OsCKX2 and OsDEP1 gene, respectively. The T-DNA construct consisted of a TALEN expression cassette driven by the maize Ubiquitin 1 promoter and a hygromycin selection marker gene driven by the 35S promoter. Pairs of TALEN monomers were linked by a T2A translational skipping sequence to form a complete open reading frame.

Agrobacterium-mediated rice transformation

The TALEN-encoding T-DNA binary vector (pGW3-T-OsBADH2b) was transformed into Agrobacterium tumefaciens strain AGL1 by electroporation. Agrobacterium-mediated transformation of embryogenic calli derived from rice cultivars Nipponbare and LPK was conducted according to Hiei et al. (1994). Hygromycin-containing medium were used for selection and regeneration of transgenic plants. After 3–4 months of cultivation, transgenic seedlings could be transferred to a paddy field during the rice-growing season or to a greenhouse (16-h light at 30 °C/8-h dark at 22 °C).

Biolistic rice transformation for multiplex gene knockout

Particle bombardments were performed using a PDS1000/He particle bombardment system (Bio-Rad, Hercules, CA, USA) with a target distance of 6.0 cm from the staying plate at helium pressure 1100 psi. Plasmid DNAs were mixed at 1 : 1 : 1 (pGW3-T-OsBADH2b: pGW3-T-OsCKX2: pGW3-T-OsDEP1) molar ratios prior to bombardment. One-month-old embryogenic calli (60–80 pieces) of rice cultivar Nipponbare were bombarded using a previously reported protocol (Li et al., 1993). Hygromycin-containing medium was used for selection and regeneration of transgenic plants.

Gas chromatography–mass spectrometry (GC-MS) determination of 2AP content

The basic protocols have been described previously (Chen et al., 2012). Dehulled rice grains (0.5 g) were placed in a 2-mL centrifuge tube and milled thoroughly in a high-speed shaker. The rice flour was transferred to a 5-mL jaw bottle to which was added 2 mL of extraction buffer (a 1 : 1 (v:v) solution of anhydrous ethanol and methylene chloride containing 0.5 mg/L of 2, 4, 6-trimethyl pyridine as an internal standard). The bottle was sealed and extracted at 80 °C for 3 h. It was then cooled to room temperature and centrifuged at 13 800 g for 5 min. The supernatant was pipetted into a sample bottle and 2AP measured with the GC-MS device (GC7890A-5975C MS; Agilent Technologies, Santa Clara, CA, USA). The initial temperature of the DB-5 MS capillary column (30 m × 0.25 mm × 0.25 μm) (J&W) was set to 50 °C; after 2 min at 50 °C, the temperature was first increased to 120 °C at a rate of 5 °C/min and then to 280 °C at a rate of 15 °C/min and maintained for 3 min. The mass spectrometer was operated in the electron impact (EI) mode with an ionization voltage of 70 eV and an ion source temperature of 230 °C. 2AP content was calculated from the equation:

$$C2 = (A2 \times C1 \times V \times 0.001)/(A1 \times W),$$

where C2 is 2AP content (mg/kg); C1, TMP concentration (mg/L); V, volume of sample injected; A1, peak area of TMP; and A2, peak area of 2AP.

PCR/RE assays and PCR-based genotyping of the T-DNA construct

Rice genomic DNA from approximately 0.5–1.0 g of leaf tissue was extracted with a DNA Quick Plant System (Tiangen, Beijing, China). PCR amplification was performed using EASY Taq polymerase (TransGen Biotech, Beijing, China) and 50 ng of genomic DNA. PCR amplicons were then digested by the appropriate restriction enzymes to screen the mutated plants. Undigested bands or PCR products were cloned into the TA cloning vector pUC-T (CWBIO, Beijing, China), and about 10 positive colonies were sequenced. PCR primers for the PCR/RE assay and for testing the presence of the T-DNA construct are listed in Table S2.

Acknowledgements

We thank the Core Facility for Metabolomics (IGDB, CAS) for help with the GC-MS assay and Prof. Shaoyang Lin for providing rice grains of cvs. Daohuaxiang and LPK. This work was funded by grants from the Ministry of Agriculture of China (2014ZX0801003B) and the National Natural Science Foundation of China (31271795 and 31200273).

References

Ashikari, M., Sakakibara, H., Lin, S., Yamamoto, T., Takashi, T., Nishimura, A., Angeles, E.R., Qian, Q., Kitano, H. and Matsuoka, M. (2005) Cytokinin oxidase regulates rice grain production. *Science*, **309**, 741–745.

Bhattacharjee, P., Singhal, R.S. and Kulkarni, P.R. (2002) Basmati rice: a review. *Int. J. Food Sci. Technol.* **37**, 1–12.

Bradbury, L., Fitzgerald, T.L., Henry, R.J., Jin, Q. and Waters, D.L.E. (2005) The gene for fragrance in rice. *Plant Biotechnol. J.* **3**, 363–370.

Bradbury, L., Gillies, S., Brushett, D., Waters, D. and Henry, R. (2008) Inactivation of an aminoaldehyde dehydrogenase is responsible for fragrance in rice. *Plant Mol. Biol.* **68**, 439–449.

Brunet, E., Simsek, D., Tomishima, M., DeKelver, R., Choi, V.M., Gregory, P., Urnov, F., Weinstock, D.M. and Jasin, M. (2009) Chromosomal translocations induced at specified loci in human stem cells. *Proc. Natl Acad. Sci. USA* **106**, 10620–10625.

Carroll, D. (2014) Genome engineering with targetable nucleases. *Annu. Rev. Biochem.* **83**, 409–439.

Chen, S., Yang, Y., Shi, W., Ji, Q., He, F., Zhang, Z., Cheng, Z., Liu, X. and Xu, M. (2008) *Badh2*, encoding betaine aldehyde dehydrogenase, inhibits the biosynthesis of 2-acetyl-1-pyrroline, a major component in rice fragrance. *Plant Cell*, **20**, 1850–1861.

Chen, M., Wei, X., Shao, G., Tang, S., Luo, J. and Hu, P. (2012) Fragrance of the rice grain achieved via artificial microRNA-induced down-regulation of *OsBADH2. Plant Breed.* **131**, 584–590.

Cong, L., Ran, F.A., Cox, D., Lin, S., Barretto, R., Habib, N., Hsu, P.D., Wu, X., Jiang, W., Marraffini, L.A. and Zhang, F. (2013) Multiplex genome engineering using CRISPR/Cas systems. *Science*, **339**, 819–823.

Feng, Z., Zhang, B., Ding, W., Liu, X., Yang, D.-L., Wei, P., Cao, F., Zhu, S., Zhang, F., Mao, Y. and Zhu, J.-K. (2013) Efficient genome editing in plants using a CRISPR/Cas system. *Cell Res.* **23**, 1229–1232.

Feng, Z., Mao, Y., Xu, N., Zhang, B., Wei, P., Yang, D.-L., Wang, Z., Zhang, Z., Zheng, R., Yang, L., Zeng, L., Liu, X. and Zhu, J.-K. (2014) Multigeneration analysis reveals the inheritance, specificity, and patterns of CRISPR/Cas-induced gene modifications in *Arabidopsis. Proc. Natl Acad. Sci. USA* **111**, 4632–4637.

Fine, E.J., Cradick, T.J., Zhao, C.L., Lin, Y. and Bao, G. (2013) An online bioinformatics tool predicts zinc finger and TALE nuclease off-target cleavage. *Nucleic Acids Res.* **42**, e42.

Haun, W., Coffman, A., Clasen, B.M., Demorest, Z.L., Lowy, A., Ray, E., Retterath, A., Stoddard, T., Juillerat, A., Cedrone, F., Mathis, L., Voytas, D.F. and Zhang, F. (2014) Improved soybean oil quality by targeted mutagenesis of the fatty acid desaturase 2 gene family. *Plant Biotechnol. J.* **12**, 934–940.

Hiei, Y., Ohta, S., Komari, T. and Kumashiro, T. (1994) Efficient transformation of rice (*Oryza sativa* L.) mediated by *Agrobacterium* and sequence analysis of the boundaries of the T-DNA. *Plant J.* **6**, 271–282.

Huang, X., Qian, Q., Liu, Z., Sun, H., He, S., Luo, D., Xia, G., Chu, C., Li, J. and Fu, X. (2009) Natural variation at the *DEP1* locus enhances grain yield in rice. *Nat. Genet.* **41**, 494–497.

Jao, L.-E., Wente, S.R. and Chen, W. (2013) Efficient multiplex biallelic zebrafish genome editing using a CRISPR nuclease system. *Proc. Natl Acad. Sci. USA* **110**, 13904–13909.

Jezussek, M., Juliano, B.O. and Schieberle, P. (2002) Comparison of key aroma compounds in cooked brown rice varieties based on aroma extract dilution analyses. *J. Agric. Food Chem.* **50**, 1101–1105.

Juillerat, A., Dubois, G., Valton, J., Thomas, S., Stella, S., Maréchal, A., Langevin, S., Benomari, N., Bertonati, C., Silva, G.H., Daboussi, F., Epinat, J.-C., Montoya, G., Duclert, A. and Duchateau, P. (2014) Comprehensive analysis of the specificity of transcription activator-like effector nucleases. *Nucleic Acids Res.* **42**, 5390–5402.

Kovach, M.J., Calingacion, M.N., Fitzgerald, M.A. and McCouch, S.R. (2009) The origin and evolution of fragrance in rice (*Oryza sativa* L.). *Proc. Natl Acad. Sci. USA* **106**, 14444–14449.

Li, L., Qu, R., Kochko, A., Fauquet, C. and Beachy, R. (1993) An improved rice transformation system using the biolistic method. *Plant Cell Rep.* **12**, 250–255.

Li, T., Liu, B., Spalding, M.H., Weeks, D.P. and Yang, B. (2012) High-efficiency TALEN-based gene editing produces disease-resistant rice. *Nat. Biotechnol.* **30**, 390–392.

Li, J.-F., Norville, J.E., Aach, J., McCormack, M., Zhang, D., Bush, J., Church, G.M. and Sheen, J. (2013a) Multiplex and homologous recombination-mediated genome editing in *Arabidopsis* and *Nicotiana benthamiana* using guide RNA and Cas9. *Nat. Biotechnol.* **31**, 688–691.

Li, W., Teng, F., Li, T. and Zhou, Q. (2013b) Simultaneous generation and germline transmission of multiple gene mutations in rat using CRISPR-Cas systems. *Nat. Biotechnol.* **31**, 684–686.

Li, C., Qi, R., Singleterry, R., Hyle, J., Balch, A., Li, X., Sublett, J., Berns, H., Valentine, M., Valentine, V. and Sherr, C.J. (2014) Simultaneous gene editing by injection of mRNAs encoding transcription activator-like effector nucleases (TALENs) into mouse zygotes. *Mol. Cell. Biol.* **34**, 1649–1658.

Liang, Z., Zhang, K., Chen, K. and Gao, C. (2014) Targeted mutagenesis in *Zea mays* using TALENs and the CRISPR/Cas system. *J. Genet. Genom.* **41**, 63–68.

Lorieux, M., Petrov, M., Huang, N., Guiderdoni, E. and Ghesquière, A. (1996) Aroma in rice: genetic analysis of a quantitative trait. *Theor. Appl. Genet.* **93**, 1145–1151.

Ma, S., Chang, J., Wang, X., Liu, Y., Zhang, J., Lu, W., Gao, J., Shi, R., Zhao, P. and Xia, Q. (2014) CRISPR/Cas9 mediated multiplex genome editing and heritable mutagenesis of *BmKu70* in *Bombyx mori. Sci. Rep.* **4**, 4489. doi:10.1038/srep04489.

Mussolino, C., Morbitzer, R., Lütge, F., Dannemann, N., Lahaye, T. and Cathomen, T. (2011) A novel TALE nuclease scaffold enables high genome editing activity in combination with low toxicity. *Nucleic Acids Res.* **39**, 9283–9293.

Niu, X., Tang, W., Huang, W., Ren, G., Wang, Q., Luo, D., Xiao, Y., Yang, S.,

Wang, F., Lu, B.-R., Gao, F., Lu, T. and Liu, Y. (2008) RNAi-directed downregulation of *OsBADH2* results in aroma (2-acetyl-1-pyrroline) production in rice (*Oryza sativa* L.). *BMC Plant Biol.* **8**, 100.

Ota, S., Hisano, Y., Ikawa, Y. and Kawahara, A. (2014) Multiple genome modifications by the CRISPR/Cas9 system in zebrafish. *Genes Cells*, **19**, 555–564.

Piganeau, M., Ghezraoui, H., De Cian, A., Guittat, L., Tomishima, M., Perrouault, L., René, O., Katibah, G.E., Zhang, L., Holmes, M.C., Doyon, Y., Concordet, J.-P., Giovannangeli, C., Jasin, M. and Brunet, E. (2013) Cancer translocations in human cells induced by zinc finger and TALE nucleases. *Genome Res.* **23**, 1182–1193.

Sakane, Y., Sakuma, T., Kashiwagi, K., Kashiwagi, A., Yamamoto, T. and Suzuki, K.-I.T. (2013) Targeted mutagenesis of multiple and paralogous genes in *Xenopus laevis* using two pairs of transcription activator-like effector nucleases. *Dev. Growth Differ.* **56**, 108–114.

Sakuma, T., Nishikawa, A., Kume, S., Chayama, K. and Yamamoto, T. (2014) Multiplex genome engineering in human cells using all-in-one CRISPR/Cas9 vector system. *Sci. Rep.* **4**, 5400. doi:10.1038/srep05400.

Shan, Q., Wang, Y., Chen, K., Liang, Z., Li, J., Zhang, Y., Zhang, K., Liu, J., Voytas, D.F., Zheng, X., Zhang, Y. and Gao, C. (2013) Rapid and efficient gene modification in rice and *Brachypodium* using TALENs. *Mol. Plant* **6**, 1365–1368.

Shao, G.N., Tang, A., Tang, S.Q., Luo, J., Jiao, G.A., Wu, J.L. and Hu, P.S. (2011) A new deletion mutation of fragrant gene and the development of three molecular markers for fragrance in rice. *Plant Breed.* **130**, 172–176.

Shi, W., Yang, Y., Chen, S. and Xu, M. (2008) Discovery of a new fragrance allele and the development of functional markers for the breeding of fragrant rice varieties. *Mol. Breed.* **22**, 185–192.

Shukla, V.K., Doyon, Y., Miller, J.C., DeKelver, R.C., Moehle, E.A., Worden, S.E., Mitchell, J.C., Arnold, N.L., Gopalan, S., Meng, X.D., Choi, V.M., Rock, J.M., Wu, Y.Y., Katibah, G.E., Zhifang, G., McCaskill, D., Simpson, M.A., Blakeslee, B., Greenwalt, S.A., Butler, H.J., Hinkley, S.J., Zhang, L., Rebar, E.J., Gregory, P.D. and Urnov, F.D. (2009) Precise genome modification in the crop species *Zea mays* using zinc-finger nucleases. *Nature*, **459**, 437–441.

Voytas, D.F. and Gao, C. (2014) Precision genome engineering and agriculture: opportunities and regulatory challenges. *PLoS Biol.* **12**, e1001877.

Wang, H., Yang, H., Shivalila, C.S., Dawlaty, M.M., Cheng, A.W., Zhang, F. and Jaenisch, R. (2013) One-step generation of mice carrying mutations in multiple genes by CRISPR/Cas-mediated genome engineering. *Cell*, **153**, 910–918.

Wang, Y., Cheng, X., Shan, Q., Zhang, Y., Liu, J., Gao, C. and Qiu, J.-L. (2014) Simultaneous editing of three homoeoalleles in hexaploid bread wheat confers heritable resistance to powdery mildew. *Nat. Biotechnol.* **32**, 947–951.

Wendt, T., Holm, P., Starker, C., Christian, M., Voytas, D., Brinch-Pedersen, H. and Holme, I. (2013) TAL effector nucleases induce mutations at a pre-selected location in the genome of primary barley transformants. *Plant Mol. Biol.* **83**, 279–285.

11

Golgi/plastid-type manganese superoxide dismutase involved in heat-stress tolerance during grain filling of rice

Takeshi Shiraya[1,†], Taiki Mori[2,‡], Tatsuya Maruyama[2], Maiko Sasaki[2], Takeshi Takamatsu[2], Kazusato Oikawa[1], Kimiko Itoh[2], Kentaro Kaneko[1], Hiroaki Ichikawa[3] and Toshiaki Mitsui[1,2,*]

[1]Department of Applied Biological Chemistry, Niigata University, Niigata, Japan
[2]Graduate School of Science and Technology, Niigata University, Niigata, Japan
[3]Division of Plant Sciences, National Institute of Agrobiological Sciences, Tsukuba, Japan

*Correspondence

email t.mitsui@agr.niigata-u.ac.jp
[†]Present address: Niigata Crop Research Center, Niigata Agricultural Research Institute, Nagaoka 940-0826, Japan.
[‡]Present address: Kanto Chemical Co., Inc., Tokyo 103-0022, Japan

Keywords: high-temperature tolerance, Golgi, grain quality, *Oryza sativa* L., plastid, superoxide dismutase.

Summary

Superoxide dismutase (SOD) is widely assumed to play a role in the detoxification of reactive oxygen species caused by environmental stresses. We found a characteristic expression of manganese SOD 1 (*MSD1*) in a heat-stress-tolerant cultivar of rice (*Oryza sativa*). The deduced amino acid sequence contains a signal sequence and an *N*-glycosylation site. Confocal imaging analysis of rice and onion cells transiently expressing *MSD1-YFP* showed MSD1-YFP in the Golgi apparatus and plastids, indicating that MSD1 is a unique Golgi/plastid-type SOD. To evaluate the involvement of MSD1 in heat-stress tolerance, we generated transgenic rice plants with either constitutive high expression or suppression of *MSD1*. The grain quality of rice with constitutive high expression of *MSD1* grown at 33/28 °C, 12/12 h, was significantly better than that of the wild type. In contrast, *MSD1*-knock-down rice was markedly susceptible to heat stress. Quantitative shotgun proteomic analysis indicated that the overexpression of *MSD1* up-regulated reactive oxygen scavenging, chaperone and quality control systems in rice grains under heat stress. We propose that the Golgi/plastid MSD1 plays an important role in adaptation to heat stress.

Introduction

Impairment of rice (*Oryza sativa* L.) grain filling under global warming is a major threat facing Asian countries. Daily mean temperatures above 26 °C during the early ripening period of *japonica* rice compromises yields through decreases in grain size and quality (Morita *et al.*, 2004; Peng *et al.*, 2004; Tashiro and Wardlaw, 1991). Perfect grains are fully rounded, transparent and filled with normal starch granules. A chalky appearance reduces commercial value because of increased cracking during polishing (Fitzgerald *et al.*, 2009) and poorer cooking quality (Singh *et al.*, 2003; Tsutsui *et al.*, 2013). Scanning microscope images of chalky areas of grain ripened under heat stress show loosely packed rounded starch granules (Evers and Juliano, 1976; Ishimaru *et al.*, 2009; Tashiro and Wardlaw, 1991). The air spaces among these abnormal starch granules refract light, making the grain appear white. Occasional small pits on the surface of the starch granules suggest attack by starch-degrading enzymes (Iwasawa *et al.*, 2009; Zakaria *et al.*, 2002); the suppression of α-amylase genes improved the quality of rice grains ripened under heat stress (Hakata *et al.*, 2012). It is widely recognized that heat stress lowers the activity of starch synthesis enzymes (Jiang *et al.*, 2003; Umemoto and Terashima, 2002; Yamakawa *et al.*, 2007). Mutants deficient in genes for starch synthesis enzymes exhibited dramatic changes in grain phenotype, including shape and chalkiness (Fujita *et al.*, 2011; Kubo *et al.*, 1999; Nishi *et al.*, 2001; Tanaka *et al.*, 2004). Furthermore, novel factors such as FLOURY ENDOSPERM2 (FLO2), GLUTELIN PRECURSOR MUTANT6 (GLUP6) and GLUTELIN PRECURSOR ACCUMULATION3 (GAP3) have been shown to be involved in the regulation of rice grain size and starch quality (Fukuda *et al.*, 2013; Ren *et al.*, 2014; She *et al.*, 2010). FLO2 contains a tetratricopeptide repeat motif that interacts with late-embryo-genesis and basic helix-loop-helix proteins (She *et al.*, 2010). GLUP6 is a guanine nucleotide exchange factor involved in intracellular transport from the Golgi apparatus to the protein storage vacuole, and the *glup6* mutant accumulates an abnormally large amount of proglutelin (Fukuda *et al.*, 2013). GAP3 is involved in post-Golgi vesicular traffic for vacuolar protein sorting (Ren *et al.*, 2014). In addition, redox regulation may affect seed maturation and quality (Onda and Kawagoe, 2011; Onda *et al.*, 2011). Thus, the mechanism of grain chalkiness caused by heat stress may be highly complex.

Abiotic stresses, including high light, drought, salinity and heat, lead to the accumulation of reactive oxygen species (ROS) such as superoxide (O_2^-), hydroxyl radicals (•OH) and hydrogen peroxide (H_2O_2; Apel and Hirt, 2004). ROS damage multiple cellular components, interfering with lipid peroxidation (Niki *et al.*, 2005), breaking DNA strands (Brawn and Fridovich, 1981) and inactivating enzymes (Fucci *et al.*, 1983). On the other hand, they also serve as signalling molecules, regulating processes including pathogen defence, programmed cell death and stomatal behaviour (Apel and Hirt, 2004). Although ROS are produced predominantly and continuously in chloroplasts, mitochondria and peroxisomes, the production and scavenging of ROS must be strictly controlled in the absence of stress. Enzymatic ROS scavenging mechanisms involve superoxide dismutase (SOD), ascorbate peroxidase, glutathione peroxidase and catalase (Apel and Hirt, 2004).

Superoxide dismutase catalyses the conversion of O_2^- to H_2O_2; it is responsible primarily for defence against oxidative stress. There are three classes of SODs categorized by their metal cofactor: Fe SOD, Mn SOD and Cu/Zn SOD (Fridovich, 1975). Plant SODs have different subcellular localizations. Typically, Mn SOD is localized to the mitochondria, Fe SOD to the plastids and Cu/Zn SOD to the plastids and cytosol (Bowler et al., 1992; Kliebenstein et al., 1998). Peroxisomal and extracellular Cu/Zn SODs also exist (Bueno et al., 1995; Streller and Wingsle, 1994). Numerous attempts have been made to enhance stress tolerance in plants by modifying the production of SOD enzymes. Ectopic production of cytosolic Cu/Zn SOD improved stress tolerance in tobacco (Faize et al., 2011), potato (Perl et al., 1993), sugar beet (Tertivanidis et al., 2004) and plum (Diaz-Vivancos et al., 2013). Overproduction of chloroplastic Cu/Zn SOD, Fe SOD and Mn SOD (fused to a chloroplast transit peptide) also increased stress resistance in tobacco (Badawi et al., 2004; van Camp et al., 1994, 1996; Sen Gupta et al., 1993; Slooten et al., 1995), potato (Perl et al., 1993), sugar beet (Tertivanidis et al., 2004), cotton (Payton et al., 2001) and alfalfa (McKersie et al., 2000). Transgenic rice overproducing cytosolic Cu/Zn SOD from mangrove (Avicennia marina) tolerated drought stress better than untransformed plants (Prashanth et al., 2008). Rice transformed with a yeast mitochondrial Mn SOD fused to the transit peptide of glutamine synthase conferred resistance to salt stress (Tanaka et al., 1999). Furthermore, rice transformed with pea (Pisum sativum) mitochondrial Mn SOD fused to the transit peptide of pea Cu/Zn SOD under the control of an oxidative stress-inducible promoter was more resistant to oxidative stress induced by methyl viologen or polyethylene glycol (Wang et al., 2005).

We have been searching for candidate genes involved in heat-stress tolerance during seed development to improve the formation of normal rice grains under a warming climate. In proteomic analysis, we detected a characteristic expression behaviour of Mn SOD in developing seeds of the heat-resistant cultivar Yukinkomai. This Mn SOD exhibited a unique subcellular localization that has never previously been described in the literature. Here, we report that control of the Golgi/plastid-type Mn SOD1 (MSD1) expression regulates tolerance to heat stress during grain filling of rice.

Results

Identification of Golgi/plastid-type Mn SOD (MSD1)

We examined the heat susceptibilities of three rice cultivars, Yukinkomai, Yukinosei and Todorokiwase, during seed development from 2004 to 2008. The plants were grown in paddy fields with irrigation water at either ambient temperature or 35 °C during the heading, ripening and maturity stages. The daily mean temperature at around the panicles in the warm-water field was 1.4–1.9 °C higher than that in the ambient-water field (25.4 °C). The percentage of damaged grains in Yukinkomai was about 22% in both treatments (Figure 1), indicating that Yukinkomai is tolerant to high temperatures during development. In contrast, that of Todorokiwase increased from 35% to 44%. Yukinosei was intermediate (Figure 1). To search for genes involved in the heat tolerance of Yukinkomai, we used a proteomic approach. As rice is sensitive to heat stress at an early stage of seed

development (Nagata et al., 2004; Satake and Yoshida, 1978), we separated grain proteins of Yukinkomai, Yukinosei and Todorokiwase at 4 days after flowering (DAF) by two-dimensional polyacrylamide gel electrophoresis (2D-PAGE). The separation profiles showed changes in the production of stress-responsive proteins, including heat shock proteins 70 (HSP70) and 16.9 (HSP16.9), 20S proteasome αF, ABA-inducible protein (R40g2), alcohol dehydrogenase (ADH) and MSD1 (Figure 2). In the heat-tolerant Yukinkomai, 20S proteasome αF, ADH and HSP16.9 were up-regulated and R40g2 were down-regulated under heat stress (Figure 2a,b). In the susceptible Todorokiwase, in contrast, HSP70, HSP16.9 and MSD1 were up-regulated (Figure 2e,f). Those in Yukinosei were intermediate (Figure 2c,d). We focused on MSD1, which was characteristically and highly expressed in developing seeds of Yukinkomai in both treatments (Figure 2a,b).

The RiceXPro public microarray database (http://ricexpro.dna. affrc.go.jp/) shows that the MSD1 gene (OsMSD1) is actively expressed throughout the rice plant, particularly in the embryo and endosperm of developing seeds. Our gel-based proteomic analysis of developing seeds supports the view that MSD1 is a major constituent in the seed proteome (Figure 2a,b). OsMSD1 is located in the centre of chromosome 5 (Figure 3a). The cDNA is 901 bp in length, encoding 231 amino acid residues that form a 24.9-kDa precursor protein (Figure 3b). MSD1 is mitochondrial enzyme in both monocots and dicots (Kliebenstein et al., 1998; del Río et al., 2003; White and Scandalios, 1988; Wu et al., 1999). Analyses by the PSORT algorithm (http://psort.hgc.jp/form. html) predicted an N-terminal mitochondrion-targeting sequence in the precursor proteins of MSD1 of Arabidopsis, maize, wheat and pea (Figure 3b). Indeed, pea MSD1 is localized chiefly in mitochondria (del Río et al., 2003). However, the prediction by PSORT and signalP (http://www.cbs.dtu.dk/services/SignalP/) showed that the rice MSD1 precursor's N-terminal sequence potentially acts as signal to the endoplasmic reticulum (ER; Figure 3b; Sakamoto et al., 1993).

To determine the subcellular localization of rice MSD1, we analysed the transient expression of OsMSD1 fused with a gene for yellow fluorescent protein (YFP) in rice and onion epidermal cells, using particle bombardment. In rice cells, confocal laser scanning microscopy showed that the distribution of MSD1-YFP

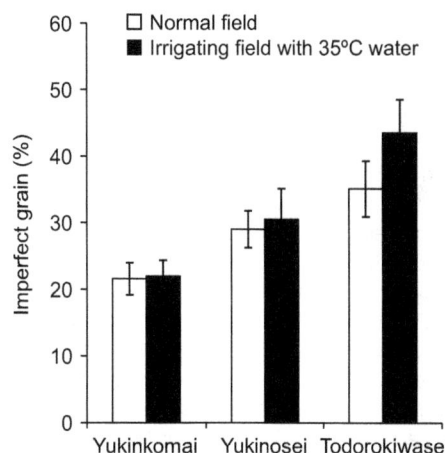

Figure 1 Proportions of imperfect grains of rice cultivars Yukinkomai, Yukinosei and Todorokiwase irrigated with water at ambient temperature (□) or 35 °C (■) from heading to maturity.

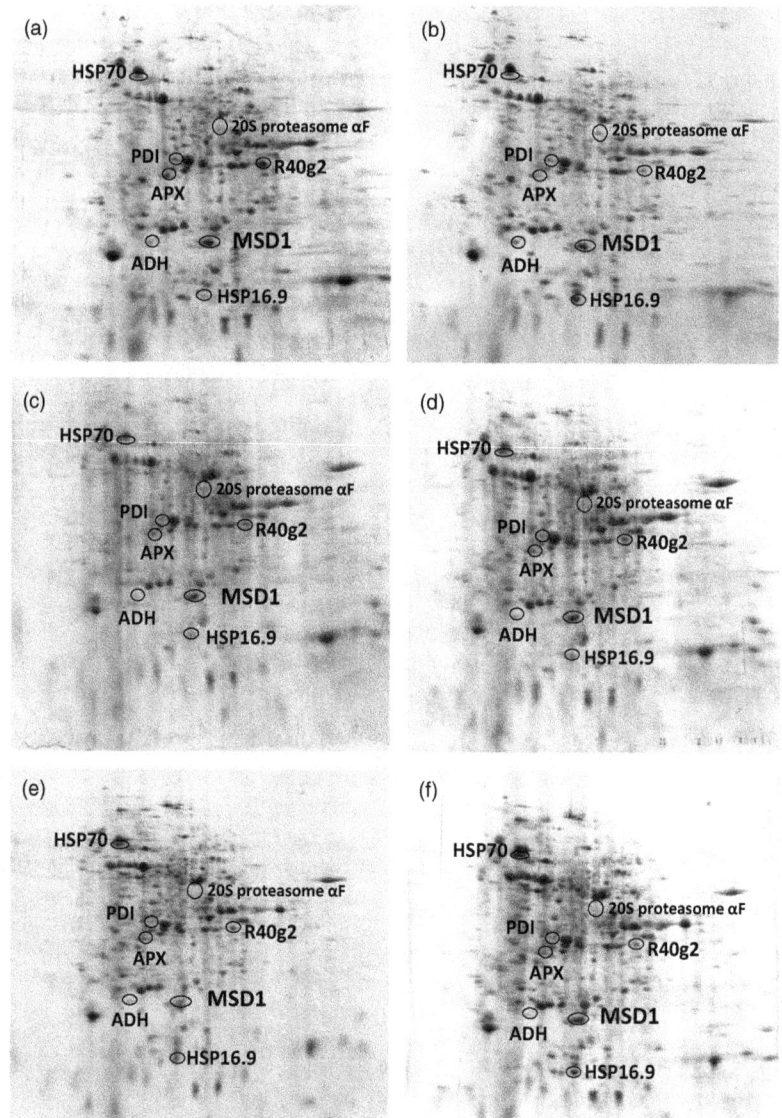

Figure 2 2D-PAGE separation profiles of proteins extracted from 4 days after flowering (DAF) grains of (a, b) Yukinkomai, (c, d) Yukinosei and (e, f) Todorokiwase grown under (a, c, e) normal or (b, d, f) heat stress treatment. Protein extracts were separated by isoelectric focusing followed by SDS-PAGE. Proteins identified included heat shock proteins 70 (HSP70; 66 kDa, pI 4.8), 20S proteasome αF (46 kDa, pI 5.7), protein disulphide isomerase (PDI; 40 kDa, pI 5.3), ABA-inducible protein (R40g2; 38 kDa, pI 6.8), ascorbate peroxidase (APX; 36 kDa, pI 5.2), alcohol dehydrogenase (ADH; 28 kDa, pI 5.0), Mn superoxide dismutase 1 (MSD1; 28 kDa, pI 5.6), HSP16.9 (18 kDa, pI 5.6) were identified in the 2D-gels.

coincided well with the autofluorescence of chloroplasts (Figure 4). In onion cells, MSD1-YFP revealed numerous particulate structures (Figure 5a,c). When *MSD1-YFP* was cobombarded with a sequence encoding a trans-Golgi marker (sialyltransferase, ST) fused at the transmembrane domain to monomeric red fluorescent protein (*ST-mRFP*) into onion cells, MSD1-YFP fluorescence overlapped well with the ST-mRFP-labelled trans-Golgi vesicles (Figure 5a). The GTPases ARF1 and SAR1 are essential for membrane trafficking between the ER and the Golgi apparatus in higher plant cells. Expression of dominant-negative ARF1 or constitutively active SAR1 mutant proteins, which are defective in GTPase cycling, prevents the ER-to-Golgi traffic (Takeuchi *et al.*, 2000, 2002). Golgi-resident proteins and secretory and vacuolar proteins are therefore retained in the ER with such mutants (Takeuchi *et al.*, 2000, 2002). We examined the effects of dominant-negative and constitutive-active mutants of ARF1 and SAR1 on the subcellular distribution of MSD1-YFP. We simultaneously expressed MSD1-YFP, the trans-Golgi marker ST-mRFP, and either AtARF1(T31N), AtARF1(Q71L), or AtSAR1(H74L) in onion cells. Both MSD1-YFP- and ST-mRFP-labelled vesicles were rearranged and remerged into tubular structures, which are

probably part of the ER network, in cells expressing the mutants (Figure 5b).

Recent investigations have revealed the dual targeting of proteins to Golgi apparatus and plastids in *Arabidopsis* (Villarejo *et al.*, 2005), rice (Asatsuma *et al.*, 2005; Chen *et al.*, 2004; Kaneko *et al.*, 2011, 2014; Kitajima *et al.*, 2009; Nanjo *et al.*, 2006) and photosynthetic micro-organisms (van Dooren *et al.*, 2001; Nowack and Grossman, 2012; Sláviková *et al.*, 2006). To test the possibility of plastid-targeting of MSD1, we cobombarded *MSD1-YFP* with a sequence encoding a plastid marker, the transit peptide of Waxy (Klösgen and Weil, 1991) fused to red fluorescent protein from *Discosoma* sp. (*WxTP–DsRed*), into onion cells. MSD1-YFP was notably colocalized with the plastids visualized by WxTP–DsRed (Figure 5c). Simultaneous expression of MSD1-YFP, the plastid marker WxTP–DsRed, and either AtARF1(T31N), AtARF1(Q71L), or AtSAR1(H74L) indicated that the plastid targeting of MSD1-YFP was inhibited in cells expressing the ARF1 and SAR1 mutant proteins (Figure 5d,f). The overall results clearly indicate that MSD1 is a multilocalizing protein that is targeted to the interior of plastids from the Golgi apparatus via the secretory pathway.

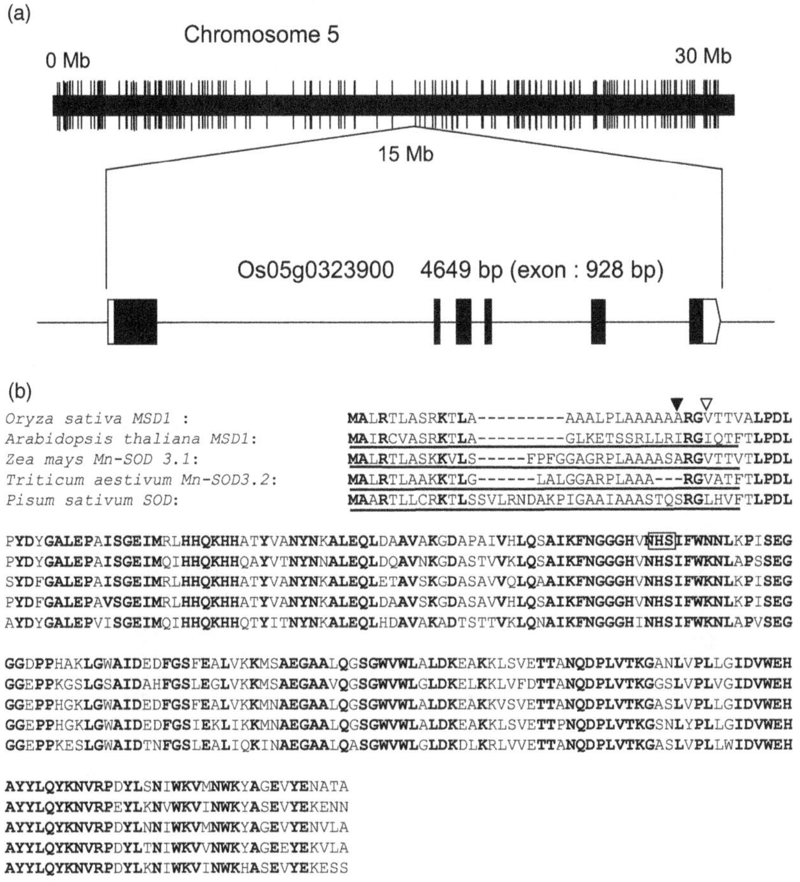

(a)

(b)

```
                                                                    ▼  ▽
Oryza sativa MSD1 :           MALRTLASRKTLA---------AAALPLAAAAAARGVTTVALPDL
Arabidopsis thaliana MSD1:    MAIRCVASRKTLA---------GLKETSSRLLRIRGIQTFTLPDL
Zea mays Mn-SOD 3.1:          MALRTLASKKVLS-----FPFGGAGRPLAAAASARGVTTVTLPDL
Triticum aestivum Mn-SOD3.2:  MALRTLAAKKTLG------LALGGARPLAAA---RGVATFTLPDL
Pisum sativum SOD:           MAARTLLCRKTLSSVLRNDAKPIGAAIAAASTQSRGLHVFTLPDL
```

```
PYDYGALEPAISGEIMRLHHQKHHATYVANYNKALEQLDAAVAKGDAPAIVHLQSAIKFNGGGHVNHSIFWNNLKPISEG
PYDYGALEPAISGEIMQIHHQKHHQAYVTNYNNALEQLDQAVNKGDASTVVKLQSAIKFNGGGHVNHSIFWKNLAPSSEG
SYDFGALEPAISGEIMRLHHQKHHATYVANYNKALEQLETAVSKGDASAVVQLQAAIKFNGGGHVNHSIFWKNLKPISEG
PYDFGALEPAVSGEIMRLHHQKHHATYVANYNKALEQLDAAVSKGDASAVVHLQSAIKFNGGGHVNHSIFWKNLKPISEG
AYDYGALEPVISGEIMQIHHQKHHQTYITNYNKALEQLHDAVAKADTSTTVKLQNAIKFNGGGHINHSIFWKNLAPVSEG
```

```
GGDPPHAKLGWAIDEDFGSFEALVKKMSAEGAALQGSGWVWLALDKEAKKLSVETTANQDPLVTKGANLVPLLGIDVWEH
GGEPPKGSLGSAIDAHFGSLEGLVKKMSAEGAAVQGSGWVWLGLDKELKKLVFDTTANQDPLVTKGGSLVPLVGIDVWEH
GGEPPHGKLGWAIDEDFGSFEALVKKMNAEGAALQGSGWVWLALDKEAKKLSVETTANQDPLVTKGASLVPLLGIDVWEH
GGEPPHGKLGWAIDEDFGSIEKLIKKMNAEGAALQGSGWVWLALDKEAKKLSVETTPNQDPLVTKGSNLYPLLGIDVWEH
GGEPPKESLGWAIDTNFGSLEALIQKINAEGAALQASGWVWLGLDKDLKRLVVETTANQDPLVTKGASLVPLLWIDVWEH
```

```
AYYLQYKNVRPDYLSNIWKVMNWKYAGEVYENATA
AYYLQYKNVRPEYLKNVWKVINWKYASEVYEKENN
AYYLQYKNVRPDYLNNIWKVMNWKYAGEVYENVLA
AYYLQYKNVRPDYLTNIWKVVNWKYAGEEYEKVLA
AYYLQYKNVRPDYLKNIWKVINWKHASEVYEKESS
```

Figure 3 *OsMSD1*. (a) Structure and position of *MSD1* (Os05g0323900) on chromosome 5. Black boxes indicate exons. (b) Alignments of predicted amino acid sequences of the deduced MSD1 proteins of possible orthologous genes from *Oryza sativa*, *Arabidopsis thaliana*, *Zea mays*, *Triticum aestivum* and *Pisum sativum*. Conserved amino acids are boldfaced. Underlines represent mitochondrion-targeting sequence predicted by PSORT. In the rice MSD1 sequence, arrowheads show possible cleavage sites of the signal peptide predicted by (▽) PSORT and (▼) signalP. An *N*-glycosylation site is boxed.

Overexpression and suppression of *MSD1* affect the grain quality of rice ripened under heat stress

To determine the possible stress-adapting function of *MSD1* in ripening seeds of rice, we generated transgenic overexpressor (OE) plants with the maize *Ubiquitin-1* promoter (P*Ubi1*) fused to *MSD1* (MSD1^{OE}) by *Agrobacterium*-mediated transformation. It was reported that P*Ubi1*-controlled genes are highly expressed in various rice tissues (Cornejo *et al.*, 1993). The expression profiles of *MSD1* mRNA in leaves, roots and developing seeds of Nipponbare wild type (WT) and MSD1^{OE} revealed a constitutive high expression of *MSD1* in MSD1^{OE} plants (Figure 6a–c). Furthermore, H$_2$O$_2$ increased in the developing seeds and young seedlings of MSD1^{OE} (Figure 7). The ratio of H$_2$O$_2$ content between MSD1^{OE} and WT seedlings under hot condition revealed that H$_2$O$_2$ formation increased in MSD1^{OE} under heat stress (Figure 7b). When plants were incubated at normal or high temperatures after heading, the ratios of perfect grains harvested were 78% (WT) and 83% (MSD1^{OE}) at 28/23 °C, 77% (WT) and 88% (MSD1^{OE}) at 30/23 °C, and 26% (WT) and 60% (MSD1^{OE}) at 33/28 °C (Figure 6d–f). Under heat stress, the grain quality of MSD1^{OE} was significantly greater than that of WT (Figures 6f and S1). To suppress the expression of *MSD1* in developing seeds, we used a 696-bp fragment of *MSD1* cDNA which contains no sequence of more than 21 nucleotides conserved with other rice SODs to construct RNA interference (RNAi) binary vectors under the control of the promoter of the developing endosperm-specific *Waxy* (P*Wx*) by arranging two identical fragments derived from *MSD1* in a tail-to-tail manner, yielding a vector generating artificial hairpin-structure transcripts (Figure S2). We generated two transgenic knock-down

(KD) rice plants transformed with P*Wx* fused to *MSD1* RNAi, designated Nipponbare MSD1^{KD} and Yukinkomai MSD1^{KD}. Both transformants were grown under heat stress after heading. The expression of *MSD1* mRNA in developing seeds decreased to 18% of WT in Nipponbare MSD1^{KD} and 53% in Yukinkomai MSD1^{KD} (Figure 6g,i), along with significant decreases in the proportion of perfect grains (to 12% and 71%, respectively; Figure 6h,j). The overall results indicate that the constitutive high expression of *MSD1* was involved in maintaining the quality of rice grains produced under heat stress during ripening.

Proteomic characterization of developing seeds of MSD1^{OE} under heat stress

To clarify how constitutive high expression of *MSD1* leads to adaptation to heat stress, we carried out quantitative shotgun proteomic analysis of ripening seeds. Proteins extracted from ripening seeds of Nipponbare WT and MSD1^{OE} grown under control (28/23 °C) and heat stress (33/28 °C) conditions at 4 and 10 DAF were labelled by iTRAQ (isobaric tag for relative and absolute quantitation), followed by tandem mass spectrometry (MS/MS) analysis. Under heat stress, 79 proteins (~6% of all identified proteins), including storage and allergen proteins, were down-regulated and 219 (~16%) were up-regulated in the ripening seeds of MSD1^{OE} relative to WT (Table S1). Under the control condition, however, the characteristic response of MSD1^{OE} did not appear. Under high temperature, scavengers of reactive oxygen species (ROS), including Cu/Zn SOD, peroxiredoxins, thioredoxin, peptide methionine sulfoxide reductase, ascorbate peroxidases, monodehydroascorbate reductase and NADH-ubiquinone oxido-reductase, were markedly up-regulated in MSD1^{OE} relative to WT

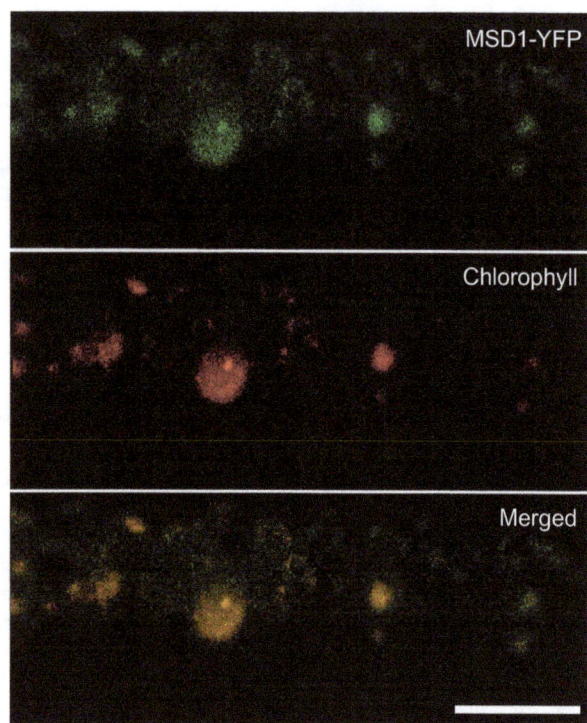

Figure 4 Expression and localization of MSD1-YFP in rice cells. Rice cells bombarded with *MSD1-YFP* were observed by laser scanning microscopy. Top: MSD1-YFP; middle: chlorophyll autofluorescence; bottom: merged. Panels are stacks of 30 images per cell, acquired from the top to the middle of the cell, every 1–2 μm. MSD1-YFP colocalized with chlorophyll autofluorescence. Bar = 10 μm.

(Figure 8 upper panel). Under the control condition, however, changes were minor. Several HSPs, chaperones, chaperonins, calreticulin, proteasome components and S-phase kinase-associated protein 1 were also up-regulated in MSD1^{OE} under heat stress (Figure 8 lower panel), but glutelin, prolamin and allergen family proteins were down-regulated (Figure S3).

Discussion

Identification of Golgi/plastid-type Mn SOD

Generally, Mn SODs are known as mitochondrial enzymes in both monocots and dicots (Kliebenstein *et al.*, 1998; del Río *et al.*, 2003; White and Scandalios, 1988; Wu *et al.*, 1999) and in eukaryotic algae (Kitayama *et al.*, 1999; Wolfe-Simon *et al.*, 2005). However, Mn SOD was localized in the chloroplasts of a marine diatom, *Thalassiosira pseudonana* (Wolfe-Simon *et al.*, 2006). The chloroplastic Mn SOD controlled by the nuclear-encoded *sodA* gene must have plastid/ER transit peptides, but typical transit peptides have not been identified (Wolfe-Simon *et al.*, 2006). As shown in Figure 3b, the deduced amino acid sequence predicted that MSD1 is an extracellular glycoprotein with an *N*-linked oligosaccharide chain. Confocal fluorescent microscopy revealed that rice MSD1 localized in multiple plastids and Golgi apparatus (Figures 4 and 5). Furthermore, the dominant-negative and constitutive-active mutants of ARF1 and SAR1 GTPases arrested the plastid-targeting of MSD1-YFP (Figure 5d), and the MSD1-YFP fluorescence was rearranged into an ER tubular network (Figure 5b). This indicates that MSD1 is transported from the Golgi apparatus via the secretory pathway to the

plastid, as are *Arabidopsis* CAH1 (Burén *et al.*, 2011; Villarejo *et al.*, 2005), *O. sativa* Amyl-1 (Asatsuma *et al.*, 2005; Kitajima *et al.*, 2009) and NPP1 (Kaneko *et al.*, 2011, 2014; Nanjo *et al.*, 2006). This is the first report of the Golgi-to-plastid traffic of noncarbohydrate metabolism-related enzyme.

The electron transport chain in chloroplasts contains several auto-oxidizable enzymes. Ferredoxin in the reduced state can react with oxygen, releasing O_2^- (Asada and Takahashi, 1987) and the aprotic interior of thylakoid membranes also produces O_2^- (Takahashi *et al.*, 1988). The outer layer tissues of developing rice seeds, namely the pericarp and the endosperm, contain chloroplasts during grain filling. Large amounts of starch molecules are synthesized and accumulated in the amyloplasts of endosperm cells. Thus, there is a need for ROS scavenging in the plastids of developing rice seeds. In addition to starches, proteins are also actively synthesized, assembled and stored in developing seeds. Storage proteins such as glutelins are synthesized in the ER and transported via the Golgi apparatus to the protein storage vacuoles (Ren *et al.*, 2014; Washida *et al.*, 2012). The production of H_2O_2 from O_2^- resulting from the maturation of glutelin in the endomembrane system (Onda *et al.*, 2009) suggests the existence of an endomembranous ROS scavenging system.

Overproduction of MSD1 improves quality grain ripened under heat stress

Ectopic production of Golgi/plastid-type MSD1 significantly improved the quality of rice grain ripened under heat stress (Figure 6f). On the other hand, suppression of MSD1 reduced the normal formation of rice grains (Figure 6h,j). These results indicate that the constitutive high expression of Golgi/plastid-type MSD1 is effective for maintaining the formation of perfect grains under heat stress during grain filling. The introduction of yeast *MnSOD* and pea mitochondrial *MnSOD* into chloroplasts of rice conferred tolerance to salt and oxidative stress (Tanaka *et al.*, 1999; Wang *et al.*, 2005). In addition, transgenic rice transformed with mangrove cytosolic *CuIZnSOD* showed better tolerance to drought (Prashanth *et al.*, 2008). We found that enhancement of *OsMSD1* conferred significant tolerance to high temperatures during rice grain filling.

OsMSD1 is located in the centre of chromosome 5 (Figure 3a). Quantitative trait loci (QTLs) controlling grain appearance quality have been identified in populations derived from crosses between *japonica* cultivars (Ebitani *et al.*, 2005; Kobayashi *et al.*, 2007; Tabata *et al.*, 2007), between *japonica* and *indica* cultivars (He *et al.*, 1999; Wan *et al.*, 2005) and between *O. sativa* and *Oryza glaberrima* (Li *et al.*, 2004). The identification of a grain chalkiness QTL (qAPG5-1, Ebitani *et al.*, 2008) close to the position of *MSD1* (Ebitani *et al.*, 2005; Yamakawa *et al.*, 2008) suggests that MSD1 is a determinant of chalkiness.

Proteomic characterization of developing seeds of MSD1^{OE} under heat stress

Quantitative proteomic analysis of ripening seeds of MSD1^{OE} and WT grown in normal and heat-stress conditions at 4 and 10 DAF revealed that 79 proteins were down-regulated and 219 were up-regulated in MSD1^{OE} under heat stress in comparison with WT (Table S1). The ROS scavenging system, molecular chaperones, chaperonins, calreticulin and proteasome components were markedly up-regulated in MSD1^{OE} under high temperature (Figure 8). In contrast, glutelin, prolamin and allergen family proteins were strongly down-regulated (Figure S3). We detected an increase in APX 1, 2 and 4 in the developing seeds of MSD1^{OE}. Monodehydroascorbate reductase, which regenerates ascorbate

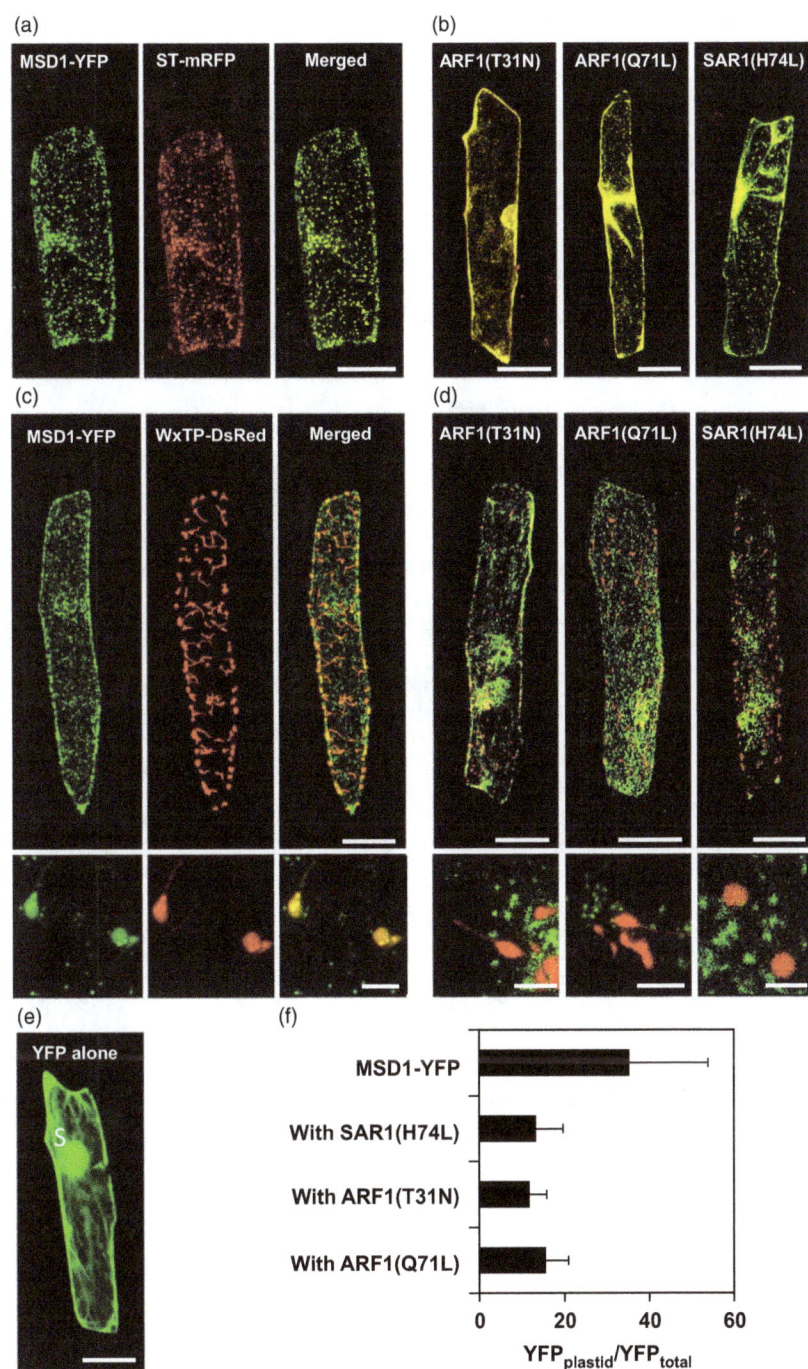

Figure 5 Expression and localization of MSD1-YFP in onion epidermal cells. (a) Onion cells expressing MSD1-YFP and ST-mRFP. Left: MSD1-YFP (green); middle: ST-mRFP (red); right: merged. Panels are stacks of 30 images per cell, acquired from the top to the middle of the cell, every 1–2 μm. MSD1-YFP colocalized with Golgi marker ST-mRFP. Bar = 100 μm. (b) Effects of AtARF1 (T31N), AtARF1(Q71L) and AtSAR1(H74L) on the distribution of MSD1-YFP and ST-mRFP. All images merge YFP with mRFP. Both MSD1-YFP and ST-mRFP were rearranged and remerged into ER tubular structures. (c) Onion cells expressing MSD1-YFP and WxTP-DsRed. Left: MSD1-YFP (green); middle: WxTP-DsRed (red); right: merged. MSD1-YFP overlapped with the plastid marker WxTP-DsRed. Bar = 100 μm. Bottom panels show close-up views of plastids; bar = 5 μm. (d) Effects of AtARF1(T31N), AtARF1(Q71L) and AtSAR1 (H74L) on the distribution of MSD1-YFP and WxTP-DsRed. All images merge YFP with DsRed. MSD1-YFP and WxTP-DsRed were distributed separately in cells. Bars = 100 μm (top) and 5 μm (bottom). (e) Onion cell expressing YFP alone. (f) Proportion of plastid localization of MSD1-YFP in the presence of AtSAR1(H74L), AtARF1(T31N) or AtARF1(Q71L). Values are means ± SD ($n = 8$–11) of ratios of the fluorescence intensity of YFP in the plastid to YFP in the whole cell ($YFP_{plastid}/YFP_{total}$, %).

from monodehydroascorbate, was also up-regulated (Figure 8 upper panel). The enhancement of APX production in rice (Lu et al., 2007; Tanaka et al., 1999) and other plants (Diaz-Vivancos et al., 2013; Faize et al., 2011) confers abiotic stress tolerance. In addition, a series of peroxiredoxins (thioredoxin peroxidases), including 2-Cys peroxiredoxin, were up-regulated in MSD1^{OE} (Figure 8 upper panel). Yeast transformed with O. sativa 2-Cys peroxiredoxin showed increased stress tolerance and fermentation capacity (Kim et al., 2013). Moreover, an HSP was increased in MSD1^{OE} under heat stress (Figure 8 lower panel). In rice (Sato and Yokoya, 2008) and Arabidopsis (Mu et al., 2013), overexpression of small HSPs enhanced tolerance to drought, salt and heat. Overall, these proteomic results and the literature strongly

support the conclusion that MSD1^{OE} rice showed improved adaptability to heat stress.

How is MSD1 involved in the adaptation of MSD1^{OE} to heat stress? We considered that the constitutive high expression of MSD1 immediately converts O_2^- to H_2O_2 under heat stress, and H_2O_2 probably serves as a trigger for enhancing the expression of the ROS scavenging system and HSP genes, as the level of H_2O_2 was higher in MSD1^{OE} than in WT (Figure 7). H_2O_2 is one of the most abundant ROS and is both highly reactive and toxic. However, H_2O_2 also functions as a signalling molecule and activates the MAPK cascade (Apel and Hirt, 2004; Neill et al., 2002). For example, H_2O_2 induced ascorbate peroxidase in embryos of germinating rice (Morita et al., 1999), in Arabidopsis

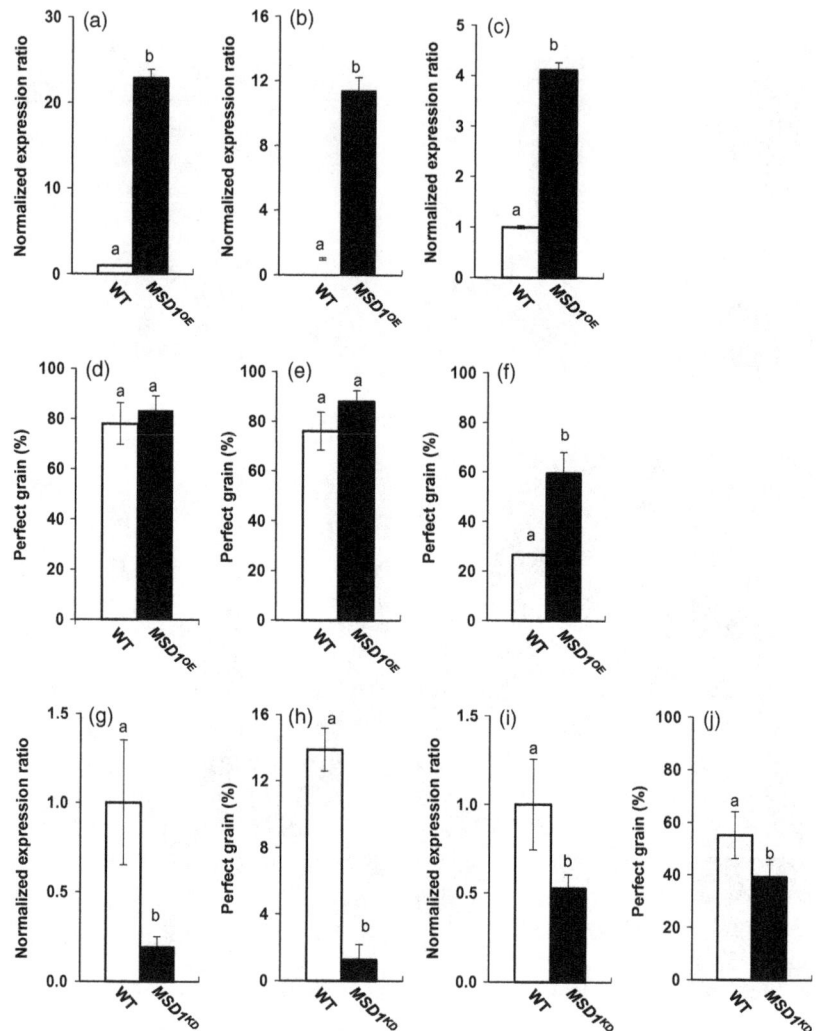

Figure 6 Evaluation of heat-stress tolerance during grain filling of rice with overexpression of *MSD1* (MSD1^{OE}) or developing endosperm-specific suppression of *MSD1* (MSD1^{KD}). (a–c) Expression profiles of *MSD1* mRNA in different organs of Nipponbare wild type (WT) and MSD1^{OE}. (a) Leaf blades and (b) roots at vegetative stage and (c) developing grains at 5 days after flowering (DAF) were harvested and used for fluorescence-based quantitative real-time PCR. Values are means ± SD ($n = 3$). (d–f) Nipponbare WT and MSD1^{OE} plants were incubated under (d) 28/23 °C (12/12 h), (e) 30/23 °C (12/12 h) or (f) 33/28 °C (12/12 h) after heading, and the appearance quality of harvested grains was evaluated. Values are means ± SD ($n = 3–7$) of proportions of perfect grains. (g, h) Nipponbare WT and MSD1^{KD} plants were incubated under 33/28 °C after heading; *MSD1* mRNA in developing grains at 5 DAF was quantified and appearance quality was evaluated ($n = 5$). (i, j) Yukinkomai WT and MSD1^{KD} plants were incubated under 33/28 °C after heading; *MSD1* mRNA in developing grains at 5 DAF was quantified, and appearance quality was evaluated ($n = 3–7$). The ratio of *MSD1* mRNA to 18S rRNA in each WT was set to 1. Columns with the same letter are not significantly different ($P < 0.05$, Student's *t*-test).

leaves (Karpinski *et al.*, 1999) and in tobacco leaves (Gupta *et al.*, 1993) and induced peroxiredoxin in mammalian thyroid cells (Kim *et al.*, 2000). Therefore, induced peroxiredoxin and ascorbate peroxidase likely work as the main regulators of intracellular H_2O_2 concentrations in MSD1^{OE}. Furthermore, heat-stress-induced H_2O_2 was involved in the early stage of activation of heat shock factor (HSF) in *Arabidopsis* cell culture (Volkov *et al.*, 2006). In rice leaves, H_2O_2 treatment induced the production of a chloroplastic small HSP (Lee *et al.*, 2000). Thus, H_2O_2 formed by Golgi/plastid-type MSD1 is the key factor that confers heat tolerance on MSD1^{OE}.

Storage and allergen family proteins were down-regulated in the early developing seeds of MSD1^{OE} under heat stress (Figure S3). The formation of protein bodies in developing seed cells of heat-susceptible Todorokiwase was brought forward by higher temperature (T. M., unpublished data). We infer that the constitutive high expression of Golgi/plastid-type MSD1 controls the redox state in the endomembrane system, leading to the normal programmed formation of protein bodies. Further studies will be needed to confirm this hypothesis.

In conclusion, we found a novel Golgi/plastid-type Mn SOD in developing rice seeds. The ectopic expression of *MSD1* dramatically induced the expression of ROS scavengers, molecular chaperones and the quality control system in developing seeds under heat stress. We consider that the constitutive high

expression of *MSD1* maintains normal grain filling and the production of perfect grains of rice under heat stress.

Experimental procedures

Plasmids

The plasmids used in this study and references describing how they were constructed are listed in Table S2.

Plant materials and growth conditions

Seeds of rice cultivars Yukinkomai, Yukinosei, Todorokiwase and Nipponbare (a model cultivar used for transformant experiments) were obtained from the Niigata Agricultural Research Institute Crop Research Center (Nagaoka city, Niigata, Japan). Transgenic lines of rice (cv. Nipponbare) overexpressing *MSD1* under the control of maize *Ubiquitin-1* constitutive promoter (MSD1^{OE}) were obtained from the full-length cDNA overexpressor (FOX) lines of rice (Nakamura *et al.*, 2007).

Transgenic plants with suppression of the *MSD1* gene in developing seeds were generated as follows: *MSD1* cDNA (bp 1–696) which contains no sequence of more than 21 nucleotides conserved with other rice SODs was amplified by PCR from pOsMSD1 (accession no. AK104160) with a primer set (Table S2) and introduced into pESWA (Islam *et al.*, 2005) to construct the RNAi vector pWX-WB-MSD1-RNAi in combination with the *Wx*

(a)

(b)

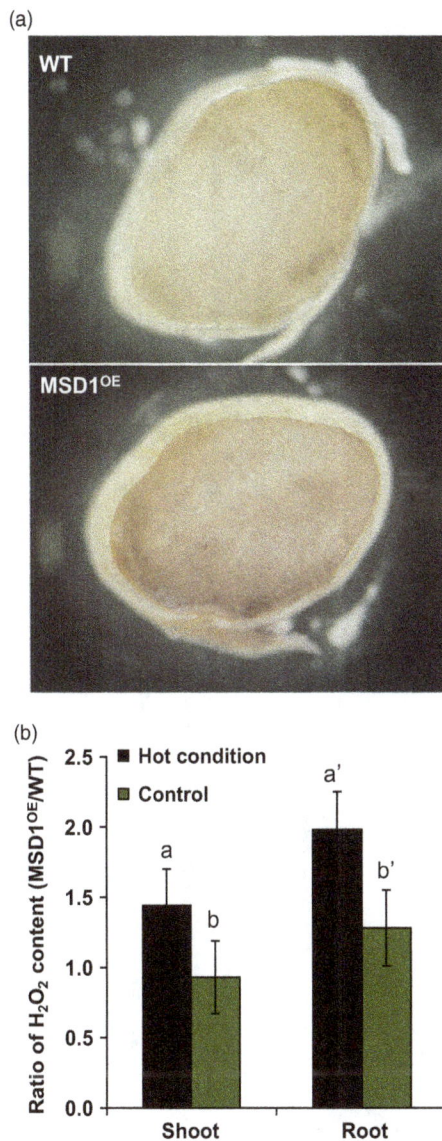

Figure 7 Increase of H_2O_2 formation in developing and germinating seeds of MSD1^{OE}. (a) Developing seeds of Nipponbare WT and MSD1^{OE} at 10 days after flowering (DAF) were stained with diaminobenzidine. (b) H_2O_2 contents in the shoots and roots of WT and MSD1^{OE} seedlings at 7 days after imbibition. Hot condition = 33/28 °C (12/12 h); control condition = 28/23 °C (12/12 h). Values are means ± SD (n = 3–4). Columns with the same letter are not significantly different ($P < 0.05$, Student's t-test).

promoter (Figure S2) using a pENTR Directional TOPO Cloning Kit and the Gateway LR Clonase Enzyme mix (Thermo Fisher Scientific, Waltham, MA). The binary RNAi vector was transformed into *Agrobacterium tumefaciens* strain EHA101 (Hood *et al.*, 1986), and *Agrobacterium*-mediated transformation of rice plants was performed as described by Hiei *et al.* (1994). We generated two transgenic knock-down (KD) rice lines transformed with pWX-WB-MSD1-RNAi, designated Nipponbare MSD1^{KD} and Yukinkomai MSD1^{KD}.

Yukinkomai, Yukinosei and Todorokiwase plants were grown in paddy fields of the Crop Research Center from 2004 to 2008 with ambient temperature or warm water. During the heading,

ripening and maturity stages, the warm water was supplied at 35 °C at a flow rate of 80 L/min, making the daily mean temperature at around the ear 1.4–1.9 °C higher than that of the ambient temperature field (25.4 °C).

Transgenic and wild-type (Nipponbare) plants were grown under 28/23 °C (12 h at 20 000 lx/12 h dark) in a growth chamber (CFH-415; Tomy Seiko, Tokyo, Japan). Grain quality (chalky or translucent) was determined with a rice grain grader (RGQI20A; Satake, Hiroshima, Japan).

Microscopy studies

Yellow fluorescent protein (YFP) is a genetic mutant of green fluorescent protein from *Aequorea victoria*. We constructed pH35GY-OsMSD1-YFP to determine the subcellular localization of rice MSD1. We PCR-amplified *MSD1* from pOsMSD1 (primers in Table S2) and introduced it into pH35GY (Funakoshi Corp, Tokyo, Japan; Kubo *et al.*, 2005) to create pH35GY-OsMSD1-YFP. To obtain pH35GY-(AAGCTT)-YFP (YFP vector alone), we PCR-amplified an *OsEMP70* fragment (bp 1083–1584) from pOsEMP70 (primers in Table S2) and introduced it into pH35GY. The pH35GY-ΔOsEMP70-YFP construct was digested with HindIII to remove the *OsEMP70* fragment and ligated with a Mighty Mix DNA ligation kit (Takara Bio, Ohtsu, Japan).

Construction of pWxTP-DsRed (red fluorescent protein from *Discosoma* sp.; Kitajima *et al.*, 2009), pST-mRFP (monomeric red fluorescent protein; Matsuura-Tokita *et al.*, 2006), pMT121-ARF1 T31N, pMT121-ARF1 Q71L and pMT121-SAR1 H74L (Takeuchi *et al.*, 2000, 2002) were described elsewhere.

To introduce plasmid DNA into rice and onion (*Allium cepa*) epidermal cells, we used the particle bombardment method, using a helium-driven particle accelerator, as described previously (Kitajima *et al.*, 2009). Confocal laser-scanning microscopes (FV300 and FV1000; Olympus, Tokyo, Japan) were used for imaging YFP, DsRed and chlorophyll autofluorescence in rice and onion cells (Kitajima *et al.*, 2009). The FV300 uses an Ar laser at 488 nm to excite YFP and a green He/Ne laser at 543 nm to excite DsRed and chlorophyll. Fluorescence was detected at 510–530 nm through BA510IF and BA530RIF emission filters with an SDM570 emission dichroic mirror (YFP) and at >565 nm through a BA565IF emission filter (DsRed and chlorophyll). The FV1000 uses an Ar laser at 488 nm to excite YFP, and at 559 nm to excite DsRed and chlorophyll. Fluorescence was detected at 510 nm through an SDM560 emission dichroic mirror (YFP) and at 581 nm with an emission dichroic mirror (DsRed and chlorophyll). Images were observed through 40× air-objective (UApo/340, NA 0.90; Olympus) and 100× oil-objective lenses (UPlanSApo, NA 1.40 Oil; Olympus). The fluorescence intensity in plastids and in whole cells was determined using Lumina Vision imaging software. The background was always set at the maximum fluorescence intensity of an area in which no structural image was present. Areas identified by either chlorophyll autofluorescence or WxTP–DsRed were defined as plastids. To evaluate the plastid-targeting abilities of YFP-labelled proteins, we determined the ratio of the fluorescent intensity of YFP in the plastid to that in the whole cell (YFP$_{plastid}$/YFP$_{total}$; Kitajima *et al.*, 2009).

Assay and diaminobenzidine staining for H_2O_2

H_2O_2 assays were carried out according to the procedure of Rao *et al.* (2000) and Xiong *et al.* (2007). Shoots and roots from rice seedlings at 7 days after imbibition were frozen and ground to a powder. Each sample (100 mg) was suspended in 0.5 mL of

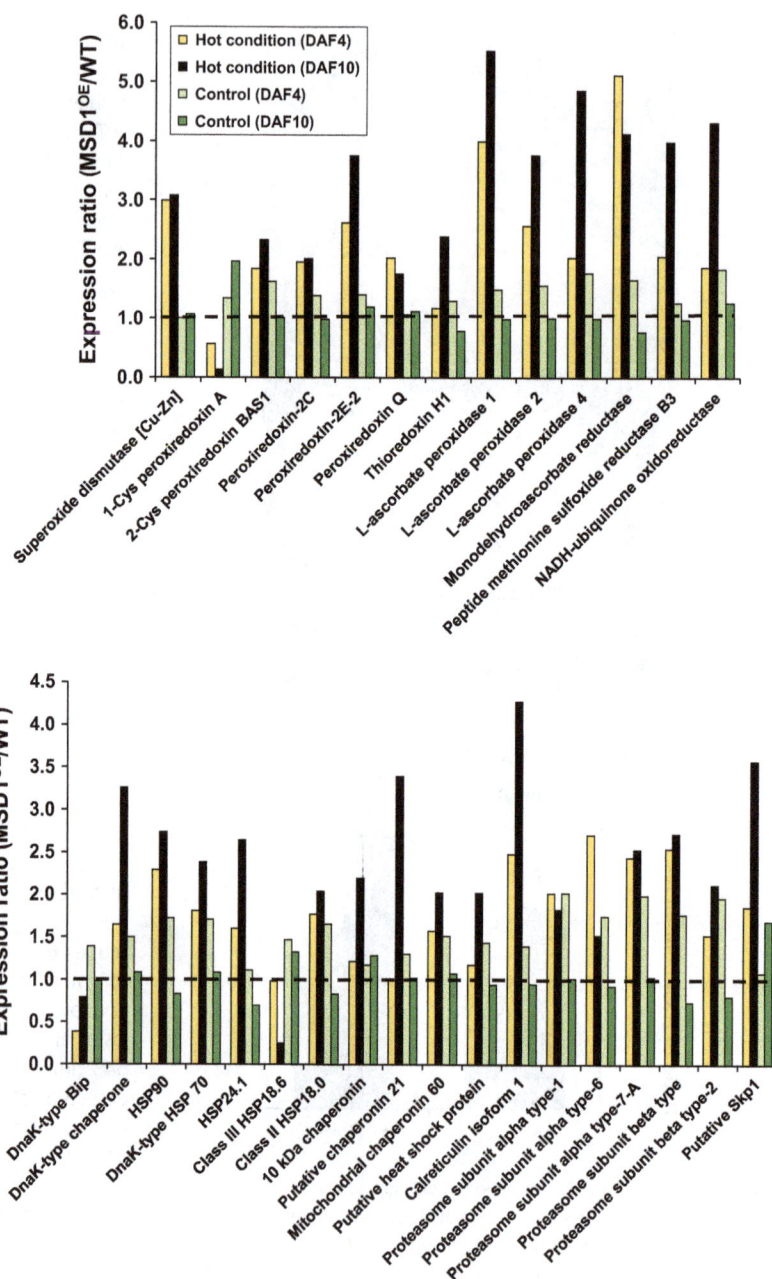

Figure 8 Increase in ROS scavenger, chaperone, quality control and programmed proteolysis systems in the developing seeds of MSD1^{OE} under heat stress. Developing seeds at 4 and 10 days after flowering (DAF) were analysed by quantitative shotgun proteomic analysis with iTRAQ labelling. WT, Nipponbare wild type; MSD1^{OE}, Nipponbare *MSD1* overexpressor. Hot condition = 33/28 °C (12/12 h); control condition = 28/23 °C (12/12 h).

0.2 M HClO$_4$, incubated on ice for 5 min and then centrifuged at 14 000 **g** for 10 min at 4 °C. The supernatant was neutralized with 0.2 M NH$_4$OH (pH 9.5) and centrifuged at 3000 **g** for 2 min. The neutralized extracts were passed through Sep-Pak Light Accell Plus QMA Carbonate columns (Nihon Waters, Tokyo, Japan) and were eluted with 0.5 mL water. H$_2$O$_2$ in the extracts was quantified using an Amplex Red Hydrogen Peroxide–Peroxidase Assay kit (Life Technology Japan, Tokyo, Japan) following the manufacturer's directions. Fluorescence was measured with an RF-5300PC spectrofluorophotometer (Shimadzu Corp., Kyoto, Japan) using excitation at 530 nm and fluorescence detection at 590 nm. For histochemical analysis, developing seeds (14 DAF) were sliced into 1-mm transverse sections and immersed in 20 mM Tris·HCl buffer (pH 6.5) containing 1% (w/v) 3,3'-diaminobenzidine. After vacuum infiltration for 30 min, the samples were incubated at room temperature for 20 h in the

dark to develop a dark-brown colour of diaminobenzidine oxidized by H$_2$O$_2$.

Gel-based proteomics

Grains of Yukinkomai, Yukinosei and Todorokiwase (100 mg) at 4 DAF were extracted with 8 M urea, 1% (w/v) CHAPS detergent, 10 mM ethylene diamine tetraacetic acid (EDTA) and 5 mM phenylmethylsulfonyl fluoride and centrifuged at 10 000 **g** for 10 min at 4 °C. The supernatants were precipitated with 10% (w/v) trichloroacetic acid and resolved with 9 M urea, 3% (w/v) IGEPAL detergent, and 2% (v/v) 2-mercaptoethanol and then used for gel-based proteomics. The procedures of 2D polyacrylamide gel electrophoresis (2D-PAGE) and matrix-assisted laser desorption ionization time-of-flight MS (MALDI-TOF-MS) were essentially identical to the previous reports (Kaneko *et al.*, 2011; Nanjo *et al.*, 2004). In 2D-PAGE, the 1st dimension used isoelectric focusing with ampholine (pH 3.5–10) and the 2nd dimension used sodium dodecyl sulphate (SDS)-PAGE with 16% separating gel. The 2D gels were stained with Coomassie brilliant blue R-250 (Nanjo et al., 2004). The protein spots excised from the gels were digested by trypsin using standard procedures (Awang et al., 2010). MALDI-TOF-MS was carried out with a matrix of α-cyano-4-hyrdoxycinnamic acid in an AXIMA-CFR mass spectrometer (Shimadzu Corp.) and an Autoflex III TOF/TOF mass spectrometer (Bruker BioSpin, Yokohama, Japan; Kaneko et al., 2011).

Quantitative shotgun proteomics

At 4 and 10 DAF, developing seeds of WT and MSD1^{OE} grown under hot (33/28 °C, 12/12 h) or control conditions (28/23 °C, 12/12 h) were used in quantitative shotgun proteomic analysis with iTRAQ labelling. The seeds (0.2 g) were ground in liquid nitrogen to a fine powder and suspended in extraction buffer consisting of 100 mM Tris·HCl (pH 8.0), 20% (w/v) glycerol, 2% (w/v) Triton X-100, 20 mM dithiothreitol, 3 M urea, 2 M thiourea and 3% (w/v) CHAPS. The homogenates were centrifuged at 10 000 g at 4 °C for 5 min. The supernatant was collected and centrifuged again. The supernatants were mixed with 1/10 volume of 100% (w/v) trichloroacetic acid, incubated on ice for 15 min and centrifuged at 10 000 g at 4 °C for 15 min. The resulting protein precipitates were washed 3 times in ice-cold acetone and resuspended in 0.5 M triethylammonium bicarbonate buffer (pH 8.5) containing 0.1% SDS. Protein concentration was determined by the Bradford method (Bio-Rad Laboratories, Hercules, CA) using bovine serum albumin as a standard. Proteins (50 μg) were reduced with tris-(2-carboxyethyl) phosphine at 37 °C for 60 min and then alkalized with methylmethanethiosulfonate at 25 °C for 60 min. Samples were digested in 10 μL of trypsin (1 μg/μL) at 37 °C for 16 h and labelled with 4-plex iTRAQ tags (Thermo Fisher Scientific) according to Fukao et al. (2011), and the resultant 4 iTRAQ-labelled peptide samples were mixed.

For quantitative proteomics, we used a combined KYA DiNA-A (KYA Tech., Tokyo, Japan) and LTQ-Orbitrap XL (Thermo Fisher Scientific) liquid chromatography-MS/MS system. The ionization voltage and capillary transfer temperature at the electrospray ionization nano-stage were set to 1.7–2.5 kV and 200 °C. iTRAQ-labelled peptides were separated in a HiQ sil C18W column (75 μm i.d. ×50 mm, 3 μm particle size; KYA Tech.), using buffers A (0.1% [v/v] acetic acid and 2% [v/v] acetonitrile in water) and B (0.1% [v/v] acetic acid and 80% [v/v] acetonitrile in water). A linear gradient from 0% to 33% B for 240 min, 33% to 100% B for 10 min and back to 0% B over 15 min was applied, and peptides eluted from the column were introduced directly into an LTQ-Orbitrap XL mass spectrometer (Thermo Fisher Scientific, Bremen, Germany) at a flow rate of 300 nL min^{-1} and a spray voltage of 4.5 kV.

Liquid chromatography-MS/MS data were acquired in data-dependent acquisition mode using Xcalibur 2.0 software (Thermo Fisher Scientific). The mass range selected for MS scan was set to 350–1600 m/z, and the top three peaks were subjected to MS/MS analysis. Full MS scan was detected in the Orbitrap, while the MS/MS scans were detected in the linear ion trap and Orbitrap. The normalized collision energy for MS/MS was set to 35 eV for collision-induced dissociation (CID) and 45 eV for higher energy C-trap dissociation (HCD). The resolution of the mass spectrometer (FTMS) was set to 60 000. Divalent or trivalent ions were subjected to MS/MS analysis in dynamic exclusion mode, and proteins were identified with Proteome Discoverer v. 1.1 software

and the SEQUEST search tool (Thermo Fisher Scientific) using the UniProt (http://www.uniprot.org/) O. sativa subsp. japonica database (63 535 proteins) with the following parameters: enzyme, trypsin; maximum missed cleavages site, 2; peptide charge, 2+ or 3+; MS tolerance, 10 ppm; MS/MS tolerance, ±0.8 Da; dynamic modification; carboxymethylation (C); oxidation (H, M, W); iTRAQ 4-plex (K, Y, N-terminus). False discovery rates were <5%.

mRNA analysis

Sample tissues (0.1 g) were ground in liquid nitrogen to fine powder and suspended in an extraction buffer consisting of 2% (w/v) cetyl trimethyl ammonium bromide, 100 mM Tris·HCl (pH 8.0), 20 mM EDTA and 1.4 M NaCl. The homogenates were mixed with 1/2 volumes of phenol and chloroform/isoamyl alcohol (25:24:1, v/v), centrifuged at 10 000 g at 4 °C for 5 min, and total RNA was extracted with RNeasy Plant Mini Kit (Qiagen, Tokyo, Japan) according to the manufacturer's instructions. Ten ng of total RNA was applied to a real-time quantitative reverse transcription PCR using SsoFast Eva Green Supermix (Bio-Rad) and CFX96 real time PCR system/C1000TM Thermal Cycler (Bio-Rad) with the PCR primer sets listed in Table S3. The mRNA contents in each sample were normalized against those of constitutive 18S rRNA gene (Accession no. AK059783).

Acknowledgements

This research was supported by a Grant for Promotion of KAAB Projects (Niigata University), Scientific Research on Innovative Areas (22114507) and Grants-in-Aid for Scientific Research (B) (22380186) from the Ministry of Education, Culture, Sports, Science and Technology, Japan to T. M. This work was also supported in part by a grant from the Ministry of Agriculture, Forestry and Fisheries of Japan (Genomics for Agricultural Innovation, AMR-0001) to H. I. We are indebted to Dr. A. Nakano (The University of Tokyo, Japan) for providing pST-mRFP, pMT121-ARF1 T31N, pMT121-ARF1 Q71L and pMT121-SAR1 H74L.

References

Apel, K. and Hirt, H. (2004) Reactive oxygen species: metabolism, oxidative stress, and signal transduction. Annu. Rev. Plant Biol. 55, 373–399.

Asada, K. and Takahashi, M. (1987) Production and scavenging of active oxygen in chloroplasts. In Photoinhibition (Kyle, D.J., Osmond, C.B. and Arntzen, C.J., eds), pp. 227–287. Amsterdam: Elsevier.

Asatsuma, S., Sawada, C., Itoh, K., Okito, M., Kitajima, A. and Mitsui, T. (2005) Involvement of α-amylase I-1 in starch degradation in rice chloroplasts. Plant Cell Physiol. 46, 858–869.

Awang, A., Karim, R. and Mitsui, T. (2010) Proteomic analysis of Theobroma cacao pod husk. J. Appl. Glycosci. 57, 245–264.

Badawi, G.H., Yamauchi, Y., Shimada, E., Sasaki, R., Kawano, N., Tanaka, K. and Tanaka, K. (2004) Enhanced tolerance to salt stress and water deficit by overexpressing superoxide dismutase in tobacco (Nicotiana tabacum) chloroplasts. Plant Sci. 166, 919–928.

Bowler, C., Van Montagu, M. and Inzé, D. (1992) Superoxide dismutase and stress tolerance. Annu. Rev. Plant Physiol. Plant Mol. Biol. 43, 83–116.

Brawn, K. and Fridovich, I. (1981) DNA strand scission by enzymatically generated oxygen radicals. Arch. Biochem. Biophys. 206, 414–419.

Bueno, P., Varela, J., Gimenez, G.G. and Del Rio, L.A. (1995) Peroxisomal copper, zinc superoxide dismutase: characterization of the isoenzyme from watermelon cotyledons. Plant Physiol. 108, 1151–1160.

Burén, S., Ortega-Villasante, C., Blanco-Rivero, A., Martínez-Bernardini, A., Shutova, T., Shevela, D., Messinger, J., Bako, L., Villarejo, A. and Samuelsson, G. (2011) Importance of post-translational modifications for functionality of a chloroplast-localized carbonic anhydrase (CAH1) in *Arabidopsis thaliana*. *PLoS ONE*, **6**, 1–15.

van Camp, W., Willekens, H., Bowler, C., Van Montagu, M. and Inze, D. (1994) Elevated levels of superoxide dismutase protect transgenic plants against ozone damage. *Bio/Technology*, **12**, 165–168.

van Camp, W., Capiau, K., Van Montagu, M., Inze, D. and Slooten, L. (1996) Enhancement of oxidative stress tolerance in transgenic tobacco overproducing Fe-superoxide dismutase in chloroplasts. *Plant Physiol.* **112**, 1703–1714.

Chen, M.H., Huang, L.F., Li, H.M., Chen, Y.R. and Yu, S.M. (2004) Signal peptide-dependent targeting of a rice α-amylase and cargo proteins to plastids and extracellular compartments of plant cells. *Plant Physiol.* **135**, 1367–1377.

Cornejo, M.J., Luth, D., Blankenship, K.M., Anderson, O.D. and Blechl, A.E. (1993) Activity of a maize ubiquitin promoter in transgenic rice. *Plant Mol. Biol.* **23**, 567–581.

Diaz-Vivancos, P., Faize, M., Barba-Espin, G., Faize, L., Petri, C., Hernández, J.A. and Burgos, L. (2013) Ectopic expression of cytosolic superoxide dismutase and ascorbate peroxidase leads to salt stress tolerance in transgenic plums. *Plant Biotechnol. J.* **11**, 976–985.

van Dooren, G.G., Schwartzbach, S.D., Osafune, T. and McFadden, G.I. (2001) Translocation of proteins across the multiple membranes of complex plastids. *Biochim. Biophys. Acta*, **1541**, 34–53.

Ebitani, T., Yamamoto, Y., Yano, M. and Funane, M. (2005) Analysis of quantitative trait loci for grain appearance, Kasalath alleles increase ratio of whole grains against Koshihikari, in rice. *Jpn. J. Crop Sci.* **74**, 290–291.

Ebitani, T., Yamamoto, Y., Yano, M. and Funane, M. (2008) Identification of quantitative trait loci for grain appearance using chromosome segment substitution lines in rice. *Breed. Res.* **10**, 91–99.

Evers, A.D. and Juliano, B.O. (1976) Varietal differences in surface ultrastructure of endosperm cells and starch granules of rice. *Starch*, **28**, 160–166.

Faize, M., Burgos, L., Faize, L., Piqueras, A., Nicolas, E., Barba-Espin, G., Clemente-Moreno, M.J., Alcobendas, R., Artlip, T. and Hernández, J.A. (2011) Involvement of cytosolic ascorbate peroxidase and Cu/Zn-superoxide dismutase for improved tolerance against drought. *J. Exp. Bot.* **62**, 2599–2613.

Fitzgerald, M.A., McCouch, S.R. and Hall, R.D. (2009) Not just a grain of rice: the quest for quality. *Trends Plant Sci.* **14**, 133–139.

Fridovich, I. (1975) Superoxide dismutases. *Annu. Rev. Biochem.* **44**, 147–149.

Fucci, L., Oliver, C.N., Coon, M.J. and Stadtman, E.R. (1983) Inactivation of metabolic enzymes by mixed-function oxidation reaction: possible implication in protein turnover and ageing. *Proc. Natl Acad. Sci. USA*, **80**, 1521–1525.

Fujita, N., Satoh, R., Hayashi, A., Kodama, M., Itoh, R., Aihara, S. and Nakamura, Y. (2011) Starch biosynthesis in rice endosperm requires the presence of either starch synthase I or IIIa. *J. Exp. Bot.* **62**, 4819–4831.

Fukao, Y., Ferjani, A., Tomioka, R., Nagasaki, N., Kurata, R., Nishimori, Y., Fujiwara, M. and Maeshima, M. (2011) iTRAQ analysis reveals mechanisms of growth defects due to excess zinc in *Arabidopsis*. *Plant Physiol.* **155**, 1893–1907.

Fukuda, M., Wen, L., Satoh-Cruz, M., Kawagoe, Y., Nagamura, Y., Okita, T.W., Washida, H., Sugino, A., Ishino, S., Ishino, Y., Ogawa, M., Sunada, M., Ueda, T. and Kumamaru, T. (2013) A guanine nucleotide exchange factor for Rab5 proteins is essential for intracellular transport of the proglutelin from the Golgi apparatus to the protein storage vacuole in rice endosperm. *Plant Physiol.* **162**, 663–674.

Gupta, A.S., Webb, R.P., Holaday, A.S. and Allen, R.D. (1993) Overexpression of superoxide dismutase protects plants from oxidative stress (Induction of ascorbate peroxidase in superoxide dismutase-overexpressing plants). *Plant Physiol.* **103**, 1067–1073.

Hakata, M., Kuroda, M., Miyashita, T., Yamaguchi, T., Kojima, M., Sakakibara, H., Mitsui, T. and Yamakawa, H. (2012) Suppression of α-amylase genes improves quality of rice grain ripened under high temperature. *Plant Biotechnol. J.* **10**, 1110–1117.

He, P., Li, S.G., Qian, Q., Ma, Y.Q., Li, J.Z., Wang, W.M., Chen, Y. and Zhu, L.H. (1999) Genetic analysis of rice grain quality. *Theor. Appl. Genet.* **98**, 502–508.

Hiei, Y., Ohta, S., Komari, T. and Kumashiro, T. (1994) Efficient transformation of rice (*Oryza sativa* L.) mediated by *Agrobacterium* and sequence analysis of the boundaries of the T-DNA. *Plant J.* **6**, 271–282.

Hood, E.E., Helmer, G.L., Fraley, R.T. and Chilton, M.D. (1986) The hypervirulence of *Agrobacterium tumefaciens* A281 is encoded in a region of pTiBo542 outside of T-DNA. *J. Bacteriol.* **168**, 1291–1301.

Ishimaru, T., Horigane, A.K., Ida, M., Iwasawa, N., San-oh, Y.A., Nakazono, M., Nishizawa, N.K., Masumura, T., Kondo, M. and Yoshida, M. (2009) Formation of grain chalkiness and changes in water distribution in developing rice caryopses grown under high-temperature stress. *J. Cereal Sci.* **50**, 166–174.

Islam, S.M.S., Miyazaki, T., Tanno, F. and Itoh, K. (2005) Dissection of gene function by RNA silencing. *Plant Biotechnol.* **22**, 443–446.

Iwasawa, N., Umemoto, T., Hiratsuka, M., Nitta, Y., Matsuda, T. and Kondo, M. (2009) Structural characters of milky-white rice grains caused by high temperature and shading during grain-filling. *Jpn. J. Crop Sci.* **78**, 322–323.

Jiang, H., Dian, W. and Wu, P. (2003) Effect of high temperature on fine structure of amylopectin in rice endosperm by reducing the activity of the starch branching enzyme. *Phytochemistry*, **63**, 53–59.

Kaneko, K., Yamada, C., Yanagida, A., Koshu, T., Umezawa, Y., Itoh, K., Hori, H. and Mitsui, T. (2011) Differential localizations and functions of rice nucleotide pyrophosphatase/phosphodiesterase isozymes 1 and 3. *Plant Biotechnol.* **28**, 69–76.

Kaneko, K., Inomata, T., Masui, T., Koshu, T., Umezawa, Y., Itoh, K., Pozueta-Romero, J. and Mitsui, T. (2014) Nucleotide pyrophosphatase/phosphodiesterase 1 exerts a negative effect on starch accumulation and growth in rice seedlings under high temperature and CO_2 concentration conditions. *Plant Cell Physiol.* **55**, 320–332.

Karpinski, S., Reynolds, H., Karpinska, B., Wingsle, G., Creissen, G. and Mullineaux, P. (1999) Systemic signaling and acclimation in response to excess excitation energy in *Arabidopsis*. *Science*, **284**, 654–657.

Kim, H., Lee, T.H., Park, E.S., Suh, J.M., Park, S.J., Chung, H.K., Kwon, O.Y., Kim, Y.K., Ro, H.K. and Shong, M. (2000) Role of peroxiredoxins in regulating intracellular hydrogen peroxide and hydrogen peroxide-induced apoptosis in thyroid cells. *J. Biol. Chem.* **275**, 18266–18270.

Kim, I.-S., Kim, Y.-S. and Yoon, H.-S. (2013) Expression of salt-induced 2-Cys peroxiredoxin from *Oryza sativa* increases stress tolerance and fermentation capacity in genetically engineered yeast *Saccharomyces cerevisiae*. *Appl. Microbiol. Biotechnol.* **97**, 3519–3533.

Kitajima, A., Asatsuma, S., Okada, H., Hamada, Y., Kaneko, K., Nanjo, Y., Kawagoe, Y., Toyooka, K., Matsuoka, K., Takeuchi, M., *et al.* (2009) The rice α-amylase glycoprotein is targeted from the Golgi apparatus through the secretory pathway to the plastids. *Plant Cell*, **21**, 2844–2858.

Kitayama, K., Kitayama, M., Osafune, T. and Togasaki, R.K. (1999) Subcellular localization of iron and manganese superoxide dismutase in *Chlamydomonas reinhardtii* (*Chlorophyceae*). *J. Phycol.* **35**, 136–142.

Kliebenstein, D.J., Monde, R.A. and Last, R.L. (1998) Superoxide dismutase in Arabidopsis: an eclectic enzyme family with disparate regulation and protein localization. *Plant Physiol.* **118**, 637–650.

Klösgen, R.B. and Weil, J.H. (1991) Subcellular location and expression level of a chimeric protein consisting of the maize waxy transit peptide and the beta-glucuronidase of *Escherichia coli* in transgenic potato plants. *Mol. Gen. Genet.* **225**, 297–304.

Kobayashi, A., Genliang, B., Shenghai, Y. and Tomita, K. (2007) Detection of quantitative trait loci for white-back and basal-white kernels under high temperature stress in japonica rice varieties. *Breed. Sci.* **57**, 107–116.

Kubo, A., Fujita, N., Harada, K., Matsuda, T., Satoh, H. and Nakamura, Y. (1999) The starch-debranching enzymes isoamylase and pullulanase are both involved in amylopectin biosynthesis in rice endosperm. *Plant Physiol.* **121**, 399–409.

Kubo, M., Udagawa, M., Nishikubo, N., Horiguchi, G., Yamaguchi, M., Ito, J., Mimura, T., Fukuda, H. and Demura, T. (2005) Transcription switches for protoxylem and metaxylem vessel formation. *Genes Dev.* **19**, 1855–1860.

Lee, B.H., Won, S.H., Lee, H.S., Miyao, M., Chung, W.I., Kim, I.J. and Jo, J. (2000) Expression of the chloroplast-localized small heat shock protein by oxidative stress in rice. *Gene*, **245**, 283–290.

Li, J., Xiao, J., Grandillo, S., Jiang, L., Wan, Y., Deng, Q., Yuan, L. and McCouch, S.R. (2004) QTL detection for rice grain quality traits using an interspecific backcross population derived from cultivated Asian (*O. sativa* L.) and African (*O. glaberrima* S.) rice. *Genome*, **47**, 697–704.

Lu, Z., Liu, D. and Liu, S. (2007) Two rice cytosolic ascorbate peroxidases differentially improve salt tolerance in transgenic *Arabidopsis*. *Plant Cell Rep.* **26**, 1909–1917.

Matsuura-Tokita, K., Takeuchi, M., Ichihara, A., Mikuriya, K. and Nakano, A. (2006) Live imaging of yeast Golgi cisternal maturation. *Nature*, **441**, 1007–1010.

McKersie, B.D., Murnaghan, J., Jones, K.S. and Bowley, S.R. (2000) Iron superoxide dismutase in transgenic alfalfa increase winter survival without a detectable increase in photosynthetic oxidative stress tolerance. *Plant Physiol.* **122**, 1427–1437.

Morita, S., Kaminaka, H., Masumura, T. and Tanaka, K. (1999) Induction of rice cytosolic ascorbate peroxidase mRNA by oxidative stress, the involvement of hydrogen peroxide in oxidative stress signalling. *Plant Cell Physiol.* **40**, 417–422.

Morita, S., Shiratsuchi, H., Takanashi, J. and Fujita, K. (2004) Effect of high temperature on grain ripening in rice plants. Analysis of the effects of high night and high day temperatures applied to the panicle and other parts of the plant. *Jpn. J. Crop Sci.* **73**, 77–83.

Mu, C., Zhang, S., Yu, G., Chen, N., Li, X. and Liu, H. (2013) Overexpression of small heat shock protein LimHSP16.45 in *Arabidopsis* enhances tolerance to abiotic stresses. *PLoS ONE*, **8**, e82264.

Nagata, K., Takita, T., Yoshinaga, S., Terashima, K. and Fukuda, A. (2004) Effect of air temperature during the early grain-filling stage on grain fissuring in rice. *Jpn. J. Crop Sci.* **73**, 336–342.

Nakamura, H., Hakata, M., Amano, K., Miyao, A., Toki, N., Kajikawa, M., Pang, J., Higashi, N., Ando, S., Toki, S., *et al.* (2007) A genome-wide gain-of function analysis of rice genes using the FOX-hunting system. *Plant Mol. Biol.* **65**, 357–371.

Nanjo, Y., Asatsuma, S., Itoh, K., Hori, H. and Mitsui, T. (2004) Proteomic identification of α-amylase isoforms encoded by RAmy3B/3C from germinating rice seeds. *Biosci. Biotechnol. Biochem.* **68**, 112–118.

Nanjo, Y., Oka, H., Ikarashi, N., Kaneko, K., Kitajima, A., Mitsui, T., Muñoz, F.J., Rodríguez-López, M., Baroja-Fernández, E. and Pozueta-Romero, J. (2006) Rice plastidial *N*-glycosylated nucleotide pyrophosphatase/phosphodiesterase is transported from the ER-Golgi to the chloroplast through the secretory pathway. *Plant Cell*, **18**, 2582–2592.

Neill, S.J., Desikan, R., Clarke, A., Hurst, R.D. and Hancock, J.T. (2002) Hydrogen peroxide and nitric oxide as signalling molecules in plants. *J. Exp. Bot.* **53**, 1237–1247.

Niki, E., Yoshida, Y., Saito, Y. and Noguchi, N. (2005) Lipid peroxidation: mechanisms, inhibition, and biological effects. *Biochem. Biophys. Res. Commun.* **338**, 668–676.

Nishi, A., Nakamura, Y., Tanaka, N. and Satoh, H. (2001) Biochemical and genetic analysis of the effects of *amylose-extender* mutation in rice endosperm. *Plant Physiol.* **127**, 459–472.

Nowack, E.C.M. and Grossman, A.R. (2012) Trafficking of protein into the recently established photosynthetic organelles of *Paulinella chromatophora*. *Proc. Natl Acad. Sci. USA*, **109**, 5340–5345.

Onda, Y. and Kawagoe, Y. (2011) Oxidative protein folding: selective pressure for prolamin evolution in rice. *Plant Signal. Behav.* **6**, 1966–1972.

Onda, Y., Kumamaru, T. and Kawagoe, Y. (2009) ER membrane-localized oxidoreductase Ero1 is required for disulfide bond formation in the rice endosperm. *Proc. Natl Acad. Sci. USA*, **106**, 14156–14161.

Onda, Y., Nagamine, A., Sakurai, M., Kumamaru, T., Ogawa, M. and Kawagoe, Y. (2011) Distinct roles of protein disulfide isomerase and P5 sulfhydryl oxidoreductases in multiple pathways for oxidation of structurally diverse storage proteins in rice. *Plant Cell*, **23**, 210–223.

Payton, P., Webb, R., Kornyeyev, D., Allen, R. and Holaday, A.S. (2001) Protecting cotton photosynthesis during moderate chilling at high light intensity by increasing chloroplastic antioxidant enzyme activity. *J. Exp. Bot.* **52**, 2345–2354.

Peng, S., Huang, J., Sheehy, J.E., Laza, R.C., Visperas, R.M., Zhong, X., Centeno, G.S., Khush, G.S. and Cassman, K.G. (2004) Rice yields decline with higher night temperature from global warming. *Proc. Natl Acad. Sci. USA*, **101**, 9971–9975.

Perl, A., Perl-Terves, R., Galili, S., Aviv, D., Shalgi, E., Makin, S. and Galun, E. (1993) Enhanced oxidative-stress defense in transgenic potato expressing tomato Cu, Zn superoxide dismutases. *Theor. Appl. Genet.* **85**, 568–576.

Prashanth, S.R., Sadhasivam, V. and Parida, A. (2008) Over expression of cytosolic copper/zinc superoxide dismutase from a mangrove plant *Avicennia marina* in indica rice var Pusa Basmati-1 confers abiotic stress tolerance. *Transgenic Res.* **17**, 281–291.

Rao, M.V., Lee, H., Creelman, R.A., Mullet, J.E. and Davis, K.R. (2000) Jasmonic acid signaling modulates ozone-induced hypersensitive cell death. *Plant Cell*, **12**, 1633–1646.

Ren, Y., Wang, Y., Liu, F., Zhou, K., Ding, Y., Zhou, F., Wang, Y., Liu, K., Gan, L., Ma, W., *et al.* (2014) GLUTELIN PRECURSOR ACCUMULATION3 encodes a regulator of post-Golgi vesicular traffic essential for vacuolar protein sorting in rice endosperm. *Plant Cell*, **26**, 410–425.

del Río, L.A., Sandalio, L.M., Altomare, D.A. and Zilinskas, B.A. (2003) Mitochondrial and peroxisomal manganese superoxide dismutase: differential expression during leaf senescence. *J. Exp. Bot.* **54**, 923–933.

Sakamoto, A., Nosaka, Y. and Tanaka, K. (1993) Cloning and sequencing analysis of a complementary DNA for manganese-superoxide dismutase from rice (*Oryza sativa* L.). *Plant Physiol.* **103**, 1477–1478.

Satake, T. and Yoshida, S. (1978) High temperature induced sterility in indica rice at flowering. *Jpn. J. Crop Sci.* **47**, 6–17.

Sato, Y. and Yokoya, S. (2008) Enhanced tolerance to drought stress in transgenic rice plants overexpressing a small heat-shock protein, sHSP17.7. *Plant Cell Rep.* **27**, 329–334.

Sen Gupta, A., Heinen, J.L., Holaday, A.S., Burke, J.J. and Allen, R.D. (1993) Increased resistance in transgenic plants that overexpress chloroplastic Cu/Zn superoxide dismutase. *Proc. Natl Acad. Sci. USA*, **90**, 1629–1633.

She, K.C., Kusano, H., Koizumi, K., Yamakawa, H., Hakata, M., Imamura, T., Fukuda, M., Naito, N., Tsurumaki, Y., Yaeshima, M., *et al.* (2010) A novel factor FLOURY ENDOSPERM2 is involved in regulation of rice grain size and starch quality. *Plant Cell*, **22**, 3280–3294.

Singh, N., Sodhi, N.S., Kaur, M. and Saxena, S.K. (2003) Physico-chemical, morphological, thermal, cooking and textural properties of chalky and translucent rice kernels. *Food Chem.* **82**, 433–439.

Sláviková, S., Vacula, R., Fang, Z., Ehara, T., Osafune, T. and Schwartzbach, S.D. (2006) Homologous and heterologous reconstitution of Golgi to chloroplast transport and protein import into the complex chloroplasts of *Euglena*. *J. Cell Sci.* **118**, 1651–1661.

Slooten, L., Capiau, K., Van Montagu, M., Sybesma, C. and Inze, D. (1995) Factors affecting the enhancement of oxidative stress tolerance in transgenic tobacco overexpressing manganese superoxide dismutase in chloroplasts. *Plant Physiol.* **107**, 737–750.

Streller, S. and Wingsle, G. (1994) *Pinus sylvestris* L. needles contain extracellular CuZn superoxide dismutase. *Planta*, **192**, 195–201.

Tabata, M., Hirabayashi, H., Takeuchi, Y., Ando, I., Iida, Y. and Ohsawa, R. (2007) Mapping of quantitative trait loci for the occurrence of white-back kernels associated with high temperatures during the ripening period of rice (*Oryza sativa* L.). *Breed. Sci.* **57**, 47–52.

Takahashi, M., Shiraishi, T. and Asada, K. (1988) Superoxide production in aprotic interior of chloroplast thylakoids. *Arch. Biochem. Biophys.* **267**, 714–722.

Takeuchi, M., Ueda, T., Sato, K., Abe, H., Nagata, T. and Nakano, A. (2000) A dominant negative mutant of Sar1 GTPase inhibits protein transport from the endoplasmic reticulum to the Golgi apparatus in tobacco and *Arabidopsis* cultured cells. *Plant J.* **23**, 517–525.

Takeuchi, M., Ueda, T., Yahara, N. and Nakano, A. (2002) Arf1 GTPase plays roles in the protein traffic between the endoplasmic reticulum and the Golgi apparatus in tobacco and *Arabidopsis* cultured cells. *Plant J.* **31**, 499–515.

Tanaka, Y., Hibino, T., Hayashi, Y., Tanaka, A., Kishitani, S., Takabe, T., Yokota, S. and Takabe, T. (1999) Salt tolerance of transgenic rice overexpressing yeast mitochondrial Mn-SOD in chloroplasts. *Plant Sci.* **148**, 131–138.

Tanaka, N., Fujita, N., Nishi, A., Satoh, H., Hosaka, Y., Ugaki, M., Kawasaki, S. and Nakamura, Y. (2004) The structure of starch can be manipulated by changing the expression levels of starch branching enzyme IIb in rice endosperm. *Plant Biotechnol. J.* **2**, 507–516.

Tashiro, T. and Wardlaw, I.F. (1991) The effect of high temperature on kernel dimensions and the type and occurrence of kernel damage in rice. *Aust. J. Agric. Res.* **42**, 485–496.

Tertivanidis, K., Goudoula, K., Vasilikiotis, K., Hassiotou, E. and Perl-Treves, R. (2004) Superoxide dismutase transgenes in sugarbeets confer resistance to oxidative agents and the fungus *C. beticola*. *Transgenic Res.* **13**, 225–233.

Tsutsui, K., Kaneko, K., Hanashiro, I., Nishinari, K. and Mitsui, T. (2013) Characteristics of opaque and translucent parts of high temperature stressed grains of rice. *J. Appl. Glycosci.* **60**, 61–67.

Umemoto, T. and Terashima, K. (2002) Activity of granule-bound synthase is an important determinant of amylose content in rice endosperm. *Funct. Plant Biol.* **29**, 1121–1124.

Villarejo, A., Burén, S., Larsson, S., Déjardin, A., Monné, M., Rudhe, C., Karlsson, J., Jansson, S., Lerouge, P., Rolland, N., *et al.* (2005) Evidence for a protein transported through the secretory pathway en route to the higher plant chloroplast. *Nat. Cell Biol.* **7**, 1224–1231.

Volkov, R.A., Panchuk, I.I., Mullineaux, P.M. and Schöffl, F. (2006) Heat stress-induced H_2O_2 is required for effective expression of heat shock genes in *Arabidopsis*. *Plant Mol. Biol.* **61**, 733–746.

Wan, X.Y., Wan, J.M., Weng, J.F., Jiang, L., Bi, J.C., Wang, C.M. and Zhai, H.Q. (2005) Stability of QTLs for rice grain dimension and endosperm chalkiness characteristics across eight environments. *Theor. Appl. Genet.* **110**, 1334–1346.

Wang, F.-Z., Wang, Q.-B., Kwon, S.-Y., Kwak, S.-S. and Su, W.-A. (2005) Enhanced drought tolerance of transgenic rice plants expressing a pea manganese superoxide dismutase. *J. Plant Physiol.* **162**, 465–472.

Washida, H., Sugino, A., Doroshenk, K.A., Satoh-Cruz, M., Nagamine, A., Katsube-Tanaka, T., Ogawa, M., Kumamaru, T., Satoh, H. and Okita, T.W. (2012) RNA targeting to a specific ER sub-domain is required for efficient transport and packaging of α-globulins to the protein storage vacuole in developing rice endosperm. *Plant J.* **70**, 471–479.

White, J.A. and Scandalios, J.G. (1988) Isolation and characterization of a cDNA for mitochondrial manganese superoxide dismutase (SOD-3) of maize and its relation to other manganese superoxide dismutases. *Biochim. Biophys. Acta*, **951**, 61–70.

Wolfe-Simon, F., Grzebyk, D., Schofield, O. and Falkowski, P.G. (2005) The role and evolution of superoxide dismutases in algae. *J. Phycol.* **41**, 453–465.

Wolfe-Simon, F., Starovoytov, V., Reinfelder, J.R., Schofield, O. and Falkowski, P.G. (2006) Localization and role of manganese superoxide dismutase in a marine diatom. *Plant Physiol.* **142**, 1701–1709.

Wu, G., Wilen, R.W., Robertson, A.J. and Gusta, L.V. (1999) Isolation, chromosomal localization, and differential expression of mitochondrial manganese superoxide dismutase and chloroplastic copper/zinc superoxide dismutase genes in wheat. *Plant Physiol.* **120**, 513–520.

Xiong, Y., Contento, A.L., Nguyen, P.Q. and Bassham, D.C. (2007) Degradation of oxidized proteins by autophagy during oxidative stress in *Arabidopsis*. *Plant Physiol.* **143**, 291–299.

Yamakawa, H., Hirose, T., Kuroda, M. and Yamaguchi, T. (2007) Comprehensive expression profiling of rice grain filling-related genes under high temperature using DNA microarray. *Plant Physiol.* **144**, 258–277.

Yamakawa, H., Ebitani, T. and Terao, T. (2008) Comparison between locations of QTLs for grain chalkiness and genes responsive to high temperature during grain filling on the rice chromosome map. *Breed. Sci.* **58**, 337–343.

Zakaria, S., Matsuda, T., Tajima, S. and Nitta, Y. (2002) Effect of high temperature at ripening stage on the reserve accumulation in seed in some rice cultivars. *Plant Prod. Sci.* **5**, 160–168.

12

Construction of a genomewide RNAi mutant library in rice

Lei Wang[1,*], Jie Zheng[1], Yanzhong Luo[1], Tao Xu[1], Qiuxue Zhang[1], Lan Zhang[1], Miaoyun Xu[1], Jianmin Wan[2], Ming-Bo Wang[3], Chunyi Zhang[1] and Yunliu Fan[1]

[1]Biotechnology Research Institute, The National Key Facility for Crop Gene Resources and Genetic Improvement, Chinese Academy of Agricultural Sciences, Beijing, China
[2]Crop Sciences Institute, The National Key Facility for Crop Gene Resources and Genetic Improvement, Chinese Academy of Agricultural Sciences, Beijing, China
[3]CSIRO Plant Industry, Canberra, NSW, Australia

*Correspondence

email wanglei01@caas.cn

Keywords: Rice, RNAi, gene silencing, siRNA, library, hairpin RNA.

Summary

Long hairpin RNA (hpRNA) transgenes are a powerful tool for gene function studies in plants, but a genomewide RNAi mutant library using hpRNA transgenes has not been reported for plants. Here, we report the construction of a hpRNA library for the genomewide identification of gene function in rice using an improved rolling circle amplification-mediated hpRNA (RMHR) method. Transformation of rice with the library resulted in thousands of transgenic lines containing hpRNAs targeting genes of various function. The target mRNA was down-regulated in the hpRNA lines, and this was correlated with the accumulation of siRNAs corresponding to the double-stranded arms of the hpRNA. Multiple members of a gene family were simultaneously silenced by hpRNAs derived from a single member, but the degree of such cross-silencing depended on the level of sequence homology between the members as well as the abundance of matching siRNAs. The silencing of key genes tended to cause a severe phenotype, but these transgenic lines usually survived in the field long enough for phenotypic and molecular analyses to be conducted. Deep sequencing analysis of small RNAs showed that the hpRNA-derived siRNAs were characteristic of Argonaute-binding small RNAs. Our results indicate that RNAi mutant library is a high-efficient approach for genomewide gene identification in plants.

Background

Forward and reverse genetics are powerful approaches for identifying the components involved in various biological processes in plants. With the completion of genomic sequencing for plants such as *Arabidopsis*, rice, maize and soybean, deciphering the function of predicted genes, identifying novel genes and understanding the molecular basis of important agronomic traits have become a major focus. This has driven the concurrent development of large libraries of mutant lines through chemical, physical or insertional mutagenesis (Bolle *et al.*, 2011; Chang *et al.*, 2012; Wang *et al.*, 2013). T-DNA insertion has been the most successful approach for generating mutagenized populations suitable for genetic studies. T-DNA insertion libraries with millions of mutants have been generated in *Arabidopsis* and rice, and these mutant lines are widely used for gene function studies (Ahn *et al.*, 2007; Bolle *et al.*, 2011; Wang *et al.*, 2013). Transposon insertion as well as physical (e.g. γ-radiation) or chemical (e.g. ethyl methanesulphonate and sodium azide) treatments has also been successfully used in mutagenesis studies (Kolesnik *et al.*, 2004; Piffanelli *et al.*, 2007; Till *et al.*, 2007). However, all of these techniques are limited by inherent imperfections. For example, a large mutagenized population must be generated to ensure sufficient genomewide coverage of genes, as mutations occur randomly and often at intergenic and non-coding regions (http://signal.salk.edu/Source/AtTOME_Data_Source.html). Many genes exist as multigene families, and obtaining mutants in which all members of a gene family are mutated is difficult; thus, scientists must commonly use mutant crossing to achieve multiple mutations, but obtaining mutations in essential genes is problematic as a loss of function is lethal to the plant. Additionally, the mapping of mutations induced by chemical or physical means is tedious and expensive, although recent advances in high-throughput genome resequencing have made this easier than before (Austin *et al.*, 2011).

RNA silencing is a eukaryotic gene repression mechanism induced by double-stranded (ds) or hairpin RNA (hpRNA). dsRNAs or hpRNAs are processed by Dicer or Dicer-like (DCL) proteins, generating small interfering RNAs (siRNAs) 20-24 nt in length (Axtell, 2013; Jones- Rhoades *et al.*, 2006; Wang *et al.*, 2012). These siRNAs are loaded onto Argonaute (AGO) proteins to form RNA-induced silencing complexes (RISCs), which cleave the cognate target mRNA (Jones- Rhoades *et al.*, 2006). RNAi has been successfully exploited as a tool in gene function analyses. Short hairpin RNA (shRNA) constructs are particularly effective at inducing silencing in mammalian cells (Shirane *et al.*, 2004), promoting the construction of many shRNA libraries for genomewide gene function analyses. In plants, however, shRNA constructs have not proven to be highly effective; instead, long hpRNA transgenes have been widely used to induce gene silencing or confer virus resistance (Mao *et al.*, 2007; Wang and Waterhouse, 2002), with the frequency of silencing often reaching >70% in a transgenic population (Smith *et al.*, 2000). Therefore, the transformation of plants with a library of long hpRNA constructs targeting all transcripts is likely to generate populations of gene knockout or knockdown lines suited for genomewide gene function analyses.

The rolling circle amplification-mediated hpRNA (RMHR) construction technique permits the rapid, efficient and inexpensive preparation of genomewide long hpRNA expression libraries (Wang *et al.*, 2008). In the present study, we optimized our RMHR system and used it to generate a long hpRNA library targeting rice genes. Subsequently, this long hpRNA library was transformed into rice to generate RNAi mutant library, resulting in a large number of transgenic lines containing hpRNAs against various types of genes and showing various phenotypic changes in comparison with wild-type (WT) plants. This RNAi mutant library therefore provides a high-efficiency tool for genomewide gene function analyses in rice.

Results

Construction of rice hpRNA libraries

Construction of intermediate rice cDNA libraries

Total RNA was extracted separately from various rice tissues (see Materials and Methods) and then mixed in equal amount for cDNA library construction, aiming to cover most rice transcripts. To minimize problems associated with the preferential ligation of shorter cDNA fragments over longer ones, cDNAs 200-400, 400-600 and 600-1000 bp in size were gel-fractionated (Figure 1a) and ligated separately into the vector pBsa2T.

pBsa2T is a modified version of pBsa, which was described previously for hpRNA library construction using the RMHR system (Wang *et al.*, 2008). In addition to two *Bsa*I sites for generating cDNA inserts with asymmetrical termini to prevent self-ligation, pBsa2T contains two *Ahd*I sites separated by the *ccdB* lethal gene sequence (Figure 1b). *Ahd*I digestion creates two 3'-T overhangs, converting pBsa2T into a 'T' vector, which is convenient for cDNA cloning. Using the T-vector ligation strategy, three cDNA libraries were generated: pOs2, pOs4 and pOs6. pOs2 contained 7.2×10^5 clones with inserts that were 200–400 bp in length, pOs4 contained 5.3×10^5 clones with inserts that were 400–600 bp in length, and pOs6 contained 2.0×10^5 clones with inserts that were 600–1000 bp in length. Sequencing of 108

randomly selected clones from the three libraries (36 clones per library) showed that they contained sequences of 92 different genes, indicating that the libraries were of high quality with relatively low sequence redundancy and suitable for hpRNA library construction.

Construction of rice hpRNA libraries using rolling circle amplification

To create circular cDNA suitable for rolling circle amplification by Phi29 polymerase, the three cDNA libraries were digested with *Bsa*I, and the cDNA inserts were ligated with two DNA oligos (mini-hairpin 1 and mini-hairpin 2) that can self-anneal to form hairpin DNA with adhesive termini matching the two different *Bsa*I termini of pBsa2T, respectively (Figure 2a). Mini-hairpin 2 contains a 50-nt spliceable intron sequence from Os01g62100, which serves as a spacer between the sense and antisense sequences of the resulting inverted repeats to stabilize the hpRNA DNA clones in bacteria. The ligation product was amplified by rolling circle amplification to generate multimeric inverted repeat DNA (Figure 2b). This amplification product was digested with *Bam*HI and *Sac*I to release single inverted repeats (Figure 2c), which were inserted downstream of the Ubi promoter in p35S-Ubi, to form three hpRNA libraries, named OsHP2, OsHP4 and OsHP6, respectively. p35S-Ubi contains a 35S-GUS:GFP cassette (Figure 2d), which drives GUS and GFP expression in transgenic cells, allowing for the quick screening of positive transgenic lines.

Quality check of the hpRNA libraries

In total, 72 clones from the three libraries (24 clones per library) were randomly selected and analysed by restriction digestion, and all of them contained inverted repeat DNA. In total, OsHP2, OsHP4 and OsHP6 contained $\sim 1 \times 10^5$, 5×10^5 and 7×10^5 clones, respectively. We further examined the quality of the three hpRNA libraries by sequencing a subset of randomly selected clones. Of 833 clones from the OsHP2 library, 793 (95.1%) had an inverted repeat sequence >100 bp in size (with an average size of 250 bp), covering sequences from 434 genes or gene families. For OsHP4, 1367 of the 1500 clones (91.1%) sequenced had inverted repeat sequences >100 bp in length corresponding to 786 genes or gene families (with an average size of 450 bp). For OsHP6, 1305 of the 1366 clones (95.5%) sequenced had inverted repeat sequences >200 bp in length from 662 genes or gene families (with an average size of 700 bp). The genomic locations of these target genes were distributed relatively evenly across the 12 rice chromosomes (Figure S1); these genes represent a diverse range of functional categories, including biological process, cellular component and molecular function. This finding indicates that the constructed OsHP libraries were of high quality with a good level of gene coverage.

Generation of the transgenic hpRNA population

For plant transformation, the three hpRNA libraries were introduced into *Agrobacterium tumefaciens* EHA105 by electroporation, generating three corresponding *Agrobacterium* libraries (Ag-OsHP2, Ag-OsHP4 and Ag-OsHP6), each containing more than 10^6 clones. The Ag-OsHP4 library was used to transform rice plants to produce a transgenic rice hpRNA population.

More than 6000 transgenic lines were obtained. Histochemical staining of leaves showed that about 3000 transgenic lines were GUS positive (Figure S2a). The hpRNA sequences in the GUS-

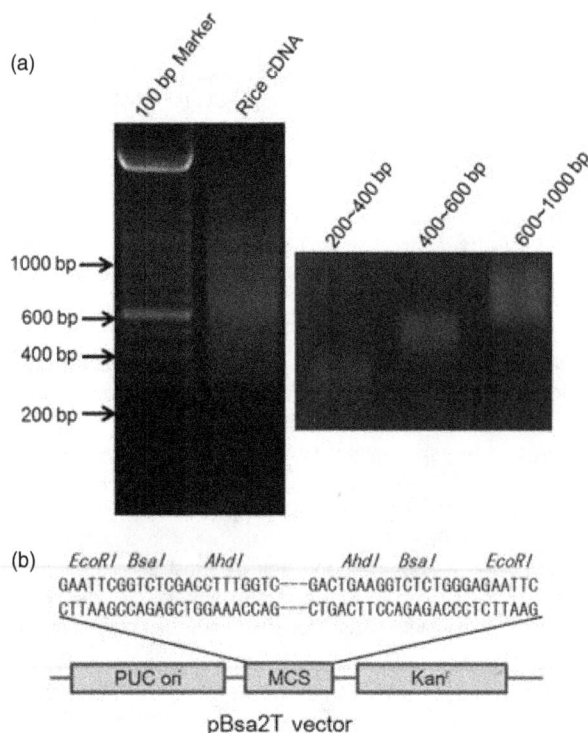

Figure 1 cDNA library construction. (a) Rice cDNAs were gel-purified and used for cDNA library construction. (b) pBsa2T can be converted to linearized, T-ended vector by *Ahd*I digestion.

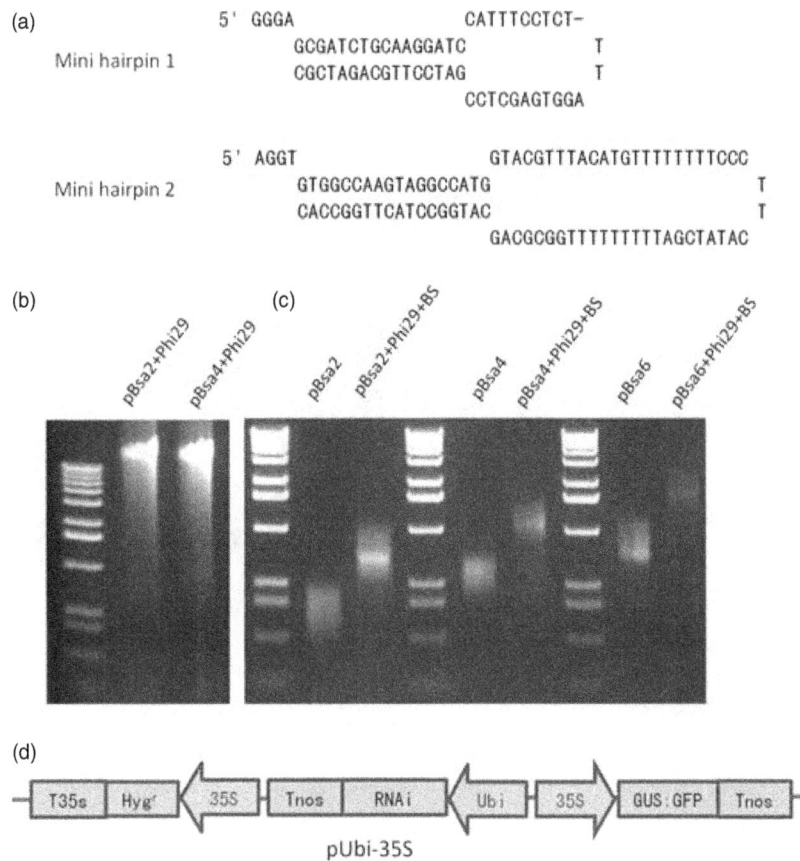

Figure 2 Construction of the hpRNA library. (a) The predicted secondary structures of the mini-hairpin 1 and mini-hairpin 2 oligonucleotides. (b) The closed circle ligation product was amplified with Phi29 DNA polymerase, producing large DNA fragments with multiple-unit inverted repeats of cDNA. (c) The amplified product was digested with *Bam*HI and *Sac*I (BS), producing a pool of single-unit inverted repeats; note that the size of the digested amplification product is approximately double that of the respective unamplified control, indicating the formation of inverted repeats. pBsa2, pBsa4 and pBsa6 represent the insert from the respective cDNA libraries. In pBsa2+Phi29+BS, the ligation product was amplified with Phi29 and digested with *Bam*HI and *Sac*I. (d) Map of the pUbi-35S vector used for hpRNA expression in rice.

positive plants were subsequently amplified by polymerase chain reaction (PCR) and sequenced, demonstrating that the hpRNA expression cassette was integrated into the rice plants (see Figure S2b for an example).

More than 30% of the GUS-positive lines showed poor growth and/or sterility. We obtained seed from about 1000 T_0 lines in ~1500 GUS-positive survival lines. T_1 seed from 211 lines was planted in the field and their phenotypes assessed. A significant proportion (~50%) of the T_1 plants exhibited various abnormal phenotypes, including reduced or increased stature; discoloration or enhanced green colour of leaves; and changes in leaf shape, tiller number or fertility (Table S2). Other phenotypes, including alterations in grain size, spikes and flowering time, as well as an altered stress response, were observed. The severities of the phenotypes varied among the transgenic lines, with the majority showing a relatively mild phenotype (Figure 3a). Many of the hpRNA lines displayed more than one phenotype, consistent with a previous report (Larmande *et al.*, 2008), suggesting that the target genes function in more than one biological process. qRT-PCR corresponding to the target genes in 8 RNAi lines were investigated, which confirmed that the level of the target mRNA was dramatically reduced (Figure 3b).

The proportion of hpRNA lines with an observable phenotype (Table S2; 47.9%) was greater than that of a typical T-DNA insertion lines (3.5%), in which most mutations occur in intergenic (35.4%) and 500-bp upstream (17.51%) regions (Wang *et al.*, 2013). This suggests that the hpRNA library effectively silenced the target genes. It also implies that this type of silencing often does not lead to the complete knockout of target genes, allowing for the recovery of phenotypes associated with essential genes that may not be recoverable using chemical, physical or insertional mutagenesis. Furthermore, compared with the mutants produced via these approaches, the target gene sequences of hpRNA lines can be easily identified by amplification of the hpRNA sequence through PCR (e.g. Figure S2b). Thus, our RNAi mutant library is potentially a powerful tool for gene identification in rice.

siRNAs are processed from hpRNAs to direct target gene silencing

hairpin RNAs are processed by DCLs to generate predominantly 21-nt siRNAs, which in turn guide the degradation of target mRNAs (Jones-Rhoades *et al.*, 2006; Watanabe, 2011). To determine whether siRNAs were produced in the rice hpRNA lines, we first analysed seven transgenic lines by small RNA Northern blot hybridization. Five of the plants showed a clear band ~21 nt in size, indicating that hpRNA was expressed and processed in these plants (Figure 4a). As expected, the abundance of the 21-nt band was variable among the different plants, reflecting different hpRNA expression levels among the transgenic lines.

Figure 3 Representative phenotypes of the hpRNA transgenic plants. (a). Ri2-Ri9 are 3-month-old T1 hpRNA plants. Ri2, dark green and sterile; Ri3, pale green and dwarf; Ri4, pale green with reduced tillering; Ri5, dwarf with reduced tillering; Ri6, pale green dwarf with reduced tillering; Ri7, increased tillering; Ri8, dwarf with reduced tillering; Ri9, increased leaf angle and dwarf. The bottom left corner shows other observed phenotypes, including reduced seed number and spike size, leaf lesions and changed awn length. (b) qRT-PCR analysis of target gene mRNA level in the 8 hpRNA lines shown in (a).

To investigate whether the expression of the target genes was repressed, we analysed 12 hpRNA lines, including the seven plants in (Figure 4a), using real-time reverse transcription (RT)-PCR. As shown in Figure 4b, the mRNA levels were clearly down-regulated in 10 of the plants. The target genes in lines 7-436 and 8-454 were not down-regulated, and correspondingly, siRNA was not detected in these two plants (Figure 4a). Thus, siRNAs from the hpRNAs were directly responsible for the silencing of the genes.

Distribution pattern of hpRNA-derived siRNAs

We investigated the distribution pattern of the hpRNA-derived siRNAs via deep sequencing of small RNAs isolated and pooled from seven hpRNA lines (RNAi #22, #25, #29, #30, #33, #46 and #47). We obtained a total of ~51 million reads of 18- to 30-nt sequences, which contained ~1.7 million reads derived specifically from the hpRNA transgenes.

Size distribution pattern

The size distribution of the total small RNA population was as described previously (Sunkar et al., 2005), with the 24-nt class being the most abundant followed by the 21-, 22- and 23-nt classes (Figure 5a). The size distribution pattern of hpRNA-derived siRNAs was distinct from that of the total small RNAs (Figure 5b), with 21-nt siRNAs being the most dominant (~57%), followed by the 22-nt (~27%) and 24-nt (~9.7%) classes. This is consistent with previous reports showing that hpRNAs are processed primarily by DCL4, but also by DCL2 and DCL3 (Fusaro et al.,

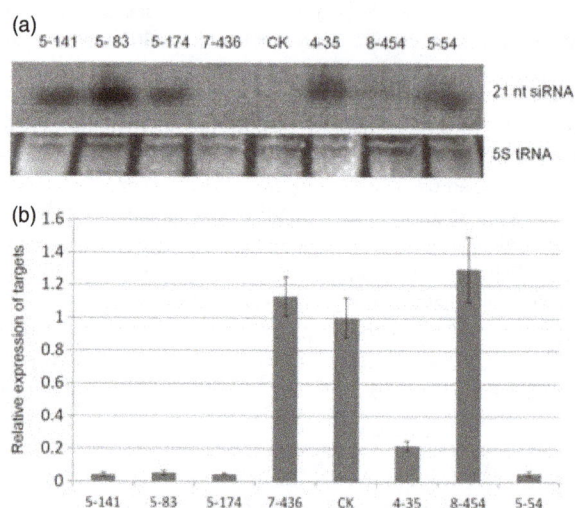

Figure 4 Northern blot analysis of small RNA accumulation (a) and qRT-PCR analysis of target gene expression (b) in transgenic hpRNA plants. CK, plant transformed with empty vector.

2006). In contrast to the total small RNA population, which contained substantial amounts of 18- to 20-, 23- and >25-nt RNA species, the abundance of hpRNA-derived small RNAs belonging to these size classes was low. This indicates that these size classes of small RNAs were not the product of in vitro RNA degradation, which may occur during RNA extraction or deep sequencing;

instead, they were produced by *in vivo* RNase processing. Examining whether these small RNA size classes are functionally important would be interesting, particularly the 20- and 23-nt classes, which are always present in significant amounts in deep sequencing data.

Distribution of siRNAs along the hpRNAs

siRNAs derived from all seven hpRNA lines overlapped with the dsRNA arm of the hpRNA, with few siRNAs corresponding to the loop region (Figure 6a). The siRNAs were unevenly distributed along the dsRNA arm, with some regions showing high abundance (siRNA hot spot) and others with low levels of siRNAs (Figure 6a). The sense siRNA hot spots often did not overlap with the antisense siRNA hot spots (Figure 6a), suggesting that these hot spots were not created by the preferential Dicer processing of specific dsRNA regions, which would otherwise generate hot spots with equal amounts of sense

and antisense siRNAs (Figure 6b). The nucleotide composition analysis showed the 20- to 22-nt siRNAs were enriched for uridine (U) at the 5′ terminus, while the 24-nt siRNAs were enriched for adenine (A) at the 5′ terminus, which is typical of small RNAs associated with AGO1 and AGO4, respectively (Wei *et al.*, 2012). This suggests that the siRNA hot spots resulted from preferential binding with AGO proteins, which may have stabilized the siRNAs.

For hpRNA lines 46, 29 and 30, the numbers of siRNAs with sense and antisense polarities were close in number. However, for the other four lines, siRNAs with a sense polarity were more abundant than those with an antisense polarity (Figure 6c). This was not due to the sequence orientation of the inverted repeats in the hpRNA constructs, of which some had a sense orientation in the front near the promoter, while others had an antisense orientation in the front (Figure 6b). A possible explanation for this is that the presence of the target mRNA increased the turnover

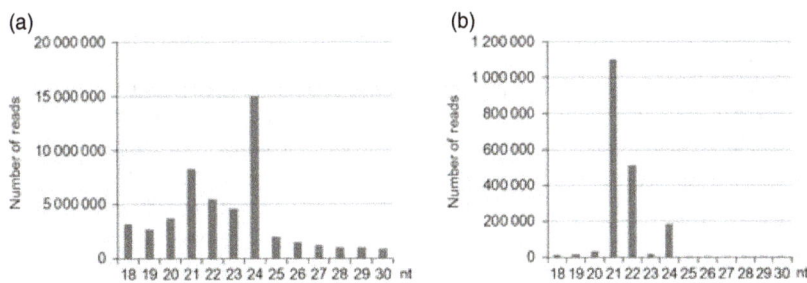

Figure 5 Size distribution of the 18- to 30-nt small RNA population (a) and hpRNA-derived small RNAs (b) in transgenic hpRNA lines obtained using deep sequencing.

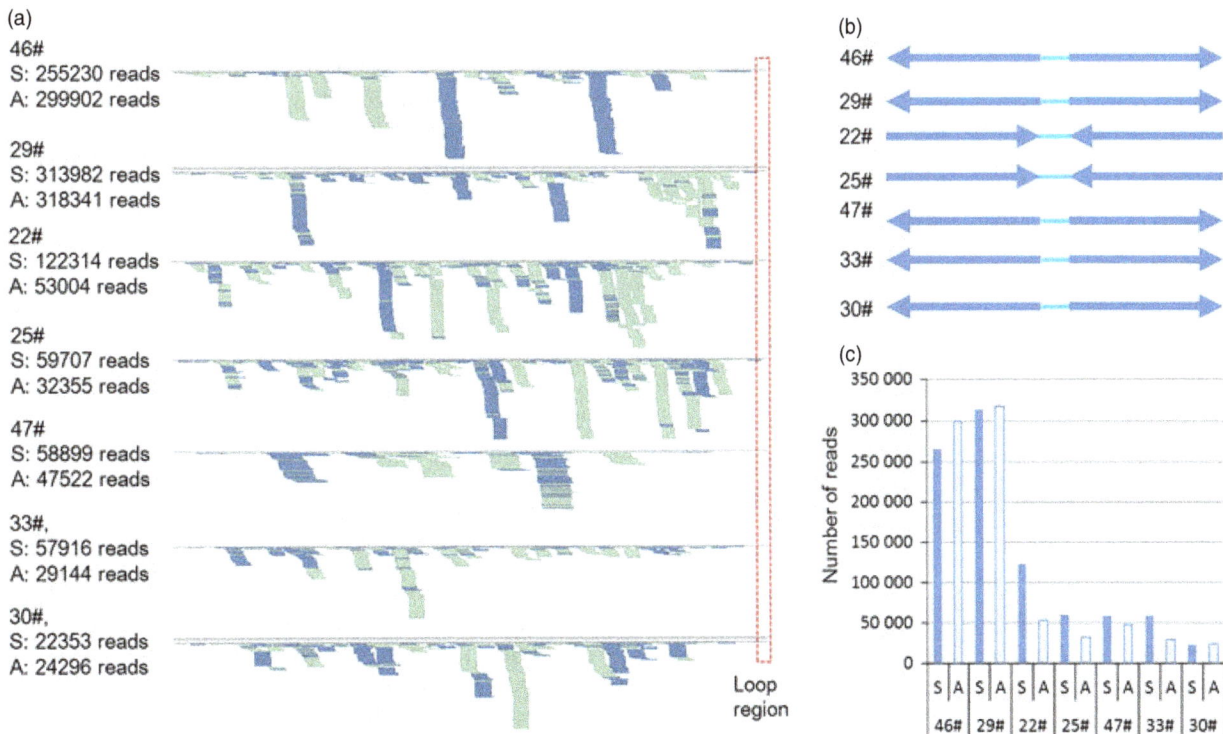

Figure 6 The siRNAs were unevenly distributed along the dsRNA arm of the hpRNA in seven independent hpRNA lines (a). The green lines represent siRNAs with a sense polarity, while the blue lines represent siRNAs with an antisense polarity that have the potential to silence target genes. The loop region, which is framed by a red dashed line, contains a small number of siRNAs. The sequence orientation of the inverted repeats in the seven hpRNA lines analysed was investigated (b). The read numbers of the sense and antisense siRNAs were calculated (c). S, sense siRNA; A, antisense siRNA.

rate of the corresponding small RNAs (i.e. the antisense siRNAs), resulting in the overrepresentation of sense siRNAs. Consistent with this, a recent study on human cells showed that the miRNA decay rate was dramatically enhanced by the presence of highly expressed target genes (Baccarini et al., 2011).

Silencing of genes in multigene families

hpRNA lines 4-68 and 7-120 contained the same hpRNA construct with a 698-bp sequence derived from the open reading frame (ORF) of Os08g0430500, a member of a four-membered gene family that includes Os0280580300, Os11g0546900 and Os03g0710800. The relative nucleotide sequence identities of the ORFs between Os08g0430500 and the other three members are 78.3%, 70.5% and 76.3%, respectively. A close look at the sequence identities revealed perfectly matched sequences between Os08g0430500 and Os0280580300 (two sequences, >21 bp in length [22 and 32 bp]), Os03g0710800 (one sequence, 23 bp in length) and Os11g0546900 (one sequence, 29 bp in length; Figure S3).

The expression level of all four members was investigated in the two hpRNA lines. Our results show that Os08g0430500, Os0280580300 and Os11g0546900 were down-regulated at different levels in the two hpRNA lines (Figure 7a), indicating that the hpRNA derived from Os08g0430500 induced silencing of itself and the related family members Os0280580300 and Os11g0546900. However, the remaining member, Os03g0710800, was not down-regulated. Instead, it was significantly up-regulated (Figure 7a). A possible explanation for this is that the hpRNA did not target Os11g0546900 mRNA effectively due to the existence of only a short stretch (23 bp) of a perfectly matched sequence and that its transcription may be up-regulated in the absence of the other three members.

The hpRNA in line 22 (Figure 8) matched perfectly with the sequence of Os04g44924 from 1 to 748 nt (Figure S4), which is conserved in the four-membered short-chain dehydrogenase/reductase family. The relative nucleotide identities in this region between Os04g44924 and the other three members of the gene family (Os04g44950, Os04g45000 and Os04g44980) are 88.6%, 82% and 66%, respectively. Although many perfectly matched sequences >21 nt in length were identified (Figure S4), only

Os04g44924 was significantly down-regulated. An examination of the small RNA deep sequencing data showed that a total of 157 700 reads mapped to the Os04g44924 transcript, with 20% of these siRNAs (~33 500) being antisense siRNAs with the potential to direct mRNA silencing. In contrast, a much smaller number of antisense siRNAs matched with the Os04g44950 (~6500 reads), Os04g45000 (~900 reads, but this gene showed no expression in leaves) and Os04g44980 (15 reads) transcripts. Thus, for the Os04g44924 gene family, effective silencing may require a threshold level of siRNAs matching the respective transcripts.

Silencing of key genes

Obtaining knockout mutants of some key genes using chemical, physical and insertional mutagenesis is problematic due to the strong deleterious and often lethal phenotypes of the homozygous mutants. In analysing our hpRNA transgenic population, we identified some lines in which key genes for development were targeted. As expected, these plants tended to show severe phenotypes; however, they survived, allowing phenotypic analyses that provided clues as to their function.

hpRNA lines 5-141 and 5-174 contained the same hpRNA construct targeting Os08g0558600, which is homologous to a protein essential for seed maturation (vesicle-associated membrane protein 727) (Ebine et al., 2008). The expression of the gene was dramatically down-regulated in both of the lines (Figure 4b). Correspondingly, the plants showed similarly strong phenotypes, including a reduced tiller number with no heading. Although the plants yielded no seed, they survived in the greenhouse for more than 6 months and could be propagated clonally, allowing for phenotypic and molecular analyses.

The hpRNA in lines 4-68 and 7-120 targeted the four-membered 14-3-3 gene family, resulting in strong repression of three family members (Figure 7a). These hpRNA lines showed strong phenotypes, including the inability to sprout a head, diminished tiller number and infertility. In plants, 14-3-3 proteins have been compared with spiders in a web of phosphorylation (De Boer et al., 2013). These phenotypes suggest that the 14-3-3 gene family is required for rice growth and development.

Discussion

Long hpRNA technology is effective at inducing gene silencing in plants and hence is useful in gene function studies; however, genomewide gene function analyses require a large number of hpRNA constructs targeting >25 000 genes in a typical plant species. Such a large number is difficult to produce using conventional cloning methods involving the PCR amplification of each specific gene sequence followed by single or multistep DNA ligation (Smith et al., 2000). We therefore developed the RMHR method, which can be used to produce a large number of hpRNA constructs simultaneously from cDNA libraries. Using RMHR, we successfully generated a long hpRNA library from an Arabidopsis cDNA population containing known and unknown genes (Wang et al., 2008). To make the RMHR system more efficient, in this study, we introduced two AhdI restriction sites into the intermediate cDNA cloning vector pBsa2T, turning it into a T-ended vector, and a ccdB lethal gene between the two AhdI sites to prevent the recovery of insert-free Escherichia coli clones. This improved vector, called pBsa2T, was highly efficient for cDNA ligation. We also found that the use of loop-specific primers (20 nt) for the rolling circle amplification reaction

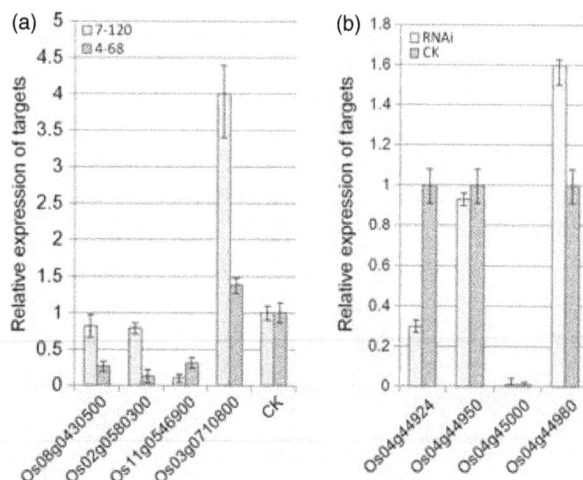

Figure 7 qRT-PCR analysis of target gene expression in the hpRNA lines. (a) The expression of a four-membered gene family was investigated in two RNAi plants. (b) Expression of the short-chain dehydrogenase/reductase family, which consists of four members, was analysed.

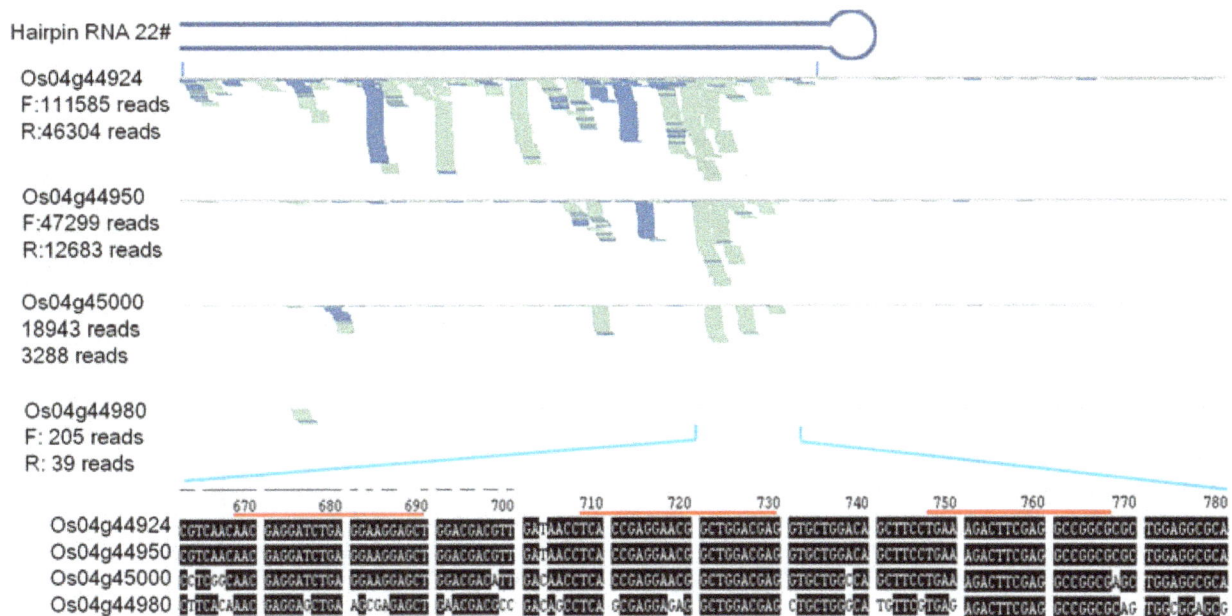

Figure 8 siRNAs derived from the hpRNA of the Os04g44924 sequence in line #22 partially match the other three members of the gene family. The red lines above the sequence alignment on the bottom indicate the three siRNA-rich regions in the four members.

eliminated the production of noninverted repeats, which sometimes occurs when 6-nt random primers are used.

Using this RMHR system, we generated a long hpRNA library with good coverage of rice genes and produced thousands of transgenic hpRNA lines. More than 50% of these transgenic lines displayed visible phenotypes, which is a much higher proportion than that reported for a typical T-DNA insertion or EMS-mutagenized population, of which only ~3.5% usually show visible phenotypes (Hirochika et al., 2004). This result is likely due to several factors. First, all of the hpRNA constructs targeted exon sequences because they were derived from cDNA libraries; this is in contrast to T-DNA insertions or EMS mutations, of which the vast majority occurs in nonexon sequences. Second, a hpRNA derived from a single member of a multigene family may direct the silencing of all or most members, as was the case for the Os08g0430500 family; this resulted in strong phenotypic changes, but such multigene repression is unlikely to occur through T-DNA or EMS mutagenesis. Furthermore, hpRNA-induced gene silencing may be partial in some members of the transgenic population, allowing for the recovery of hpRNA lines in which housekeeping or lethal genes are targeted. The hpRNA constructs used in this study were driven by the maize ubiquitin promoter, which is a strong constitutive promoter. A tissue-specific or inducible promoter could be used in future studies to further improve the recovery of hpRNA lines targeting housekeeping genes.

However, it is worth noting that although hpRNA-induced RNAi is effective for knocking down gene expression, a small amount of targeted mRNA is expected to exist in transgenic cells due to the nature of post-transcriptional gene silencing. Therefore, RNAi may not be suitable for disrupting the function of genes, which requires only a low level of mRNA accumulation. Furthermore, as the use of hpRNA libraries relies on tissue culture and plant transformation, care is needed to distinguish hpRNA-induced phenotypes from those of tissue culture-induced somaclonal variation (Jeong et al., 2002). In the current study,

we observed dominant mutant phenotypes not only in transgenic hpRNA lines but sometimes also in untransformed lines regenerated through tissue culture. However, somaclonal variations can be differentiated from hpRNA-induced phenotypes through examination of cosegregation between a specific phenotype and a hpRNA transgene in a segregating transgenic population.

Experimental procedures

Construction of pBsa2T and pUbi-35S

To produce pBsa2T, two oligonucleotides were synthesized: Bsa-Ahd Fw (5'-GGGAATTCGGTCTCGACCTTTGGTCCAAGCT TTAAATGGTTACTAAAAGCCAGA) and Bsa-Ahd Rv (5'-CCGA ATTCTCCCAGAGACCTTCAGTCTATATTCCCCAGAACAT). Polymerase chain reaction was performed in a total volume of 50 μL consisting of 1× PCR buffer, 0.2 mM dNTPs, 0.25 μM each primer, 2 U of Taq DNA polymerase and 1 ng of pENTR1A, using pENTR1A (Qiagen, Hilden, Germany) as template. The amplification programme consisted of 1 min at 95 °C, 25 cycles of 30 s at 94 °C, 30 s at 56 °C and 30 s at 72 °C, followed by 10 min at 72 °C. The product was digested with EcoRI and cloned into the EcoRI site of pENTR1A to form the intermediate vector pBsa2T. For cDNA cloning, pBsa2T was digested with AhdI, and the larger fragment was gel-purified, generating linearized pBsa2T with T overhangs.

To prepare pUbi-35S, pBI121 was digested with HindIII and EcoRI, and the 35S-GUS-Nos cassette was gel-purified and inserted into the HindIII and EcoRI sites in pCAMBIA1303, generating pC-35S. The Ubi-GUS expression cassette in pAHC25 was excised by HindIII and SacI digestion and inserted into the same sites in pC-35S, forming pUbi-35S.

Rice cDNA library construction

Rice plants (Oryza sativa cv. Nipponbare) were grown in the field. Various tissues were collected, including whole seedlings, roots, stems and leaves at the tillering stage; young panicles and stems at the heading stage; panicles at the flowering stage; and anthers and panicles at the filling stage and at the yellow ripening stage. Total RNA was extracted separately from the tissues using TRIzol reagent (Invitrogen, Carlsbad, CA) and mixed in equal amounts. mRNA was isolated from the total RNA using PolyATtract mRNA Isolation System III (Promega, Madison, WI).

A cDNA library was constructed following the SMART cDNA library manual (Clontech Laboratories Inc., Mountain View, CA) with the following exceptions: LD PCR products were digested with BamHI, SacI, BsaI and XbaI and then treated with T4 DNA polymerase to produce blunt ends. An 'A' tail was added to the 3' terminus using Taq DNA polymerase. The DNA was fractionated using preparative DNA gel electrophoresis after digestion, and fragments 200–400, 400–600 and 600–1000 bp in length were purified and ligated separately into linearized pBsa2T. Ligations with pBsa2T were performed immediately following cDNA size fractionation. Three libraries, pOs2, pOs4 and pOs6, were generated with 1.45×10^6 clones in total.

hpRNA library construction

To generate closed circle cDNA, the pOs2, pOs4 and pOs6 cDNA libraries were digested with BsaI, and the insert was gel-purified and ligated separately with the mini-hairpin 1 (5'- GGGAGCGATC TGCAAGGATCCATTTCCTCTTTAGGTGAGCTCCGATCCTTGCAG ATCGC) and mini-hairpin 2 (5'-GTGGCCAAGTAGGCCATGCTGC GCCAAAAAAAAATCGATATGAAGGGAAAAAAAACATGTAAAC GTACCATGGCCTACTTGGCCACACCT) oligonucleotides, which were preannealed and phosphorylated. The circular DNAs were amplified by rolling circle amplification using loop-specific primers (F: 5'-GGAAAAAAAACATGTAAACG and R: 5'-TTCATATCGATTT TTTTTGG), yielding linear concatemers of inverted repeat DNA (Wang et al., 2008). Following digestion with BamHI and SacI, purified fragments from the three libraries were inserted into pUbi-35S, generating three hpRNA libraries, OsHP2, OsHP4 and OsHP6.

Rice transformation

Rice (O. sativa L. cv. Kitaake) was transformed using seed-derived callus tissue based on the protocol of Hiei et al. (Hiei and Komari, 2008) with modifications. Surface-sterilized seeds were germinated and cultured on NBI medium (NB + 2 mg/L 2,4-D; PhytoTechnology Laboratories, Shawnee Mission, KS). After 7 days of incubation at 30 °C in the dark, the calli were transferred to fresh NBI medium and cultured for 25–30 days at 30 °C. The calli were then harvested and cultured on fresh NBI medium for another 3–4 days. The calli were then incubated for 20 min in A. tumefaciens resuspended ($OD_{600} = 0.3$–0.4) in liquid NBASS medium (NBI + 200 μM acetosyringone + 10 g/L sucrose), dried on sterile filter paper and cocultivated with A. tumefaciens for 3 days at 25 °C on NBASS medium. After cocultivation, the rice calli were rinsed thoroughly (four times) in sterile water containing 50 mg/L timentin and transferred to selective medium (NB + 40 mg/L hygromycin-B, 500 mg/L cefotaxime and 2 mg/L 2,4-D) for 4 weeks at 28 °C. Vigorously growing callus pieces 0.5–1 mm in size were transferred to regeneration medium (NB + 0.5 mg/L NAA, 4 mg/L 6-BA and 40 mg/L hygromycin-B) for 2–3 weeks at 28 °C under 5000 Lux of light. Shoots arising from the calli on regeneration medium were transferred to rooting media (½ MS salt + MS vitamin solution + 30 g/L sucrose and 40 mg/L hygromycin-B). Once roots formed, the plants were transferred to soil.

Identification of hpRNA transformants

Putative transgenic plants were screened first by GUS histochemical staining (Gao et al., 2010). GUS-positive plants were further analysed by PCR and sequencing using primers to amplify the Nos terminator and Ubi promoter sequences, respectively: OsPin fw1 (5'-TGTAAACGTACCATGGCCTAC) and Nos60 (5'-CAACAG GAT TCA ATC TTA AGA AAC, to amplify the Nos terminator sequence) and OsPin rv1 (5'-CATATCGATTTTTTTTTTGGCG) and OsUbiFw (5'-TATGCAGCAGCTATATGTGG for the ubiquitin promoter sequence).

Small RNA Northern blot hybridization

Northern blot hybridization was performed based on a published protocol (Wang et al., 2007) with slight modifications. Total RNA was extracted from leaves (~400 mg) using TRIzol reagent (Invitrogen) and dissolved in 20 μL of sterile water; 15 μg was separated in a 17% denaturing polyacrylamide gel and then transferred to a Hybond+ membrane. About 30 ng of DNA for each probe was labelled with ^{32}P-dATP for ~15 h at room temperature using a Prime-a-Gene Kit (Promega). Church buffer was used for hybridization at 40 °C overnight.

Real-time PCR

Total RNA was treated with DNase I, and cDNA was synthesized using a reverse transcription system (Promega) following the manufacturer's instructions. qRT-PCR was performed using an Applied Biosystems Prism 7500 analyzer and GoTaq qPCR Master Mix (Promega). For each plant, three independent biological replicates were analysed. Sample comparisons were performed using the $2^{(-\Delta\Delta CT)}$ method (Xu et al., 2012). The primers used are listed in Table S1.

Small RNA deep sequencing and bioinformatic analysis

Total RNAs were extracted from the leaves of transgenic plants. Small RNAs were then enriched as described previously (Wang et al., 2007). Small RNA sequence reads were generated using an Illumina Genome Analyzer II (Illumina Inc., San Diego, CA). The raw data were preprocessed using the NGSQCToolkit (http://59. 163.192.90:8080/ngsqctoolkit/) pipeline to remove low-quality (Patel and Jain, 2012) and contaminated reads and to clip adapter sequences. The clean reads were then filtered for siRNA analysis (Hardcastle et al., 2012). Small RNAs ranging from 18 to 30 nt in length were collected and mapped to the hpRNA sequences using Bowtie 2 (http://bowtie-bio.sourceforge.net/bowtie2/index.shtml) (Langmead and Salzberg, 2012). The analysed data are presented using Tablet (http://bioinf.scri.ac.uk/tablet/) (Milne et al., 2010).

Acknowledgements

This work was supported by the National Natural Science Foundation of China (grant number 31271801) and National High Technology Research and Development Program of China (grant numbers 2012AA10A306) and the National Special Program for Genetically Modified Organisms Development (grant number 2009ZX08009-16B).

References

Ahn, J.H., Kim, J., Yoo, S.J., Yoo, S.Y., Roh, H., Choi, J.H., Choi, M.S., Chung, K.S., Han, E.J., Hong, S.M., Jung, S.H., Kang, H.J., Kim, B.K., Kim, M.D., Kim, Y.K., Kim, Y.H., Lee, H., Park, S.H., Yang, J.H., Yang, J.W., Yoo, D.H., Yoo,

S.K. and Lee, J.S. (2007) Isolation of 151 mutants that have developmental defects from T-DNA tagging. *Plant Cell Physiol.* **48**, 169–178.

Austin, R.S., Vidaurre, D., Stamatiou, G., Breit, R., Provart, N.J., Bonetta, D., Zhang, J., Fung, P., Gong, Y., Wang, P.W., McCourt, P. and Guttman, D.S. (2011) Next-generation mapping of *Arabidopsis* genes. *Plant J.* **67**, 715–725.

Axtell, M.J. (2013) Classification and Comparison of Small RNAs from Plants. *Annu. Rev. Plant Biol.* **64**, 137–159.

Baccarini, A., Chauhan, H., Gardner, T.J., Jayaprakash, A.D., Sachidanandam, R. and Brown, B.D. (2011) Kinetic analysis reveals the fate of a microRNA following target regulation in mammalian cells. *Curr. Biol.* **21**, 369–376.

Bolle, C., Schneider, A. and Leister, D. (2011) Perspectives on systematic analyses of gene function in *Arabidopsis thaliana*: new tools, topics and trends. *Curr. Genomics*, **12**, 1–14.

Chang, Y., Long, T. and Wu, C. (2012) Effort and contribution of T-DNA Insertion mutant library for rice functional genomics research in China: review and perspective. *J. Integr. Plant Biol.* **54**, 953–966.

De Boer, A.H., van Kleeff, P.J. and Gao, J. (2013) Plant 14-3-3 proteins as spiders in a web of phosphorylation. *Protoplasma*, **250**, 425–440.

Ebine, K., Okatani, Y., Uemura, T., Goh, T., Shoda, K., Niihama, M., Morita, M.T., Spitzer, C., Otegui, M.S., Nakano, A. and Ueda, T. (2008) A SNARE complex unique to seed plants is required for protein storage vacuole biogenesis and seed development of *Arabidopsis thaliana*. *Plant Cell*, **20**, 3006–3021.

Fusaro, A.F., Matthew, L., Smith, N.A., Curtin, S.J., Dedic-Hagan, J., Ellacott, G.A., Watson, J.M., Wang, M.B., Brosnan, C., Carroll, B.J. and Waterhouse, P.M. (2006) RNA interference-inducing hairpin RNAs in plants act through the viral defence pathway. *EMBO Rep.* **7**, 1168–1175.

Gao, F., Su, Q., Fan, Y. and Wang, L. (2010) Expression pattern and core region analysis of AtMPK3 promoter in response to environmental stresses. *Sci. China Life Sci.* **53**, 1315–1321.

Hardcastle, T.J., Kelly, K.A. and Baulcombe, D.C. (2012) Identifying small interfering RNA loci from high-throughput sequencing data. *Bioinformatics*, **28**, 457–463.

Hiei, Y. and Komari, T. (2008) *Agrobacterium*-mediated transformation of rice using immature embryos or calli induced from mature seed. *Nat. Protoc.* **3**, 824–834.

Hirochika, H., Guiderdoni, E., An, G., Hsing, Y.I., Eun, M.Y., Han, C.D., Upadhyaya, N., Ramachandran, S., Zhang, Q., Pereira, A., Sundaresan, V. and Leung, H. (2004) Rice mutant resources for gene discovery. *Plant Mol. Biol.* **54**, 325–334.

Jeong, D.H., An, S., Kang, H.G., Moon, S., Han, J.J., Park, S., Lee, H.S., An, K. and An, G. (2002) T-DNA insertional mutagenesis for activation tagging in rice. *Plant Physiol.* **130**, 1636–1644.

Jones- Rhoades, M.W., Bartel, D.P. and Bartel, B. (2006) MicroRNAS and their regulatory roles in plants. *Annu. Rev. Plant Biol.* **57**, 19–53.

Kolesnik, T., Szeverenyi, I., Bachmann, D., Kumar, C.S., Jiang, S., Ramamoorthy, R., Cai, M., Ma, Z.G., Sundaresan, V. and Ramachandran, S. (2004) Establishing an efficient Ac/Ds tagging system in rice: large-scale analysis of Ds flanking sequences. *Plant J.* **37**, 301–314.

Langmead, B. and Salzberg, S.L. (2012) Fast gapped-read alignment with Bowtie 2. *Nat. Methods*, **9**, 357–359.

Larmande, P., Gay, C., Lorieux, M., Perin, C., Bouniol, M., Droc, G., Sallaud, C., Perez, P., Barnola, I., Biderre-Petit, C., Martin, J., Morel, J.B., Johnson, A.A., Bourgis, F., Ghesquiere, A., Ruiz, M., Courtois, B. and Guiderdoni, E. (2008) Oryza Tag Line, a phenotypic mutant database for the Genoplante rice insertion line library. *Nucleic Acids Res.* **36**, D1022–D1027.

Mao, Y.B., Cai, W.J., Wang, J.W., Hong, G.J., Tao, X.Y., Wang, L.J., Huang, Y.P. and Chen, X.Y. (2007) Silencing a cotton bollworm P450 monooxygenase gene by plant-mediated RNAi impairs larval tolerance of gossypol. *Nat. Biotechnol.* **25**, 1307–1313.

Milne, I., Bayer, M., Cardle, L., Shaw, P., Stephen, G., Wright, F. and Marshall, D. (2010) Tablet–next generation sequence assembly visualization. *Bioinformatics*, **26**, 401–402.

Patel, R.K. and Jain, M. (2012) NGS QC Toolkit: a toolkit for quality control of next generation sequencing data. *PLoS ONE*, **7**, e30619.

Piffanelli, P., Droc, G., Mieulet, D., Lanau, N., Bes, M., Bourgeois, E., Rouviere, C., Gavory, F., Cruaud, C., Ghesquiere, A. and Guiderdoni, E. (2007) Large-scale characterization of Tos17 insertion sites in a rice T-DNA mutant library. *Plant Mol. Biol.* **65**, 587–601.

Shirane, D., Sugao, K., Namiki, S., Tanabe, M., Iino, M. and Hirose, K. (2004) Enzymatic production of RNAi libraries from cDNAs. *Nat. Genet.* **36**, 190–196.

Smith, N.A., Singh, S.P., Wang, M.B., Stoutjesdijk, P.A., Green, A.G. and Waterhouse, P.M. (2000) Total silencing by intron-spliced hairpin RNAs. *Nature*, **407**, 319–320.

Sunkar, R., Girke, T. and Zhu, J.K. (2005) Identification and characterization of endogenous small interfering RNAs from rice. *Nucleic Acids Res.* **33**, 4443–4454.

Till, B.J., Cooper, J., Tai, T.H., Colowit, P., Greene, E.A., Henikoff, S. and Comai, L. (2007) Discovery of chemically induced mutations in rice by TILLING. *BMC Plant Biol.* **7**, 19.

Wang, M.B. and Waterhouse, P.M. (2002) Application of gene silencing in plants. *Curr. Opin. Plant Biol.* **5**, 146–150.

Wang, L., Wang, M.B., Tu, J.X., Helliwell, C.A., Waterhouse, P.M., Dennis, E.S., Fu, T.D. and Fan, Y.L. (2007) Cloning and characterization of microRNAs from *Brassica napus*. *FEBS Lett.* **581**, 3848–3856.

Wang, L., Luo, Y.Z., Zhang, L., Jiao, X.M., Wang, M.B. and Fan, Y.L. (2008) Rolling circle amplification-mediated hairpin RNA (RMHR) library construction in plants. *Nucleic Acids Res.* **36**, e149.

Wang, M.B., Masuta, C., Smith, N.A. and Shimura, H. (2012) RNA silencing and plant viral diseases. *Mol. Plant Microbe Interact.* **25**, 1275–1285.

Wang, N., Long, T., Yao, W., Xiong, L., Zhang, Q. and Wu, C. (2013) Mutant resources for the functional analysis of the rice genome. *Mol. Plant*, **6**, 596–604.

Watanabe, Y. (2011) Overview of plant RNAi. *Methods Mol. Biol.* **744**, 1–11.

Wei, W., Ba, Z., Gao, M., Wu, Y., Ma, Y., Amiard, S., White, C.I., Rendtlew, D.J., Yang, Y.G. and Qi, Y. (2012) A role for small RNAs in DNA double-strand break repair. *Cell*, **149**, 101–112.

Xu, M.Y., Dong, Y., Zhang, Q.X., Zhang, L., Luo, Y.Z., Sun, J., Fan, Y.L. and Wang, L. (2012) Identification of miRNAs and their targets from *Brassica napus* by high-throughput sequencing and degradome analysis. *BMC Genomics*, **13**, 421.

13

A whole-genome SNP array (RICE6K) for genomic breeding in rice

Huihui Yu[1,†], Weibo Xie[2,†], Jing Li[1], Fasong Zhou[1,*] and Qifa Zhang[2,*]

[1]Life Science and Technology Center, China National Seed Group Co., Ltd, Wuhan, China
[2]National Key Laboratory of Crop Genetic Improvement, National Center of Crop Molecular Breeding, Huazhong Agricultural University, Wuhan, China

*Correspondence

emails zhoufasong@
sinochem.com; qifazh@mail.hzau.edu.cn
[†]These authors contributed equally to this work.
The authors H. Yu, J. Li and F. Zhou have commercial interest in RICE6K as employees of China National Seed Group Co., Ltd.

Keywords: RICE6K, SNP chip, genomic breeding, functional markers, rice (*Oryza sativa* L.).

Summary

The advances in genotyping technology provide an opportunity to use genomic tools in crop breeding. As compared to field selections performed in conventional breeding programmes, genomics-based genotype screen can potentially reduce number of breeding cycles and more precisely integrate target genes for particular traits into an ideal genetic background. We developed a whole-genome single nucleotide polymorphism (SNP) array, RICE6K, based on Infinium technology, using representative SNPs selected from more than four million SNPs identified from resequencing data of more than 500 rice landraces. RICE6K contains 5102 SNP and insertion–deletion (InDel) markers, about 4500 of which were of high quality in the tested rice lines producing highly repeatable results. Forty-five functional markers that are located inside 28 characterized genes of important traits can be detected using RICE6K. The SNP markers are evenly distributed on the 12 chromosomes of rice with the average density of 12 SNPs per 1 Mb and can provide information for polymorphisms between *indica* and *japonica* subspecies as well as varieties within *indica* and *japonica* groups. Application tests of RICE6K showed that the array is suitable for rice germplasm fingerprinting, genotyping bulked segregating pools, seed authenticity check and genetic background selection. These results suggest that RICE6K provides an efficient and reliable genotyping tool for rice genomic breeding.

Introduction

In the history of plant breeding, cross-pollination and transgenic technologies have twice revolutionized the way of crop improvement in the beginning and the end of last century, respectively. Plant breeders realized gene recombination within species through controlled pollination and introduced genes for alien traits, such as insect resistance and herbicide tolerance, into crops via transformation. The advances of genomic research in the last decade may have afforded tools and resources for a third technology breakthrough in plant breeding, which may be termed 'genomic breeding'. By genomic breeding, plant breeders can explore the genomic information including DNA sequences and gene functions to design ideal genotypes and conduct selection to modify the whole genome for varietal improvement. In practice, genomic breeding would select the genes of interest for target traits using molecular markers and optimize genetic background based on genome-wide DNA polymorphisms. Compared to field selection based only on phenotype in conventional breeding, whole-genome selection can integrate the target genes into a better-defined genetic background with greatly improved efficiency.

Two main types of high-throughput genotyping platforms are now available in technology market that can be adopted for genomic breeding: DNA sequencing (Davey *et al.*, 2011) and DNA array (Gupta *et al.*, 2008). New sequencing technologies have been widely applied in genetic studies (Metzker, 2010; Varshney *et al.*, 2009). In rice (*Oryza sativa* L.), Huang *et al.* (2009) reported a high-throughput resequencing method to genotype a recombinant inbred line (RIL) population (Huang

et al., 2009). Xie *et al.* (2010) developed a genotype-imputing method to construct haplotypes using low-coverage genome resequencing data (Xie *et al.*, 2010). Both methods were validated in genetic studies for gene discovery (Wang *et al.*, 2011; Yu *et al.*, 2011). Diverse germplasm collections of rice have been resequenced for genome-wide association studies of agronomic traits (Huang *et al.*, 2010, 2011). Many valuable linkage disequilibria, quantitative trait loci (QTLs) and genes have been identified from the analyses, and millions of DNA polymorphisms were detected in the genomic data. Such unprecedented large amounts of information have laid a solid foundation for platform development for genomic breeding.

Various DNA array-based genotyping platforms have been developed and tested for genetic studies and germplasm characterization (Borevitz *et al.*, 2003; Jaccoud *et al.*, 2001; McNally *et al.*, 2009; Miller *et al.*, 2007a,b; Wang *et al.*, 2010; Xie *et al.*, 2009). For crops such as rice that has a complete genome sequence available, single nucleotide polymorphism (SNP) array has been taken as the preferred technique because of its high-density, assay accuracy, simple data analysis and easy data exchange between research programmes. In rice, three major SNP assay platforms built on different assay principles are available. Affymetrix gene-chip detects SNPs based on differential hybridization efficiency between DNA probes and template sequences; a Rice 44K SNP genotyping array was developed and applied in genome-wide association studies (Famoso *et al.*, 2011; McCouch *et al.*, 2010; Tung *et al.*, 2010; Zhao *et al.*, 2011). Illumina GoldenGate SNP Chip detects SNPs based on DNA extension and differential ligation; various SNP chips of this type were also developed for rice and used in different genetic analysis and breeding projects (Boualaphanh *et al.*, 2011; Chen *et al.*, 2011; Nagasaki *et al.*, 2010; Thomson *et al.*, 2012; Yamamoto *et al.*, 2010; Zhao *et al.*, 2010). As the most recent SNP assay technology, Illumina Infinium SNP array is based on differential single nucleotide extension. Data of genome screening obtained in human, animals and plants demonstrated that this technique enjoyed the advantages of high specificity, reproducibility and call rate (Oliphant *et al.*, 2002; Steemers and Gunderson, 2007). It has now been used in human disease diagnosis, and genetic and breeding studies in crops including maize (Cook *et al.*, 2012; Ganal *et al.*, 2011), wheat and barley (Miedaner and Korzun, 2012), and oilseed rape (Snowdon and Iniguez Luy, 2012). These molecular marker assay platforms have provided options for breeding applications.

Genotyping technologies for genetic studies and breeding share some commonalities but at the same time have significant distinctions. Both need the capacity to detect molecular markers evenly distributed in the genome. Genetic studies aim at revealing unknowns by discovering new genes or QTLs, whereas breeders need genotyping tools that can quickly and reliably find the known targets to identify the expected genotype. Moreover, high-throughput molecular marker assay platform for breeding application has to be cost-effective, time-saving, technically reliable, easy to use and widely adaptable for various breeding applications.

Here, we report on our effort in developing a SNP array for rice genomic breeding. We extracted millions of SNPs based on data from resequencing of rice germplasm collections and designed a rice whole-genome SNP array using Infinium technology. We showed that this platform is useful for a range of applications.

Results

Design of RICE6K

A rice whole-genome SNP array was designed essentially for efficient progeny screening in rice breeding with two considerations, genetic background selection and genotyping of target genes. Illumina BeadArray technology and Infinium SNP assay platform were chosen for the SNP array fabrication because of its demonstrated high specificity, reproducibility and accuracy in SNP call (Oliphant *et al.*, 2002; Steemers and Gunderson, 2007). Two sorts of DNA variations were considered in designing the array: (i) SNPs with adequate coverage and representation of the genome diversity judged on the basis of resequencing diverse germplasm collections (Huang *et al.*, 2010) and (ii) allelic variations of characterized functional genes controlling important breeding traits (Jiang *et al.*, 2011).

Probes for genome diversity

The selection of the probes took several steps. In the first step, raw sequences of 4 236 029 SNP sites were identified from low-coverage (~1×) genome sequences of 520 rice accessions (Huang *et al.*, 2010). In doing so, the released assembly version 6.1 of genomic pseudomolecules of *japonica* cv. Nipponbare (http://rice.plantbiology.msu.edu/) was used as the reference genome. Sequence reads of all accessions were aligned to the reference genome using software MAQ (Li *et al.*, 2008). SNPs were identified using custom PERL scripts from output of MAQ. The following criteria were applied in the processes of sequence comparison and SNP identifications. The mapping quality of sequence reads must be ≥20. At each SNP site, there are at least ten sequence reads showing consensus to each of the two

polymorphic nucleotides, of which at least five reads each had mapping quality >40 and the corresponding base quality >20. Because the average sequence coverage after sequence alignment was ~400×, we limited the total number of sequence reads covering a SNP site between 50 and 1000 to avoid possible repeat sequences.

In a predominantly self-pollinating species like rice, a plant from a germplasm collection is expected to be highly homozygous. Thus, a SNP site would be removed from the candidate list if more than five germplasm accessions showed heterozygous genotype. A SNP site would also be removed if: (i) more than 800 or less than 80 sequence reads covering the SNP site were obtained or (ii) the added frequency of a minor allele in *indica* group (374 accessions), in *japonica* group (146 accessions) and in all germplasm accessions is less than 0.2 or (iii) there are other SNP sites within the flanking 50-bp sequences. This process reduced the number of candidate SNP sites to 1 559 745.

In the following step, the 50-bp flanking sequences on both sides of each selected SNP site were extracted and aligned against to the genome sequence of Nipponbare using BLAST program (Kent, 2002), and the SNP site was removed from the candidate list if more than one matched hits were found with identity >85% of either side flanking sequence. The 50-bp flanking sequences on both sides of the SNP site were extracted from genome sequences of Zhenshan 97 and Minghui 63, two typical *indica* varieties, and compared to the reference Nipponbare sequence. The SNP was excluded if the 50-bp flanking sequences in both Zhenshan 97 and Minghui 63 were different from those in Nipponbare at both sides. This screen kept 1 055 959 candidate SNP sites in the list, of which 35.5% show polymorphism between *japonica* and *indica* (with an allele frequency >0.9 in one subspecies and <0.1 in the other), 42.1% between two random *indica* accessions and 16.9% between two *japonica* accessions.

To further reduce the number of candidate SNPs, the distribution of the selected 1 055 959 SNPs on chromosomes was displayed in windows of 100 kb. Linkage disequilibrium between closely linked SNP sites was evaluated by squared correlation coefficient (r^2) with the threshold value set at 0.64. Two SNP sites with the $r^2 \geq 0.64$ were placed in the same group using a greedy algorithm (Carlson *et al.*, 2004), which resulted in a total of 86 075 groups. With one or two SNPs selected from each group, a total of 187 284 SNPs were used as probe candidates. These selected SNPs (called tag-SNPs) and the corresponding flanking sequences were submitted to Illumina Inc. (http://www.illumina.com/) for probe screen. After removing the tag-SNPs with a design score <0.6, a total of 115 740 SNP sites met the Illumina Infinium probe designing criteria.

To select the final set of SNPs, we defined an In/Ja SNP such that it could differentiate between the main alleles (>90% frequency) of *indica* and *japonica*. Since *indica–japonica* differentiation represents most of the genome diversity in rice germplasms, we chose two In/Ja SNPs in each 100-kb region. Other types of SNPs were added when there were less than two In/Ja SNPs in an 100-kb region. In Infinium SNP assays, two bead types are used to detect an A/T or G/C SNP (Infinium I type SNPs), while only one bead type is needed for other types of SNPs, such as A/G, A/C, T/G, T/C (Infinium II type SNPs). In order to put as many SNPs as possible on the chip with a total of 6000 bead types, we defined an empirical scoring system: $S = \text{MAF} + T*3.5$, where MAF is the minor allele frequency (%) of the SNP site, and $T = 1$ for Infinium II SNPs (non-A/T or G/C SNPs) or $T = 0$ for Infinium I SNPs. A SNP site with $S \geq 33$ would be selected.

Eventually a total of 5556 SNP sites were selected from the 115 740 tag-SNPs, which together with the corresponding flanking sequences were used for synthesizing the probes.

Probes for functional genes

More than 600 rice genes controlling important agronomic traits and biological processes had been identified and characterized at the time when the array was designed (Jiang *et al.*, 2011). To incorporate these genes in the array, we identified SNP/InDel sequences inside 40 functional genes for important agronomic traits that were isolated by map-based cloning. A gene-specific probe (functional probe), either a SNP or an InDel sequence, that represents a characterized function or phenotype was designed for each selected gene.

To put functional markers (FMs) on RICE6K array, we selected genes isolated via map-based cloning that are functionally important to agronomy traits. First, different alleles of 40 functionally characterized genes were identified by searching publications, and the related allele sequences were downloaded from the public DNA database (http://www.ncbi.nlm.nih.gov). These sequences were aligned together and subsequently the polymorphic SNP/InDel markers were developed. These markers were then converted into Infinium probes. If the functionally characterized polymorphic site of a gene is a single SNP and has no other SNPs or mutations in the 50-bp flanking sequence of one side, the conserved 50-bp sequence next to the SNP was directly used as a candidate probe. If the functional polymorphic site was an InDel, two strategies were taken to develop functional probes (Figure S1). One strategy was to convert the InDel marker to a SNP marker: the probe was developed from the conserved side of the insertion/deletion, and the nucleotide to be detected was the first base of the insertion sequence or the base following the deletion if a polymorphism between functional and non-functional alleles existed. The other strategy was to directly use the specific insertion sequence as a probe, and thus, the genomic sequence of the insertion allele could be detected with a strong signal, but the counterpart allele would show very low signal. The first-type FM is codominant and the second type is dominant.

In total, 80 functional probes for 40 genes controlling traits like grain yield, grain quality, heading date, hybrid fertility, biotic and abiotic stress resistance were included in the array.

Quality assay of the RICE6K array

A total of 5636 markers including 5556 SNPs for genetic diversity and 80 for specific gene functions were synthesized and put on RICE6K chip (Table S1). Genotyping accuracy and SNP call rate of RICE6K were tested following the recommended protocols. In order to establish an accurate genotyping procedure, RICE6K was used to genotype rice varieties and a F2 population derived from a cross between Balilla, a typical *japonica* variety, and Nanjing 11, a typical *indica* variety. Genotyping results from assay of 181 rice samples including 112 rice inbred lines, 2 F1 hybrids and 67 F2 plants were used to define SNP genotype clusters. Of the 5636 SNPs on RICE6K chip, 5102 (90.5%) passed bead representation and decoding quality metrics, of which 5034 were genetic diversity SNPs and 68 were gene functional ones. The called SNPs with the following characteristics were considered of high quality: (i) genotypes were clearly grouped into three clusters, AA, AB and BB, in the F2 population, or into two clusters, AA and BB, in case of inbred lines; (ii) less than 80% of 112 inbred lines were genotyped as 'NC' (no call, that means missing genotypes); (iii) less than 5% of 112 inbred lines were called to be heterozygous and

at least one line was called as one of the two homozygous genotypes, AA or BB. Among the 5034 genetic diversity SNPs detected on the array, 4428 were considered to be of high quality (Table 1). To test the reliability of RICE6K in the identification of functional genes, the 112 inbred lines were genotyped using the SNP array. Forty-five functional markers of 28 genes performed well in the assay and were able to report different functional alleles of the corresponding genes (Table S2, Table S3). Furthermore, Balilla (sample ID: P5.Balilla) and Nanjing 11 (sample ID: P1.NJ11) are taken as representative varieties of *japonica* and *indica* to test the subspecies-related functional alleles. Different functional alleles of seven genes including plant height (*Sd-1*), grain number (*Gn1a*), plant architecture (*TAC1*), hybrid fertility (*S5* and *Sa*) and grain size (*GW2* and *qSW5/GW5*) were detected in the two varieties using RICE6K. The detected genotypes were consistent with the corresponding phenotypes (Table 2). Additionally, the 12 SNP/InDel markers were heterozygous in the F1 and segregated in the F2 populations derived from a cross between Balilla and Nanjing 11. In other cases, assay of RICE6K showed that the *japonica* variety Kongyu 131 (sample ID: Y3) has a short allele of *GS3* (Fan *et al.*, 2006; Mao *et al.*, 2010) and wide alleles of *GW2* (Song *et al.*, 2007) and *qSW5/GW5* (Shomura *et al.*, 2008; Weng *et al.*, 2008), in agreement with the phenotype of short and round grains. Our RICE6K assay also showed that Daohuaxiang (sample ID: Y6) and Yuexiangzhan (sample ID: Y7), two aromatic varieties, indeed had the mutant allele of *BADH2* (*fgr*) conditioning the fragrance in the grain (Chen *et al.*, 2008). Additionally, the functional marker, Os07g15770.2, could distinguish three alleles of the pleiotropic gene *Ghd7*. The RICE6K array detected three genotypes in our tested lines, 'C', 'A' and 'NC' (no call due to low detecting signal), corresponding to the functional allele (e.g. Minghui 63) and two non-functional alleles, *Ghd7-0a*, a premature termination in the predicted coding region (e.g. Mudanjiang 8), and *Ghd7-0*, with the *Ghd7* locus completely deleted (e.g. Zhenshan 97) (Xue *et al.*, 2008). In total, these assays identified 4473 high-quality markers including 45 functional markers that are evenly distributed on the 12 chromosomes with an average of 12 markers per Mb (Figure S2).

We also tested reproducibility of the RICE6K array in genotyping by assaying four independent DNA samples of *indica* variety '93-11'. Among the 5102 markers of the RICE6K array, only 0–5 SNPs (<0.1%) showed different genotypes between any two tested samples and no difference was detected in the results from the 4473 high-quality markers, indicating high reproducibility of the array in genotyping.

The genotyping data of the 106 unique accessions selected from the 112 tested inbred lines (six duplicates identified in this work were removed in the following analysis; Table S3) were used to predict the number of polymorphic markers between any two varieties. The 106 rice varieties were clearly clustered into two groups based on the genotypes of 4473 high-quality

Table 1 Markers on the RICE6K array

Types	Synthesized markers	Detected markers	High-quality markers	%*
Genetic diversity	5556	5034	4428	87.96
Gene function	80	68	45	66.18
Total	5636	5102	4473	87.67

*Percentage of high-quality markers = High-quality markers/Detected markers × 100%.

Table 2 Genotypes of Balilla and Nanjing11 at gene functional markers detected by RICE6K array

Gene	MSU locus	Probe	Genotype Balilla	Nanjing 11
Sd-1	LOC_Os01g66100	ID01g00SD1.1	Semi-dwarf plant	High plant
		ID01g00SD1.2	Semi-dwarf plant	High plant
		Os01g66100.1	Semi-dwarf plant	High plant
TAC1	LOC_Os09g35980	Os09g35980.1	Compact plant	Spread-out plant
Gn1a	LOC_Os01g10110	Os01g10110.1	Small spike	Big spike
		Os01g10110.2	Small spike	Big spike
SaF	LOC_Os01g39670	Os01g39670.1	Japonica type	Indica type
S5	LOC_Os06g11010	Os06g11010.1	Japonica type	Indica type
		Os06g11010.2	Japonica type	Indica type
GW2	LOC_Os02g14720	Os02g14720.2	Wide grain	Narrow grain
qSW5/GW5	(GeneBank: AB433345)	Os05g00GW5.1	Wide grain	Narrow grain
		ID05g00GW5.3	Wide grain	Narrow grain

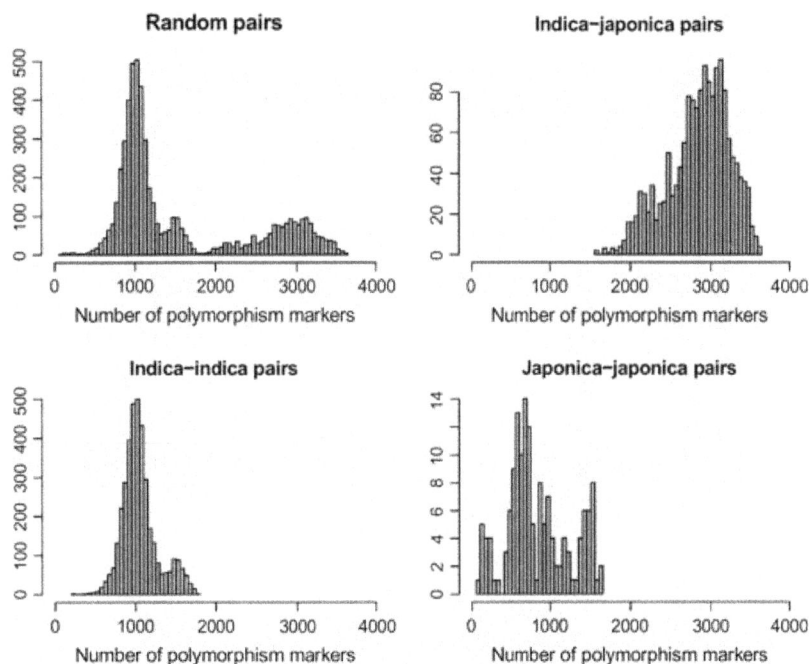

Figure 1 Polymorphism marker number distributions between two varieties. For each histogram, x-axis shows the number of polymorphism marker between two varieties and y-axis shows the number of pairs.

SNP/InDel markers, and all 18 japonica varieties were clustered in one group and the rest 88 indica ones in the other. Additionally, subgroups in japonica and indica were also defined with a fine resolution (Figure S3). The average high-quality polymorphic markers between any two tested varieties are 1559, and the ones between two tested indica varieties, two tested japonica varieties or between tested indica and japonica are 1053, 824 and 2853, respectively (Figure 1). This result suggests that RICE6K array can be widely used to genotype different populations derived from different crosses, not only for inter-subspecies between indica and japonica, but also for intra-subspecies.

Applications of the RICE6K

We validated the usefulness of the RICE6K array in a range of applications including genomic breeding and genetic analysis.

Genetic background selection in breeding process

The improvement of a specific trait by backcrossing is an important breeding strategy, aiming to transfer a desired trait from a donor germplasm into the genome of an elite variety without disturbing the genetic background. It is critical to be able to track the DNA fragments from the different genomes. Kongyu 131, a japonica cultivar widely grown in north-east China in the last decade, has become highly susceptible to rice fungal blast (Magnaporthe grisea) in recent years. A genomics-based introgression of Pi1 (Hua et al., 2012) and Pi2 (Zhou et al., 2006) from the donors into Kongyu 131 has been implemented using SSR markers. In BC4F1, 29 plants with the introduced Pi1 or Pi2 genes were examined using the RICE6K array, which provided an unambiguous graphic genotype for each individual (Figure 2). The results showed that the genetic backgrounds of several tested plants were similar to the recurrent parent Kongyu 131 (e.g. L16, L22, L24, L25 and L28), but the genomic regions containing *Pi1* and *Pi2* had large dragged fragments from the donor parents, which may have potential adverse genetic effects (Figure 2). Indeed, a flowering gene *Hd1* (9.3 Mb on Chr06) (Yano *et al.*, 2000) was linked to *Pi2* (10.4 Mb on Chr06). Transferring late flowering allele into Kongyu 131 is undesirable for the improvement of this variety to be planted in targeted area. This result suggested selection for recombination must be performed in early generations of backcrossing as suggested by Chen *et al.* (2000). Nonetheless, it demonstrated that the RICE6K array can provide a powerful tool for genotype selection.

Genotyping biparental segregating populations

We genotyped individual lines/plants and their parents in several biparental cross-populations including (i) an F2 population from a cross between an *indica* variety Nanjing 11 and a *japonica* variety Balilla, (ii) a RIL population derived from a cross between two *indica* varieties, (iii) a RIL population from a cross between an *indica* and a *japonica* variety and (iv) CSSL (chromosomal segmental substitution lines) from three different crosses (Table 3). Among the 4473 high-quality SNP/InDel markers on

the RICE6K array, the number of markers detected to be homozygous and polymorphic for the two parents of each tested population was ranged from 1336 to 3775: more than 3000 in inter-subspecific populations and more than 1000 in *indica* populations (Table 3). For each population, the genotypes of the called SNPs were assigned as 'AA' (female parental genotype), 'BB' (male parental genotype) or 'AB' (heterozygous genotype). As a result, the called high-quality polymorphic SNPs could provide high-density graphical genotypes for each individual (Figure 3), which can be used for genetic investigations. For example, genotyping the 197 lines in the ZX-RIL population using the RICE6K array resulted in a high-density genetic linkage map consisting of 1495 recombination bins covering 1591.2 cM with average length of 1.1 cM per bin (Tan *et al.*, 2013). The total length of the genetic map was similar to the ones reported in previous studies using sequence-based genotyping method (Huang *et al.*, 2009; Yu *et al.*, 2011). These tests indicated that RICE6K SNP array is robust and efficient in population genotyping.

Varietal identity and purity tests

Identity and purity of seeds are always of main concern in crop production and seed industry. In China, a set of 24 SSR markers

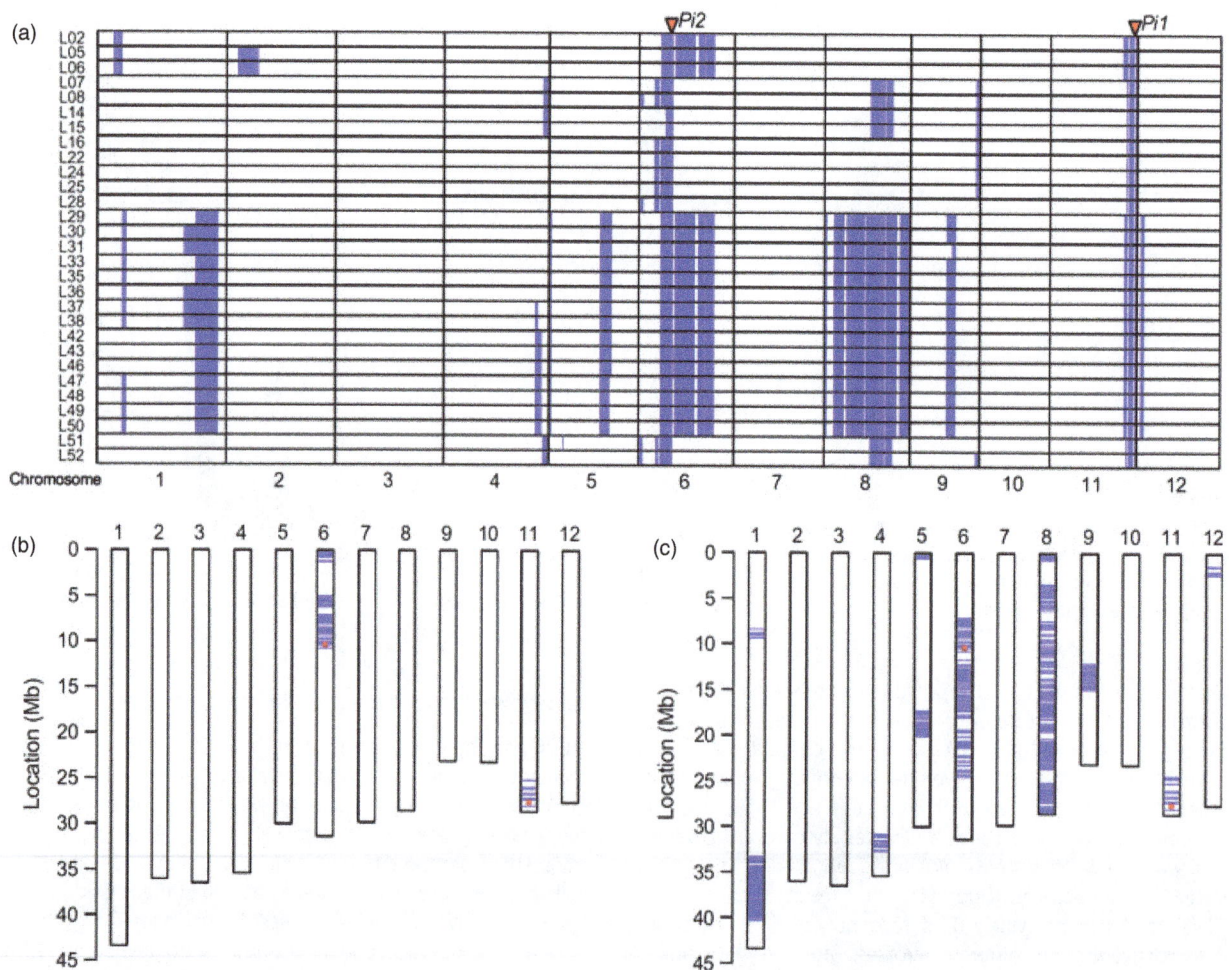

Figure 2 Genetic background screen using RICE6K array. (a) The genetic background of all the 29 plants in BC4F1. (b) The detailed genotyping map of the plant L28. (c) The detailed genotyping map of the plant L50. Twelve chromosomes of rice are labelled from 1 to 12. The reference genome is Nipponbare (rice TIGR6.1). The triangles and the dots indicate the positions of the two target genes, *Pi1* on chromosome 11 and *Pi2* on chromosome 6, respectively. The blue lines indicated the positions of the single nucleotide polymorphism (SNP)s with heterozygous genotypes where genomic fragments of the donor parent were introgressed, and the genotypes of the rest genomic regions were the same as the recurrent parent Kongyu 131.

Table 3 Population tested on RICE6K array

Population	Cross	Type	Polymorphism markers	%*
ZM-RIL	ZS 97 (R) × MH 63 (R)	*Indica × Indica*	1336	26.18
ZX-RIL	Zhenshan 97 × Xizang 2	*Indica × Japonica*	3362	65.90
BN-F2	Balilla × Nanjing 11	*Indica × Japonica*	3775	73.99
ZN-CSSL	Zhenshan 97 × Nipponbare	*Indica × Japonica*	3709	72.70
ZM-CSSL	Zhenshan 97 × Minghui 63	*Indica × Indica*	1342	26.30
ZI-CSSL	Zhenshan 97 × *Oryza rufipogon* (IRGC-105491)	*Indica × Oryza rufipogon*	1789	35.06

RIL, recombinant inbred line; CSSL, chromosomal segmental substitution line.

*Percentage of polymorphism markers detected by the RICE6K array.

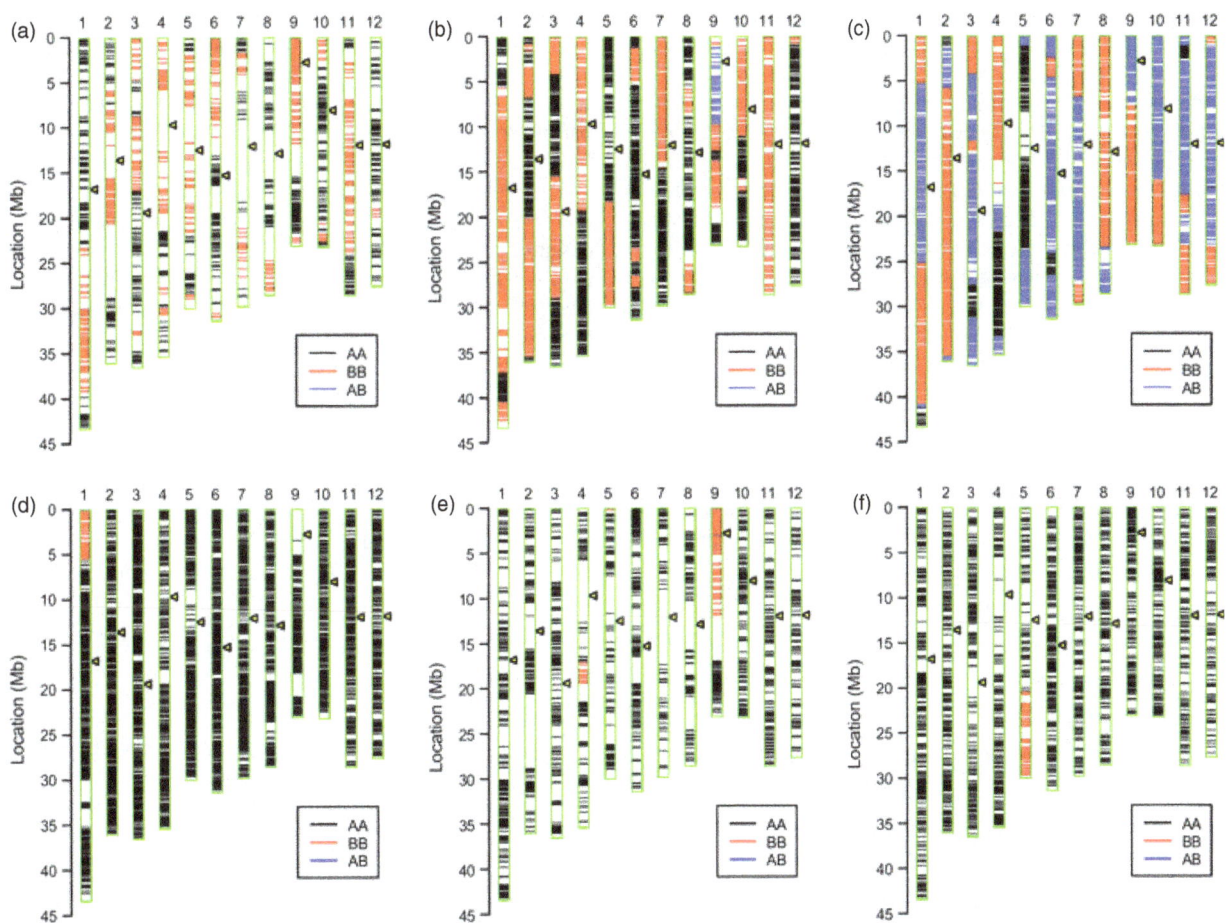

Figure 3 Haplotype maps of example lines/plants from different populations detected by RICE6K array. Each map shows one example line/plant from one of six populations as described in Table 3. (a) ZM-RIL population, (b) ZX-RIL population, (c) BN-F2 population, (d) ZN-CSSL population, (e) ZM-CSSL population and (f) ZI-CSSL population. Each short line at the chromosomes indicates the position of a single nucleotide polymorphism (SNP), and the triangle arrows indicate the centromeres of rice 12 chromosomes. The physical position of each marker is based on rice TIGR6.1. 'AA' represents female parental homozygous genotype, 'BB' represents male parental homozygous genotype and 'AB' represents heterozygous genotype.

has been officially used for testing identity and purity of rice varieties (Chinese Agricultural Industry Standard, NY/T 1433-2007), in which two samples are regarded as different varieties if two or more markers show polymorphisms, as related varieties if one marker is different, and as the same variety if no polymorphism is detected in all the 24 markers. Although the markers were carefully selected with even distribution on the 12 chromosomes and showing high polymorphisms in tested varie-

ties (Zhuang *et al.*, 2006), these 24 SSR markers cover only a very small partition of the rice genome. We used the RICE6K array to fingerprint rice varieties that produced results that challenge the notion of 'varieties'. For example, YU is not only an elite inbred rice variety in China, but also the parent of an elite hybrid, which has been widely planted for more than 5 years. A sample of bulked seeds of YU was fingerprinted using RICE6K, which revealed several genomic regions that were heterogeneous and thus still segregating (Figure 4). The two main heterogeneous regions were 9–11.5 Mb on Chr02 and 0–7 Mb on Chr09. Using one of the plants designated YU001 as a reference, four major types of heterogeneous plants were identified from the population. However, none of the heterogeneous regions could be identified using the 24 SSR markers. These results indicated that genetic heterogeneity would be retained in a variety for a long period, which might be the cause of variety deterioration. This makes selection and subsequent field tests necessary a few years after variety release, which also suggests that genotyping the selected lines using a method like the RICE6K before varietal release may help maintaining the quality and purity of the varieties.

Bulked segregant analysis

Bulked segregant analysis (BSA) is an efficient method for rapidly identifying molecular markers linked to any specific genes or genomic regions (Michelmore et al., 1991). We tested the application of RICE6K array in BSA by mapping the fertility-restorer gene-controlling cytoplasmic male sterility (CMS). The three-line hybrid F1 plants, derived from a cross between CMS line JN 2A and restorer line JH 3, were self-pollinated to generate an F2 population of about 2000 plants. The F2 population was planted in a field nursery in Sanya, China, and 94 plants in the population were examined for spikelet fertility. The ratio of the fertile to sterile plants was $74 : 20$ [x^2 (3 : 1) = 0.695, $P = 0.405$], indicating that the fertility was controlled by a single locus. DNA bulks from ten fertile plants and 20 sterile plants were separately prepared and assayed using the RICE6K array. The result showed that the main difference between the two bulks was in region 18.1–19.9 Mb on Chr10, where the bulk from fertile plants was heterozygous and the bulk of sterile plants was homozygous (Figure 5). There are two characterized rice fertility-restorer genes in this region, one is Rf1/Rf1a/Rf5 (18.8 Mb) and the other is Rf1b (18.9 Mb) (Akagi et al., 2004; Hu et al., 2012;

Wang et al., 2006), and some other fertility-restorer genes have been also mapped to this region, such as Rf4 and Rf6 (Ahmadikhah and Karlov, 2006; Liu et al., 2004). Thus, the RICE6K can be used to quickly locate the gene to the genomic region. Moreover, the polymorphic SNP markers in the region identified by RICE6K array can be used for fine mapping of this gene.

In conclusion, the results show that the RICE6K chip provides a robust tool for a range of applications in genotyping.

Discussion

One of the biggest challenges that plant breeders have to face is to stack multiple target genes in a favourable genetic background. This requires the breeders to be able to track many functional genes in a segregating population in a short time and reliable manner at a reasonable cost. RICE6K accommodated 80 functional markers covering 40 rice genes. In our tests using various genetic materials, at least 45 functional markers representing 28 genes, 1–5 markers per gene, performed well. The parallel interrogation of dozens functional genes can greatly facilitate gene stacking.

Some of the genes have several functional mutations, for example, Hd1 for heading date (Takahashi et al., 2009). We thus designed multiple probes in the RICE6K to detect these alleles, which detected several haplotypes for determining functional type of the gene. For functional mutations located inside repeated motifs of a gene, for example, Pi2/Pi9/Piz-t (Qu et al., 2006; Zhou et al., 2006), a locus for blast resistance with multiple alleles, it is difficult to design a probe to differentiate the alleles. In such case, we also recommend to use multiple SNP probes differentiating functional allele haplotypes for the screening.

Agronomically important traits like hybrid vigour, grain yield and quality are affected by many genes/QTLs each with a small effect on the trait. Superior varieties or hybrids are the results of gradual accumulation of many favourite alleles through repeated

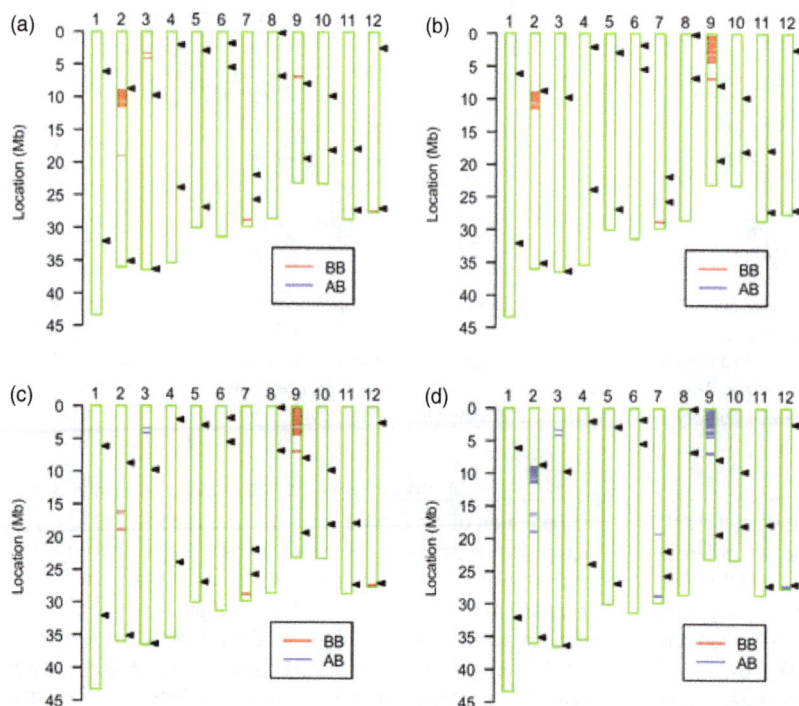

Figure 4 Comparative genotyping of different plants in 'YU' population. The plant 'YU001' is the reference genotype. The figure shows the maps of different plants compared with the reference plant 'YU001' genotyped by the RICE6K array. The single nucleotide polymorphism (SNP) genotype is assigned to 'AA' when the detected genotype of the plant is the same as the referent plant 'YU001' and is not shown. The SNP genotype is assigned to 'BB' when the genotypes of the plant and the reference are different and both of them are homologous. The SNP genotype is assigned to 'AB' when the genotype of the reference is homologous and one of the plants is heterozygous. (a) Genotyping map of the plant 'YU003', (b) genotyping map of the plant 'YU010', (c) genotyping map of the plant 'YU033' and (d) genotyping map of mixture sample harvested from random 20 plants of the population. The triangle arrows indicate the positions of 24 SSR markers recommended by the Chinese Agricultural Industry Standard (NY/T 1433-2007).

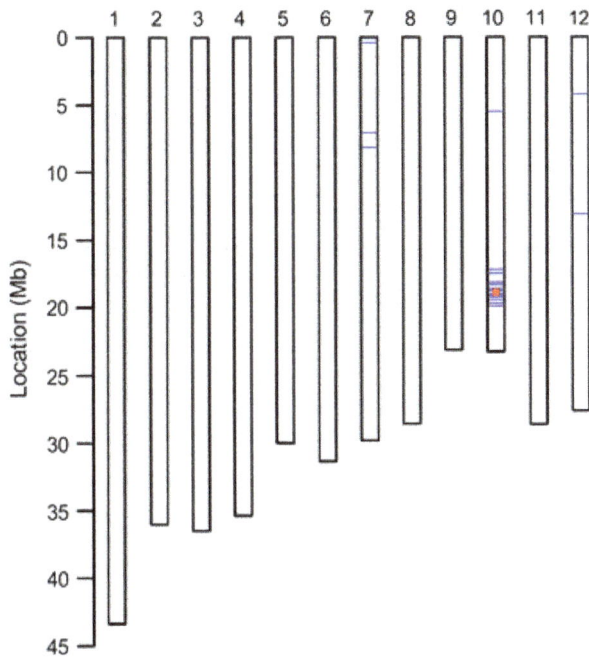

Figure 5 Bulked segregant analysis of a fertility-restorer gene using RICE6K array. The map shows different genotypes between the DNA bulk sample from ten fertile plants and the sample from 20 sterile plants. The blue short lines on the chromosomes represent the single nucleotide polymorphism (SNP) sites with different genotypes, at which the fertility bulk sample was heterozygous and the sterility bulk sample was homozygous. The red dot indicates the positions of the cloned fertility-restorer gene *Rf1/Rf1a/Rf5*.

crossing and phenotype selections in the long breeding history. These varieties probably represent the best combinations of genes controlling yield, quality and adaptation to climate and cultivation conditions. However, they will inevitably become susceptible to diseases or other biotic stresses because of emergence of new strains of the pathogens. In this case, a quick backcrossing scheme is usually performed to incorporate a specific gene to improve the resistance with minimal disturbance of the genetic background. In previous reports, breeders used linked markers (RFLP, SSR or AFLP) to perform selection for recombination between the target gene and the genetic background, and a set of unlinked makers for genetic background screening (Chen et al., 2000; Jorasch, 2005; Liu et al., 2003), which produced variable results. Chip-based high-density genome fingerprinting can greatly improve the efficiency of backcrossing by selecting for precise recombinations and the background of the recurrent parent. We showed that RICE6K array could provide detailed information of the genomic composition of the selected progeny at resolution of <100 kb. This may lead to highly accurate prediction of the performance of the selected individuals. Thus, the RICE6K array would be particularly useful for backcross breeding to introgress genes into elite backgrounds. In addition, knowledge of the genes in the array would also help predict the performance of the selected individuals in other types of breeding programmes.

Genomic information including genome sequences and functional genes is still accumulating on accelerated pace (Jiang et al., 2011), which provides practically unlimited resources for improving genotyping tools like the RICE6K array. Ideally, the genotyping information should provide adequate for detecting

polymorphisms in both inter-subspecific (*indica/japonica*) and intra-subspecific (*indica/indica* and *japonica/japonica*) crosses of rice. This may be readily addressed with the present results from the already available sequencing data as well as the still ongoing efforts. It is also desired that the genotyping tools can incorporate information on functional genes as they become available. These will entail economically practical higher-density SNP arrays, which should now be placed in the pipeline for future development.

Experimental procedures

The assay protocol of the RICE6K

First, rice genomic DNA samples were extracted and their quality was examined. Two types of tissues, either seed or leaf, were used for DNA extraction. For seed, about 20 dry seeds were dehulled, mixed and grinded after freezing with liquid nitrogen, and the genomic DNA was extracted using CoWin SurePlant DNA Kit (Beijing CoWin Bioscience Co., Ltd., Beijing, China). For leaf, fresh young seedling leaflet of 3–5 cm in length were harvested, grinded after freezing with liquid nitrogen, and the genomic DNA was extracted using Wizard Magnetic 96 DNA Plant System Kit (Promega Corporation, Madison, WI). DNA quality was checked by 1%–1.5% agarose gel electrophoresis. The DNA samples with high quality (>10-kb fragments) and appropriate concentration (10–50 ng/μL) were used for SNP assays.

DNA amplification, fragmentation and chip hybridization, washing, and staining were performed according to Infinium assay standard protocol (Infinium HD Assay Ultra Protocol Guide,http://www.illumina.com/). HiScan scanner (Illumina Inc., San Diego, CA) was used for chip scanning, and GenomeStudio software was used for raw data analysis. R platform was employed for further analysis, for example, genotype identification, comparison and map drawing (R Development Core Team, 2011).

Acknowledgements

This work was supported by the Introduction of International Advanced Agricultural Science and Technology Program of China (948 Program, Grant No. 2012-G2), the National High Technology Research and Development Program of China (863 Program, Rice Functional Genomics Research Project, Grant No. 2012AA10A304), the National Natural Science Foundation of China (Grant No. 31100962), the Research Fund for the Doctoral Program of Higher Education of China (Grant No. 20110146120013). The lines/plants used in this study for RICE6K SNP chip tests were provided by Yuqing He, Sibin Yu, Xingming Lian and Yongzhong Xing in Huazhong Agricultural University. We highly appreciate the generous support.

References

Ahmadikhah, A. and Karlov, G. (2006) Molecular mapping of the fertility-restoration gene *Rf4* for WA-cytoplasmic male sterility in rice. *Plant Breed.* **125**, 363–367.

Akagi, H., Nakamura, A., Yokozeki-Misono, Y., Inagaki, A., Takahashi, H., Mori, K. and Fujimura, T. (2004) Positional cloning of the rice *Rf-1* gene, a restorer of BT-type cytoplasmic male sterility that encodes a mitochondria-targeting PPR protein. *Theor. Appl. Genet.* **108**, 1449–1457.

Borevitz, J.O., Liang, D., Plouffe, D., Chang, H.S., Zhu, T., Weigel, D., Berry, C.C., Winzeler, E. and Chory, J. (2003) Large-scale identification of single-feature polymorphisms in complex genomes. *Genome Res.* **13**, 513–523.

Boualaphanh, C., Daygon, V.D., Calingacion, M.N., Sanitchon, J., Jothityangkoon, D., Mumm, R., Hall, R.D. and Fitzgerald, M.A. (2011) Use of new generation single nucleotide polymorphism genotyping for rapid development of near-isogenic lines in rice. *Crop Sci.* **51**, 2067–2073.

Carlson, C.S., Eberle, M.A., Rieder, M.J., Yi, Q., Kruglyak, L. and Nickerson, D.A. (2004) Selecting a maximally informative set of single-nucleotide polymorphisms for association analyses using linkage disequilibrium. *Am. J. Hum. Genet.* **74**, 106–120.

Chen, S., Lin, X.H., Xu, C.G. and Zhang, Q. (2000) Improvement of bacterial blight resistance of 'Minghui 63', an elite restorer line of hybrid rice, by molecular marker-assisted selection. *Crop Sci.* **40**, 239–244.

Chen, S., Yang, Y., Shi, W., Ji, Q., He, F., Zhang, Z., Cheng, Z., Liu, X. and Xu, M. (2008) *Badh2*, encoding betaine aldehyde dehydrogenase, inhibits the biosynthesis of 2-acetyl-1-pyrroline, a major component in rice fragrance. *Plant Cell*, **20**, 1850–1861.

Chen, H., He, H., Zou, Y., Chen, W., Yu, R., Liu, X., Yang, Y., Gao, Y.M., Xu, J.L., Fan, L.M., Li, Y., Li, Z.K. and Deng, X.W. (2011) Development and application of a set of breeder-friendly SNP markers for genetic analyses and molecular breeding of rice (*Oryza sativa* L.). *Theor. Appl. Genet.* **123**, 869–879.

Cook, J.P., McMullen, M.D., Holland, J.B., Tian, F., Bradbury, P., Ross-Ibarra, J., Buckler, E.S. and Flint-Garcia, S.A. (2012) Genetic architecture of maize kernel composition in the nested association mapping and inbred association panels. *Plant Physiol.* **158**, 824–834.

Davey, J.W., Hohenlohe, P.A., Etter, P.D., Boone, J.Q., Catchen, J.M. and Blaxter, M.L. (2011) Genome-wide genetic marker discovery and genotyping using next-generation sequencing. *Nat. Rev. Genet.* **12**, 499–510.

Famoso, A.N., Zhao, K., Clark, R.T., Tung, C.-W., Wright, M.H., Bustamante, C., Kochian, L.V. and McCouch, S.R. (2011) Genetic architecture of aluminum tolerance in rice (*Oryza sativa*) determined through genome-wide association analysis and QTL mapping. *PLoS Genet.* **7**, e1002221.

Fan, C., Xing, Y., Mao, H., Lu, T., Han, B., Xu, C., Li, X. and Zhang, Q. (2006) *GS3*, a major QTL for grain length and weight and minor QTL for grain width and thickness in rice, encodes a putative transmembrane protein. *Theor. Appl. Genet.* **112**, 1164–1171.

Ganal, M.W., Durstewitz, G., Polley, A., Bérard, A., Buckler, E.S., Charcosset, A., Clarke, J.D., Graner, E.M., Hansen, M. and Joets, J. (2011) A large maize (*Zea mays* L.) SNP genotyping array: development and germplasm genotyping, and genetic mapping to compare with the B73 reference genome. *PLoS ONE*, **6**, e28334.

Gupta, P.K., Rustgi, S. and Mir, R.R. (2008) Array-based high-throughput DNA markers for crop improvement. *Heredity*, **101**, 5–18.

Hu, J., Wang, K., Huang, W., Liu, G., Gao, Y., Wang, J., Huang, Q., Ji, Y., Qin, X., Wan, L., Zhu, R., Li, S., Yang, D. and Zhu, Y. (2012) The rice pentatricopeptide repeat protein RF5 restores fertility in Hong-Lian cytoplasmic male-sterile lines via a complex with the glycine-rich protein GRP162. *Plant Cell*, **24**, 109–122.

Hua, L., Wu, J., Chen, C., Wu, W., He, X., Lin, F., Wang, L., Ashikawa, I., Matsumoto, T. and Pan, Q. (2012) The isolation of *Pi1*, an allele at the *Pik* locus which confers broad spectrum resistance to rice blast. *Theor. Appl. Genet.* **125**, 1047–1055.

Huang, X., Feng, Q., Qian, Q., Zhao, Q., Wang, L., Wang, A., Guan, J., Fan, D., Weng, Q. and Huang, T. (2009) High-throughput genotyping by whole-genome resequencing. *Genome Res.* **19**, 1068–1076.

Huang, X., Wei, X., Sang, T., Zhao, Q., Feng, Q., Zhao, Y., Li, C., Zhu, C., Lu, T., Zhang, Z., Li, M., Fan, D., Guo, Y., Wang, A., Wang, L., Deng, L., Li, W., Lu, Y., Weng, Q., Liu, K., Huang, T., Zhou, T., Jing, Y., Lin, Z., Buckler, E.S., Qian, Q., Zhang, Q.F., Li, J. and Han, B. (2010) Genome-wide association studies of 14 agronomic traits in rice landraces. *Nat. Genet.* **42**, 961–967.

Huang, X., Zhao, Y., Wei, X., Li, C., Wang, A., Zhao, Q., Li, W., Guo, Y., Deng, L., Zhu, C., Fan, D., Lu, Y., Weng, Q., Liu, K., Zhou, T., Jing, Y., Si, L., Dong, G., Huang, T., Lu, T., Feng, Q., Qian, Q., Li, J. and Han, B. (2011) Genome-wide association study of flowering time and grain yield traits in a worldwide collection of rice germplasm. *Nat. Genet.* **44**, 32–39.

Jaccoud, D., Peng, K., Feinstein, D. and Kilian, A. (2001) Diversity arrays: a solid state technology for sequence information independent genotyping. *Nucleic Acids Res.* **29**, E25.

Jiang, Y., Cai, Z., Xie, W., Long, T., Yu, H. and Zhang, Q. (2011) Rice functional genomics research: progress and implications for crop genetic improvement. *Biotechnol. Adv.* **30**, 1059–1070.

Jorasch, P. (2005) Intellectual property rights in the field of molecular marker analysis. In *Molecular Marker Systems in Plant Breeding and Crop Improvement* (Lorz, H. and Wenzel, G., eds), pp. 433–471. Berlin, Germany: Springer.

Kent, W.J. (2002) BLAT – the BLAST-like alignment tool. *Genome Res.* **12**, 656–664.

Li, H., Ruan, J. and Durbin, R. (2008) Mapping short DNA sequencing reads and calling variants using mapping quality scores. *Genome Res.* **18**, 1851–1858.

Liu, S.P., Li, X., Wang, C.Y., Li, X.H. and He, Y.Q. (2003) Improvement of resistance to rice blast in Zhenshan 97 by molecular marker-aided selection. *Acta Bot. Sin.* **45**, 1346–1350.

Liu, X.Q., Xu, X., Tan, Y.P., Li, S.Q., Hu, J., Huang, J.Y., Yang, D.C., Li, Y.S. and Zhu, Y.G. (2004) Inheritance and molecular mapping of two fertility-restoring loci for Honglian gametophytic cytoplasmic male sterility in rice (*Oryza sativa* L.). *Mol. Genet. Genomics*, **271**, 586–594.

Mao, H., Sun, S., Yao, J., Wang, C., Yu, S., Xu, C., Li, X. and Zhang, Q. (2010) Linking differential domain functions of the GS3 protein to natural variation of grain size in rice. *Proc. Natl Acad. Sci. USA*, **107**, 19579–19584.

McCouch, S.R., Zhao, K., Wright, M., Tung, C.W., Ebana, K., Thomson, M., Reynolds, A., Wang, D., DeClerck, G. and Ali, M.L. (2010) Development of genome-wide SNP assays for rice. *Breed. Sci.* **60**, 524–535.

McNally, K.L., Childs, K.L., Bohnert, R., Davidson, R.M., Zhao, K., Ulat, V.J., Zeller, G., Clark, R.M., Hoen, D.R., Bureau, T.E., Stokowski, R., Ballinger, D.G., Frazer, K.A., Cox, D.R., Padhukasahasram, B., Bustamante, C.D., Weigel, D., Mackill, D.J., Bruskiewich, R.M., Ratsch, G., Buell, C.R., Leung, H. and Leach, J.E. (2009) Genomewide SNP variation reveals relationships among landraces and modern varieties of rice. *Proc. Natl Acad. Sci. USA*, **106**, 12273–12278.

Metzker, M.L. (2010) Sequencing technologies – the next generation. *Nat. Rev. Genet.* **11**, 31–46.

Michelmore, R.W., Paran, I. and Kesseli, R.V. (1991) Identification of markers linked to disease-resistance genes by bulked segregant analysis: a rapid method to detect markers in specific genomic regions by using segregating populations. *Proc. Natl Acad. Sci. USA*, **88**, 9828–9832.

Miedaner, T. and Korzun, V. (2012) Marker-assisted selection for disease resistance in wheat and barley breeding. *Phytopathology*, **102**, 560–566.

Miller, M.R., Atwood, T.S., Eames, B.F., Eberhart, J.K., Yan, Y.L., Postlethwait, J.H. and Johnson, E.A. (2007a) RAD marker microarrays enable rapid mapping of zebrafish mutations. *Genome Biol.* **8**, R105.

Miller, M.R., Dunham, J.P., Amores, A., Cresko, W.A. and Johnson, E.A. (2007b) Rapid and cost-effective polymorphism identification and genotyping using restriction site associated DNA (RAD) markers. *Genome Res.* **17**, 240–248.

Nagasaki, H., Ebana, K., Shibaya, T., Yonemaru, J. and Yano, M. (2010) Core single-nucleotide polymorphisms – a tool for genetic analysis of the Japanese rice population. *Breed. Sci.* **60**, 648–655.

Oliphant, A., Barker, D.L., Stuelpnagel, J.R. and Chee, M.S. (2002) BeadArray technology: enabling an accurate, cost-effective approach to high-throughput genotyping. *Biotechniques*, **32**, S56–S61.

Qu, S., Liu, G., Zhou, B., Bellizzi, M., Zeng, L., Dai, L., Han, B. and Wang, G.L. (2006) The broad-spectrum blast resistance gene *Pi9* encodes a nucleotide-binding site-leucine-rich repeat protein and is a member of a multigene family in rice. *Genetics*, **172**, 1901–1914.

R Development Core Team (2011) *R: A Language and Environment for Statistical Computing*. R Foundation for Statistical Computing, Vienna, Austria. ISBN 3-900051-07-0, URL http://www.R-project.org.

Shomura, A., Izawa, T., Ebana, K., Ebitani, T., Kanegae, H., Konishi, S. and Yano, M. (2008) Deletion in a gene associated with grain size increased yields during rice domestication. *Nat. Genet.* **40**, 1023–1028.

Snowdon, R.J. and Iniguez Luy, F.L. (2012) Potential to improve oilseed rape and canola breeding in the genomics era. *Plant Breed.* **131**, 351–360.

Song, X.J., Huang, W., Shi, M., Zhu, M.Z. and Lin, H.X. (2007) A QTL for rice grain width and weight encodes a previously unknown RING-type E3 ubiquitin ligase. *Nat. Genet.* **39**, 623–630.

Steemers, F.J. and Gunderson, K.L. (2007) Whole genome genotyping technologies on the BeadArray platform. *Biotechnol. J.* **2**, 41–49.

Takahashi, Y., Teshima, K.M., Yokoi, S., Innan, H. and Shimamoto, K. (2009) Variations in Hd1 proteins, *Hd3a* promoters, and *Ehd1* expression levels contribute to diversity of flowering time in cultivated rice. *Proc. Natl Acad. Sci. USA*, **106**, 4555–4560.

Tan, C., Han, Z., Yu, H., Zhan, W., Xie, W., Chen, X., Zhao, H., Zhou, F. and Xing, Y. (2013) QTL scanning for rice yield using a whole genome SNP array. *J. Genet. Genomics*, doi:10.1016/j.jgg.2013.06.009.

Thomson, M.J., Zhao, K., Wright, M., McNally, K.L., Rey, J., Tung, C.-W., Reynolds, A., Scheffler, B., Eizenga, G. and McClung, A. (2012) High-throughput single nucleotide polymorphism genotyping for breeding applications in rice using the BeadXpress platform. *Mol. Breed.* **29**, 875–886.

Tung, C.-W., Zhao, K., Wright, M.H., Ali, M.L., Jung, J., Kimball, J., Tyagi, W., Thomson, M.J., McNally, K. and Leung, H. (2010) Development of a research platform for dissecting phenotype–genotype associations in rice (*Oryza* spp.). *Rice*, **3**, 205–217.

Varshney, R.K., Nayak, S.N., May, G.D. and Jackson, S.A. (2009) Next-generation sequencing technologies and their implications for crop genetics and breeding. *Trends Biotechnol.* **27**, 522–530.

Wang, Z., Zou, Y., Li, X., Zhang, Q., Chen, L., Wu, H., Su, D., Chen, Y., Guo, J., Luo, D., Long, Y., Zhong, Y. and Liu, Y.G. (2006) Cytoplasmic male sterility of rice with boro II cytoplasm is caused by a cytotoxic peptide and is restored by two related PPR motif genes via distinct modes of mRNA silencing. *Plant Cell*, **18**, 676–687.

Wang, J., Yu, H., Xie, W., Xing, Y., Yu, S., Xu, C., Li, X., Xiao, J. and Zhang, Q. (2010) A global analysis of QTLs for expression variations in rice shoots at the early seedling stage. *Plant J.* **63**, 1063–1074.

Wang, L., Wang, A., Huang, X., Zhao, Q., Dong, G., Qian, Q., Sang, T. and Han, B. (2011) Mapping 49 quantitative trait loci at high resolution through sequencing-based genotyping of rice recombinant inbred lines. *Theor. Appl. Genet.* **122**, 327–340.

Weng, J., Gu, S., Wan, X., Gao, H., Guo, T., Su, N., Lei, C., Zhang, X., Cheng, Z. and Guo, X. (2008) Isolation and initial characterization of GW5, a major QTL associated with rice grain width and weight. *Cell Res.* **18**, 1199–1209.

Xie, W., Chen, Y., Zhou, G., Wang, L., Zhang, C., Zhang, J., Xiao, J., Zhu, T. and Zhang, Q. (2009) Single feature polymorphisms between two rice cultivars detected using a median polish method. *Theor. Appl. Genet.* **119**, 151–164.

Xie, W., Feng, Q., Yu, H., Huang, X., Zhao, Q., Xing, Y., Yu, S., Han, B. and Zhang, Q. (2010) Parent-independent genotyping for constructing an ultrahigh-density linkage map based on population sequencing. *Proc. Natl Acad. Sci. USA*, **107**, 10578–10583.

Xue, W., Xing, Y., Weng, X., Zhao, Y., Tang, W., Wang, L., Zhou, H., Yu, S., Xu, C., Li, X. and Zhang, Q. (2008) Natural variation in Ghd7 is an important regulator of heading date and yield potential in rice. *Nat. Genet.* **40**, 761–767.

Yamamoto, T., Nagasaki, H., Yonemaru, J., Ebana, K., Nakajima, M., Shibaya, T. and Yano, M. (2010) Fine definition of the pedigree haplotypes of closely related rice cultivars by means of genome-wide discovery of single-nucleotide polymorphisms. *BMC Genomics*, **11**, 267.

Yano, M., Katayose, Y., Ashikari, M., Yamanouchi, U., Monna, L., Fuse, T., Baba, T., Yamamoto, K., Umehara, Y., Nagamura, Y. and Sasaki, T. (2000) *Hd1*, a major photoperiod sensitivity quantitative trait locus in rice, is closely related to the Arabidopsis flowering time gene *CONSTANS*. *Plant Cell*, **12**, 2473–2484.

Yu, H., Xie, W., Wang, J., Xing, Y., Xu, C., Li, X., Xiao, J. and Zhang, Q. (2011) Gains in QTL detection using an ultra-high density SNP map based on population sequencing relative to traditional RFLP/SSR markers. *PLoS ONE*, **6**, e17595.

Zhao, K., Wright, M., Kimball, J., Eizenga, G., McClung, A., Kovach, M., Tyagi, W., Ali, M.L., Tung, C.W., Reynolds, A., Bustamante, C.D. and McCouch, S.R. (2010) Genomic diversity and introgression in *O. sativa* reveal the impact of domestication and breeding on the rice genome. *PLoS ONE*, **5**, e10780.

Zhao, K., Tung, C.W., Eizenga, G.C., Wright, M.H., Ali, M.L., Price, A.H., Norton, G.J., Islam, M.R., Reynolds, A., Mezey, J., McClung, A.M., Bustamante, C.D. and McCouch, S.R. (2011) Genome-wide association mapping reveals a rich genetic architecture of complex traits in *Oryza sativa*. *Nat. Commun.* **2**, 467.

Zhou, B., Qu, S., Liu, G., Dolan, M., Sakai, H., Lu, G., Bellizzi, M. and Wang, G.L. (2006) The eight amino-acid differences within three leucine-rich repeats between Pi2 and Piz-t resistance proteins determine the resistance specificity to Magnaporthe grisea. *Mol. Plant Microbe Interact.* **19**, 1216–1228.

Zhuang, J., Shi, Y., Ying, J., Ezg, Z.R.Z., Chen, J. and Zhu, Z. (2006) Construction and testing of primary microsatellite database of major rice varieties in China. *Chinese J. Rice Sci.* **20**, 460–468.

14

The CRISPR/Cas9 system produces specific and homozygous targeted gene editing in rice in one generation

Hui Zhang[1], Jinshan Zhang[1,2], Pengliang Wei[1,2], Botao Zhang[1], Feng Gou[1,2], Zhengyan Feng[1,2,3], Yanfei Mao[1], Lan Yang[1], Heng Zhang[1], Nanfei Xu[1,]* and Jian-Kang Zhu[1,4,]*

[1]*Shanghai Center for Plant Stress Biology, Chinese Academy of Sciences, Shanghai, China*
[2]*University of Chinese Academy of Sciences, Shanghai, China*
[3]*Institute of Plant Physiology and Ecology, Shanghai Institutes for Biological Sciences, Chinese Academy of Sciences, Shanghai, China*
[4]*Department of Horticulture and Landscape Architecture, Purdue University, West Lafayette, IN, USA*

*Correspondence

e-mails nfxu@sibs.ac.cn and
jkzhu@purdue.edu
Hui Zhang, Jinshan Zhang, Pengliang Wei
and Botao Zhang contributed equally to this
work.

Summary

The CRISPR/Cas9 system has been demonstrated to efficiently induce targeted gene editing in a variety of organisms including plants. Recent work showed that CRISPR/Cas9-induced gene mutations in Arabidopsis were mostly somatic mutations in the early generation, although some mutations could be stably inherited in later generations. However, it remains unclear whether this system will work similarly in crops such as rice. In this study, we tested in two rice subspecies 11 target genes for their amenability to CRISPR/Cas9-induced editing and determined the patterns, specificity and heritability of the gene modifications. Analysis of the genotypes and frequency of edited genes in the first generation of transformed plants (T0) showed that the CRISPR/Cas9 system was highly efficient in rice, with target genes edited in nearly half of the transformed embryogenic cells before their first cell division. Homozygotes of edited target genes were readily found in T0 plants. The gene mutations were passed to the next generation (T1) following classic Mendelian law, without any detectable new mutation or reversion. Even with extensive searches including whole genome resequencing, we could not find any evidence of large-scale off-targeting in rice for any of the many targets tested in this study. By specifically sequencing the putative off-target sites of a large number of T0 plants, low-frequency mutations were found in only one off-target site where the sequence had 1-bp difference from the intended target. Overall, the data in this study point to the CRISPR/Cas9 system being a powerful tool in crop genome engineering.

Keywords: CRISPR/Cas9, targeted gene editing, rice.

Introduction

The lion's share of our food supply is directly or indirectly from plants. Efficient production and high yield of crops are hallmarks of advancement in human societies. Crop improvement either by simple selection or by breeding has been largely based on genetic diversity accumulated through a long history of natural mutations. As our demand on food accelerates, the pressure on crop breeding continues to build, and the need to create greater and targeted gene diversity becomes more urgent. In the last few decades, mutagenesis of plant genes by means of chemical treatment and radiation has been extensively used to create new traits in plants (Jacobsen and Schouten, 2007). Although these agents can greatly speed up the mutation process and led to more than 2250 new plant varieties (Jacobsen and Schouten, 2007), the resulting changes are random in the genes mutated and the direction of the mutations. For these reasons, plant biologists have been searching for ways to easily edit specific plant genes of interest to desired outcome.

Although targeted genome modification in plants was reported in the 1980s (Paszkowski *et al.*, 1988), for a long time, it remained far from practical breeding applications, waiting for improvement in efficiency. Later advances in the engineering of

site-specific nucleases, such as zinc-finger nucleases (ZFNs) and transcription activator-like effector nucleases (TALENs), have allowed more reliable targeted plant gene editing (Bogdanove and Voytas, 2011; Carroll, 2011). Several publications have reported successful modifications in native plant genes, but these methods are relatively tedious, expensive and with only a small number of genes modified to date (Chen and Gao, 2014). A more recently developed CRISPR/Cas system employing a Cas9 endonuclease and a guide RNA complex has shown much higher efficiency in targeted gene editing and sparked great enthusiasm in the research of this new method. It has been successfully used for efficient genome editing in human cell lines, zebra fish and mouse (Cong *et al.*, 2013; Hwang *et al.*, 2013; Wang *et al.*, 2013) and recently applied to gene modification in plants (Feng *et al.*, 2013; Jiang *et al.*, 2013; Li *et al.*, 2013; Mao *et al.*, 2013; Miao *et al.*, 2013; Nekrasov *et al.*, 2013; Shan *et al.*, 2013). These methods rely on engineered sequence-specific DNA-binding proteins (zinc finger or TALE) or small RNAs to target DNA nucleases to genes of interests to create double-stranded breaks (DSBs), which can lead to gene mutations due to nonhomologous end-joining (NHEJ) repair or gene replacement or correction as a result of homologous recombination-based repair (HR). Recent work in our laboratory has demonstrated that the mutations created at the target sites as a result of the CRISPR/Cas9 activities were stable and passed to next generations in Arabidopsis by the classic Mendelian inheritance, although the majority of mutations were somatic in nature in the early generation of Arabidopsis transformed with CRISPR/Cas9 (Feng *et al.*, 2014). It remains unclear whether the targeted gene editing by CRISPR/Cas9 is heritable and specific in a crop plant. Here, we show that the CRISPR/Cas9-induced mutations in targeted endogenous plant genes are readily detectable in T0 rice plant, a cereal crop, and most of the mutations are inherited in the progenies. With the exception of low-frequency off-target mutations detected in one site, our deep sequencing and other analysis did not detect any off-target mutations in multiple CRISPR/Cas9 transgenic lines, indicating that the mutagenesis effect of CRISPR/Cas9 is reasonably specific in rice.

Results

Efficiency of CRISPR/Cas9 system in causing target gene mutations in rice

To extensively evaluate the efficacy of the CRISPR/Cas9 system in making double-strand breaks (DSB) and hence allowing target gene mutation in rice, we tested a list of genes, *OsPDS, OsPMS3, OsEPSPS, OsDERF1, OsMSH1, OsMYB5, OsMYB1, OsROC5, OsSPP* and *OsYSA*, with diverse functions (Table S1). Some of these genes targeted, such as *OsDERF1* and *OsMYB1*, do not play major roles in plant development, and the knockout mutations are not expected to result in easily identifiable phenotype. Other selected targets, such as *OsPDS*, carry critical functions in pigment synthesis, and disruption of their functions leads to obvious phenotypes, such albino plants (Fang *et al.*, 2008). Rice has only one copy of the *OsEPSPS* gene that encodes a key enzyme in the synthesis of aromatic amino acids; if this gene is knocked out on both sets of chromosomes, the cells cannot survive and a mutant plant cannot be developed (Xu *et al.*, 2002).

After transformation of the CRISPR/Cas9 constructs in rice callus and regeneration of transgenic plants, many Cas9-positive T0 plants were identified for each of the targets and these plants were analysed to detect mutations in the targeted sequence regions (Table 1). DSB must have happened in nearly half of the T0 plants tested because 44.4% of them carried mutations.

Mutation rate varied in a wide range from target gene to target gene, that is, from 21.1% to 66.7%. For the two target sites in the same *OsYSA* gene, the difference was relatively small (43.1% and 66.7%).

It has been reported that the % GC content, targeting strand and targeting context of sgRNA targeting sequences may influence sgRNA efficacy (Wang *et al.*, 2014). In our study, in all, except one, of the 11 sgRNAs examined, those having a higher GC% content showed a higher editing efficiency (Table S2).

Homozygous mutations were found in the T0 plants of several targets, accounting for 3.8% of all T0 plants (Table 1) or 7.7% in plants carrying targeted mutations. There seems to be a tendency that targets that showed higher mutation rates were more likely to have homozygous mutations at T0. However, some targets did not have any homozygous mutations. This may be related to the importance of the target gene in plant growth and development, as discussed later.

Simultaneous targeting of two genes was also tested for several pairs of targets (Table 2). Mutation rates for most of the individual target sites in this experiment were similar to those in the experiments where only one site was targeted (compare Tables 2 and 1). The percentages of T0 plants that carried mutations at both target sites were similar to the expected double mutation rates, that is, the mutation rate at one site times the rate at the other site (Table 2). These data suggested that mutations at the two sites targeted by one construct likely occurred independently of each other and that the levels of Cas9 and guide RNAs were not limiting in the transgenic rice plants.

Variety and frequency of mutations

In a detailed study in Arabidopsis (Feng *et al.*, 2014), the NHEJ mutations resulting from CRISPR/Cas9 activities were predominantly short indels. In the current study in rice, mutation types and frequencies were followed in 11 target sites (Figure 1a). More than half of the mutations were insertions, and all insertions are of 1 bp. Deletions rated second only to insertions and were found in nearly half of the T0 transgenic plants with target gene mutations. The majority of deletions were short, although there were longer deletions that ranged from tens to hundreds of bps. There were about 4% of base substitutions in the targets, and all

Table 1 Percentage of T0 plants found with mutations in the target sequence

Target gene	Guide RNA	No. of plants examined	No. of plants with mutations	Mutation rate (%)	Putative homozygous mutations	
					Number	%
OsMYB1	SgRNA	36	24	66.7	4	11.1
OsYSA	sgRNA1	54	36	66.7	2	3.7
OsROC5	SgRNA	43	28	65.1	3	7.0
OsDERF1	SgRNA	55	28	50.9	3	5.5
OsYSA	sgRNA2	72	37	51.4	0	0
OsPDS	SgRNA	43	18	41.9	0	0
OsMSH1	SgRNA	27	10	37.0	1	3.7
OsMYB5	SgRNA	47	15	31.9	0	0
OsSPP	SgRNA	38	11	28.9	0	0
OsPMS3	SgRNA	38	10	26.3	4	10.5
OsEPSPS	SgRNA	38	8	21.1	0	0
Average		44.6	20.5	44.4	1.5	3.8

Table 2 Mutation rates in T0 plants with 2 sites targeted

Target 1	Target 2	No. of T0 plants examined	No. of plants with mutations			Mutation frequency (%)			Expected double mutation frequency (%)*
			Site 1	Site 2	Both sites	Site 1	Site 2	Both sites	
OsMSH1	OsDERF1	30	15	17	10	50.0	56.7	33.3	28.3
OsMSH1	OsPDS	55	22	31	18	40.0	56.4	32.7	22.5
OsPDS	OsPMS3	35	11	6	3	31.4	17.1	8.6	5.4
OsPDS	OsDERF1	35	13	6	2	37.1	17.1	5.7	6.4

*Expected double mutation rate if mutation at each site is independent of each other (product of mutation rates at 2 individual sites).

of those substitutions involved just 1 bp. Overall, about 3/4 of all mutations only changed 1 bp. Cas9 cleaves double-strand DNA at a position three base pairs upstream of the PAM sequence (Jinek et al., 2012). All but one 1-bp deletions occurred right upstream of this DSB position, at the 4th base from the PAM site, and 100% of the 1-bp insertions were also at this position (Table S3). When the base composition of the 1-bp insertions was examined, most of them were A (44.8%) or T (43.4%) insertions (Figure 1b).

Considering targets and mutation types, some targets showed different mutation types than most other targets (Figure 1c), although the number of events tested for some targets was limited. For example, all detected OsEPSPS mutations were of 1 bp. In contrast, nearly all of the mutations in the OsMSH1 target region were deletions, and the length of those deletions spread from 1 bp to more than 3 bp, with the majority 3 bp or longer. Although both insertions and deletions were abundant in

OsMYB1 and OsPDS, the distribution of mutation length was similar to that in OsMSH1.

Genotypes and segregations of mutants

Rice is a diploid with two copies of each gene, one copy on each chromosome in a chromosome pair. When the CRISPR/Cas9 components are inserted in the genome of a rice cell and begin to function, one or both copies of the target genes may be cleaved and mutated. If both copies of the target are mutated in an embryogenic cell before the cell undergoes division, the genotype of the cell may be homozygous if the mutations are the same, or bi-allelic if different mutations occurred on the two copies of the target gene. If only one copy of the target gene is mutated, a heterozygote is expected. If mutations happen after division of the transformed embryogenic cell, the plant regenerated may be chimeric (Figure 2). To investigate the timing of the mutations and their distribution in the regenerated T0 transgenic of the

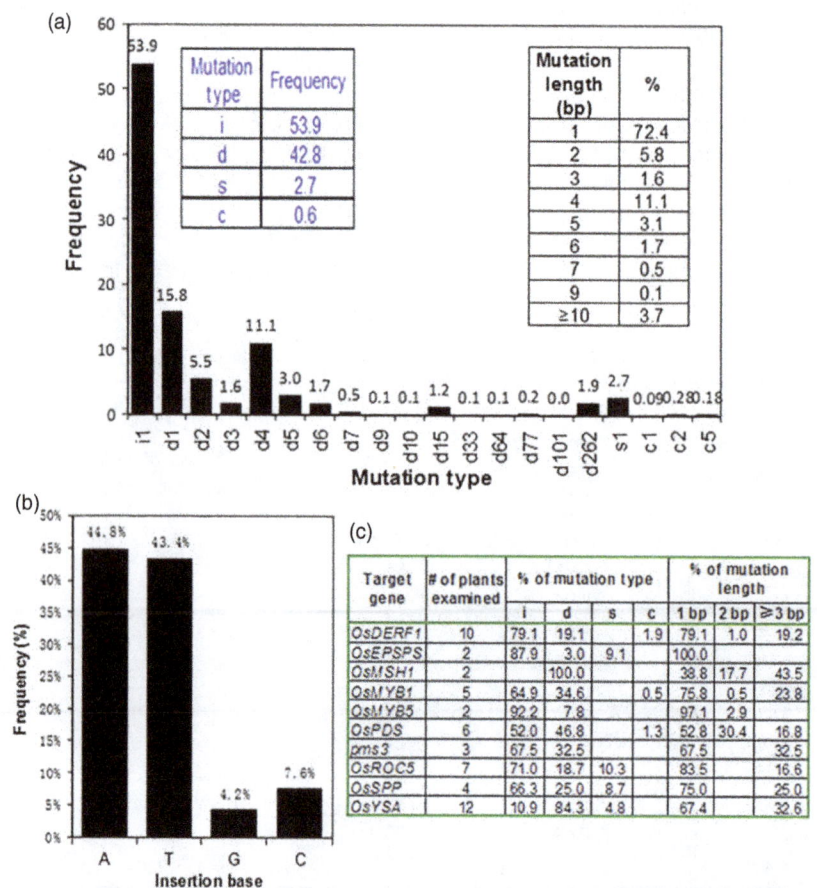

Figure 1 CRISPR/Cas9-induced mutation types and frequency. Graph (a), mutation types and frequency. Graph (b) and Table (c) are from combined data of 11 different targets at T0 generation. Left insert in (a), occurrence of deletion (d), insertion (i), substitution (s) and combined (c) mutation types. Right insert in (a), frequency of different mutation length. In x-axis: d#, # of base pair (bp) deleted from target site; i#, # of bp inserted at target site, c#, combined mutation. (b), percentage of different bases in the 1-bp insertion mutants. (c), mutation type and length for each target gene.

CRISPR/Cas9 system, the putative genotypes from a leaf sample from many T0 plants were examined and the data are summarized in Table 3. The segregation in the genotypes in T1 plants was also followed to confirm the genotypes at T0 (Table 3). Detailed sequencing information is in Table S3.

Of all of the T0 plants examined in Table 3, 10.5% (6/57) were putatively homozygous and 15.8% (9/57) were putatively bi-allelic, giving a total of 26.3%. Most T0 homozygotes carry 1-bp insertion (i1), and 1 T0 homozygote carries 1-bp deletion (d1), indicating that these two types of mutations happened at high frequency at the early stage of T0 plant development. Bi-alleles also predominantly carried the combination of i1 and d1 mutations. There were similar numbers of heterozygotes (12/57) as homozygotes and bi-alleles combined. However, some of these putative heterozygotes could be chimeras with some cells in the tissue sample homozygous with one mutation and other cells wild type. The most abundant T0 genotype was chimera, 40.4% (23/57). Seven of the 57 T0 plants were found to have no mutation at the target genes; none of the T1 descendants of these WT plants showed any mutation in the targets even in the presence of the Cas9 transgene, indicating that the CRISPR/Cas9 system was not functional in these transgenic events, possibly due to silencing of the *Cas9* and/or guide RNA transgenes.

A number of T1 plants from each of the T0 plants in Table 3 were examined for the genotype at the target sites. Given the large number of samples, only one sequencing run was performed on each sample; therefore, homozygotes and wild type could be clearly identified, but bi-alleles, heterozygotes and chimeras could not be distinguished and are marked as 'h' in Table 3. All T1 plants from 4 T0 homozygotes were homozygous for the same mutations, indicating that the mutations in these transgenic lines were stably passed to the next generation as expected by Mendelian law. Two of the homozygous T0 lines (T0-13 of *OsROC5* and T0-25 of *OsYSA*, marked by *) showed some unexpected genotypes in T1; this will be addressed in a later section. If the bi-allelic genotypes are inherited normally, segregation of 1xx:2xy:1yy is expected in T1. Although we did not confirm in every bi-allelic line that the 'h' genotype is the expected 'xy' bi-allele in T1 (marked * in Table 3), all the mutations, except one [d5 in *OsMYB1*, line T0-20(2)], in the 9 bi-allele T0 plants were found as homozygotes in T1 generation, and the ratios between the two mutations in a bi-allele were in the range of 1 : 1 by chi-square test. Similarly, homozygotes were found in T1 plants for all but one mutation in heterozygous T0 transgenic lines. As expected, the T1 segregation patterns of T0 chimeras were more diverse and less predictable, but homozygotes were found in T1 for the majority of the mutations detected in T0 chimeras.

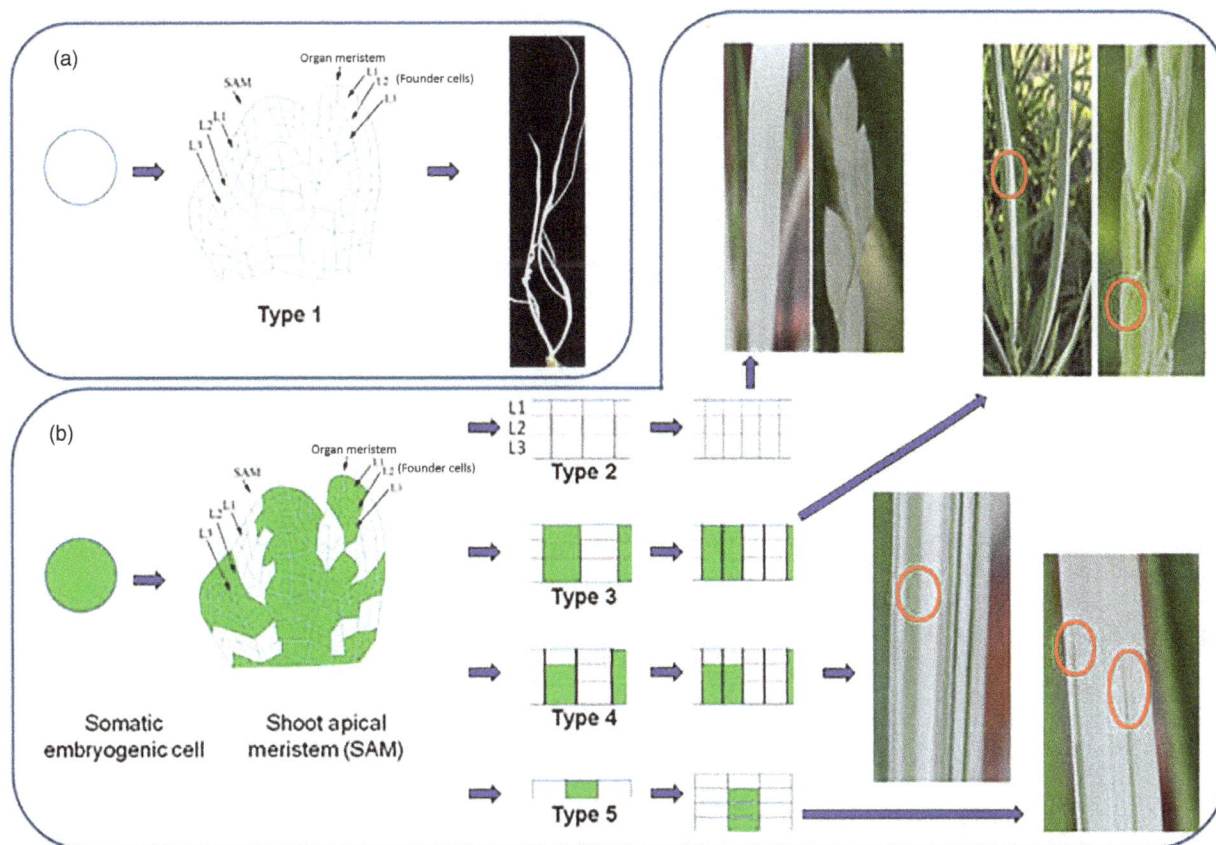

Figure 2 Diagramatic models of mutations in the cells transformed with the CRISPR/Cas9 system. The pictures shown were from T0 plants of the *OsPDS* target. Typical patterns of different mosaic types are marked by red circles. (a) When both copies of the *OsPDS* gene are simultaneously mutated, that is, in homozygous or bi-allelic plants, all the cells are albino (Type 1). (b) A cell with one or two copies of wild-type *OsPDS* gene is green. So, when two copies of the *OsPDS* are mutated after the division of the first embryogenic cell, the resulting cells have different genotypes and the regenerated plants exhibit different chimeric phenotypes (Types 2–5) depending on when the *OsPDS* gene is fully mutated.

Table 3 Segregation of CRISPR/Cas9-induced mutations in target genes

			T0			T1	
Target gene	sgRNA	Line*	Zygosity†	Genotype‡	Cas9§	Segregation ratio	Cas9
OsDERF1	sgRNA	T0-11	Homozygote	i1i1	+	8i1i1	nt
OsDERF1	sgRNA	T0-32	Homozygote	i1i1	+	21i1i1	nt
OsMYB1	sgRNA	T0-8	Homozygote	i1i1	+	20i1i1	nt
OsROC5	sgRNA	T0-13	Homozygote	10i1i1	+	9i1i1 : 2h : 9WT¶	20+
OsYSA	sgRNA1	T0-25	Homozygote	12d1d1	+	20d1d1 : 2h¶	22+
OsYSA	sgRNA1	T0-23**	Homozygote	i1i1	+	6i1i1	nt
OsYSA	sgRNA1	T0-20**	Bi-allele	5i1,6d1	+	3i1 : 1h	nt
OsROC5	sgRNA	T0-14**	Bi-allele	4i1a,5i1b	+	1i1ai1a : 3i1bi1b : 5h	nt
OsMSH1	sgRNA	T0-6	Bi-allele	11d1,7d4	+	7d1d1 : 5d4d4 : 9d1d4	19+ : 1−
OsMYB1	sgRNA	T0-10	Bi-allele	10i1,9d1	+	7i1i1 : 8d1d1 : 12i1d1	nt
OsMYB1	sgRNA	T0-20(1)	Bi-allele	6i1,5d5	+	3i1ai1a : 2d5d5 : 1i1bi1b : 14h¶	17+ : 3−
OsMYB1	sgRNA	T0-20(2)	Bi-allele	5i1,6d5	+	1i1ai1a : 2i1bi1b : 6h : 2WT¶	8+ : 2−
OsMYB5	sgRNA	T0-10	Bi-allele	8i1,7d1	+	8i1i1 : 6d1d1 : 15i1d1	nt
OsROC5	sgRNA	T0-8	Bi-allele	8d262,5i1	+	3i1i1 : 1d262d262 : 3h : 3WT¶	11+
OsYSA	sgRNA1	T0-35	Bi-allele	8d1a,5d1b	+	12d1ad1a : 5d1bd1b¶	10+ : 7−
OsYSA	sgRNA1	T0-35**	Heterozygote	6d1,5WT	+	3d1d1 : 8h : 2WT	nt
OsDERF1	sgRNA	T0-2(1)	Heterozygote	8i1,4WT	+	4i1i1 : 12h : 4WT	20+
OsDERF1	sgRNA	T0-2(2)	Heterozygote	7i1,4WT	+	4i1i1 : 11h : 1WT	15+ : 1−
OsDERF1	sgRNA	T0-7	Heterozygote	8i1,9WT	+	4i1i1 : 12h : 4WT	18+ : 2−
OsDERF1	sgRNA	T0-29	Heterozygote	7i1,13WT	+	4i1i1 : 20h : 3WT	nt
OsDERF1	sgRNA	T0-30(1)	Heterozygote	9d4,3WT	+	4d4d4 : 1d9d9 : 7h : 1WT	13+
OsDERF1	sgRNA	T0-30(2)	Heterozygote	6d4,5WT	+	3d4d4 : 1d9d9 : 7h : 1WT	10+ : 2−
OsMYB5	sgRNA	T0-1	Heterozygote	6i1,9WT	+	4i1i1 : 9h : 4WT	nt
OsPDS	sgRNA	T0-2(2)	Heterozygote	7i1,8WT	+	5i1i1 : 10h	nt
OsSPP	sgRNA	T0-17	Heterozygote	5i1,9WT	+	1h : 13WT	14+
OsYSA	sgRNA1	T0-38	Heterozygote	10d1,3WT	+	2d1d1 : 14h : 6WT	19+ : 3−
OsPMS3	sgRNA	T0-7	Heterozygote	2i1,12WT	+	5i1i1 : 6h : 7WT	nt
OsEPSPS	sgRNA	T0-16	Chimera	6i1a,1i1b,1s1,7WT	+	7h : 11WT	nt
OsEPSPS	sgRNA	T0-23	Chimera	4i1a,2i1b,1s1,8WT	+	8h : 18WT	nt
OsMSH1	sgRNA	T0-26	Chimera	3d2,4d15,3WT	+	17h : 2WT	19+
OsPDS	sgRNA	T0-2(1)	Chimera	8i1a,1i1b,4WT	+	3i1ai1a : 11h	nt
OsPDS	sgRNA	T0-2(3)	Chimera	5i1,1d4,6WT	+	6i1i1 : 7h : 1WT	nt
OsPDS	sgRNA	T0-2(4)	Chimera	11i1,1d6,3WT	+	3i1i1 : 6h : 2WT	nt
OsPDS	sgRNA	T0-4	Chimera	2d2,1d33,11WT	+	1d2d2 : 15h : 16WT	nt
OsPDS	sgRNA	T0-11	Chimera	7d2,1d6,3WT	+	11d2d2 : 21h : 2WT	nt
OsROC5	sgRNA	T0-28	Chimera	6i1a,5i1b,1i1c	+	20i1ai1a : 3i1bi1b : 1h	24+
OsROC5	sgRNA	T0-30	Chimera	1d1,1i1a,1i1b,1r1a,1r1b,6WT	+	24WT	24+
OsROC5	sgRNA	T0-43	Chimera	7i1,2s1,2WT	+	3i1i1 : 6h : 4WT	11+ : 2
OsROC5	sgRNA	T0-44	Chimera	1d7,5i1a,1i1b,1r1,7WT	+	1d1d1 : 4h : 18WT	23+
OsSPP	sgRNA	T0-5	Chimera	1d4,2d3,1s1,9WT	+	24WT	24+
OsSPP	sgRNA	T0-12	Chimera	2i1a,1i1b,9WT	+	24WT	24+
OsYSA	sgRNA1	T0-20	Chimera	1d1,1r1,10WT	+	2h : 20WT	24+
OsYSA	sgRNA1	T0-31	Chimera	5d1,1d3,1i1,4WT	+	12h : 12WT	18+ : 6−
OsYSA	sgRNA2	T0-9	Chimera	7i1,1r1,7WT	+	5d3d3 : 1d4d4 : 1i1i1 : 10h : 5WT	18+ : 4−
OsYSA	sgRNA2	T0-29	Chimera	1d4,1d3,1r1,9WT	+	13h : 17WT	25+ : 5−
OsYSA	sgRNA2	T0-31	Chimera	12d4,1d3,1d1	+	16d4d4 : 1d3d3 : 2d1d1 : 3h	24+
OsYSA	sgRNA2	T0-35	Chimera	5d4,4d6a,1d6b,1d6c,2i1	+	4d6cd6c : 12h	16+
OsYSA	sgRNA2	T0-47	Chimera	1d4,1r1,9WT	+	4h : 14WT	15+ : 3−
OsPMS3	sgRNA	T0-31	Chimera	1d77,14WT	+	9h : 13WT	nt
OsPMS3	sgRNA	T0-32	Chimera	2i1,1d3,12WT	+	6h : 21WT	nt
OsDERF1	sgRNA	T0-13	WT	WT	+	17WT	nt
OsDERF1	sgRNA	T0-15	WT	WT	+	20WT	20+

Table 3 Continued

| Target gene | sgRNA | Line* | T0 | | | Cas9§ | T1 | |
			Zygosity†	Genotype‡			Segregation ratio	Cas9
OsMYB1	sgRNA	T0-22	WT	WT		+	11WT	nt
OsROC5	sgRNA	T0-41	WT	WT		+	17WT	12+ : 5−
OsSPP	sgRNA	T0-14	WT	WT		+	10WT	8+ : 2−
OsYSA	sgRNA1	T0-17	WT	WT		+	24WT	20+ : 4−
OsYSA	sgRNA2	T0-50	WT	WT		+	24WT	15+ : 9−

*Line name is in the format of T0-#(#) with (#) specifying the different tillers from the same T0 plant. For example, T0-20(1) refers to tiller No. 1 of T0 plant line No. 20.
†The zygosity of homozygote, bi-allele, heterozygote in T0 plants is putative.
‡WT, wild-type sequence with no mutation detected.
§Presence of Cas9 sequence; +, Cas9 detected; −, Cas9 not detected; nt, not tested.
¶More data are needed to fully explain the T1 phenotypes. d#, # of bp deleted from target site; d#a, same number of deletion at one site; d#b, same number of deletion at other sites; i#, # of bp inserted at target site; i#a, same number of insertion at one site; i#b, same number of insertion of different nucleotide at the same site; c#, combined mutation; h, heterogeneous, more than one sequence detected in the sample.
**Lines in Kasalath background.

Table 4 Extended examination of genotypes of some transgenic plants

| Target gene | Line* | T0 genotype | | T1 segregation ratio | Sequences of 1 nonhomozygous plant |
		Leaf sample	Mixed sample*		
OsDERF1	T0-2(2)	7i1,4WT	8i1,1c1,11WT	4i1i1 : 11h : 1WT	5i1,2d4,5WT
OsDERF1	T0-30(1)	9d4,3WT	13d4,1d9,1c1a,1c1b,3WT	4d4d4 : 1d9d9 : 7h : 1WT	3i1,3d4,1d9,5WT
OsEPSPS	T0-23	4i1a,2i1b,1s1,8WT	1d1,18WT	8h : 18WT	3i1,3d2,5d15,1WT
OsPDS	T0-4	2d2,1d33,11WT	5i1,1d2,13WT	1d2d2 : 15h : 16WT	11d2,8d6
OsPDS	T0-11	7d2,1c1,3WT	11d2,1d4,7WT	11d2d2 : 21h : 2WT	9i1,6d2,1WT

See Table 3 for line naming, genotype denotation and T0 data collection. The DNA in the target region of one of the T1 plants that showed heterogeneous target sequences was PCR amplified, cloned, sequenced, and the data are listed in the last column.
*Mixed sample consists of shoots, panicle branches and glume.

The WT copy of the target gene in heterozygous and chimeric plants could continue to mutate either in T0 or T1 generation. If new mutations happened in T0, they are expected to be found in other parts of the T0 plants. To further investigate these possibilities, different tissues were sampled from some T0 plants and the sequences of one of the T1 plant that was heterogeneous in first round sequencing were fully distinguished (Table 4). New mutations were found in other parts of all the T0 plants so examined. Mutations that were not detected in any of the T0 tissues were also found in T1 plants [e.g. d4 in line T0-2(2) of OsDERF1 and i1 in line T0-11 of OsPDS] (Table 4).

Off-target analysis

To evaluate the potential off-target effects of CRISPR/Cas9 in rice, we carried out whole genome sequencing on 7 T0 rice lines, three in Kasalath (Kas) (Feng et al., 2013) and four in Nipponbare (Nip) background, targeting OsYSA, OsROC5 and OsSPP, and 2 T1 lines in Nip targeting OsDERF1 and OsMYB1 (Table S4). The Nip and Kas wild-type plants were also included as controls in genome resequencing and subsequent analysis (Table S4). The sequencing results of the lines in Nip background were compared with the GenBank japonica reference and those in Kas to the indica reference, to detect SNPs and indels. As shown in Table S5, the number of SNPs and indels detected in the CRISPR/Cas9 transgenic plants was comparable with their respective wild-type

control plants (Table S5). The slight difference in the number of indels and SNPs for the Kas wild-type sample was likely due to sequencing depth.

To identify the candidate off-target sites, we used a previously developed biased workflow (Feng et al., 2014). Two groups of putative target sites of CRISPR/Cas9 were searched by the BLASTN algorithm using the full 20-bp target sequence and the critical last 12-bp sequence (i.e. 9- to 20-bp fragment of the 20-bp target) as queries, respectively. Any of the putative off-target sites that bare a SNP or indel is listed in the overlap column of Table S6. The sequence reads at each of these overlaps were then manually examined. It was found that the putative off-target sites on this short list were either the target itself or sites that lack a NGG sequence critical for Cas9 activity, and there was no mutation in any of these latter sites. In the end, our analysis found no off-target site for all the samples sequenced in rice (Table S6). In contrast, the expected mutations in the intended target sites were easily identified for all samples (Figures S1–S9). We also selected several putative off-target loci that have a PAM sequence and high sequence similarity to the OsDERF1 and OsMYB1 target sites and to the sgRNA2 site of OsYSA, to further examine potential off-target effects (Table 5). DNA was extracted from a large number of transgenic plants, and the putative off-target sites PCR amplified and sequenced. Among all putative off-targets examined, mutations were detected in 7 of 72 plants at

Table 5 Mutations detected in the putative CRISPR/Cas9 off-target sites

Target	Name of putative off-target site	Putative off-target locus	Sequence of the putative off-target site	No. of mismatching bases	No. of plants sequenced*	No. of plants with mutations
OsDERF1	H1	Chr3:21146571-21146590	CCGGCTAGCGGCGGCGGCGATGG	5	20	0
	H2	Chr3:14051263-14051282	GTCATTAGCGGCGGCGGCGGCGG	3	20	0
	H3	Chr3:21685763-21685782	AGAATTTGCGGCGGCGGCGGCGG	3	20	0
	H4	Chr5:26767168-26767187	CGTCATAGCGGCGGCGGCGATGG	4	20	0
	H5	Chr9:15561391-15561410	ACCATTAGCGGCGGCGGCGGTGG	4	20	0
OsMYB1	H6	Chr3:14449191-14449210	CAGCGGCGTCGAGGCGTTGGCGG	5	20	0
	H7	Chr5:4662212-4662231	GGCGCGCGTCGAGGCGCAGGCGG	3	20	0
	H8	Chr12:2404144-2404163	CAGCGGCGTCGAGGCGTTGGCGG	5	20	0
OsYSA sgRNA2	H9	Chr11:1535478-1535497	CCGCGACAAGCACCTTCATGAGG	1	72	7
	H10	Chr6:26468378-26468397	CTGCATCGAGCACCTTCATTGGG	4	72	0
	H11	Chr11:8377872-8377891	CAAACCATAGCACCTTCATGAGG	7	72	0
	H12	Chr12:19162972-19162991	TCGGATCAAGCACCTTCATAGGG	6	72	0
	H13	Chr11:22996586-22996605	CAGACAAATGCACCTTCATGAGG	6	71	0

The PAM motif (NGG) is marked by a grey box; mismatching bases are shown in red color.

*For targets OsDERF1 and OsMYB1, T0 plants from 10 events and T1 plants from another 10 events were used in this test. For the OsYSA-sgRNA2 targeting, T0 plants from all available events were used. All the T0 or T1 plants listed in this table were Cas9 positive, and all of the lines targeting OsDERF1 and OsMYB1 harbored CRISPR/Cas9 induced mutations at the target sites.

one locus with a 1-bp mismatch outside the seed region of the OsYSA target site (Table 5). The result agrees with the reported Cas9/sgRNA specificity, which is determined by the position and number of mismatches in sgRNA target pairing (Jinek et al., 2012). Thus, a well-designed specific sgRNA, like the other sgRNAs in this study, would not target undesired sites.

In four homozygous T0 lines tested that had 1-bp mutation, no new mutation or WT target sequence was found in any of the T1 plants examined (Table 3). To test whether there is any remutation in the bi-alleles, more extensive sequencing was performed on both T0 and T1 plants for five of the transgenic lines listed in Table 3. The new data are summarized in Table 6. For the first three transgenic lines in this table, the mutation genotypes in all other tissues on T0 plants were the same as the genotype originally detected in the one-leaf sample, and segregations found in T1 plants all followed classic Mendelian inheritance (1x:2xy:1y) with no new mutations or reversion (Table 6). The lack of new mutations in the progenies of homozygous or bi-allelic mutant plants suggests that the mutated genes which have one mismatch with the sgRNAs cannot be targeted by the sgRNAs. This is consistent with the notion that sequences

proximate to PAM are important for cleavage by Cas9 (Jinek et al., 2012).

New mutations and WT sequence were found in some T1 plants of the other two putative bi-allelic T0 lines, but they were accounted for in other tissues in the T0 plants. Therefore, even though the one-leaf sample indicated that these 2 T0 plants were bi-alleles, they were actually chimeras; the new mutations and WT genotypes were likely from the chimeric parts of the T0 plants. Similarly, the unexpected genotypes found in T1 plants of line T0-13 of OsROC5 and those in T0-25 of OsYSA (Table 3) were likely because that the T0 plants of these two putative homozygous lines were actually chimeric.

CRISPR-Cas acts early during the regeneration of T0 plants

Rice is a diploid plant, and there are two copies of each target gene in the genome. If both copies of the target gene are mutated before the first division of the embryogenic cell, the regenerated T0 plant is expected to be either homozygous or bi-allelic. Because the vast majority of mutations were i1 with a significant percentage of d1 (Figure 1), the majority of homo-

Table 6 Extended examination of genotypes of some transgenic plants

Target gene	Line	T0 genotype			T1 segregation ratio
		Putative zygosity	Leaf sample	Mixed sample*	
OsMSH1	T0-6	Bi-allele	11d1,7d4	14d1,6d4	7d1d1 : 5d4d4 : 9d1d4
OsMYB1	T0-10	Bi-allele	10i1,9d1	10i1,8d1	7i1i1 : 8d1d1 : 12i1d1
OsMYB5	T0-10	Bi-allele	8i1,7d1	9i1,10d1	8i1i1 : 6d1d1 : 15i1d1
OsMYB1	T0-20(2)	Bi-allele	5i1a,6d5	4i1b,1d5,3d7,10WT	1i1ai1a : 2i1bi1b : 6h : 2WT
OsMYB1	T0-20(1)	Bi-allele	6i1a,5d5	4i1b,1d5,3d7,10WT	3i1ai1a : 2d5d5 : 1i1bi1b : 14h

See Table 3 for line naming and genotype denotation. The putative zygosity of T0 plants was based on the genotyping of the leaf sample only. For the T0 plants in this table, besides leaf samples, samples were also taken from other tissues to examine their mutations. For the first 3 transgenic lines in the table, the heterogeneous sequences (h in Table 3) were cloned and sequenced.

*Mixed sample consisted of shoots, panicle branches and glume.

Table 7 Mutation types in the *OsMYB1* and *OsDERF1* targets in T0 plants

Target gene	No. of plants examined	No. of plants with mutations	Mutation rate (%)	Putative homozygous		Putative bi-allelic		Putative heterozygous		Chimeric	
				Number	%	Number	%	Number	%	Number	%
OsMYB1	36	24	66.7	4	11.1	11	30.6	4	11.1	5	13.9
OsDERF1	55	28	50.9	3	5.5	2	3.6	8	14.5	15	27.3

Table 8 Extended examination of genotypes in different tissues of T0 plants

Target gene	Line number	Zygosity*	Genotype	
			Leaf	Mixed tissues†
OsDERF1	T0-11	Homozygote	i1i1	i1i1
OsDERF1	T0-32	Homozygote	i1i1	i1i1
OsMYB1	T0-8	Homozygote	i1i1	i1i1
OsMYB5	T0-10	Bi-allele	8i1,7d1	9i1,10d1
OsMSH1	T0-6	Bi-allele	11d1 7d4	14d1,6d4
OsMYB1	T0-10	Bi-allele	10i1,9d1	10i1,8d1
OsMYB1	T0-20(2)	Bi-allele	5i1a,6d5	4i1b,1d5, 3d7,10WT
OsMYB1	T0-20(1)	Bi-allele	6i1a,5d5	4i1b,1d5, 3d7,10WT
OsPDS	T0-11	Chimera	7d2,1d6,3WT	11d2,1d4,7WT
OsPDS	T0-2(4)	Chimera	11i1,1d6,3WT	12i1,7WT
OsDERF1	T0-30(1)	Chimera	9d4,3WT	13d4,1d9,1c5, 1c5a,3WT
OsEPSPS	T0-23	Chimera	4i1a,2i1b,6c, 19WT	1d1,18WT
OsPMS3	T0-32	Chimera	2i1,1d3,12WT	2i1,1d3,14WT
OsPDS	T0-4	Chimera	2d2,1d33, 11WT	5i1,1d2,13WT
OsDERF1	T0-2(2)	Chimera	7i1,4WT	8i1,1c2,11WT
OsMYB5	T0-1	Chimera	6i1,9WT	8i1,1d2,7WT
OsDERF1	T0-29	Heterozygote	7i1,13WT	7i1,3WT
OsDERF1	T0-7	Heterozygote	8i1,9WT	8i1,4WT

Discrepancy between different tissues from the same plant is underlined.

*The zygosity of homozygote, bi-allele, heterozygote in T0 plants is putative.

†Mixed sample consisted of shoots, panicle branches and glume.

zygotes are expected to be i1 with some d1. Also, the bi-allelic genotypes should mostly involve i1 and d1 mutations. The data in Table 3 match this prediction very well, demonstrating that the formation of homozygotes and bi-alleles followed the simple statistic rules and indicating that the mutation of the copies of the target genes is largely independent of each other. Because putative homozygotes and bi-alleles accounted for 26.3% of all T0 plants (Table 3), it can be calculated that 51.3% (square root of 26.3%) of the transformed rice embryogenic cells experienced target gene mutations before their first cell division. Some of the WT genotypes were not followed up in T1 for the data collection in Table 3, and the mutation rate may therefore be inflated. To have a more accurate estimate on the early mutation rate, the different mutation types in many transgenic lines of two targets were carefully examined and the data are summarized in Table 7. For *OsMYB1*, 11.1% homozygotes and 30.6% bi-alleles gave a

total of 41.7% fully mutated genotypes, which indicated a mutation rate of 64.6% in the first embryogenic cells (Table 7). Similar calculation of data shows a 30.2% mutation rate in the first embryogenic cells for the *OsDERF1* target. Such high mutation rates in such a short-time window suggested a high efficiency of the CRISPR/Cas9 system in rice. Some of the DSB could have been repaired without causing mutations; therefore, the DSB rate in those rice cells must be even higher.

Mutation types were examined in different tissues, that is, leaves, shoots, panicle branches and glum, in some transgenic lines. In homozygous and bi-allelic T0 lines, the mutations detected in leaves were the same as those found in other tissues mixed together (Table 8). In heterozygous and chimeric lines that still had wild-type target sequences, same mutations were in both the leaf sample and the mixed sample from the same plant, but different mutations were detected in the other tissues in some lines (Table 8). These results indicated that most of the mutations in different parts of a T0 plant came from a same source, likely the shoot apical meristem (SAM) because cells in different tissues all originated from the SAM. We also examined mutations in different tillers of the same plants (Table 9). Identical mutations were found in all tillers in each of the two transgenic lines (Table 9), and the tillers in the third line also shared the same mutations but also had small numbers of different mutations in some tillers. This result further supported the notion that the majority of the mutations detected in different parts of a T0 plant had their origin from SAM. These results were also consistent with the high rate of homozygous or bi-allelic mutations observed in T0 plants.

OsPDS encodes a phytoene desaturase; disruption of the function of this gene by mutation results in malfunction in pigment synthesis, thus creating albino tissues. This feature can be used to estimate the timing and pattern of mutations of this gene in cells with the CRISPR/Cas9 system. Based on the phenotypes observed in the T0 plants targeting the *OsPDS* gene, five types of DSB patterns and timing can be formulated (Figure 2).

When both copies of the *OsPDS* genes were mutated before the division of the first embryogenic cell, all the cells in the regenerated T0 plant should have the same genotype of the target gene, shown as Type 1 in Figure 2. Completely albino T0 plants were found in the transgenic events targeting *OsPDS* gene, indicating that this type of early mutation did occur. When some cells become homozygotes or bi-alleles after the first division of the embryogenic cell, the SAM developed from this cell, and thus the plant developed thereafter, would be chimeric. If the mutations happened shortly after the first cell division, large portions of the SAM and whole organs of the regenerated plant may have the same mutation genotype; this constitutes the Type 2 pattern in Figure 2. Plants of this mutation type would have albino strips in leaves, albino organs, panicle branches or whole

Table 9 Extended examination of genotypes in different tillers of the same plant

Target gene	Lines and tillers*	Zygosity[†]	Genotype[‡]
OsDERF1	T0-30(1)	Heterozygote	9d4,3WT
	T0-30(2)	Heterozygote	6d4,5WT
	T0-30(3)	Heterozygote	6d4,3WT
	T0-30(4)	Heterozygote	5d4,3WT
OsDERF1	T0-2(1)	Heterozygote	8i1,4WT
	T0-2(2)	Heterozygote	7i1,4WT
	T0-2(3)	Heterozygote	6i1,5WT
	T0-2(4)	Heterozygote	6i1,4WT
OsPDS	T0-2(1)	Chimera	8i1a,1i1b,4WT
	T0-2(2)	Heterozygote	7i1a,8WT
	T0-2(3)	Chimera	5i1a,1d4,6WT
	T0-2(4)	Chimera	11i1a,1d6,3WT

*The number in the bracket shows the different tillers from the same T0 plant.
[†]The zygosity of homozygote, bi-allele, heterozygote in T0 plants is putative.
[‡]Different mutations found in different tillers from the same T0 plant are underlined.

tillers. The other mutation types involved mutations in different L layers. Type 3 are those plants that had mutations in one or 2-L layers before periclinal cell division, to form uniform strips of different colours in the leaves or panicles. When mutations take place after some anticlinal cell division, there may be longitudinal difference in the genotype and colour in the strips of the leaves or other organs. This was observed as Type 5 as shown in Figure 2. Combination of Type 3 and Type 5 gave Type 4 phenotypes where both longitudinal and horizontal differences were noted.

Discussion

Our data suggested a high efficiency of the CRISPR/Cas9 system in rice. A significant percentage of T0 plants were homozygous or bi-allelic for the edited/mutated genes. However, not all target sequences were mutated in T0 plants as evidenced by an abundance of heterozygotes and chimeras. WT target sequences were also detected in T1 plants. This continued presence of nonmutated target sequence appears inconsistent with the high efficiency indicated from the large percentage of homozygous and bi-allelic T0 plants. One of the reasons for this discrepancy may be that there is a wide range in the activity of the CRISPR/Cas9 system in the different transgenic events, depending on where in the host genome the *CRISPR/Cas9* transgene is inserted. DSB and mutations may be slow or even very slow in some events where the activity of CRISPR/Cas9 system is low. Alternatively, it is possible that there is a high percentage of faultless repairs after DSB.

The mutation types and frequency seemed to be related to the target gene. The *OsPMS3* gene codes a noncoding RNA, and the target sequence is not within the 21 nt small RNA region critical for the function of this gene. The mutations of this target were predominantly insertions and deletion, with a significant percentage of the mutations involved three or more bps (Figure 1c). In contrast, *OsEPSPS* is a key enzyme in the pathway for aromatic amino acid synthesis. The loss of function in this gene results in lack of some essential amino acids, leading to plant death. Simultaneous mutation on both copy of the gene, homozygous or bi-allelic, may render the plant unsurvivable, hence not appear in T0 plants. Supporting this argument, no T1 *OsEPSPS* homozygous or bi-allelic mutants were recovered (Table 3). The data

for *OsPDS* were similar, although some T1 homozygous mutants were found for this target gene (Table 3). However, the number of transgenic lines available for this kind of comparison was small for some targets, and the relationships between target genes and mutation types need further investigation.

Fewer mutation types were detected in CRISPR/Cas9 transgenic rice in this study compared to those in Arabidopsis (Feng *et al.*, 2014). For example, only 1-bp insertions were found in rice, while many insertion lengths were reported in Arabidopsis. This could be caused by intrinsic DNA repair differences between these two species, the different transformation methods or the different sets of target genes. Some mutations may be less detrimental to the function of the gene so that the plant carry these mutations had better chance to survive. It could be that there is no significant difference in mutation types among different target genes, but only certain mutations for some targets can be tolerated and therefore selected through plant propagation.

Inheritance of the mutations in homozygous, bi-allelic and heterozygous T0 plants to form homozygotes in T1 suggested that most, if not all, of the mutations as a result of CRISPR/Cas9 activity are highly stable and can be inherited to next generations in the same way other normal genes follow. It is understandable that some of the mutations found in somatic cells only are lost because the cells did not contribute to the germ-line development, but plenty of mutations did end up in germ-line cells and most, if not all, mutation types found in T0 rice plants were also found in T1 (Table 3).

Conclusion

In summary, the CRISPR/Cas9 system was highly efficient in rice for generating DSB-induced target gene editing. Because of the high efficiency and action early during plant regeneration, homozygotes of edited target genes were readily found in T0 plants, the first generation. This is in contrast to the lack of any homozygotes or bi-alleles in the T1 generation of CRISPR/Cas9 transgenic Arabidopsis plants (Feng *et al.*, 2014). The presence of homozygotes of edited genes in CRISPR/Cas9 transgenic rice presents the fastest possible scenario in inducing changes to a crop plant genome, which greatly saves breeding time compared with traditional breeding or gene transformation. The edited genes were passed to the next generation following classic Mendelian law, without any detectable new mutation or reversion. Even with extensive searches, we could not find any evidence of large-scale off-targeting in rice for any of the many targets tested in this study, and only found one instance of low-frequency mutation in an off-target site that had 1-bp difference from the intended target. All the data in this study pointed to the CRISPR/Cas9 system being a powerful tool in crop improvement through targeted gene modification.

Experimental procedures

Plant materials, growth and generation of transgenic plants

The rice variety Nipponbare (*Oryza sativa* L. ssp. *japonica*) and Kasalath (*Oryza sativa* L. ssp. *indica*) were used as hosts in *Agrobacterium*-mediated transformation as previously described (Hiei *et al.*, 1997). The transgenic rice lines were grown in the paddy field in Shanghai (30_N, 121_E), China, during normal rice-growing seasons. Mature seeds were collected from T0 plants, dried and geminated for 2 days at 37 °C in the dark. The germinated seeds were then planted in soil, and the seedlings were grown under standard greenhouse conditions (16-h light at 30 °C/8-h night at 22 °C).

Vector construction

The sgRNA-Cas9 plant expression vectors were constructed as previously described (Feng et al., 2013; Mao et al., 2013). The oligos used in constructing the sgRNA vectors for OsDERF1, OsEPSPS, OsMYB5, OsPMS3 and OsPDS are listed in Table S8. Those for OsMYB1 can be found in Mao et al. (2013) and for OsROC5, OsYSA and OsSPP in Feng et al. (2013).

Genotyping

Total DNA was isolated using the cetyltrimethyl ammonium bromide (CTAB) method from transgenic plants (Rowland and Nguyen, 1993). PCR was performed to amplify the genomic region surrounding the CRISPR target sites using the specific primers (Table S8). The primer sequences used for amplifying targeted regions of OsMYB1, OsROC5, OsYSA and OsSPP were listed in Mao et al. (2013) and Feng et al. (2013). The PCR fragments were directly sequenced or cloned into the pMD18-T (TaKaRa, Dalian, China) vector and then sequenced by Sanger method to identify mutations.

Whole Genome Sequencing and bioinformatics analysis

The total DNA was extracted from about 100 mg leaf samples of rice. Libraries were constructed and used for high-throughput sequencing on Illumina Hiseq 2500 (Illumina, San Diego, CA, USA) according to the standard protocols. The data were analysed to identify the SNP and indels using the SAM tools software. To identify the potential off-target sites, the BLASTN algorithm was performed using the full 20-bp target sequence and the 12-bp seed sequence, respectively, as queries. Putative off-target mutations were searched by overlapping the aforementioned SNP/indels with the potential off-target sites. These putative off-target mutations were manually examined to see whether there were indeed any mutations and whether they were the targets themselves or whether they have a PAM sequence at the 3′ end.

Acknowledgements

The authors declare that they have no conflicts of interest with respect to this work. This work was supported by the Chinese Academy of Sciences.

References

Bogdanove, A.J. and Voytas, D.F. (2011) TAL effectors: customizable proteins for DNA targeting. Science, **333**, 1843–1846.

Carroll, D. (2011) Genome engineering with zinc-finger nucleases. Genetics, **188**, 773–782.

Chen, K. and Gao, C. (2014) Targeted genome modification technologies and their applications in crop improvements. Plant Cell Rep. **33**, 575–583.

Cong, L., Ran, F.A., Cox, D., Lin, S., Barretto, R., Habib, N., Hsu, P.D., Wu, X., Jiang, W., Marraffini, L.A. and Zhang, F. (2013) Multiplex genome engineering using CRISPR/Cas systems. Science, **339**, 819–823.

Fang, J., Chai, C., Qian, Q., Li, C., Tang, J., Sun, L., Huang, Z., Guo, X., Sun, C. and Liu, M. (2008) Mutations of genes in synthesis of the carotenoid precursors of ABA lead to pre-harvest sprouting and photo-oxidation in rice. Plant J. **54**, 177–189.

Feng, Z., Zhang, B., Ding, W., Liu, X., Yang, D.-L., Wei, P., Cao, F., Zhu, S., Zhang, F. and Mao, Y. (2013) Efficient genome editing in plants using a CRISPR/Cas system. Cell Res. **23**, 1229.

Feng, Z.Y., Mao, Y.F., Xu, N.F., Zhang, B.T., Wei, P.L., Wang, Z., Zhang, Z.J., Yang, D.L., Yang, L., Zeng, L., Liu, X.D. and Zhu, J-K. (2014) Multi-generation analysis reveals the inheritance, specificity and patterns of CRISPR/Cas induced gene modifications in Arabidopsis. Proc. Natl Acad. Sci. USA, **111**, 4632–4637.

Hiei, Y., Komari, T. and Kubo, T. (1997) Transformation of rice mediated by Agrobacterium tumefaciens. Plant Mol. Biol. **35**, 205–218.

Hwang, W.Y., Fu, Y., Reyon, D., Maeder, M.L., Tsai, S.Q., Sander, J.D., Peterson, R.T., Yeh, J.R.J. and Joung, J.K. (2013) Efficient genome editing in zebrafish using a CRISPR-Cas system. Nat. Biotechnol. **31**, 227–229.

Jacobsen, E. and Schouten, H.J. (2007) Cisgenesis strongly improves introgression breeding and induced translocation breeding of plants. Trends Biotechnol. **25**, 219–223.

Jiang, W., Zhou, H., Bi, H., Fromm, M., Yang, B. and Weeks, D.P. (2013) Demonstration of CRISPR/Cas9/sgRNA-mediated targeted gene modification in Arabidopsis, tobacco, sorghum and rice. Nucleic Acids Res. **41**, e188.

Jinek, M., Chylinski, K., Fonfara, I., Hauer, M., Doudna, J.A. and Charpentier, E. (2012) A programmable dual-RNA-guided DNA endonuclease in adaptive bacterial immunity. Science, **337**, 816–821.

Li, J.-F., Norville, J.E., Aach, J., McCormack, M., Zhang, D., Bush, J., Church, G.M. and Sheen, J. (2013) Multiplex and homologous recombination-mediated genome editing in Arabidopsis and Nicotiana benthamiana using guide RNA and Cas9. Nat. Biotechnol. **31**, 688–691.

Mao, Y., Zhang, H., Xu, N., Zhang, B., Gou, F. and Zhu, J.-K. (2013) Application of the CRISPR–Cas System for Efficient Genome Engineering in Plants. Mol. Plant, **6**, 2008–2011.

Miao, J., Guo, D., Zhang, J., Huang, Q., Qin, G., Zhang, X., Wan, J., Gu, H. and Qu, L.-J. (2013) Targeted mutagenesis in rice using CRISPR-Cas system. Cell Res. **23**, 1233.

Nekrasov, V., Staskawicz, B., Weigel, D., Jones, J.D.G. and Kamoun, S. (2013) Targeted mutagenesis in the model plant Nicotiana benthamiana using Cas9 RNA-guided endonuclease. Nat. Biotechnol. **31**, 691–693.

Paszkowski, J., Baur, M., Bogucki, A. and Potrykus, I. (1988) Gene targeting in plants. EMBO J. **7**, 4021–4026.

Rowland, L.J. and Nguyen, B. (1993) Use of polyethylene glycol for purification of DNA from leaf tissue of woody plants. Biotechniques, **14**, 734–736.

Shan, Q., Wang, Y., Li, J., Zhang, Y., Chen, K., Liang, Z., Zhang, K., Liu, J., Xi, J.J. and Qiu, J.-L. (2013) Targeted genome modification of crop plants using a CRISPR-Cas system. Nat. Biotechnol. **31**, 686–688.

Wang, H., Yang, H., Shivalila, C.S., Dawlaty, M.M., Cheng, A.W., Zhang, F. and Jaenisch, R. (2013) One-step generation of mice carrying mutations in multiple genes by CRISPR/Cas-mediated genome engineering. Cell, **153**, 910–918.

Wang, T., Wei, J.J., Sabatini, D.M. and Lander, E.S. (2014) Genetic screens in human cells using the CRISPR-Cas9 system. Science, **343**, 80–84.

Xu, J., Feng, D., Li, X., Chang, T. and Zhu, Z. (2002) Cloning of genomic DNA of rice 5-enolpyruvylshikimate 3-phosphate synthase gene and chromosomal localization of the gene. Sci. China, C, Life Sci. **45**, 251–259.

15

Down-regulation of *OsSPX1* caused semi-male sterility, resulting in reduction of grain yield in rice

Kang Zhang[†], Qian Song[†], Qiang Wei[†], Chunchao Wang, Liwei Zhang, Wenying Xu* and Zhen Su*

State Key Laboratory of Plant Physiology and Biochemistry, College of Biological Sciences, China Agricultural University, Beijing, China

*Correspondence
emails zhensu@cau.
edu.cn and x_wenying@yahoo.com
[†]These authors contributed equally to this work.

Keywords: *OsSPX1*, rice, pollen fertility, grain yield, expression profiling.

Summary

OsSPX1, a rice SPX domain gene, involved in the phosphate (Pi)-sensing mechanism plays an essential role in the Pi-signalling network through interaction with *OsPHR2*. In this study, we focused on the potential function of *OsSPX1* during rice reproductive phase. Based on investigation of *OsSPX1* antisense and sense transgenic rice lines in the paddy fields, we discovered that the down-regulation of *OsSPX1* caused reduction of seed-setting rate and filled grain number. Through examination of anthers and pollens of the transgenic and wild-type plants by microscopy, we found that the antisense of *OsSPX1* gene led to semi-male sterility, with lacking of mature pollen grains and phenotypes with a disordered surface of anthers and pollens. We further conducted rice whole-genome GeneChip analysis to elucidate the possible molecular mechanism underlying why the down-regulation of *OsSPX1* caused deficiencies in anthers and pollens and lower seed-setting rate in rice. The down-regulation of *OsSPX1* significantly affected expression of genes involved in carbohydrate metabolism and sugar transport, anther development, cell cycle, etc. These genes may be related to pollen fertility and male gametophyte development. Our study demonstrated that down-regulation of *OsSPX1* disrupted rice normal anther and pollen development by affecting carbohydrate metabolism and sugar transport, leading to semi-male sterility, and ultimately resulted in low seed-setting rate and grain yield.

Introduction

Rice (*Oryza sativa*), one of the major food staples for the world's population, is a model monocot plant for molecular biological study and a model crop for agronomical improvement. The rice grain yield is affected by many genetic and environmental factors, such as photosynthesis ability, nutrient efficiency, processes of pollination and fertilization, biotic and abiotic stresses, etc. Pollen fertility is a critical factor for rice yield. Dozens of genes are involved in anther and pollen development in rice and *Arabidopsis*, including anther cell differentiation, meiosis and pollen development (Wilson and Zhang, 2009). The development of pollen and anther requires nutrients such as sugars and lipids from source organs to support pollen development and maturation (Goetz *et al.*, 2001). Carbon Starved Anther (*CSA*), encoding a MYB transcription factor, is involved in sugar partitioning and the *csa* mutant showed low carbohydrate level in later anthers with male sterility (Zhang *et al.*, 2010). Defective Pollen Wall (*DPW*) is a fatty acyl reductase, and the mutant *dpw* showed defective anther development and degenerated pollen grains with an irregular exine (Shi *et al.*, 2011). The tapetum degeneration triggered by a programmed cell death (PCD) process provides cellular contents supporting pollen wall formation (Wu and Cheun, 2000). The *TDR* (Tapetum Degeneration Retardation) encodes a putative transcription factor with a bHLH domain. In the *tdr* mutant, the tapetum PCD was retarded in the anther with aborted pollen development (Li *et al.*, 2006). The *Arabidopsis* ortholog *AMS* (Aborted Microspores) showed similar function (Xu *et al.*, 2010). UDP-glucose pyrophosphorylase (UGPase) is a key enzyme in carbohydrate metabolism, producing UDP-glucose for sucrose synthesis in leaves. In

Arabidopsis pho1 mutants, *Ugp* was found to be up-regulated under conditions of phosphate deficiency (Ciereszko *et al.*, 2001). Rice *Ugp1* is essential for pollen callose deposition, and the *Ugp1*-silenced plants showed thermosensitive male sterility (Chen *et al.*, 2007). Rice *OsUgp2* is a pollen-preferential gene and plays a critical role in starch accumulation during pollen maturation (Mu *et al.*, 2009).

Phosphorus is one of the major mineral nutrients for plant growth and development. Phosphate (Pi) has regulatory function in reactions of photosynthetic carbon metabolism and is involved in the photosynthetic carbon assimilation and carbon partitioning processes between starch and sucrose (Rao, 1996), through the operation of the Pi translocator to facilitate a rapid exchange of Pi, triose-P and 3-phosphoglyceric acid (PGA) (Flugge, 1995). The Pi concentration inside and outside the chloroplast could affect the photosynthetic carbon reduction and control the balance between starch in chloroplast and sucrose in the cytosol (Rao, 1996). The dynamic interactions between sink and source tissues affected the response of photosynthesis to phosphate limitation (Pieters *et al.*, 2001).

There exists close relationship between phosphate signalling and rice reproductive development. Rice plants can accumulate abundant Pi in vegetative organs such as leaves at the early developmental stage and transport the Pi stored in the leaves to reproductive organs such as panicle at the late developmental stage (Marschner, 1995). The rice phosphate transporter gene *OsPT8* is involved in Pi translocation from vegetative organs to reproductive organs in rice. The suppression of *OsPT8* resulted in lower seed-setting rate, higher phosphorus content in the panicle axis and decreased phosphorus content in unfilled grain hulls (Jia *et al.*, 2011). Under high Pi level, overexpressing *OsPHR2* up-regulated some Pi transporter genes and suppressed the growth parameters during harvest stage, for example lower seed-setting rate, lower tiller number and grain number. In particular, the *OsPHR2-Ov1* transgenic line showed disordered male reproductive organs such as twisted anther structures, few pollen grains and low pollen viability (Zhou et al., 2008). The *ltn1* (*OsPHO2*) mutant significantly reduced tiller number and fertility compare to WT (Hu et al., 2011). *OsSPX1*, as one of the Pi-dependent inhibitors of *OsPHR2* activity in rice (Wang et al., 2014), is involved in the Pi starvation signalling network related to *OsPHR2* and *OsPHO2* (Liu et al., 2010), but there is no any report about the effect of *OsSPX1* on pollen development and grain yield.

In plants, many SPX domain proteins (with SPX domain, named after the SYG1/Pho81/XPR1 proteins) were identified to be involved in the phosphate-related signal transduction pathway and regulation pathways. For example, the AtPHO1 (At3g23430) protein is involved in ion transport in *Arabidopsis* (Hamburger et al., 2002; Stefanovic et al., 2007; Wang et al., 2004, 2008); the AtSPX family proteins were considered as part of the phosphate-signalling pathways controlled by PHR1 and SIZ1 (Duan et al., 2008); AtSPX1 was identified as the inhibitor of PHR1 and the SPX1/PHR1 interaction was Pi-dependent (Puga et al., 2014); the *OsSPX1* is involved in the Pi-sensing mechanism (Liu et al., 2010); OsSPX1 and OsSPX2 are Pi-dependent inhibitors of OsPHR2 through the protein-protein interaction and are involved in the Pi-sensing process of rice (Wang et al., 2014); *OsSPX3* and *OsSPX5* are redundant genes negatively regulating root-to-shoot Pi translocation and restored phosphate balance under phosphate starvation (Shi et al., 2014); OsSPX4 protein is responsive to Pi concentration and regulates the activity of OsPHR2 with the protein–protein interaction (Lv et al., 2014). Besides the essential role of SPX domain proteins involved in phosphate signalling, some of them have other key functions. For example, a PHO1 family protein, SHB1, contains an N-terminal SPX domain and a C-terminal EXS domain and was reported to specifically regulate blue-light responses and/or possibly red and far-red light responses in *Arabidopsis* (Kang and Ni, 2006). Both SPX and EXS domains likely anchor SHB1 to a protein complex, and the SPX domain is critical for SHB1 signalling, which plays dual roles in photoperiodic and autonomous flowering (Zhou and Ni, 2009; 2010). Furthermore, SHB1 was identified as a positive regulator of *Arabidopsis* seed development that affected both cell size and number (Zhou et al., 2009).

Our previous study reported that constitutive overexpression of *OsSPX1* in tobacco and *Arabidopsis* plants caused the improvement of cold tolerance with decreasing total leaf Pi (Zhao et al., 2009) and down-regulation of *OsSPX1* caused transgenetic rice high sensitivity to cold and oxidative stresses in seedling stage (Wang et al., 2013). In the generated *OsSPX1* transgenic rice lines, we observed the *Ubi::OsSPX1*-antisense (down-regulation of *OsSPX1*) lines showed significantly lower seed-setting rate in the reproductive stage, which may be correlated with semi-male sterility in the *Ubi::OsSPX1*-antisense lines. In this study, we focused on the effect of *OsSPX1* on pollen development and grain yield with the *OsSPX1* antisense and sense transgenic rice lines. We conducted rice whole-genome GeneChip to elucidate the possible molecular mechanism and identified the downstream key genes involved in the relationship between *OsSPX1* and rice pollen development and grain yield. This work on *OsSPX1* may aid understanding of the possible novel functions of *OsSPX1* and

be greatly beneficial for improving plant growth and crop grain yield.

Results

Antisense of *OsSPX1* caused reduction of rice grain yield in paddy fields

To investigate the traits during reproductive stages, the *Ubi::OsSPX1-antisense* transgenic, *Ubi::OsSPX1-sense* transgenic and wild-type (WT) plants were planted in paddy fields. The construction of *OsSPX1* transgenic lines was described in our previous work (Wang et al., 2013). The expression levels of *OsSPX1* in the mature leaves of WT and transgenic rice lines were shown in Figure S1. *OsSPX1* were significantly suppressed in *Ubi::OsSPX1-antisense* transgenic lines (lines A1 and A2) compared to WT plants (left chart in Figure S1), and constitutively and more strongly expressed in *Ubi::OsSPX1-sense* transgenic lines (S1 and S2) than WT plants (right chart in Figure S1).

During the ripening phases, we compared the traits related to grain yield among the *Ubi::OsSPX1-antisense* transgenic lines, *Ubi::OsSPX1-sense* transgenic lines and WT plants. The *Ubi::OsSPX1-antisense* transgenic lines (A1 and A2) exhibited lower seed-setting rate and filled grain number (Figure 1). The panicles in lines A1 and A2 were straight while those in *Ubi::OsSPX1-sense* transgenic lines (S1 and S2) and WT plants were bent (Figure 1a). The harvested panicles of A1 and A2 lines were not mature, and the panicles of A1 were much smaller than those of WT, S1 and S2 (Figure 1b, the empty seeds in lines A1 and A2 were highlighted by white arrows). We separated the filled and unfilled grains of individual plants of each lines, and noticed the seed-setting ratio was significantly lower in A1 and A2 than those in WT, S1 and S2 (Figure 1c). The seed-setting rate in A1 and A2 lines was at least 50% lower compared to WT, S1 and S2 lines (the *t*-test results were significant, *P*-values were lower than 0.01). The lower seed-setting rate led to a reduction of grain yield in the *Ubi::OsSPX1-antisense* transgenic rice plants (Figure 1d), and the phenotype was stable for later generations (Figure S2).

Antisense of *OsSPX1* caused lower pollen viability with disordered pollen and anther

The pollen fertility is a critical factor for rice grain yield. We proposed that down-regulation of *OsSPX1* may affect the anther and pollen development in rice. We discovered that the *Ubi::OsSPX1-antisense* transgenic lines showed semi-male sterility during anther and pollen development. Figure 1e,f showed the anther phenotype of *Ubi::OsSPX1-antisense* transgenic lines, *Ubi::OsSPX1-sense* transgenic lines and WT plants in paddy fields. During the heading stage, the spikelets from *Ubi::OsSPX1-antisense* transgenic lines had smaller pale-yellow anthers compared to the normal yellow anthers from *Ubi::OsSPX1-sense* transgenic lines and WT (Figure 1e,f). Unlike mature pollen of WT and *Ubi::OsSPX1-sense* transgenic lines, large proportion of pollen of *Ubi::OsSPX1-antisense* transgenic lines could not be stained by Alexander's solution (Figure 1g,h). These results indicated that the *Ubi::OsSPX1-antisense* transgenic lines lacked normal mature pollen grains and this might cause the lower seed-setting rate and filled grain number. The pollen viability of different transgenic lines and WT rice plants in the heading stage was correlated with the seed-setting rate at the harvest stage (Figure 1d,h).

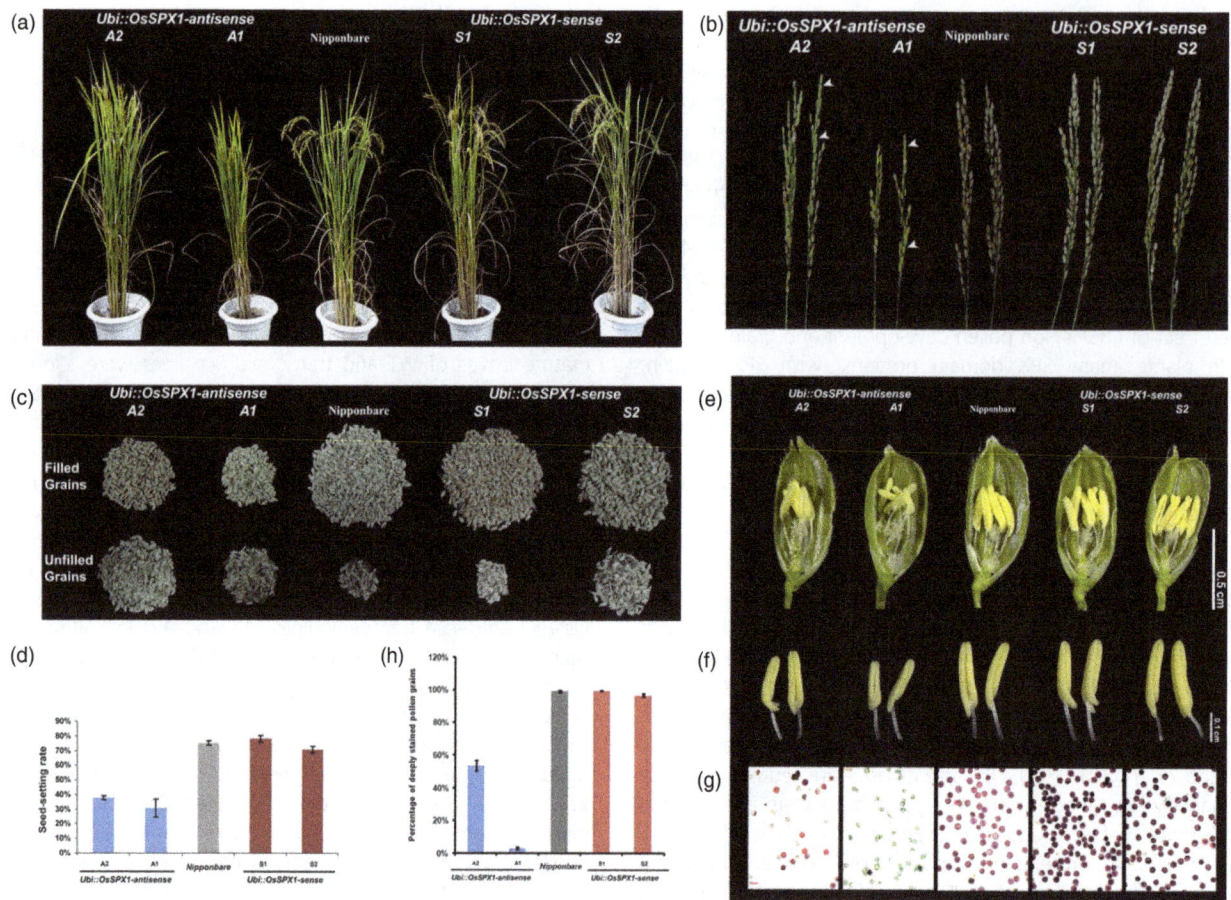

Figure 1 Agricultural traits and anther phenotypes of WT, *Ubi::OsSPX1-antisense* and *Ubi::OsSPX1-sense* transgenic rice lines. (a) Comparison of *Ubi:: OsSPX1-antisense* transgenic lines (lines A1, and A2, on the left), *Ubi::OsSPX1-sense* transgenic lines (lines S1, and S2, on the right) with the wild-type (WT) rice at the grain-filling stage grown in paddy fields. (b) The harvested panicles of *Ubi::OsSPX1-antisense* transgenic lines (lines A1, and A2, on the left), *Ubi::OsSPX1-sense* transgenic lines (lines S1, and S2, on the right) and the wild-type (WT) rice. (white arrows indicate empty seeds). (c) The total grains of the plant of *Ubi:: OsSPX1-antisense* transgenic lines (lines A1, and A2, on the left), *Ubi::OsSPX1-sense* transgenic lines (lines S1, and S2, on the right) and the wild-type (WT) rice. (d) Comparison of seed-setting rate of *Ubi::OsSPX1-antisense* transgenic lines (in blue bars), *Ubi::OsSPX1-sense* transgenic lines (in red bars) with the WT (in grey bars) rice. (e) Comparison of spikelets after removing half of the lemma and palea, Bar = 0.5 cm. (f) Comparison of anthers from *Ubi::OsSPX1-antisense* transgenic lines (lines A1, and A2, on the left), *Ubi::OsSPX1-sense* transgenic lines (lines S1, and S2, on the right) and the wild-type (WT) rice, Bar = 0.1 cm. (g) Comparison of pollen grains stained with Alexander's solution, Bar = 50 μm. (h) Percentage of stained pollen grains.

Furthermore, we examined the anther development by light microscopy and analysed the anther and pollen by scanning electron microscopy (SEM) and transmission electron microscopy (TEM) (Figure 2). At stage 12 of anther development (Zhang *et al.*, 2011), for WT and *Ubi::OsSPX1-sense* transgenic line S1, the pollen grains were full of starch and lipids and the tapetums were almost degenerated; whereas in the anther of *Ubi::OsSPX1-antisense* transgenic line A1, the microspores were hardly visible and the anther locule was almost empty (Figure 2a–c).

The cuticle on the exterior of the anthers of *Ubi::OsSPX1-antisense* transgenic line was not as well formed as those on the anther outer surface of the WT and *Ubi::OsSPX1-sense* transgenic line (Figure 2d–f). The Ubisch bodies on the inner locule surface were also different (Figure 2g–i), granular for WT and *Ubi:: OsSPX1-sense* transgenic line S1 and shrunken for *Ubi::OsSPX1-antisense* transgenic line A1. In addition, the pollen grains of WT and S1 had a smooth and particulate exine pattern, whereas the pollen grains of A1 appeared severely shrunken and empty (Figure 2j–l).

Further TEM observations showed consistent phenomena. At stage 12, the pollen grains of WT and *Ubi::OsSPX1-sense* transgenic line S1 were full of storage materials, with more starch granules and lipids in S1 and WT, whereas the *Ubi:: OsSPX1-antisense* transgenic line A1 pollen grains were collapsed with almost no accumulated storage materials (Figure 2m–o), the storage materials in the pollen of A1, WT and S1 were correlated with the Alexander staining results for their pollen (Figure 1g). Compared with those of the WT and S1, the pollen wall of A1 showed thicker exine layer (both tectum and nexines) and almost no intine layer (Figure 2p–r).

Transcriptome map of anthers from *OsSPX1* transgenic lines and WT plants

We conducted rice whole-genome GeneChip to elucidate the possible molecular mechanism underlying the down-regulation of *OsSPX1* in causing deficiencies in anther and pollen and lower seed-setting rate in rice. There were nine anther samples in total: three independent biological samples each were

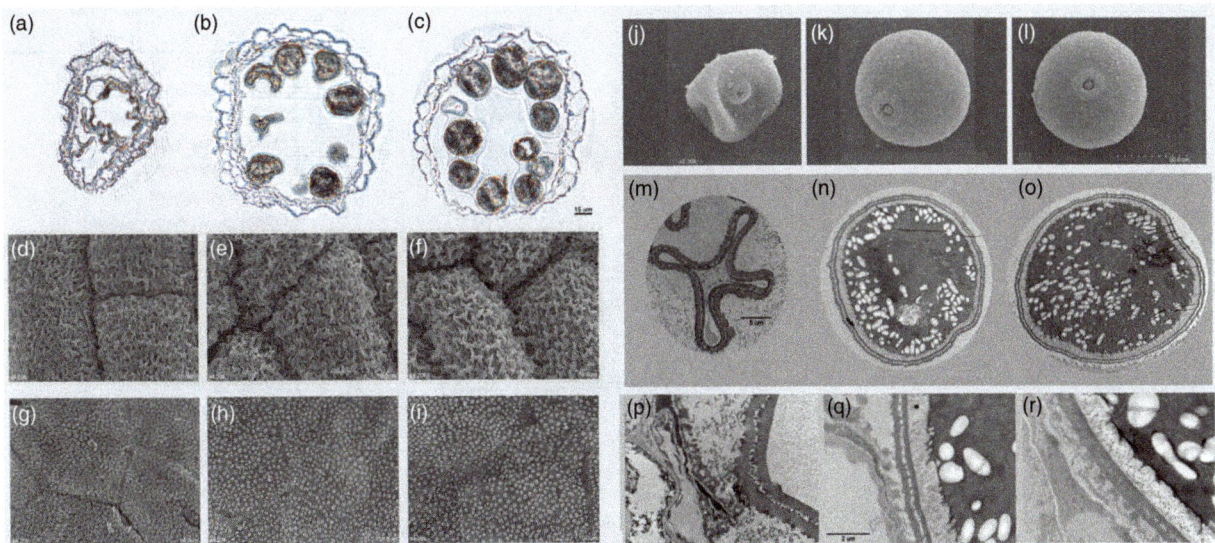

Figure 2 Pollen phenotypes among WT, *Ubi::OsSPX1-antisense* and *Ubi::OsSPX1-sense* transgenic rice lines. (a–c) Cross section of a single locule in stage 12 for A1, WT and S1, Bar = 15 μm. (d–f) Scanning electron microscopy analysis of the anther surface of A1, WT and S1 at stage 12, Bar = 10 μm. (g–i) Scanning electron microscopy analysis of the inner surface of the anther wall layers at stage 12 for A1, WT and S1, Bar = 10 μm. (j–l) Scanning electron microscopy analysis of the pollen grain at stage 12 for A1, WT and S1, Bar = 20 μm. (m–o) Transmission electron micrograph of the pollen grain at stage 12 for A1, WT and S1, Bar = 5 μm. (p–r) Transmission electron micrograph of the pollen wall at stage 12 for A1, WT and S1, Bar = 2 μm.

collected from *Ubi::OsSPX1-antisense* transgenic line A1, *Ubi::OsSPX1-sense* transgenic line S1 and WT (Nipponbare) plants during heading stage. We mainly focused on the differentially expressed probe sets between line A1 (*Ubi::OsSPX1-antisense*) and line S1 (*Ubi::OsSPX1-sense*), and between line A1 and WT. With ANOVA test ($P \leq 0.05$) and 1.5-fold change as cut-off, there were 867 probe sets significantly higher expressed in anthers of line S1 than A1, and 803 probe sets were significantly lower expressed in line S1 than A1; 333 probe sets were significantly higher expressed in WT than in line A1 anthers, and 1062 probe sets significantly lower expressed in WT than A1 (shown in Figure 3a, the detailed information for each probe set was listed in Table S1). The Venn diagram in Figure 3a illustrated the intersection of probe sets between each groups, there were 237 overlapped probe sets more highly expressed both in line S1 and WT plants compared to line A1 anthers, and 543 overlapped probe sets were less expressed both in line S1 and WT plants compared to A1.

We classified these differentially expressed genes using GeneBins (Goffard and Weiller, 2007). Figure 3b highlighted several classes (BINs) of genes significantly more highly expressed in line S1 than in A1, including sugar porter (SP) family, starch and sucrose metabolism, etc. As to the genes significantly more highly expressed in line A1, the enriched BINs included chitinase, lipid-transfer protein, etc. Further analysis with MapMan tool (Thimm et al., 2004) showed that there were several biological metabolism processes and large gene families in these differentially expressed genes (shown in Figure 3c). The genes related to cell wall, starch–sucrose, cell cycle and sugar transporter were highly expressed in line S1, whereas some transcription factor genes (such as WRKY, BHLH, ZIM, etc.) and hormone response genes (related to GA, ABA, ethylene, etc.) were highly expressed in line A1.

To further identify the co-expressed probe sets with similar expression patterns in the transgenic and WT plants, both self-organized mapping (SOM) and hierarchical methods were used

for clustering the 2403 differentially expressed probe sets among line S1, line A1 and WT (listed in Table S1). A colour heat map represented the relative expression level of each probe set across the nine anther samples (shown in Figure 4, from left to right, three replicates each of line A1, WT and line S1). These probe sets could be grouped into multiple clusters, and we focused on two of them: the top one representing the probe sets lower expressed in line A1 anther samples, and the bottom one representing the probe sets higher expressed in line A1. We applied Gene Ontology (GO) analysis-using agriGO (Du et al., 2010) and REVIGO (Supek et al., 2011)-to the selected clusters (shown in the right side of Figure 4). For the top cluster representing the probe sets lower expressed in line A1 anther samples, the enriched GO terms were mainly related to sucrose and starch metabolism (FDR P-value: 3.40E-16 and 7.20E-10, respectively), carbohydrate transport (FDR P-value: 4.10E-05), sugar:hydrogen symporter (FDR P-value: 6.10E-04), etc. As to the probe sets highly expressed in line A1 anther samples, the enriched GO terms included defence response (FDR P-value: 9.50E-05), chitinase activity (FDR P-value: 2.40E-07), etc.

Finally, we highlighted several enriched GO terms and functional groups for either lower or higher expressed in *Ubi::OsSPX1-antisense* transgenic lines (Figures 3 and 4) based on the results including gene cluster analysis, GeneBins, MapMan, and GO enrichment analysis. Some important genes were also highlighted in Table 1 and Table S3, were involved in sucrose and starch metabolism, sugar transporter, anther development, cell cycle and microtubule-based process, chitinase, and phenylalanine metabolism.

There were 15 genes selected for real-time RT-PCR validation (shown in Figure 5) based on the functional enrichment analysis of differentially expressed genes. Additional biological replicate anther samples were collected for real-time RT-PCR validation, and more than 90% of the tested genes confirmed the GeneChip results, especially for the seven sugar transporter genes (Figure 5). Compared to the GeneChip data, the fold change of

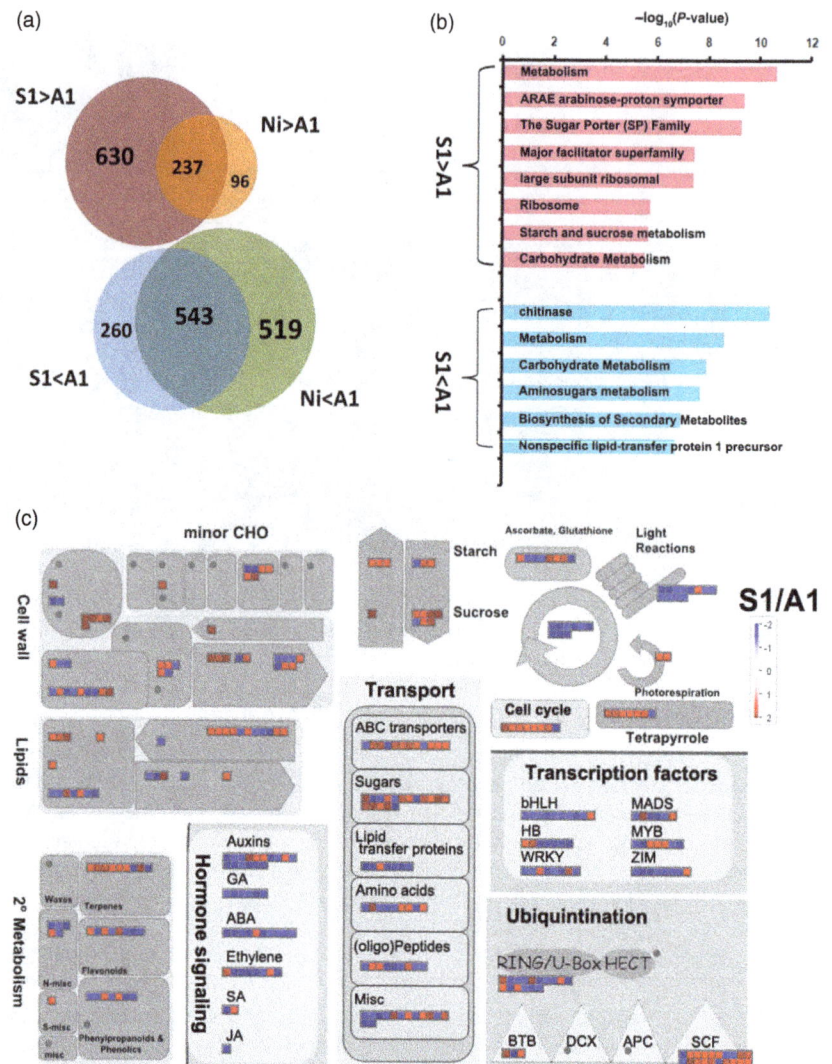

Figure 3 Venn diagrams analysis and MapMan view of differential expression probe sets in the anthers between line A1 and S1, and between line A1 and WT(Ni). (a) Venn diagrams illustrate the differential expression probe sets in the anthers among transgenic line A1, S1 and Ni (WT). (b) GeneBins analysis of the differential expression probe sets in the anthers between line A1 and S1. (c) MapMan view of the differential expression probe sets in the anthers between line A1 and S1. Fold changes are shown in colour, red boxes indicate up-regulated in S1, and blue boxes indicate up-regulated in line A1.

genes analysed by real-time RT-PCR was not exactly same, but the change trends were similar.

Discussion

Phosphorus is one of the major mineral nutrients for plant growth and development. Rice plants can accumulate abundant Pi in vegetative organs such as leaves at the early developmental stage and transport the Pi stored in the leaves to reproductive organs such as panicle at the late developmental stage. *OsSPX1* is involved in the Pi-signalling network related to *OsPHR2* and *OsPHO2*. *OsPHR2-Ov1* transgenic line showed disordered male reproductive organs. The *ltn1* (*OsPHO2*) mutant had significantly reduced tiller number and fertility compared to WT. We studied the potential function of *OsSPX1* during rice reproductive phase, and discovered that the *OsSPX1* antisense transgenic rice lines had lower seed-setting rate. Thus, we further investigated the anther and pollen development of the transgenic and wild-type plants and found that the pollen fertility was affected by antisense of *OsSPX1* gene, possibly through influence on pollen fertility. We conducted rice whole-genome GeneChip analysis to compare the gene expression profiling between wild-type and transgenic rice lines. GO and GeneBins analysis results

showed that the down-regulated genes in the *OsSPX1* antisense lines were significantly enriched in the following biological processes, including starch and sucrose metabolism, carbohydrate metabolism and sugar transport, anther development, cell cycle, etc.

During the pollen maturation, starch and lipids accumulate and the supply of photosynthetic assimilates (including sugar and lipids) from source organs is required (Goetz et al., 2001). In the functional enrichment analysis for the probe sets that had significantly lower expression in *Ubi::OsSPX1-antisense* transgenic line A1, GeneBins, MapMan and GO analyses all highlighted the sucrose and starch metabolism and sugar transporter categories. Starch biosynthesis is critical during pollen maturation and sterile pollen are normally starch deficient (Datta et al., 2002). In maize, several key sugar metabolic genes were lower expressed in late pollen stage of CMS male-sterile genotype than that of male-fertile genotypes, including SPP (sucrose 6-phosphate phosphohydrolase), IVR (invertase), HXK (Hexokinase), hexose transporter, etc. (Datta et al., 2002). In our results, two HXK genes (*LOC_Os05g09500* and *LOC_Os05g3111*) and one SPP homolog (*LOC_Os05g05270*) were significantly lower expressed in *Ubi::OsSPX1-antisense* transgenic line A1 (Table 1). Moreover, a dozen of sugar transporter genes

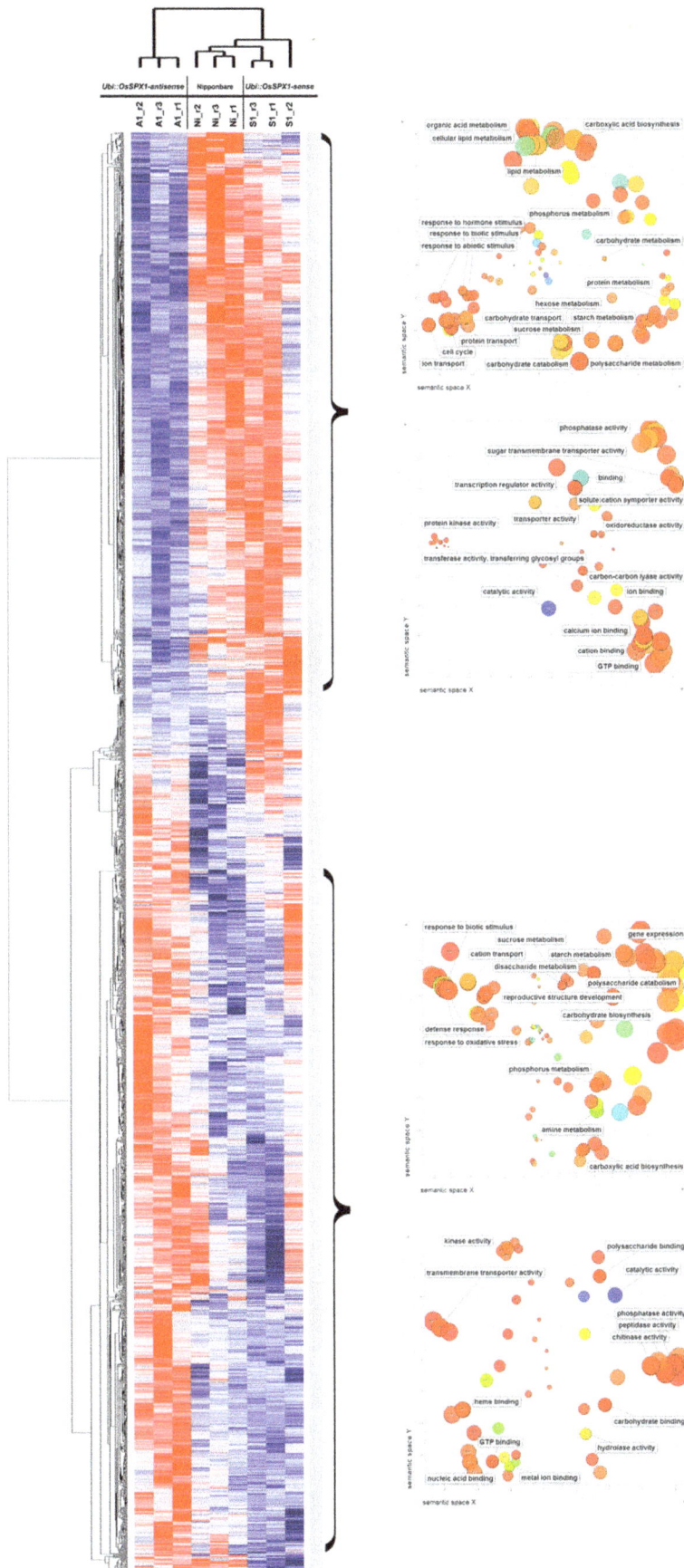

Figure 4 Cluster and gene ontology (GO) analysis of the differentially expressed probe sets in the anthers among transgenic line A1, S1 and WT(Ni). The overview hierarchical cluster result of 2403 probe sets showing differential expression in anthers among *Ubi::OsSPX1-sense* transgenic line (S1), WT(Ni) and *Ubi::OsSPX1-antisense* transgenic lines (A1); the red (high) and blue (low) colours represent the relative expression level across the samples. The marked groups represent these probe sets specifically down-regulated in line A1 (upper group) or up-regulated in line A1 (lower group). The charts on the right represent the enriched GO terms in the probe sets belonging to the marked group.

Table 1 Selected differentially expressed probe sets related to enriched function terms

| Probe set ID | S1 vs. A1 | | Ni vs. A1 | | Locus ID | Annotation |
	P-value	Fold change	P-value	Fold change		
Starch and sucrose metabolism						
Os.11216.1.S1_at	1.70E-02	1.62	3.07E-02	1.54	LOC_Os03g55090	Alpha-1,4 glucan phosphorylase, L isozyme
Os.57438.1.S1_at	3.57E-02	1.76	3.70E-02	1.76	LOC_Os02g52700	Alpha-amylase precursor
Os.52873.1.S1_a_at	3.30E-02	1.54	4.02E-02	1.51	LOC_Os05g25550	ATPase 7, plasma membrane-type
Os.52968.1.S1_at	9.61E-03	1.78	3.56E-02	1.56	LOC_Os11g19160	Beta-D-xylosidase
Os.50337.1.S1_at	4.43E-02	2.12	5.41E-01	1.29	LOC_Os04g33720	Beta-fructofuranosidase, insoluble isoenzyme 3 precursor
OsAffx.26489.1.S1_at	6.59E-03	1.81	4.12E-02	1.52	LOC_Os04g46760	Conserved hypothetical protein
OsAffx.28245.1.S1_at	2.32E-02	1.54	2.66E-02	1.52	LOC_Os07g03260	CSLC10—cellulose synthase-like family C
Os.52482.1.S1_at	3.96E-02	1.70	8.68E-02	1.55	LOC_Os07g36630	CSLF8—cellulose synthase-like family F; beta1,3;1,4 glucan synthase
Os.37822.2.S1_at	3.94E-02	1.60	3.90E-01	1.21	LOC_Os01g12030	Endoglucanase 1 precursor
Os.54770.1.S1_at	3.90E-03	1.97	1.89E-02	1.68	LOC_Os02g03120	Endoglucanase 1 precursor
Os.54812.1.S1_at	1.76E-02	2.61	2.98E-01	1.57	LOC_Os09g36060	Endoglucanase 1 precursor
OsAffx.3061.1.S1_x_at	4.89E-02	1.71	1.31E-01	1.50	LOC_Os02g54030	Endo-polygalacturonase precursor
OsAffx.17383.1.S1_at	1.73E-02	1.66	6.61E-01	1.09	LOC_Os08g37750	Exo-1,3-beta-glucanase
Os.5670.1.S1_at	4.15E-02	1.66	7.43E-02	1.55	LOC_Os08g23790	Exopolygalacturonase precursor
Os.8324.1.S1_a_at	3.32E-02	1.61	3.40E-02	1.61	LOC_Os01g66940	Fructokinase-1
Os.12780.1.S1_at	3.31E-02	1.60	1.57E-01	1.35	LOC_Os08g02120	Fructokinase-2
Os.24051.1.S1_at	5.27E-03	1.78	3.47E-01	1.19	LOC_Os04g33640	Glucan endo-1,3-beta-glucosidase 7 precursor
Os.24051.1.S1_x_at	4.54E-04	1.58	2.45E-02	1.25	LOC_Os04g33640	Glucan endo-1,3-beta-glucosidase 7 precursor
Os.24051.2.S1_at	4.89E-04	1.58	8.32E-02	1.18	LOC_Os04g33640	Glucan endo-1,3-beta-glucosidase 7 precursor
Os.33745.1.S1_at	9.96E-03	2.11	3.24E-02	1.83	LOC_Os01g58730	Glucan endo-1,3-beta-glucosidase GVI precursor
Os.53364.1.S1_at	3.53E-02	2.26	7.75E-01	1.14	LOC_Os07g40740	Heparanase-like protein 3 precursor
Os.6114.1.S1_at	1.79E-02	2.06	5.28E-02	1.79	LOC_Os05g09500	Hexokinase-1
OsAffx.4483.1.S1_at	1.25E-02	1.66	6.50E-02	1.42	LOC_Os05g31110	Hexokinase-1
Os.3414.1.A1_at	2.74E-02	1.65	2.07E-01	1.32	LOC_Os08g40930	Isoamylase
OsAffx.29137.1.S1_at	2.41E-02	1.53	1.73E-01	1.28	LOC_Os08g10604	Pectinesterase-1 precursor
Os.18244.1.S1_at	1.45E-02	1.87	4.68E-01	1.20	LOC_Os03g53790	Periplasmic beta-glucosidase precursor
Os.9731.1.S1_at	3.06E-02	1.68	1.71E-01	1.38	LOC_Os12g36810	Polygalacturonase
Os.4879.1.S1_a_at	2.50E-02	1.52	2.56E-01	1.22	LOC_Os01g47550	Ribokinase
Os.26441.1.S1_s_at	3.29E-02	1.70	2.69E-02	1.74	LOC_Os11g45710	SFR2
Os.12725.1.S1_at	3.09E-02	1.53	1.05E-01	1.36	LOC_Os06g06560	Soluble starch synthase 1, chloroplast precursor
Os.49091.1.S1_at	2.52E-02	2.05	6.12E-02	1.82	LOC_Os05g05270	Sucrose phosphate synthase
Os.25677.1.S1_at	4.21E-02	1.91	3.69E-01	1.34	LOC_Os03g28330	Sucrose synthase 2
Os.57465.1.S1_x_at	4.79E-02	1.52	2.95E-01	1.24	LOC_Os03g28330	Sucrose synthase 2
Os.49763.1.S1_s_at	1.03E-02	2.22	2.44E-02	1.99	LOC_Os03g17230	UDP-glucuronic acid decarboxylase 1
Membrane transport including sugar porter and ABCG						
Os.26786.1.S1_at	3.76E-02	1.56	1.79E-01	1.32	LOC_Os09g15330	Sugar transport protein 14
Os.50123.1.S1_x_at	1.05E-02	1.89	4.97E-02	1.59	LOC_Os07g10590	Sugar transport protein 8
Os.45939.1.S1_at	8.66E-03	1.81	2.24E-02	1.65	LOC_Os01g04190	Arabinose-proton symporter
Os.45939.1.S1_x_at	2.72E-02	1.55	7.65E-02	1.40	LOC_Os01g04190	Arabinose-proton symporter
Os.6624.1.S1_s_at	2.96E-02	1.60	6.09E-02	1.49	LOC_Os01g04190	arabinose-proton symporter
Os.27138.1.S1_at	2.32E-02	1.81	2.28E-01	1.36	LOC_Os10g21590	Carbohydrate transporter/sugar porter
Os.54757.1.S1_at	3.41E-02	1.71	1.11E-01	1.49	LOC_Os03g05610	Inorganic phosphate transporter 1-2
Os.50503.1.S1_at	2.47E-02	1.70	7.45E-02	1.51	LOC_Os03g04360	Inorganic phosphate transporter 1-7
OsAffx.30403.1.S1_at	1.65E-02	2.16	2.33E-01	1.47	LOC_Os03g03680	Major facilitator superfamily protein
OsAffx.30403.1.S1_s_at	3.49E-02	1.91	4.03E-01	1.30	LOC_Os03g03680	Major facilitator superfamily protein
Os.53018.2.S1_x_at	3.65E-03	1.73	1.46E-02	1.53	LOC_Os04g43210	Proton myo-inositol cotransporter
Os.31838.1.S1_at	2.10E-02	1.79	6.71E-02	1.57	LOC_Os03g24870	Solute carrier family 2, facilitated glucose transporter member 8
Os.26932.1.S1_at	1.38E-02	1.88	3.65E-01	1.25	LOC_Os07g37320	Sugar carrier protein C
Os.27613.1.A1_at	1.77E-02	1.74	1.81E-02	1.74	LOC_Os05g07870	Triose phosphate/phosphate translocator
Os.4704.1.S1_at	6.41E-02	1.49	5.61E-02	1.51	LOC_Os03g07480	Sucrose Transporter
Os.45486.1.S1_x_at	4.19E-02	1.82	9.03E-02	1.64	LOC_Os01g03144	ABC-2 type transporter family protein
Os.46480.1.S1_at	4.86E-02	1.65	1.97E-01	1.38	LOC_Os10g35180	ATP-binding cassette sub-family G member 2
Os.11454.2.S1_at	3.22E-02	1.58	9.25E-02	1.42	LOC_Os01g34970	Multidrug resistance protein 8

Table 1 Continued

Probe set ID	S1 vs. A1		Ni vs. A1		Locus ID	Annotation
	P-value	Fold change	P-value	Fold change		
Os.33212.2.S1_at	2.13E-02	1.53	9.72E-02	1.34	LOC_Os01g67580	Multidrug resistance-associated protein 9
Os.51901.1.S1_at	7.71E-03	1.91	2.19E-02	1.71	LOC_Os02g21750	Multidrug resistance protein 4
Os.409.1.S1_at	5.28E-03	1.58	2.32E-02	1.41	LOC_Os07g15460	Metal transporter Nramp6
Os.25771.1.S1_at	1.70E-02	1.68	1.88E-02	1.67	LOC_Os01g65000	Ammonium transporter 2
Os.2678.1.S1_at	3.94E-02	1.61	5.11E-01	1.16	LOC_Os02g13870	Aquaporin NIP1.2
OsAffx.22646.1.S1_at	4.66E-02	1.63	4.84E-02	1.62	LOC_Os03g61290	ATCHX19
OsAffx.26700.2.S1_at	5.33E-04	1.89	3.47E-04	1.96	LOC_Os05g02870	ATPase, coupled to transmembrane movement of substances
OsAffx.26700.2.S1_x_at	3.42E-04	1.86	2.07E-04	1.94	LOC_Os05g02870	ATPase, coupled to transmembrane movement of substances
Os.46553.1.S1_at	4.92E-02	1.73	7.97E-02	1.62	LOC_Os10g13830	ATPase, coupled to transmembrane movement of substances
Os.46553.2.S1_x_at	3.69E-02	1.76	4.96E-02	1.70	LOC_Os10g13830	ATPase, coupled to transmembrane movement of substances
Os.24908.1.S1_at	1.06E-02	1.83	3.31E-02	1.63	LOC_Os08g39950	Potassium transporter 17
Os.25736.1.S1_at	4.22E-02	1.58	2.57E-01	1.28	LOC_Os09g31486	Heat shock 70 kDa protein, Mitochondrial precursor
Os.38164.1.S1_at	4.57E-02	1.64	1.63E-01	1.40	LOC_Os12g38180	heat shock cognate 70 kDa protein 2
Anther development						
Os.49681.1.S1_at	3.60E-02	1.65	2.62E-01	1.30	LOC_Os03g07140	Male sterility protein 2, DPW
Os.18429.1.S1_x_at	4.16E-02	1.61	2.97E-01	1.27	LOC_Os01g63580	Glycerol-3-phosphate acyltransferase 8
Os.49822.1.S1_at	1.51E-02	1.75	2.75E-02	1.64	LOC_Os06g11970	MADS-box protein AGL66; MADS63
Os.53212.1.S1_at	1.20E-02	1.70	4.93E-02	1.48	LOC_Os04g21660	26S protease regulatory subunit 6A
Os.46480.1.S1_at	4.86E-02	1.65	1.97E-01	1.38	LOC_Os10g35180	ATP-binding cassette subfamily G member 2
Os.20530.1.S1_at	2.01E-02	1.59	2.19E-01	1.26	LOC_Os08g44530	Dihydroxy-acid dehydratase
Os.41468.1.S1_at	1.17E-02	2.01	7.91E-02	1.60	LOC_Os01g47050	Kelch motif family protein
Os.46553.1.S1_at	4.92E-02	1.73	7.97E-02	1.62	LOC_Os10g13830	ATPase, coupled to transmembrane movement of substances
Os.46553.2.S1_x_at	3.69E-02	1.76	4.96E-02	1.70	LOC_Os10g13830	ATPase, coupled to transmembrane movement of substances
OsAffx.12789.1.S1_s_at	9.09E-03	2.35	8.85E-02	1.72	LOC_Os03g08754	MADS-box transcription factor 47
Os.50337.1.S1_at	4.43E-02	2.12	5.41E-01	1.29	LOC_Os04g33720	Beta-fructofuranosidase, insoluble isoenzyme 3 precursor
Os.33948.1.S1_at	4.50E-02	1.69	2.89E-01	1.32	LOC_Os04g41110	N terminus of Rad21-/Rec8-like protein

were also lower expressed in this line (Table 1). Expression patterns of seven sugar transporter genes were validated by real-time RT-PCR (Figure 5). For example, *LOC_Os10g21590*, a carbohydrate transporter/sugar porter gene has its *Arabidopsis* homologs, *AtPMT1* and *AtPMT2* more highly expressed in mature or germinating pollen grains, as well as in growing pollen tubes (Klepek *et al.*, 2010). Analyses of reporter genes performed with promoter sequences showed expression in hydathodes and young xylem cells (both genes). For another gene of a sugar transport protein, *LOC_Os07g10590* – of, its *Arabidopsis* homologs, *AtSTP6* was only expressed during the late stages of pollen (Scholz-Starke *et al.*, 2003) and *AtSTP9* was specifically expressed in the male gametophyte (Schneidereit *et al.*, 2003). In addition, several *ABCG* (ATP-binding cassette transporter) genes were lower expressed in line A1 (Table 1), including *LOC_Os10g13830*, homolog of *Arabidopsis ABCG31*. It was reported that the many pollen grains in double mutant *abcg9/abcg31* were shrivelled up and collapsed when exposure to dry air (Schneidereit *et al.*, 2003). These results indicate that antisense of *OsSPX1*, possibly through regulation of phosphate homoeostasis, affects the expression levels of some key genes related to carbohydrate metabolism and sugar transport, and

then influences the transport of nutrients from source organs like flag leaves to sink organs like anthers.

Many genes were reported to be related to anther and pollen development both in rice and *Arabidopsis* (Wilson and Zhang, 2009). For example, both rice and *Arabidopsis* MIKC* type *MADS*-box genes showed conserved expression in the gametophyte, while *OsMADS62*, *OsMADS63* and *OsMADS68* were all specifically expressed late in pollen development (Liu *et al.*, 2013). In our result, the *OsMADS63* (*LOC_Os06g11970*) was significantly lower expressed in *Ubi::OsSPX1-antisense* transgenic line A1 (Table 1). Another male sterility gene, *DPW* (*LOC_Os03g07140*), was also lower expressed in line A1. The mutant *dpw* was reported to show defective anther development and degenerated pollen grains (Shi *et al.*, 2011).

Cell cycle process also plays an important role in the male gametophyte development (McCormick, 2004). The SPX protein in yeast, Pho81, is involved cyclin–cdk complex as a CDK inhibitor (Lee *et al.*, 2000; Lenburg and O'Shea, 1996). There were several cyclin genes significantly lower expressed in *Ubi:: OsSPX1-antisense* transgenic line A1, including members of cyclins A, B and D, as well as some important regulator genes

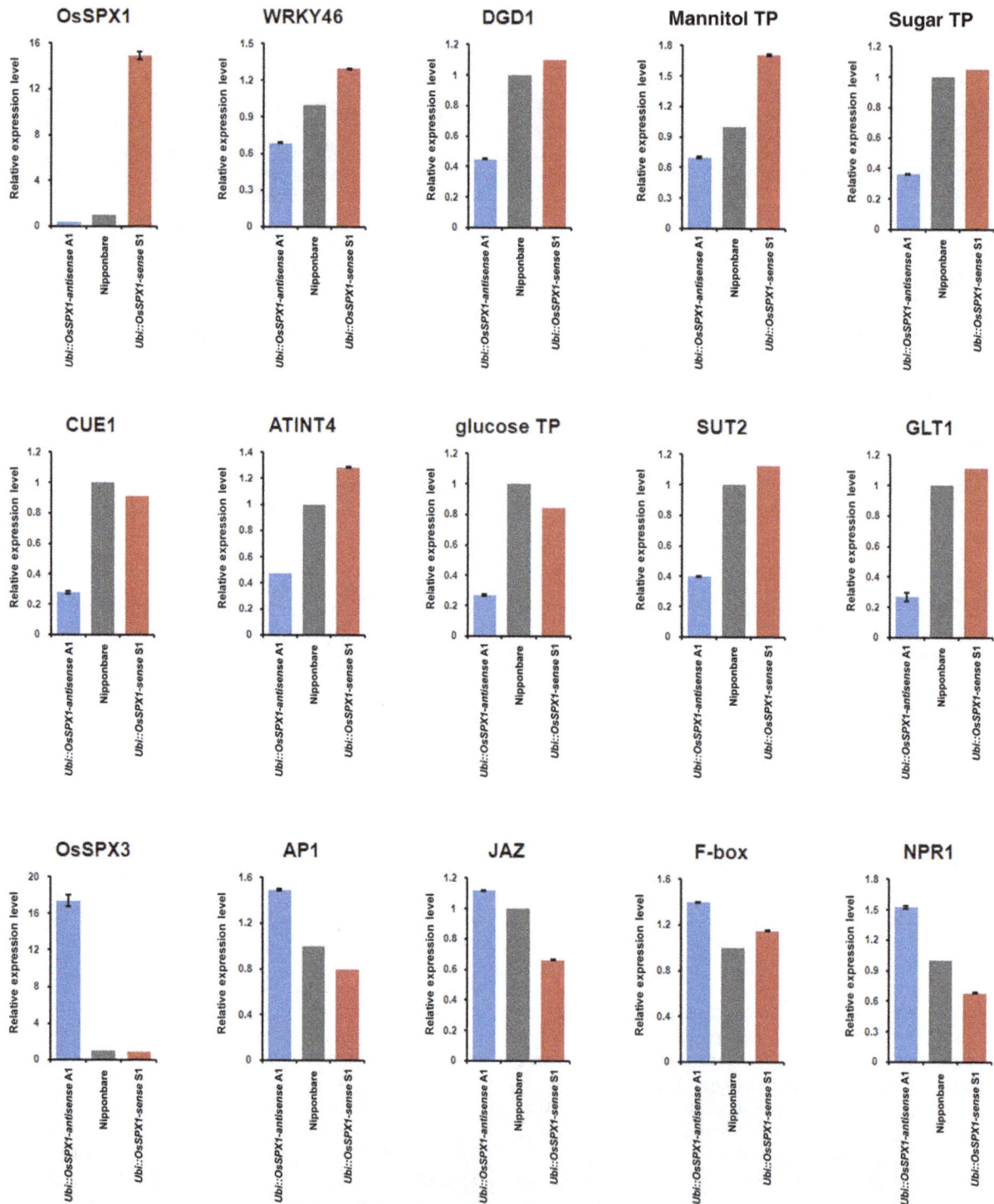

Figure 5 Real-time RT-PCR validation for selected probe sets in anthers. The probe sets were selected for real-time RT-PCR to validate the expression patterns among *Ubi::OsSPX1-sense* transgenic line (red bar), WT (grey bar) and *Ubi::OsSPX1-antisense* transgenic line (blue bar), the error bars represent the standard deviations of three replicates. The transcripts are as follows (**t**he primers for each probe set are listed in Table S2): OsSPX1—LOC_Os06g40120; IDS4-like protein; SPX domain containing protein; WRKY46—LOC_Os12g02440; WRKY transcription factor 46; DGD1—LOC_Os04g34000; digalactosyldiacylglycerol synthase 1; mannitol TP—LOC_Os10g21590; carbohydrate transporter/sugar porter; sugar TP—LOC_Os07g10590; sugar transport protein 8; CUE1—LOC_Os05g07870; triose phosphate/phosphate translocator; ATINT4—LOC_Os04g43210; proton myo-inositol cotransporter; glucose TP—LOC_Os03g24870; solute carrier family 2, facilitated glucose transporter member 8; SUT2—LOC_Os03g07480; sucrose transporter; GLT1—LOC_Os01g04190; arabinose-proton symporter; OsSPX3—LOC_Os10g25310; SPX domain containing protein; AP1—LOC_Os07g01820; MADS-box transcription factor 15; JAZ—LOC_Os07g42370; pnFL-2; F-box—LOC_Os10g41838; F-box protein interaction domain containing protein; NPR1—LOC_Os01g09800; regulatory protein NPR1; BTBA1—Bric-a-Brac, Tramtrack, Broad Complex BTB domain with Ankyrin repeat region.

(Table S3). For example, in the transgenic RNAi rice lines of *LOC_Os02g40450*, reduced expression level of the *ROCK-N-ROLLERS* gene resulted in reduced fertility with partially sterile flowers and defective pollens (Chang *et al.*, 2009). These data might indicate a connection between Pi starvation and cell cycle process, and their roles in male gametophyte development.

In addition, some phenylalanine metabolism pathway genes were affected by down-regulation of *OsSPX1* and were highly expressed in line A1, including phenylalanine ammonia-lyase (PAL), peroxidise, tropinone reductase, etc. (Table S3). Phenylalanine metabolism pathway is involved in the pollen development process. The transition of phenylpropanoids to flavonoids is considered as essential condition for viable pollen (Wiermann, 1970) and the *PAL* in tapetum cells of anthers might play an important role in pollen development (Kehrel and Wiermann, 1985). The activity of PAL protein was related to the number of fertile pollen grains at the flowering stage of broccoli (Kishitani *et al.*, 1993).

In brief, we discovered a novel role of *OsSPX1* in rice pollen fertility and grain yield using the *OsSPX1* antisense and sense transgenic rice lines. Our results showed that antisense of *OsSPX1* caused rice semi-male sterility and lower seed-setting rate. We further conducted rice whole-genome GeneChip analysis to elucidate the possible molecular mechanism and found that the enriched functional groups related to starch and sucrose metabolism, sugar porter, cell cycle, anther development, phenylalanine metabolism pathway, etc. Several genes related to male sterility and male gametophyte development were also lower expressed in *Ubi::OsSPX1-antisense* transgenic lines, such as *DPW* and *ROCK-N-ROLLERS*. These results may help us to understand the possible novel functions of *OsSPX1* involved in rice reproductive development and grain yield.

Experimental procedures

Plant materials

Seeds of rice (Nipponbare as WT, and *Ubi::OsSPX1-antisense* and *Ubi::OsSPX1-sense* transgenic lines) were surface-sterilized in 5% (w/v) sodium hypochlorite for 20 min and then washed in distilled water three or four times, then germinated in water for 2 day at room temperature and 1 day at 37 °C. The seedlings were planted in the paddy fields during the growing season in Beijing, China.

For phenotype evaluation: The spikelets of rice (Nipponbare as WT, and *Ubi::OsSPX1-antisense* and *Ubi::OsSPX1-sense* transgenic lines) were randomly collected at heading stage.

For RNA isolation: Anther samples were harvested from rice plants during heading stage under natural conditions in the paddy fields.

Characterization of anther and pollen phenotypes

Anthers of the sampled flowers were dissected and immersed in Alexander's solution (Alexander, 1969). Stained pollen grains were released from anthers and observed under light microscopy (Zeiss, A1, Thuringia, Germany). For SEM, fresh anthers were coated with palladium-gold in a sputter coater (Hummer), then observed and photographed by Hitachi S-3400N scanning electron microscope (Hitachi, Japan). For TEM observation, the anthers were fixed in formaldehyde acetic acid using standard plastic sections and Hitachi JEM-1230 (HC) transmission electron microscope (Hitachi, Japan) was used.

RNA isolation and real-time RT-PCR

All anther and flag leaf samples from transgenic lines and WT were homogenized in liquid nitrogen before isolation of the RNA. Total RNA was isolated using TRIZOL® reagent (Invitrogen, Carlsbad, CA) and purified using Qiagen RNeasy columns (Qiagen, Hilden, Germany). Reverse transcription was performed using Moloney murine leukaemia virus (M-MLV; Invitrogen). We heated 10 μL samples containing 2 μg of total RNA, and 20 pmol of random hexamers (Invitrogen) at 70 °C for 2 min to denature the RNA and then chilled the samples on ice for 2 min. We added reaction buffer and M-MLV to a total volume of 20 μL containing 500 μm dNTPs, 50 mm Tris–HCl (pH 8.3), 75 mm KCl, 3 mm $MgCl_2$, 5 mm dithiothreitol, 200 units of M-MLV and 20 pmol random hexamers. The samples were then heated at 42 °C for 1.5 h. The cDNA samples were diluted to 8 ng/μL for real-time RT-PCR analysis.

For real-time RT-PCR, triplicate quantitative assays were performed on 1 μL of each cDNA dilution using the SYBR Green Master Mix (PN 4309155; Applied Biosystems) with an ABI 7900 sequence detection system according to the manufacture's protocol (Applied Biosystems, Carlsbad, CA). The gene-specific primers were designed using PRIMER3 (http://frodo.wi.mit.edu/primer3/input.htm). The amplification of 18S rRNA was used as an internal control to normalize all data (forward primer, 5′-CGGCTACCACATCCAAGGAA-3′; reverse primer, 5′- TGTCAC-TACCTCCCCGTGTCA-3′). Gene-specific primers were listed in Table S2. The relative quantification method (ΔΔCT) was used to evaluate quantitative variation between replicates examined.

Affymetrix GeneChip analysis

For each sample, 8 μg of total RNA was used for making biotin-labelled cRNA targets. All the processes about cDNA and cRNA synthesis, cRNA fragmentation, hybridization, washing and staining, and scanning followed the GeneChip Standard Protocol (Eukaryotic Target Preparation). In this experiment, Poly-A RNA Control Kit and the One-Cycle cDNA Synthesis kit were applied. Affymetrix GCOS software was used to do data normalization and comparative analysis.

In order to map the probe set ID to the locus ID in the rice genome, the consensus sequence of each probe set was compared by BLAST (Basic Local Alignment and Search Tool) against the TIGR Rice Genome version 5. The cut-off e-value was set as 1e−20. The singular enrichment analysis (SEA) tool in agriGO (Du *et al.*, 2010) was applied for functional enrichment analysis of selected gene list, with the default parameters for Affymetrix Rice Genome Array. The functional enrichment analysis result was presented by REVIGO (Supek *et al.*, 2011) tool with its default parameter for *Oryza sativa* GO term background. The gene function categorization was based on the functional classification BINs of *Oryza sativa* from MapMan (Thimm *et al.*, 2004) and GeneBins (Goffard and Weiller, 2007).

Acknowledgements

We thank Qunlian Zhang for rice breeding; we thank Liqin Wei, Junzhen Jia, Haihong Liu and Yan Liang for their technical support with microscopy; we also thank Hong Yan for support on GeneChip experiment. This work was supported by grants from the Ministry of Science and Technology of China (31371291 and 2013CBA01402). The authors declare that they have no conflict of interest.

References

Alexander, M.P. (1969) Differential staining of aborted and nonaborted pollen. *Stain Technol.* **44**, 117–122.

Chang, L., Ma, H. and Xue, H.W. (2009) Functional conservation of the meiotic genes SDS and RCK in male meiosis in the monocot rice. *Cell Res.* **19**, 768–782.

Chen, R., Zhao, X., Shao, Z., Wei, Z., Wang, Y., Zhu, L., Zhao, J. *et al.* (2007) Rice UDP-glucose pyrophosphorylase1 is essential for pollen callose deposition and its cosuppression results in a new type of thermosensitive genic male sterility. *Plant Cell,* **19**, 847–861.

Ciereszko, I., Johansson, H., Hurry, V. and Kleczkowski, L.A. (2001) Phosphate status affects the gene expression, protein content and enzymatic activity of UDP-glucose pyrophosphorylase in wild-type and pho mutants of Arabidopsis. *Planta,* **212**, 598–605.

Datta, R., Chamusco, K.C. and Chourey, P.S. (2002) Starch biosynthesis during pollen maturation is associated with altered patterns of gene expression in maize. *Plant Physiol.* **130**, 1645–1656.

Du, Z., Zhou, X., Ling, Y., Zhang, Z. and Su, Z. (2010) agriGO: a GO analysis toolkit for the agricultural community. *Nucleic Acids Res.* **38**, W64–70.

Duan, K., Yi, K., Dang, L., Huang, H., Wu, W. and Wu, P. (2008) Characterization of a sub-family of Arabidopsis genes with the SPX domain reveals their diverse functions in plant tolerance to phosphorus starvation. *Plant J.* **54**, 965–975.

Flugge, U.-L. (1995) Phosphate translocation in the regulation of photosynthesis. *J. Exp. Bot.* **46**, 1317–1323.

Goetz, M., Godt, D.E., Guivarc'h, A., Kahmann, U., Chriqui, D. and Roitsch, T. (2001) Induction of male sterility in plants by metabolic engineering of the carbohydrate supply. *Proc. Natl Acad. Sci. USA,* **98**, 6522–6527.

Goffard, N. and Weiller, G. (2007) GeneBins: a database for classifying gene expression data, with application to plant genome arrays. *BMC Bioinformatics,* **8**, 87.

Hamburger, D., Rezzonico, E., MacDonald-Comber Petetot, J., Somerville, C. and Poirier, Y. (2002) Identification and characterization of the Arabidopsis PHO1 gene involved in phosphate loading to the xylem. *Plant Cell,* **14**, 889–902.

Hu, B., Zhu, C., Li, F., Tang, J., Wang, Y., Lin, A., Liu, L. *et al.* (2011) LEAF TIP NECROSIS1 plays a pivotal role in the regulation of multiple phosphate starvation responses in rice. *Plant Physiol.* **156**, 1101–1115.

Jia, H., Ren, H., Gu, M., Zhao, J., Sun, S., Zhang, X., Chen, J. *et al.* (2011) The phosphate transporter gene OsPht1;8 is involved in phosphate homeostasis in rice. *Plant Physiol.* **156**, 1164–1175.

Kang, X. and Ni, M. (2006) Arabidopsis SHORT HYPOCOTYL UNDER BLUE1 contains SPX and EXS domains and acts in cryptochrome signaling. *Plant Cell,* **18**, 921–934.

Kehrel, B. and Wiermann, R. (1985) Immunochemical localization of phenylalanine ammonia-lyase and chalcone synthase in anthers. *Planta,* **163**, 183–190.

Kishitani, S., Yomoda, A., Konno, N. and Tanaka, Y. (1993) Involvement of phenylalanine ammonia-lyase in the development of pollen in broccoli (*Brassica oleracea* L.). *Sex. Plant Reprod.* **6**, 244–248.

Klepek, Y.S., Volke, M., Konrad, K.R., Wippel, K., Hoth, S., Hedrich, R. and Sauer, N. (2010) *Arabidopsis thaliana* POLYOL/MONOSACCHARIDE TRANSPORTERS 1 and 2: fructose and xylitol/H+ symporters in pollen and young xylem cells. *J. Exp. Bot.* **61**, 537–550.

Lee, M., O'Regan, S., Moreau, J.L., Johnson, A.L., Johnston, L.H. and Goding, C.R. (2000) Regulation of the Pcl7-Pho85 cyclin-cdk complex by Pho81. *Mol. Microbiol.* **38**, 411–422.

Lenburg, M.E. and O'Shea, E.K. (1996) Signaling phosphate starvation. *Trends Biochem. Sci.* **21**, 383–387.

Li, N., Zhang, D.S., Liu, H.S., Yin, C.S., Li, X.X., Liang, W.Q., Yuan, Z. *et al.* (2006) The rice tapetum degeneration retardation gene is required for tapetum degradation and anther development. *Plant Cell,* **18**, 2999–3014.

Liu, F., Wang, Z., Ren, H., Shen, C., Li, Y., Ling, H.Q., Wu, C. *et al.* (2010) OsSPX1 suppresses the function of OsPHR2 in the regulation of expression of OsPT2 and phosphate homeostasis in shoots of rice. *Plant J.* **62**, 508–517.

Liu, Y., Cui, S., Wu, F., Yan, S., Lin, X., Du, X., Chong, K. *et al.* (2013) Functional conservation of MIKC*-Type MADS box genes in Arabidopsis and rice pollen maturation. *Plant Cell,* **25**, 1288–1303.

Lv, Q., Zhong, Y., Wang, Y., Wang, Z., Zhang, L., Shi, J., Wu, Z. *et al.* (2014) SPX4 Negatively regulates phosphate signaling and homeostasis through its interaction with PHR2 in rice. *Plant Cell,* **26**, 1586–1597.

Marschner, H. (1995) *Mineral Nutrition of Higher Plants.* London: ACADEMIC PRESS Harcourt Brace & Company, Publishers.

McCormick, S. (2004) Control of male gametophyte development. *Plant Cell,* **16** (Suppl.), S142–S153.

Mu, H., Ke, J., Liu, W., Zhuang, C. and Yip, W. (2009) UDP-glucose pyrophosphorylase2 (OsUgp2), a pollen-preferential gene in rice, plays a critical role in starch accumulation during pollen maturation. *Chin. Sci. Bull.* **54**, 234–243.

Pieters, A.J., Paul, M.J. and Lawlor, D.W. (2001) Low sink demand limits photosynthesis under P(i) deficiency. *J. Exp. Bot.* **52**, 1083–1091.

Puga, M.I., Mateos, I., Charukesi, R., Wang, Z., Franco-Zorrilla, J.M., de Lorenzo, L., Irigoyen, M.L. *et al.* (2014) SPX1 is a phosphate-dependent inhibitor of Phosphate Starvation Response 1 in Arabidopsis. *Proc. Natl Acad. Sci. USA,* **111**, 14947–14952.

Rao, I. (1996) Role of phosphorus in photosynthesis. In *Handbook of Photosynthesis.* (Pessarakli, M., ed), pp. 173–193. New York: Marcel Dekker.

Schneidereit, A., Scholz-Starke, J. and Buttner, M. (2003) Functional characterization and expression analyses of the glucose-specific AtSTP9 monosaccharide transporter in pollen of Arabidopsis. *Plant Physiol.* **133**, 182–190.

Scholz-Starke, J., Buttner, M. and Sauer, N. (2003) AtSTP6, a new pollen-specific H+-monosaccharide symporter from Arabidopsis. *Plant Physiol.* **131**, 70–77.

Shi, J., Tan, H., Yu, X.H., Liu, Y., Liang, W., Ranathunge, K., Franke, R.B. *et al.* (2011) Defective pollen wall is required for anther and microspore development in rice and encodes a fatty acyl carrier protein reductase. *Plant Cell,* **23**, 2225–2246.

Shi, J., Hu, H., Zhang, K., Zhang, W., Yu, Y., Wu, Z. and Wu, P. (2014) The paralogous SPX3 and SPX5 genes redundantly modulate Pi homeostasis in rice. *J. Exp. Bot.* **65**, 859–870.

Stefanovic, A., Ribot, C., Rouached, H., Wang, Y., Chong, J., Belbahri, L., Delessert, S. *et al.* (2007) Members of the PHO1 gene family show limited functional redundancy in phosphate transfer to the shoot, and are regulated by phosphate deficiency via distinct pathways. *Plant J.* **50**, 982–994.

Supek, F., Bosnjak, M., Skunca, N. and Smuc, T. (2011) REVIGO summarizes and visualizes long lists of gene ontology terms. *PLoS One,* **6**, e21800.

Thimm, O., Blasing, O., Gibon, Y., Nagel, A., Meyer, S., Kruger, P., Selbig, J. *et al.* (2004) MAPMAN: a user-driven tool to display genomics data sets onto diagrams of metabolic pathways and other biological processes. *Plant J.* **37**, 914–939.

Wang, Y., Ribot, C., Rezzonico, E. and Poirier, Y. (2004) Structure and expression profile of the Arabidopsis PHO1 gene family indicates a broad role in inorganic phosphate homeostasis. *Plant Physiol.* **135**, 400–411.

Wang, Y., Secco, D. and Poirier, Y. (2008) Characterization of the PHO1 gene family and the responses to phosphate deficiency of Physcomitrella patens. *Plant Physiol.* **146**, 646–656.

Wang, C., Wei, Q., Zhang, K., Wang, L., Liu, F., Zhao, L., Tan, Y. *et al.* (2013) Down-regulation of OsSPX1 causes high sensitivity to cold and oxidative stresses in rice seedlings. *PLoS One,* **8**, e81849.

Wang, Z., Ruan, W., Shi, J., Zhang, L., Xiang, D., Yang, C., Li, C. *et al.* (2014) Rice SPX1 and SPX2 inhibit phosphate starvation responses through interacting with PHR2 in a phosphate-dependent manner. *Proc. Natl Acad. Sci. USA,* **111**, 14953–14958.

Wiermann, R. (1970) Synthesis of phenylpropanes during pollen development. *Planta.* **95**, 133–145.

Wilson, Z.A. and Zhang, D.B. (2009) From Arabidopsis to rice: pathways in pollen development. *J. Exp. Bot.* **60**, 1479–1492.

Wu, H.M. and Cheun, A.Y. (2000) Programmed cell death in plant reproduction. *Plant Mol. Biol.* **44**, 267–281.

Xu, J., Yang, C., Yuan, Z., Zhang, D., Gondwe, M.Y., Ding, Z., Liang, W. *et al.* (2010) The ABORTED MICROSPORES regulatory network is required for postmeiotic male reproductive development in Arabidopsis thaliana. *Plant Cell,* **22**, 91–107.

Zhang, H., Liang, W., Yang, X., Luo, X., Jiang, N., Ma, H. and Zhang, D. (2010) Carbon starved anther encodes a MYB domain protein that regulates sugar partitioning required for rice pollen development. *Plant Cell*, **22**, 672–689.

Zhang, D., Luo, X. and Zhu, L. (2011) Cytological analysis and genetic control of rice anther development. *J. Genet. Genomics*, **38**, 379–390.

Zhao, L., Liu, F., Xu, W., Di, C., Zhou, S., Xue, Y., Yu, J. *et al.* (2009) Increased expression of OsSPX1 enhances cold/subfreezing tolerance in tobacco and Arabidopsis thaliana. *Plant Biotechnol. J.* **7**, 550–561.

Zhou, Y. and Ni, M. (2009) SHB1 plays dual roles in photoperiodic and autonomous flowering. *Dev. Biol.* **331**, 50–57.

Zhou, Y. and Ni, M. (2010) SHORT HYPOCOTYL UNDER BLUE1 truncations and mutations alter its association with a signaling protein complex in Arabidopsis. *Plant cell*, **22**, 703–715.

Zhou, J., Jiao, F., Wu, Z., Li, Y., Wang, X., He, X., Zhong, W. *et al.* (2008) OsPHR2 is involved in phosphate-starvation signaling and excessive phosphate accumulation in shoots of plants. *Plant Physiol.* **146**, 1673–1686.

Zhou, Y., Zhang, X., Kang, X., Zhao, X., Zhang, X. and Ni, M. (2009) SHORT HYPOCOTYL UNDER BLUE1 associates with MINISEED3 and HAIKU2 promoters *in vivo* to regulate Arabidopsis seed development. *Plant cell*, **21**, 106–117.

16

Pedigree-based analysis of derivation of genome segments of an elite rice reveals key regions during its breeding

Degui Zhou[1,†], Wei Chen[2,3,†], Zechuan Lin[2,†], Haodong Chen[2,4,†], Chongrong Wang[1,†], Hong Li[1], Renbo Yu[2,4], Fengyun Zhang[5], Gang Zhen[2,4], Junliang Yi[5], Kanghuo Li[1], Yaoguang Liu[6], William Terzaghi[7], Xiaoyan Tang[4,8], Hang He[2,4,8,*], Shaochuan Zhou[1,5,*] and Xing Wang Deng[2,4,8,*]

[1]Guangdong Provincial Key Laboratory of New Technology in Rice Breeding, Rice Research Institute, Guangdong Academy of Agricultural Sciences, Guangzhou, China

[2]State Key Laboratory of Protein and Plant Gene Research, Peking–Tsinghua Center for Life Sciences, School of Advanced Agricultural Sciences and College of Life Sciences, Peking University, Beijing, China

[3]School of Information and Engineering, Wenzhou Medical University, Wenzhou, China

[4]Shenzhen Institute of Crop Molecular Design, Shenzhen, China

[5]Agricultural College, Hunan Agricultural University, Changsha, China

[6]College of Life Sciences, South China Agricultural University, Guangzhou, China

[7]Department of Biology, Wilkes University, Wilkes-Barre, PA, USA

[8]Frontier Laboratories of Systems Crop Design Co., Ltd., Beijing, China

*Correspondence
email hehang@pku.edu.cn
email xxs123@163.com
email: deng@pku.edu.cn
†These authors contributed equally to this work.

Keywords: genetic improvement, resequencing, single nucleotide polymorphisms, pedigree, artificial selection, breeding.

Summary

Analyses of genome variations with high-throughput assays have improved our understanding of genetic basis of crop domestication and identified the selected genome regions, but little is known about that of modern breeding, which has limited the usefulness of massive elite cultivars in further breeding. Here we deploy pedigree-based analysis of an elite rice, Huanghuazhan, to exploit key genome regions during its breeding. The cultivars in the pedigree were resequenced with 7.6× depth on average, and 2.1 million high-quality single nucleotide polymorphisms (SNPs) were obtained. Tracing the derivation of genome blocks with pedigree and information on SNPs revealed the chromosomal recombination during breeding, which showed that 26.22% of Huanghuazhan genome are strictly conserved key regions. These major effect regions were further supported by a QTL mapping of 260 recombinant inbred lines derived from the cross of Huanghuazhan and a very dissimilar cultivar, Shuanggui 36, and by the genome profile of eight cultivars and 36 elite lines derived from Huanghuazhan. Hitting these regions with the cloned genes revealed they include numbers of key genes, which were then applied to demonstrate how Huanghuazhan were bred after 30 years of effort and to dissect the deficiency of artificial selection. We concluded the regions are helpful to the further breeding based on this pedigree and performing breeding by design. Our study provides genetic dissection of modern rice breeding and sheds new light on how to perform genomewide breeding by design.

Introduction

Rice (Oryza sativa L.) is one of the most important food crops. The breeding of semi-dwarf cultivars and the application of hybrid technology have dramatically improved grain yield, quality and resistance to biotic and abiotic stresses of rice during 1960s–1990s. Recently, the rapid population growth and the decrease in available farmland, plus suffering from heavy damage by pest and disease during cultivation, improvement of grain yield as well as biotic and abiotic resistance remains a severe task in rice breeding. However, breeding improvement has been limited because of ineffective selection in quantitative traits including grain yield which can be addressed by molecular breeding, breeding by design, genome selection or gene modification (Peleman and van der Voort, 2003; Wang et al., 2005).

Breeding by design was considered to be an important way to continue improvement breeding (Peleman and van der Voort, 2003) which involved two steps: identifying the best parents and simplest crossing scheme(s) that can produce, through recombination, desirable targeted multilocus genotype(s) in breeding populations, and second, developing the most efficient and effective selection scheme(s) that can lead to the quickest identification and fixation of the targeted genotype(s) from the breeding populations (Li and Zhang, 2013). Although marker-assisted breeding has been widely used in rice breeding, it has been limited to the selection of one or a few loci. Breeding by design has not yet produced a perfect case since it was proposed 10 years ago. The bottleneck is that our knowledge of functional genomics is still limited to process the information needed for breeding by design (Chen et al., 2013).

Sequencing of rice genome, which has been followed by tremendous advances on functional genome studies, has been made in rice in the past decade (Goff et al., 2002; Yu et al., 2002; IRGS Project., 2005). More than 900 rice genes have been cloned and characterized (Chen et al., 2013; Jiang et al., 2012; Li and Zhang, 2013; Yamamoto et al., 2012), and many QTLs have been mapped, but about 32 000 genes have been predicted in rice genome. Therefore, available genetic/molecular information on rice, however, is still insufficient for designing the genotypes of whole genomes. In addition, in the long history of rice domestication and artificial selection, a series of varied alleles have been distributed to different varieties in order to adapt to environment and cropping season. Meanwhile, most rice agronomical important traits are quantitative traits governed by QTLs. Genetic regulation mechanisms underlying quantitative traits are complex netlike and systematically organized, which have made them uncontrollable and difficult to be predicted. At this case, gathering elite allele of genes for agronomical important traits after gene mapping would be a precondition for designing idea genotype, but this process requires tremendous effort, which seems rather difficult to be addressed recently.

Elite cultivars, such as Huanghuazhan which has been grown over 4.5 million ha in southern China, should have excellent allele combinations of important genes for desirable traits. Huanghuazhan, an elite *indica* rice bred by Zhou *et al.* (2008, 2009, 2010), is a semi-dwarf, super high yield, good eating quality and highly adapt to vast areas of nine provinces in China, and twenty cultivars have been derived from it. Most significantly, a clear breeding history with a carefully recorded pedigree of Huanghuazhan (Figure S1) makes it an ideal candidate for dissecting genetic basis of modern breeding, which would uncover selected genome regions and characteristics of artificial selection, potentially providing a priori knowledge towards the selection of favourable breeding parents and breeding by design (Ge *et al.*, 2012).

In this study, 20 cultivars in the Huanghuazhan pedigree were selected for high-throughput sequencing (Table 1). Using this sequence information, we were able to understand the dynamics of chromosomal recombination during the breeding process and clarify the regions or genes which were artificially selected. By traditional QTL analysis and hitting against the table of cloned gene, key genome regions in Huanghuazhan and dynamic gene change during its breeding and the characterization of artificial selection could be exploited. In summary, our exploration of the construction of Huanghuazhan genome during its breeding built

a bridge between traditional breeding and its genetic mechanisms, and the genetic information derived from this study will help directing further pedigree breeding, genetic improvement using Huanghuazhan and breeding by molecular design.

Results

Trait improvement of Huanghuazhan pedigree

Thirteen core cultivars were used during Huanghuazhan breeding (Figure 1). We collected 11 obtainable of them and measured 30 traits (Figures 2, S2 and S3). Cultivars were then divided into two groups based on their breeding effects: the donors which donated genome segments, and receptors which accepted genome segments and therefore have been improved. In the receptors, to the generation of Huanghuazhan, some yield and plant-type traits were changed but not improved after all, others such as panicle length were improved, and potential donor could be identified based on trait values (Figure S2, see Data S1).

Quality traits improvement is the main achievement of the pedigree (Figures 2 and S3). Except protein content, alkali spreading value, brown rice rate, milled rice rate and head rice rate, quality traits have been greatly improved, and potential donors could be found based on trait values (see Data S1). Overall, compared to original receptor, Teqing, Huanghuazhan has been extremely improved in grain quality, whereas grain yield has been reduced but been kept in a high level.

Pedigree sequencing and SNP calling

To determine the artificially selected genome regions of Huanghuazhan, 19 core cultivars including its 11 progenitors and eight cultivars derived from it were collected and resequenced with 7.6× depth on average (Table 1). Additionally, Shuanggui 36, another parental line which has been crossed with Huanghuazhan to generate a set of recombinant inbred lines, was also resequenced. All sequence reads were aligned against the reference Nipponbare genome (version 6.1) (Ouyang *et al.*, 2007) using BWA (Li and Durbin, 2010). On average, 88.1% of reads were mapped to the genome. The sequences of Huanghuazhan, Qingliuai and Teqing covered approximately 91.9%, while the others covered 78.6%–82.1% of the whole genome (Table 1).

We used Samtools (Li *et al.*, 2009) to call SNPs. Overall, 2 113 844 SNPs were identified in these listed 20 cultivars (Table 1), but the genotypes of many cultivars at these loci were largely missing because of sequencing gaps, which created difficulties for further analyses (Figure 3). To fill these gaps, we imputed the genotypes using a k-nearest neighbour-based imputation algorithm (Huang *et al.*, 2010, 2012). During imputation, resequencing data from another 518 traditional Chinese landraces (Huang *et al.*, 2010) were included for better accuracy. This allowed us to reduce the missing genotypes from an average of 29.5% to 1.1% (Figure 4c).

According to the genome annotation (MSU 6.1), 56.4% of all SNPs were found in intergenic regions, 13.6% in introns, 2.1% in UTRs and 14.5% in gene coding regions, containing 207 301 synonymous SNPs and 144 309 nonsynonymous SNPs (Figure 4a). Based on the MSU annotation, 9494 SNPs were expected to introduce premature stop codons, 835 SNPs disrupted stop codons, 499 SNPs altered start codons and 4740 SNPs affected splice donor or acceptor sites. These large-effect SNPs were distributed in 9018 genes, and these genes were significantly abundant (*P* values <0.05 in chi-square tests) in one of seven

Table 1 Resequencing of Shuanggui 36 and 19 cultivars in the Huanghuazhan pedigree

Name	Reads (M)	Bases (G)	Map reads (%)	Map reads	Depth (x)	Cov (%)
Texianzhan 25	26.88	2.69	87.48	23 531 737	6.32	81.07
Jingxian 89	26.21	2.62	86.88	22 761 509	6.11	80.15
Changsizhan	30.25	3.03	87.02	26 366 606	7.08	81.76
Huangruanzhan	23.28	2.33	85.83	19 997 374	5.37	79.12
Fengaizhan	22.78	2.28	86.73	19 775 049	5.31	78.65
Fengqingai	26.30	2.63	87.09	22 905 776	6.15	80.14
Huangxiuzhan	23.94	2.39	86.55	20 686 202	5.56	79.83
Huasizhan	25.62	2.56	86.90	22 246 194	5.98	80.68
Huangsizhan	26.15	2.61	87.03	22 713 595	6.10	80.45
Huangxinzhan	25.42	2.54	87.27	22 166 612	5.96	80.61
Molisimiao	27.69	2.77	87.31	24 185 645	6.50	81.85
Moliruanzhan	24.27	2.43	87.10	21 166 339	5.69	79.21
Fengxiuzhan	24.19	2.42	86.70	20 981 020	5.64	80.28
Huanglizhan	30.94	3.09	86.38	26 691 443	7.17	82.09
Fengbazhan	24.37	2.44	87.27	21 293 912	5.72	80.54
Fenghuazhan	26.79	2.68	87.58	23 470 194	6.31	81.06
Qingliuai	166.38	7.32	94.47	153 665 342	18.58	92.07
Teqing	147.66	6.5	94.68	136 755 218	16.53	92.03
Huanghuazhan	121.45	5.29	95.31	112 045 685	13.55	91.52
Shuanggui 36	26.30	2.63	86.59	22 773 087	6.12	80.31

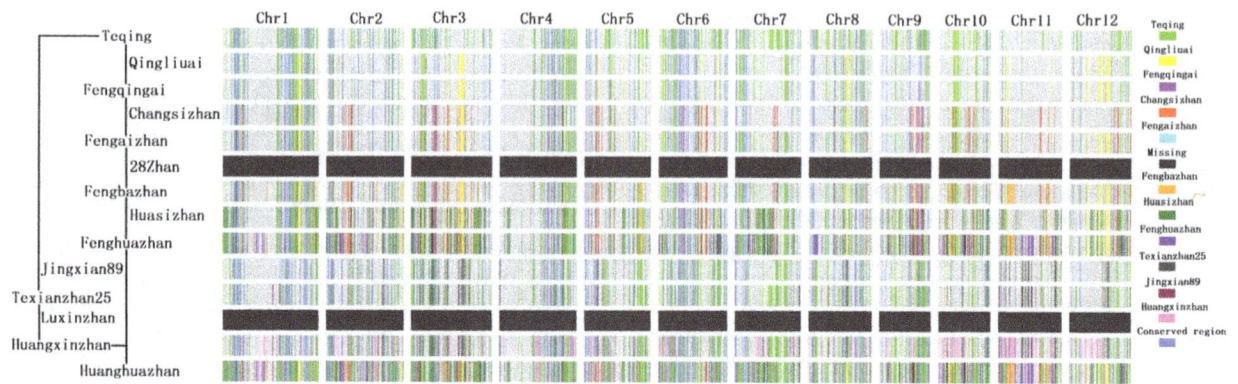

Figure 1 Genome flow of Huanghuazhan breeding. Corresponding color of each cultivar were noted in the right size. The black suggest missing regions or cultivars. From the top to the bottom is the flow of the pedigree.

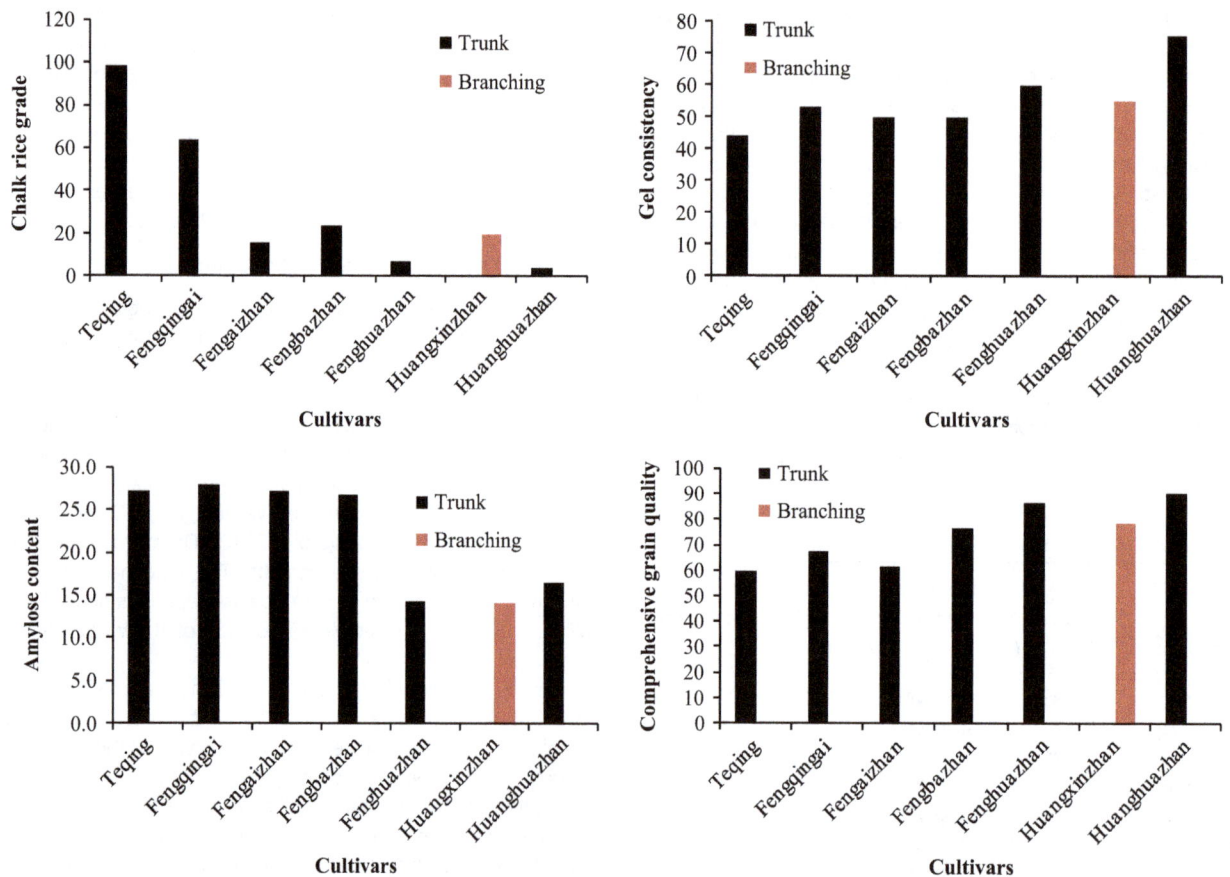

Figure 2 Quality trait improvement in Huanghuazhan pedigree.

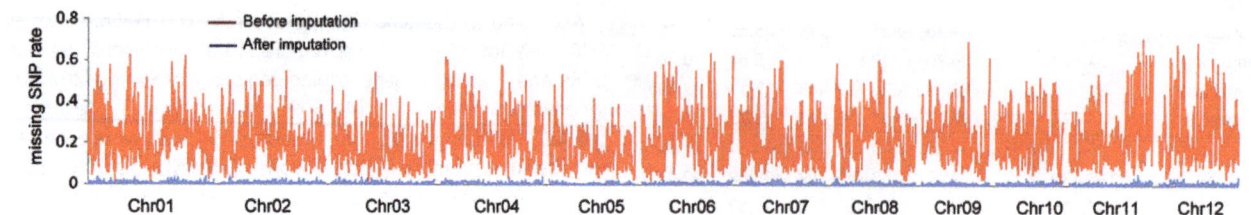

Figure 3 Proportions of missing SNPs across the chromosomes of 20 cultivars. Proportions of missing SNPs were calculated as missing SNP number divided by total SNP number per 100 kb region.

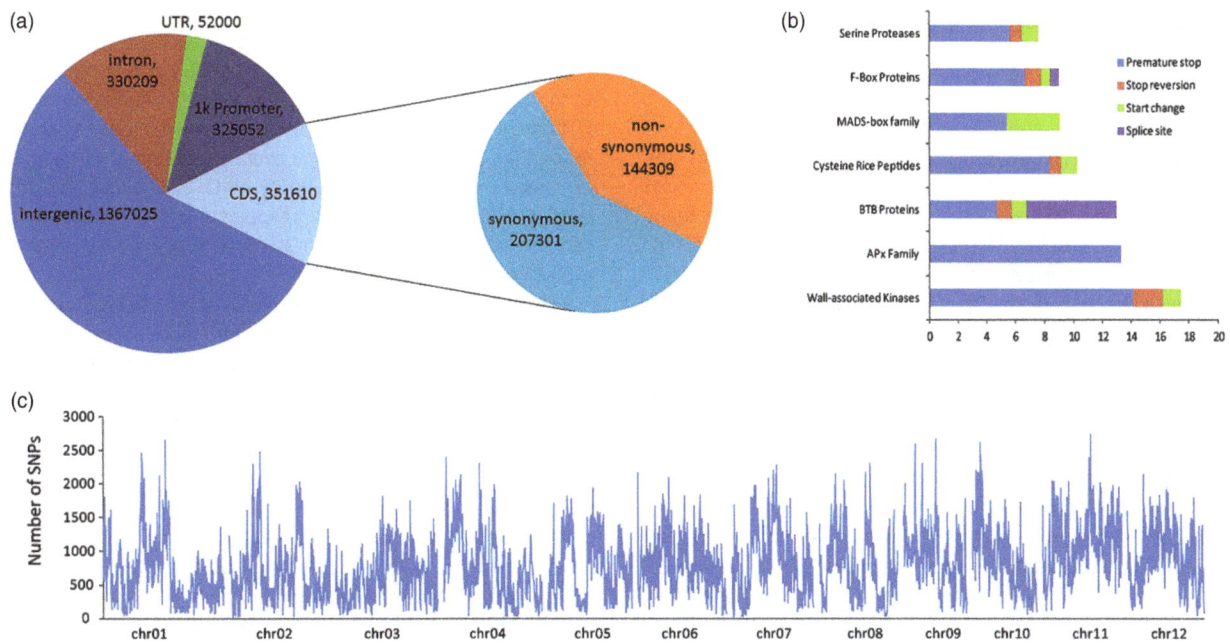

Figure 4 Distribution and annotation of SNPs in 20 cultivars. (a) Distribution of SNP localization. (b) Large-effect SNPs in annotated rice gene families. (c) Distribution of SNPs along the rice genome after imputation.

families (Figure 4b). However, hitting against recently cloned genes, we found only 31 of them have fatal SNPs, but it should be noted that they enrich in biotic resistance-related gene (Table S1), and key blast resistance genes, such as *Xa21* (Song *et al.*, 1995), *Pb-1* (Hayashi *et al.*, 2010), *Pik2* (Zhai *et al.*, 2011), *RGA5* (*PiCO39*) (Cesari *et al.*, 2013), *Pi9* (Qu *et al.*, 2006) and *Pib* (Wang *et al.*, 1999) were observed to have different kinds of large-effect SNPs. *Pb-1*, *Pik2*, *RGA5*(*PiCO39*), *Pi9* and *Pib* are important genes which related to resistance to *Magnaporthe oryzae*, but to our knowledge, the receptors have poor ability to resist this bacterium. Therefore, blast resistance genes may more frequently mutate with fatal SNPs, but it would not always give arise to the variation of resistance ability.

Genome structure of Huanghuazhan

A bin strategy with bin size of 50 kb (see Experimental Procedures) was employed to divide rice genome into 7440 adjacent blocks. Using a cut-off of more than 85% identity between ancestors and Huanghuazhan to exploit the Huanghuazhan traceable blocks (HTBs) (or inherited ancestor genome segments), we found 61.76% of them could be traced by pedigree-based block derivation analysis (see Experimental Procedures), while the remaining blocks were untraceable or have historic recombination event (Figures 1 and S4). In all blocks, 13.05% were conserved across all cultivars in the pedigree, 18.21% were derived from Teqing, 2.61% from Qingliuai, 0.26% from Fengqingai, 0.19% from Changsizhan, 0.62% from Fengaizhan, 2.76% from Fengbazhan, 8.62% from Huasizhan, 8.01% from Fenghuazhan, 0.74% from Jingxian 89, 0.65% from Texianzhan 25 and 4.40% from Huangxinzhan, respectively (Figure S4). The number of HTBs each cultivar donated to Huanghuazhan is strongly correlated with grain yield (Pearson $r^2 = 0.79$, *P* value <0.01), revealing they relate to high yield performance of Huanghuazhan. Furthermore, genomic identity among cultivars (Figure S5) and phylogenetic profiling (Figure S6)

proved the record of pedigree flow (Figure 1) was correct and could be used in the following analysis.

Selective sweeps during Huanghuazhan breeding

We first analysed LD decay rate in pedigree, but found it was consistent with that of *indica* rice landraces (Huang *et al.*, 2010, 2012). To exploit the regions selected by breeders, which are so-called selective sweeps (SSWs, see Data S1), we calculated π, $θ_w$ and Tajima's *D* (Tajima, 1989) by sliding windows (see Experimental Procedures) across the 12 chromosomes (Figure 5a, Figure 5b) with Variscan (Hutter *et al.*, 2006). Using 5% cut-offs in Tajima's *D*, we found 884 SSWs, totalling 18.12 Mb, distributing on all chromosomes except chromosome 3 and 7 (Figure S7). GO analyses indicated that 3066 putative genes in the SSWs enriched in stress-related biological processes (Figure S8). Furthermore, we found only 57.91% of SSWs were included in HTBs, indicating not all the derivation of them could be traced by HTBs analysis in the study, or possibly they showed different genetic information of the pedigree.

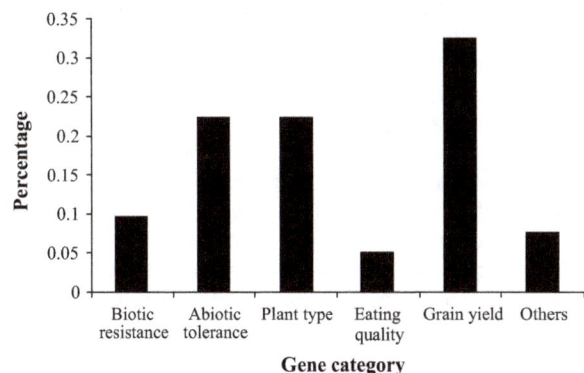

Figure 5 Functional category of cloned gene in cHTBs.

HTBs and SSWs included agronomically important genes

Identifying the function of SSWs and HTBs provides insight into the artificially selected genome regions and is a prerequisite to apply them on breeding practice. To exploit their putative function, we tested the association between SNPs and 30 agronomic traits (see Experimental Procedures, see Data S1) in the pedigree, and 86 significant loci for 10 traits were mapped to overlap with many important genes (Table S2A), for example, the one on chromosome 6 for grain amylose content, which with a P-value of 1.03E-17, overlapped with the wx locus, a genome region which was supported to critically determining eating quality and amylose/amylopectin ratio (Tian et al., 2009). These loci explained 39%–83% of phenotypic variations. Ten loci locate on SSWs (Table S3B) and 68 loci on HTBs (Table S3A), both of which included the one that has lowest P-value and overlaps with wx locus.

Expectedly, agronomic traits are inherited quantitatively and can rarely be identified via a single strategy. Therefore, a traditional QTL mapping was performed with a RIL constructed from crossing Huanghuazhan to a dissimilar cultivar, Shuanggui 36. In total, 114 QTLs were mapped for 22 traits (Table S4A) with some of which were also detected in association test. Of these 114 loci, all were found in HTBs (Table S4A), and 58 with 12 loci explained more than 10% and 4 more than 30% of phenotypic variations were found in SSWs (Table S4B).

QTL mapping and association test have totally obtained 200 loci for only 28 traits, which are only sufficiency to demonstrate the function of minute proportion of SSWs and HTBs. We then employed the catalogue of mapped genes/QTLs released by Q-TARO (Yamamoto et al., 2012) to additionally exploit the function of SSWs and HTBs. By integrating with our mapping results, we found 85 genes and 278 QTLs located in the former, whereas 675 genes and 582 QTLs located in the latter (Table S5). Many important genes such as wx (Tian et al., 2009) and $Ghd8$ (Yan et al., 2011) were found in SSWs (Table S11). Meanwhile, the greater number of important genes were found in HTBs, for example $Xa21$ (Song et al., 1995), wx (Tian et al., 2009), $GS5$ (Li et al., 2011) and $IPA1$ (Jiao et al., 2010). Furthermore, using this gene/QTL catalogue to match against our results from both association test and QTL mapping, we found 49 of them were also mapped for the same trait in other studies, and 88 overlapped with previous cloned genes (Table S6), indicating our mapping results are reliable.

Strict conserved HTBs are important functional blocks of HTBs

Huanghuazhan traceable blocks overlapped with all QTLs mapped from RILs indicated they could explain all phenotypic variations contributed by genetic variations. This together with the elite performance of Huanghuazhan, and large scale of important genes locate on HTBs, should be strong evidences supporting they are the combination of elite allele of important genes. However, 61.79% of 50K blocks are HTBs. We deduced many of them were randomly introgressed blocks which are not harmful to agronomical traits, they were inherited because of artificial selection has not ready to reject them from genome yet, and as the growth of breeding generation, they should not be found in the genome. As expected, except these conserved across all cultivars (Figure 1), we found only 26.22% of HTBs including 196 of cloned genes were conserved in at least three continued breeding generation (Table S8). These strictly conserved HTBs (cHTBs) extremely enrich in yield gene category, followed with abiotic stress resistance and plant-type category (Figure 5). cHTBs included 81.6% of QTLs mapped from RILs (Table S7), and each of them explains 11.16% of phenotypic variations on average, both of which are, respectively, higher than average number of QTLs included by HTBs which obtained from 1000 times random resampling to all HTBs and average phenotypic variations explained by them (9.15%) and by all QTLs (9.67%), indicating cHTBs included major QTLs mapped from RILs, which has demonstrated they can explain all most of phenotypic variations explained by genetic variations.

Interestingly, the fundamental genes of rice flowering system (Tsuji et al., 2011), $RFT1$, $Hd3a$ and $Ehd1$, were found in cHTBs (Table S8). The expressional variation of $Hd3a$ and $Ehd1$ contributed to the diversity of flowering time in rice (Takahashi et al., 2009). Flowering time determines the adaption of cultivars; therefore, selection towards genes involved in it is the primary step in rice breeding. Important flowering genes included by cHTBs indicated they should be the actual regions which have undergone human selection. Expectedly, many other important genes such as $Pi21$, $Pb1$, $GS5$ and ALK were also found (Table S8). Overall, we conclude that cHTBs are important functional blocks of HTBs which have been selected artificially.

Genetic architecture of Huanghuazhan breeding

The HTBs, together with analysis of sequence change of cloned genes (see Experimental Procedures), enable us to visualize how important genes and genome segments were introgressed, selected and fixed in the offspring. As a result, we found 13.05% of HTBs and 306 cloned genes were conserved across cultivars (Table S9, Figure 1). Of the conserved genes (Table S10), like these in cHTBs, they are abundant in plant type, grain yield and abiotic tolerance-related genes (Figure S11). Many key genes for plant type and yield such as $SD1$ (Sasaki et al., 2002), $IPA1$ (Jiao et al., 2010), FZP (Komatsu et al., 2003) and $Gn1a$ (Ashikari et al., 2005) conserved across cultivars in the pedigree (Table S9). We deduced these regions may have undergone selection in the breeding of Teqing. Taken the results together suggested that modern cultivars have a proportion of genome regions/genes lacking of polymorphism, and these are indispensible elements of modern cultivars to maintaining high yield, and grain yield, abiotic stress resistance and plant type are the primary goals of artificial selection in modern breeding.

Teqing donated 18.21% of Huanghuazhan blocks, 4.84% by trunk, 10.09% by branch and 3.28% by either trunk or branch of the pedigree (Figure 1). Analysis of sequence change of cloned genes revealed that 33.54%–55.10% of traceable genes in each gene category originated from Teqing (Table S10, Figure S12). Like HTBs conserved across cultivars, they enriched in category of abiotic stress resistance, plant type and yield. Meanwhile, it donated important genes/QTLs such as ALK (Gao et al., 2003) for eating quality, $TAC1$ (Yu et al., 2007) for plant type and $S5$ (Yang et al., 2012) for sterility for the pedigree (Table S10). For flowering system, it donated core genes, $RFT1$ and $Hd3a$ (Komiya et al., 2008). The difference of days to flowering among receptors was 8 days (Figure S2), suggesting it has perfectly established the flowering system for downstream receptors. Therefore, Teqing is the backbone of the pedigree, contributing about half number of fundamental elements corresponding to excellent performance of downstream cultivars.

The remaining 11 cultivars in the pedigree donated 30.54% of Huanghuazhan blocks. Although each of which contributed little genome (Figure S4) for the Huanghuazhan, some of them were able to greatly improve specific important agronomic trait and donated important genes (Figures S2, S3 and Table S10). The introduction of Changsizhan has largely improved grain size and grain density, and it donated GS5 (Li et al., 2011) and GS3 (Mao et al., 2010), which related to grain size, for the pedigree (Table S10). Huasizhan extremely improved grain quality of receptor and donated wx (Tian et al., 2009), which breeders support as a key locus for eating quality, and Fengbazhan donated Xa21 (Song et al., 1995), a widely used bacterial blight resistance gene, for the pedigree. These were according with the breeding process, which always selected an elite rice as the original receptor, and improved it step by step with lines superior in one or few specific traits.

Interestingly, we found some donors, such as Jingxian 89, did not largely improve the receptor (Figures S2 and S3). Correspondingly, little amount of its genome with very few genes/QTLs have been directly inherited by Huanghuazhan, and most significantly, none of the genes/QTLs it donated were reported to sufficiently affect important traits (Table S10, Data of QTLs no shown). Taken these together suggested that not all step in the breeding is efficient to improve the receptor, although it needs equal amount of efforts as effective one. Selection of a plausible donor to improve receptor is important and able to speed up the breeding process. To understand the genetic background of candidate donors by high-density genotyping and analyse the genotypic variations of key genes of target traits among donors and candidate receptors should be helpful to recent breeding which mainly relies on artificial selection.

Discussion

HTBs included the combination of elite allele of important genes

Pedigree analysis, a strategy that can reveal the dynamic change of genome during crop breeding, has attracted the attention of breeders recently (Yamamoto et al., 2010). Resequencing of Huanghuazhan pedigree and analysing genomic variations have successfully exploited 61.76% of genomic segments that progenitors donated to Huanghuazhan. By employing traditional QTL analysis and association test, we successfully identified 200 loci for 28 traits, and totally 182 loci located on HTBs (Tables S2A, S3A and S4A); on the other hand, hitting to the table of previously cloned genes, HTBs contained important genes such as Xa21 (Song et al., 1995), wx (Tian et al., 2009), GS5 (Li et al., 2011) and IPA1 (Jiao et al., 2010) for grain yield, quality and plant type (Table S5). Given the elite performance of Huanghuazhan (Figures S2 and S3), HTBs are genome blocks containing elite alleles of important genes. The genes locating on HTBs enriched in abiotic stress resistance and grain yield gene categories, followed by plant-type category, but quality category is pretty low (Table S5). These suggested yield, abiotic stress resistance and plant type have been frequently selected in Huanghuazhan breeding programme, which is also according with the conclusion provided by cHTBs (Figure 5). The reason that abiotic stress-related genes were favoured by breeders is these genes were spontaneously selected to respond to varied environments, so that bred cultivars can maintain high yield and good quality wherever cultivated. Phenotypic variations explained by them are always allocated to these so-called gene-environment interaction (G × E) or environment variations in recent QTL analysis pipeline (Wang et al., 2012; Yang et al., 2008), which makes them rather difficult to be detected by QTL study, but play key roles in elite cultivar's genome, and should have great potential in breeding

practice. We therefore deduced these environment-specific QTLs are the class of abiotic resistance genes.

Hence, molecular breeding in rice aids to pyramid elite allele of important genes of yield, quality and resistance into elite cultivars (Chen et al., 2000; Wang et al., 2005; Zhou et al., 2003); therefore, assessment of allele performance is the prerequisite and biggest task. Recently, assessment of allele performance mainly depend on QTL analysis with genetic populations such as NILs and CSSLs (Ohsumi et al., 2011; Zong et al., 2012), which are time and money consuming. In our study, Huanghuazhan was bred after several decades of artificial selection, having conserved amount of valuable genome segments which was called HTBs in the study. As discussed above, HTBs included important genes, and some of them such as wx (Tian et al., 2009) were both detected in QTL mapping and association test (Tables S1, S3A and S10), further revealing it conserved elite allele of genes selected by breeders. Analysis of HTBs should be helpful to high-throughput assessment of allele performance in rice populations, then facilitated to hence molecular breeding. We have introgressed Huanghuazhan allele of Gn1a (Ashikari et al., 2005), which located on HTBs (Table S5), into a japonica backbone parent, and the improvement of grain number per panicle could be found in successful lines (unpublished data), indicating this method is reliable and HTBs are powerful resources for Huanghuazhan pedigree breeding and for the improvement of backbone parents using Huanghuazhan as donor.

Pedigree analysis uncovers the characteristics of artificial selection

The numbers of HTBs donated by each cultivars were strongly correlated with yield. About 13.05% of HTBs and 306 genes were conserved across pedigree (Figure 1, Table S9). Of the conserved blocks/genes, key yield and plant-type genes such as SD1 (Sasaki et al., 2002), IPA1 (Jiao et al., 2010), FZP (Komatsu et al., 2003) and Gn1a (Ashikari et al., 2005) were found (Table S9), suggesting in modern rice breeding lines, elite allele of important genes had been conserved by breeders.

By employing the table of cloned genes, candidate genes from every donor to improve specific trait of receptors were identified (Table S10). On the other hand, SNPs of the pedigree revealed 9018 SNPs causing fatal mutation (Figure 4). Although few cloned genes have fatal SNPs, they have enriched in biotic resistance genes, especially these related to resistance to M. oryzae (Table S1). However, only two of eight of them have been conserved in the pedigree (Table S10). Of the cloned genes locating in HTBs, only 2.82% have fatal SNPs; of important genes, few of which have fatal SNPs. As pretty small proportion of gene in HTBs containing fatal SNPs and the presence/absence of them would commonly result in extreme phenotype, they were not favoured by breeders.

We found many genes were donated by diverse donors (Table S12). We traced the sequence change of GS5 (Li et al., 2011) and wx (Tian et al., 2009) in the pedigree (Figure 6), which may be key gene donated by two donors, Changsizhan and Huasizhan. Of GS5, three of four SNPs including one in 5'-UTR on its annotated cDNA are sense mutation, and all of them first emerged in Changsizhan within the same haplotype and had been fixed in the pedigree since Fengaizhan. The result consists with our QTL mapping of RILs, which have mapped a large-effect QTL approaching to GS5 for grain weight, length, width and length/width (Table S3A). Similar situation was also found in wx. Interestingly, for wx, we found its favoured allele had existed in

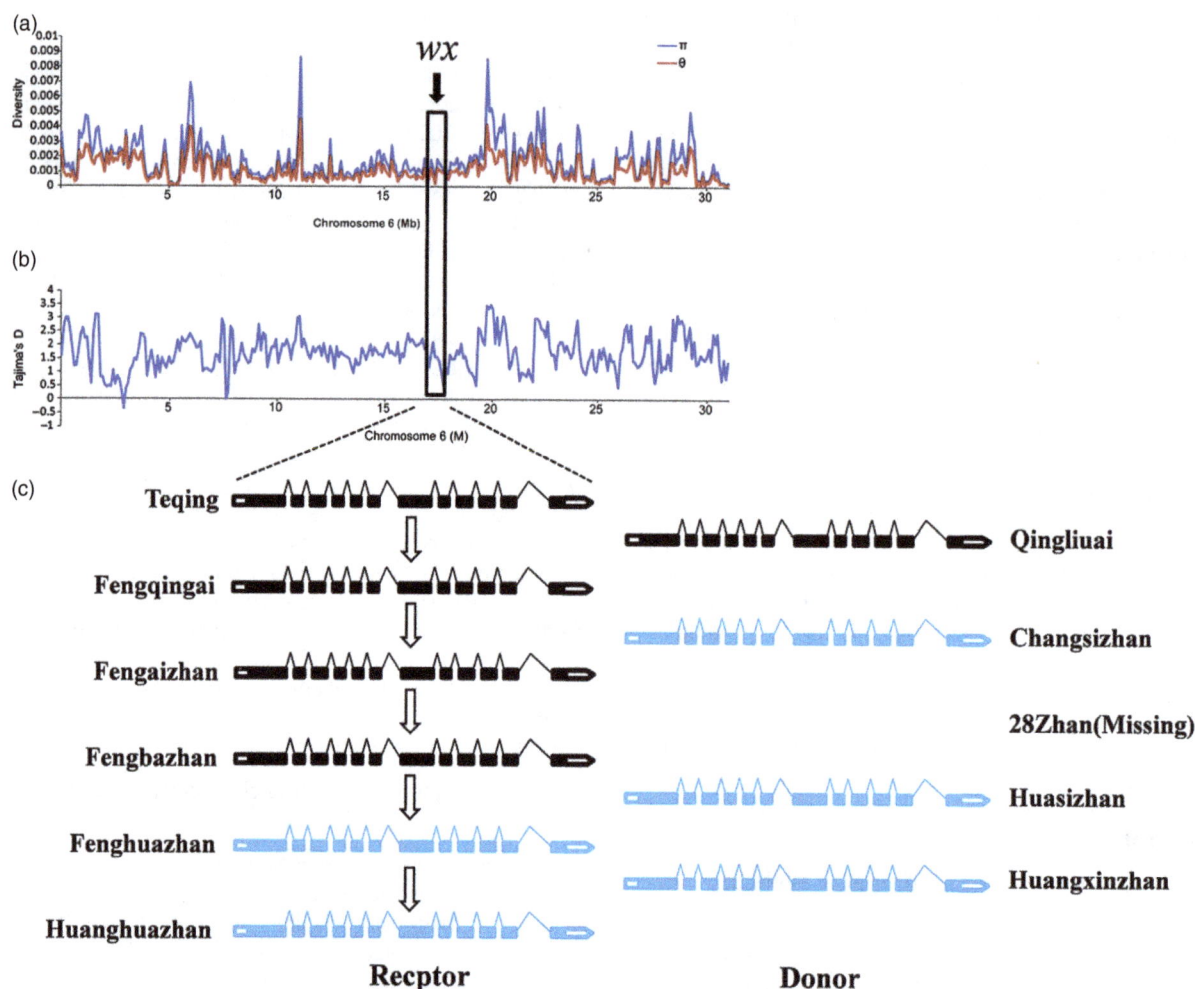

Figure 6 Whole-genome screening of SSWs (a, b) and artificial selection in *wx* locus (c) in the Huanghuazhan pedigree. Diversity π and θ_w (a), and Tajima's *D* (b) were calculated in 100 kb windows across chromosome 6. (c) Dynamic process of allele selection in *wx* gene. Favored allele (blue) has emerged before the application of Huasizhan to improve grain quality by replacing Teqing allele (black).

the donor cultivar which is in the upstream of the cultivar they derived from (Figure 6). Furthermore, analysis of sequence variations of traceable cloned genes revealed 31.61% of them are the same as *wx*. However, they have not introgressed into corresponding receptors, revealing artificial selection is unable to select large scale of elite alleles simultaneously, therefore missing proportion of valuable alleles at every breeding generation. The question may arise from the poor understanding of genetic background of breeding parents, as well as from the unbalance selection in each breeding step, where breeders only focus on several more than many traits at each breeding generation. Poor understanding of genetic background of breeding parents also lead to using ineffective lines to improve receptor, resulting in null breeding step as discussed above. These bottleneck could be overcome by breeding by design (Peleman and van der Voort, 2003; Wang *et al.*, 2007) or high-dense genotyping, where variations of key genes could be designed or identified, then selection towards to all of them would be carried out efficiently.

cHTBs were important functional blocks which have frequently selected in Huanghuazhan pedigree breeding system

That cHBTs overlapped with 81.6% of QTLs mapped from RILs, included core genes of flowering time (Table S8), have been

conserved in at least three continued breeding generations and finally have been preserved on Huanghuazhan, should have supported that they were important functional blocks of HTBs. We therefore tested the efficiency of cHTBs in breeding practice by evaluating the average selected frequency of them with 44 superior lines which developed from Huanghuazhan by artificial selection in the field. The lines were divided into two subgroups: first group includes eight lines which were resequenced with the same strategy as these in Huanghuazhan pedigree; second group includes 36 lines which were genotyped by 50K chip designed in our former study (Chen *et al.*, 2013). Genome blocks of them were obtained by the same method as that of Huanghuazhan pedigree, and they were compared to Huanghuazhan to exploit blocks directly inherited from Huanghuazhan. The first group and the second group, respectively, showed 100% and 97.00% of HTBs were conserved in at least one cultivar, suggesting almost all HTBs had been selected in the field, although the former and latter just inherited half and about 70% of Huanghuazhan genome (Figures S13 and S14), respectively.

The average selection frequency of cHTBs is 0.82 in the first group and 0.81 in the second group, which is better than all HTBs (Figure 7). Furthermore, as expected, HTBs which conserved across all cultivars in the pedigree have an average frequency of 0.93 in the first group and 0.84 in the second group, which are

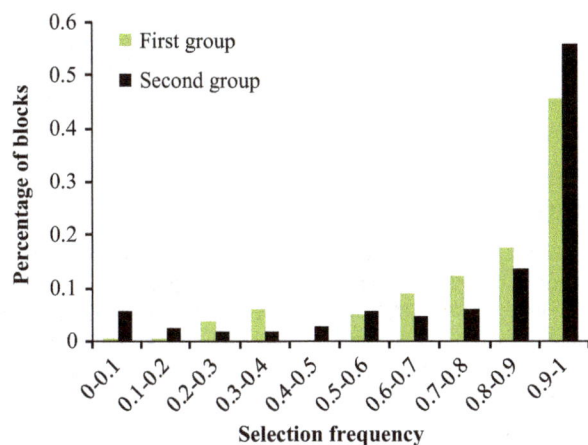

Figure 7 Selection frequency of cHTBs revealed by eight (a, first group) and 36 (b, second group) superior lines derived from Huanghuazhan.

better than these out of the region (Figure S15), further indicating all most of them conserved across our breeding lines. Lacking of polymorphism makes them be automatically inherited by progenies; therefore, they are not the target of artificial selection during breeding. On the other hand, cHTBs, which have been strictly conserved in historic breeding scheme, are the real selection target of Huanghuazhan breeding. The higher selection frequency of them confirmed that they would facilitate to improve yield, quality or disease resistance so that are favoured by breeders, which suggest we have successfully identified valuable blocks of Huanghuazhan pedigree and they would be useful in breeding practice.

Pedigree resequencing shed new light on breeding by design

In the study, we found HTBs were frequently artificially selected in the breeding programmes within Huanghuazhan pedigree, and HTBs potentially include elite alleles of genes for important traits. However, 61.76% of 50K genome blocks (4598) are HTBs, making it difficult to perform breeding by design. But by exploiting cHTBs and artificial selection of eight cultivars and 36 elite lines derived from Huanghuazhan, we found the cHTBs have high average selection frequency, suggesting they are blocks favoured by breeders, although some blocks, such as the one included *Pi21* (Fukuoka *et al.*, 2009), were less frequently selected because of resistance to disease is more difficult to be observed as the pest and disease control during field management were performed before breeders to select elite lines. Therefore, cHTBs are good candidate for performing breeding by design, which are more time and money saving than strategies developed in the past (Li and Zhang, 2013; Peleman and van der Voort, 2003). Expanding the pedigree would more accurately identify cHTBs and thereby help the design of effective breeding platform.

Experimental procedures

Plant materials and genome resequencing

Seeds of single plant of Huanghuazhan, and another parent of RILs employed in the study, Shuanggui 36, together with 11 core cultivars in the pedigree which were used to improve its traits, and eight lines developed from but with superior agronomical

traits than Huanghuazhan, were collected, and their genomic DNA was extracted using plant DNA extraction kits (Qiagen, Hilden, Germany). We constructed paired-end sequencing libraries in which at least 5 µg genomic DNA was used for each cultivar according to the manufacturer's instructions (Illumina, San Diego). 100-bp short reads at each end were generated following the manufacturer's base calling pipeline (SolexaPipeline-1.0; Illumina). Low-quality reads and adaptor sequences were removed by a Perl script with Cutadapt software (Martin, 2011), yielding a total of 909.19 million high-quality paired-end reads of 100-bp or 45-bp read lengths. We have submitted the sequences to Sequence Reads Archive (SRA), National Center for Biotechnology Information (NCBI), as Accession Number in BioProject: PRJNA271838.

SNP calling and imputation

The high-quality reads and 100-bp paired-end reads from a resequencing of 518 rice landraces reported by Huang *et al.* (2010) were aligned to reference Nipponbare genome version 6.1 using BWA (version 0.6) (Ouyang *et al.*, 2007) with default parameter setting, and SNPs were called using Samtools (version 1.8) (Li *et al.*, 2009) also with default parameters. The k-nearest neighbour-based imputation algorithm was employed to fill in missing SNP genotypes as described in Huang *et al.* (2010). The parameters were set as the combination of $w = 80$, $p = -7$, $k = 5$ and $f = 0.7$, where w represented window size of adjacent SNPs, p represented a penalty for different genotypes, the nearest k individuals were defined as neighbours, and the genotype of this individual was determined to be the same as the major allele of the nearest neighbours if the major allele frequency was higher than a threshold, f. Then SNPs with the following criteria were excluded from the SNP data set: (i) SNPs with more than 80% missing data and SNPs with minor allele frequency (MAF) <0.05 in all accessions and (ii) SNPs with missing data and SNPs with MAF <0.05 in the pedigree cultivars.

5 540 907 SNPs that emerged at least three times during imputation, but because many SNPs are tightly linked, 277 045 of them were selected and used in following analysis except exploiting selective sweeps. The genotype accuracy was tested at the basis of pedigree, that is, the SNP of child must accord to one of its parents, otherwise the SNP is set to nonconforming, which is potentially incorrect. At this criterion, 9.75% of SNPs were found to be nonconforming at least in one cultivar in the pedigree. But for all SNPs, only <0.02% of SNPs was nonconforming.

Construction of genome bins, identification of HTBs and selection sweep regions

50-kb bins were constructed by allocating 277 045 SNPs into 50-kb step-length window to form as unique marker to regenotype all resequenced cultivars. We chose 50 kb to construct the bins because of short LD in *indica* rice and casual SNP error (see Data S1). We employed the bin strategy with 85% identity as the threshold of bins damaged, that is, if identity of according bin of upstream cultivars and Huanghuazhan is larger than or equals to 0.85, then it was deemed as conserved blocks, or on other words, HTBs. This strategy would lead to a proportion of positive bias, but we found the bias could be adjusted by integrating with pedigree flow, where we can obtain the real HTBs by checking if they were continuously inherited by receptors in the order of pedigree flow. That is, based on the flow of pedigree tree, each bin of Huanghuazhan was compared to the upstream to check if it was consistently inherited from progenitors during breeding. The number of HTBs donated by each cultivars was counted and used to test the Pearson correlations between HTBs donated by cultivars and all traits.

To identify the selection sweeps (SSWs), π, θw and Tajima's D (Tajima, 1989) were calculated with sliding window of 10 kb or 100 kb across 12 chromosomes with Variscan (Hutter et al., 2006) using filtered SNPs obtained from imputation. But finally 10-kb windows were chosen because it enabled more accurate identification of small selected regions. We used 5% as a cut-off of Tajima's D test to identify selective sweeps which may display highly skewed site frequency. Adjacent SSWs were merged together to build a whole selected region. Figures of conserved blocks or selection sweep regions were drawn using Perl script with GD module (www.perl.org).

Association test and QTL mapping

Adjacent SNPs with the same segregation pattern were combined to form a marker for association test, which totally resulted in 58 629 SNP block markers covering the whole genome of rice. The markers and 30 agronomic traits were fitted to general liner model with PLINK (Purcell et al., 2007). P value threshold was set to be 10^{-5}, then 100 000 adaptive permutation tests were performed and cut-off was set to 0.0001 in PLINK.

A total of 260 recombinant inbred lines derived from the hybrid of Huanghuazhan and Shuanggui 36 by single seed descendant were genotyped using Illumina Golden Gate assay containing 384 SNPs designed from the SNPs of these two parents (Figure S9). We scored the chip data using the Illumina Genome Studio genotyping software with a no-call threshold of 0.25, which is a recommended bound for obtaining a reliable genotype call, and the SNPs with call rates lower than 90% were removed. Finally, 230 SNP markers were retained for data analyses. The linkage map was constructed using ICIMapping version 3.1 according to the order of 230 successfully genotyped SNP markers (Figure S10). QTL analyses were conducted with the inclusive composite interval mapping (ICIM) method of ICIMapping (Wang et al., 2012). Significance thresholds (P value <0.05) for QTL detection were determined with 1000 permutations.

Identification of cHTBs and simulation study

Huanghuazhan traceable blocks which have been inherited at least three breeding generation were deemed as cHTBs. Therefore, only donors/receptors before Fenghuazhan contribute cHTBs. Fengbazhan were also included because of the missing progenitor, 28 Zhan, has donated genome blocks to it, and these from 28Zhan are also over three generations. After obtaining cHTBs, the numbers were counted and used in simulation study. In simulation study, 1000 times of repeats were performed, with each time randomly resampled the equal amount of HTBs as cHTBs from all HTBs. Average number of QTLs included by resampled HTBs were counted as the total number of QTLs included in all resampled HTBs divided to 1000. Phenotypic variations explained by each cHTB and resampled HTB were count as the sum of phenotypic variations of total cHTBs/resampled HTBs divided to the number of them.

Analysis of sequence variations of cloned genes

We found about 10% cloned genes locate near HTBs (within 10 kb) more than on them. To compensate this effect, we trace the sequence change of each cloned gene. Based on the locus position of cloned genes, 5 540 907 SNPs which obtained from imputation were allocated to corresponding gene. Alleles of each cultivar of each gene then was compared each other to obtain the cultivar they derived, therefore generating the table of traceable genes.

Acknowledgements

This work was supported by grants from the National High Technology Research and Development Program of China (863 Program: 2012AA10A304, 2014AA10A602), the National Program on Key Basic Research Project of China (973 Program: 2011CB100101), the National Natural Science Foundation of China (U1031001, 31201277) and the Ministry of Agriculture of China (948 Program: 2011-G2B, 2012-G2).

References

Ashikari, M., Sakakibara, H., Lin, S., Yamamoto, T., Takashi, T., Nishimura, A., Angeles, E.R., Qian, Q., Kitano, H. and Matsuoka, M. (2005) Cytokinin oxidase regulates rice grain production. Science, **309**, 741–745.

Cesari, S., Thilliez, G., Ribot, C., Chalvon, V., Michel, C., Jauneau, A., Rivas, S., Alaux, L., Kanzaki, H. and Okuyama, Y. (2013) The rice resistance protein pair RGA4/RGA5 recognizes the Magnaporthe oryzae effectors AVR-Pia and AVR1-CO39 by direct binding. Plant Cell, **25**, 1463–1481.

Chen, S., Lin, X., Xu, C. and Zhang, Q. (2000) Improvement of bacterial blight resistance of 'Minghui 63', an elite restorer line of hybrid rice, by molecular marker-assisted selection. Crop Sci. **40**, 239–244.

Chen, H., He, H., Zhou, F., Yu, H. and Deng, X.W. (2013) Development of genomics-based genotyping platforms and their applications in rice breeding. Curr. Opin. Plant Biol. **16**, 247–254.

Fukuoka, S., Saka, N., Koga, H., Ono, K., Shimizu, T., Ebana, K., Hayashi, N., Takahashi, A., Hirochika, H. and Okuno, K. (2009) Loss of function of a proline-containing protein confers durable disease resistance in rice. Science, **325**, 998–1001.

Gao, Z., Zeng, D., Cui, X., Zhou, Y., Yan, M., Huang, D., Li, J. and Qian, Q. (2003) Map-based cloning of the ALK gene, which controls the gelatinization temperature of rice. Sci. China, C Life Sci. **46**, 661–668.

Ge, H., You, G., Wang, L., Hao, C., Dong, Y., Li, Z. and Zhang, X. (2012) Genome selection sweep and association analysis shed light on future breeding by design in wheat. Crop Sci. **52**, 1218–1228.

Goff, S.A., Ricke, D., Lan, T.H., Presting, G., Wang, R., Dunn, M., Glazebrook, J., Sessions, A., Oeller, P., Varma, H., Hadley, D., Hutchison, D., Martin, C., Katagiri, F., Lange, B.M., Moughamer, T., Xia, Y., Budworth, P., Zhong, J., Miguel, T., Paszkowski, U., Zhang, S., Colbert, M., Sun, W.L., Chen, L., Cooper, B., Park, S., Wood, T.C., Mao, L., Quail, P., Wing, R., Dean, R., Yu, Y., Zharkikh, A., Shen, R., Sahasrabudhe, S., Thomas, A., Cannings, R., Gutin, A., Pruss, D., Reid, J., Tavtigian, S., Mitchell, J., Eldredge, G., Scholl, T., Miller, R.M., Bhatnagar, S., Adey, N., Rubano, T., Tusneem, N., Robinson, R., Feldhaus, J., Macalma, T., Oliphant, A. and Briggs, S. (2002) A draft sequence of the rice genome (Oryza sativa L. ssp. japonica). Science, **296**, 92–100.

Hayashi, N., Inoue, H., Kato, T., Funao, T., Shirota, M., Shimizu, T., Kanamori, H., Yamane, H., Hayano Saito, Y. and Matsumoto, T. (2010) Durable panicle blast-resistance gene Pb1 encodes an atypical CC-NBS-LRR protein and was generated by acquiring a promoter through local genome duplication. Plant J. **64**, 498–510.

Huang, X., Wei, X., Sang, T., Zhao, Q., Feng, Q., Zhao, Y., Li, C., Zhu, C., Lu, T., Zhang, Z., Li, M., Fan, D., Guo, Y., Wang, A., Wang, L., Deng, L., Li, W., Lu, Y., Weng, Q., Liu, K., Huang, T., Zhou, T., Jing, Y., Li, W., Lin, Z., Buckler, E.S., Qian, Q., Zhang, Q., Li, J. and Han, B. (2010) Genome-wide association studies of 14 agronomic traits in rice landraces. Nat. Genet. **42**, 961–967.

Huang, X., Zhao, Y., Wei, X., Li, C., Wang, A., Zhao, Q., Li, W., Guo, Y., Deng, L., Zhu, C., Fan, D., Lu, Y., Weng, Q., Liu, K., Zhou, T., Jing, Y., Si, L., Dong, G., Huang, T., Lu, T., Feng, Q., Qian, Q., Li, J. and Han, B. (2012) Genome-wide association study of flowering time and grain yield traits in a worldwide collection of rice germplasm. *Nat. Genet.* **44**, 32–39.

Hutter, S., Vilella, A.J. and Rozas, J. (2006) Genome-wide DNA polymorphism analyses using VariScan. *BMC Bioinformatics*, **7**, 409.

IRGS Project. (2005) The map-based sequence of the rice genome. *Nature*, **436**, 793–800.

Jiang, Y., Cai, Z., Xie, W., Long, T., Yu, H. and Zhang, Q. (2012) Rice functional genomics research: progress and implications for crop genetic improvement. *Biotechnol. Adv.* **30**, 1059–1070.

Jiao, Y., Wang, Y., Xue, D., Wang, J., Yan, M., Liu, G., Dong, G., Zeng, D., Lu, Z., Zhu, X., Qian, Q. and Li, J. (2010) Regulation of *OsSPL14* by OsmiR156 defines ideal plant architecture in rice. *Nat. Genet.* **42**, 541–544.

Komatsu, M., Chujo, A., Nagato, Y., Shimamoto, K. and Kyozuka, J. (2003) *FRIZZY PANICLE* is required to prevent the formation of axillary meristems and to establish floral meristem identity in rice spikelets. *Development*, **130**, 3841–3850.

Komiya, R., Ikegami, A., Tamaki, S., Yokoi, S. and Shimamoto, K. (2008) *Hd3a* and *RFT1* are essential for flowering in rice. *Development*, **135**, 767–774.

Li, H. and Durbin, R. (2010) Fast and accurate long-read alignment with Burrows-Wheeler transform. *Bioinformatics*, **26**, 589–595.

Li, Z.K. and Zhang, F. (2013) Rice breeding in the post-genomics era: from concept to practice. *Curr. Opin. Plant Biol.* **16**, 261–269.

Li, H., Handsaker, B., Wysoker, A., Fennell, T., Ruan, J., Homer, N., Marth, G., Abecasis, G. and Durbin, R. (2009) The sequence alignment/map format and SAMtools. *Bioinformatics*, **25**, 2078–2079.

Li, Y., Fan, C., Xing, Y., Jiang, Y., Luo, L., Sun, L., Shao, D., Xu, C., Li, X., Xiao, J., He, Y. and Zhang, Q. (2011) Natural variation in *GS5* plays an important role in regulating grain size and yield in rice. *Nat. Genet.* **43**, 1266–1269.

Mao, H., Sun, S., Yao, J., Wang, C., Yu, S., Xu, C., Li, X. and Zhang, Q. (2010) Linking differential domain functions of the GS3 protein to natural variation of grain size in rice. *Proc. Natl Acad. Sci. USA*, **107**, 19579–19584.

Martin, M. (2011) Cutadapt removes adapter sequences from high-throughput sequencing reads. *EMBnet.journal*, **17**, 10–12.

Ohsumi, A., Takai, T., Ida, M., Yamamoto, T., Arai-Sanoh, Y., Yano, M., Ando, T. and Kondo, M. (2011) Evaluation of yield performance in rice near-isogenic lines with increased spikelet number. *Field. Crop. Res.* **120**, 68–75.

Ouyang, S., Zhu, W., Hamilton, J., Lin, H., Campbell, M., Childs, K., Thibaud-Nissen, F., Malek, R.L., Lee, Y., Zheng, L., Orvis, J., Haas, B., Wortman, J. and Buell, R. (2007) The TIGR rice genome annotation resource: improvements and new features. *Nucleic Acids Res.* **35**, D883–D887.

Peleman, J.D. and van der Voort, J.R. (2003) Breeding by design. *Trends Plant Sci.* **8**, 330–334.

Purcell, S., Neale, B., Todd-Brown, K., Thomas, L., Ferreira, M.A., Bender, D., Maller, J., Sklar, P., De Bakker, P.I. and Daly, M.J. (2007) PLINK: a tool set for whole-genome association and population-based linkage analyses. *Am. J. Hum. Genet.* **81**, 559–575.

Qu, S., Liu, G., Zhou, B., Bellizzi, M., Zeng, L., Dai, L., Han, B. and Wang, G.-L. (2006) The broad-spectrum blast resistance gene *Pi9* encodes a nucleotide-binding site–leucine-rich repeat protein and is a member of a multigene family in rice. *Genetics*, **172**, 1901–1914.

Sasaki, A., Ashikari, M., Ueguchi-Tanaka, M., Itoh, H., Nishimura, A., Swapan, D., Ishiyama, K., Saito, T., Kobayashi, M. and Khush, G. (2002) Green revolution: a mutant gibberellin-synthesis gene in rice. *Nature*, **416**, 701–702.

Song, W., Wang, G., Chen, L., Kim, H., Pi, L., Holsten, T., Gardner, J., Wang, B., Zhai, W. and Zhu, L. (1995) A receptor kinase-like protein encoded by the rice disease resistance gene, *Xa21*. *Science*, **270**, 1804–1806.

Tajima, F. (1989) Statistical method for testing the neutral mutation hypothesis by DNA polymorphism. *Genetics*, **123**, 585–595.

Takahashi, Y., Teshima, K.M., Yokoi, S., Innan, H. and Shimamoto, K. (2009) Variations in Hd1 proteins, *Hd3a* promoters, and *Ehd1* expression levels contribute to diversity of flowering time in cultivated rice. *Proc. Natl Acad. Sci. USA*, **106**, 4555–4560.

Tian, Z., Qian, Q., Liu, Q., Yan, M., Liu, X., Yan, C., Liu, G., Gao, Z., Tang, S., Zeng, D., Wang, Y., Yu, J., Gu, M. and Li, J. (2009) Allelic diversities in rice starch biosynthesis lead to a diverse array of rice eating and cooking qualities. *Proc. Natl Acad. Sci. USA*, **106**, 21760–21765.

Tsuji, H., Taoka, K. and Shimamoto, K. (2011) Regulation of flowering in rice: two florigen genes, a complex gene network, and natural variation. *Curr. Opin. Plant Biol.* **14**, 45–52.

Wang, Z., Yano, M., Yamanouchi, U., Iwamoto, M., Monna, L., Hayasaka, H., Katayose, Y. and Sasaki, T. (1999) The Pib gene for rice blast resistance belongs to the nucleotide binding and leucine-rich repeat class of plant disease resistance genes. *Plant J.* **19**, 55–64.

Wang, Y., Xue, Y. and Li, J. (2005) Towards molecular breeding and improvement of rice in China. *Trends Plant Sci.* **10**, 610–614.

Wang, J., Wan, X., Li, H., Pfeiffer, W.H., Crouch, J. and Wan, J. (2007) Application of identified QTL-marker associations in rice quality improvement through a design-breeding approach. *Theor. Appl. Genet.* **115**, 87–100.

Wang, J., Li, H., Zhang, L. and Meng, L. (2012) *QTL IciMapping Version 3.2.* Beijing: The Quantitative Genetics Group, Institute of Crop Science Chinese Academy of Agricultural Sciences (CAAS).

Yamamoto, T., Nagasaki, H., Yonemaru, J.-I., Ebana, K., Nakajima, M., Shibaya, T. and Yano, M. (2010) Fine definition of the pedigree haplotypes of closely related rice cultivars by means of genome-wide discovery of single-nucleotide polymorphisms. *BMC Genom.* **11**, 267.

Yamamoto, E., Yonemaru, J.-I., Yamamoto, T. and Yano, M. (2012) OGRO: the overview of functionally characterized genes in rice online database. *Rice*, **5**, 1–10.

Yan, W., Wang, P., Chen, H., Zhou, H., Li, Q., Wang, C., Ding, Z., Zhang, Y., Yu, S., Xing, Y. and Zhang, Q. (2011) A major QTL, *Ghd8*, plays pleiotropic roles in regulating grain productivity, plant height, and heading date in rice. *Mol. Plant*, **4**, 319–330.

Yang, J., Hu, C., Hu, H., Yu, R., Xia, Z., Ye, X. and Zhu, J. (2008) QTLNetwork: mapping and visualizing genetic architecture of complex traits in experimental populations. *Bioinformatics*, **24**, 721–723.

Yang, J., Zhao, X., Cheng, K., Du, H., Ouyang, Y., Chen, J., Qiu, S., Huang, J., Jiang, Y., Jiang, L., Ding, J., Xu, C., Li, X. and Zhang, Q. (2012) A killer-protector system regulates both hybrid sterility and segregation distortion in rice. *Science*, **337**, 1336–1340.

Yu, J., Hu, S., Wang, J., Wong, G.K., Li, S., Liu, B., Deng, Y., Dai, L., Zhou, Y., Zhang, X., Cao, M., Liu, J., Sun, J., Tang, J., Chen, Y., Huang, X., Lin, W., Ye, C., Tong, W., Cong, L., Geng, J., Han, Y., Li, L., Li, W., Hu, G., Li, J., Liu, Z., Qi, Q., Li, T., Wang, X., Lu, H., Wu, T., Zhu, M., Ni, P., Han, H., Dong, W., Ren, X., Feng, X., Cui, P., Li, X., Wang, H., Xu, X., Zhai, W., Xu, Z., Zhang, J., He, S., Xu, J., Zhang, K., Zheng, X., Dong, J., Zeng, W., Tao, L., Ye, J., Tan, J., Chen, X., He, J., Liu, D., Tian, W., Tian, C., Xia, H., Bao, Q., Li, G., Gao, H., Cao, T., Zhao, W., Li, P., Chen, W., Zhang, Y., Hu, J., Liu, S., Yang, J., Zhang, G., Xiong, Y., Li, Z., Mao, L., Zhou, C., Zhu, Z., Chen, R., Hao, B., Zheng, W., Chen, S., Guo, W., Tao, M., Zhu, L., Yuan, L. and Yang, H. (2002) A draft sequence of the rice genome (*Oryza sativa* L. ssp. *indica*). *Science*, **296**, 79–92.

Yu, B., Lin, Z., Li, H., Li, X., Li, J., Wang, Y., Zhang, X., Zhu, Z., Zhai, W., Wang, X., Xie, D. and Sun, C. (2007) TAC1, a major quantitative trait locus controlling tiller angle in rice. *Plant J.* **52**, 891–898.

Zhai, C., Lin, F., Dong, Z., He, X., Yuan, B., Zeng, X., Wang, L. and Pan, Q. (2011) The isolation and characterization of *Pik*, a rice blast resistance gene which emerged after rice domestication. *New Phytol.* **189**, 321–334.

Zhou, P., Tan, Y., He, Y., Xu, C. and Zhang, Q. (2003) Simultaneous improvement for four quality traits of Zhenshan 97, an elite parent of hybrid rice, by molecular marker-assisted selection. *Theor. Appl. Genet.* **106**, 326–331.

Zhou, S., Li, H., Huang, D., Lu, D., Lai, S., Zhou, D., Wang, Z., Fu, C. and Li, K. (2008) Analysis on good characteristics and breeding achievements of Huanghuazhan—the core germplasm be planted as early, middle or late rice. *J. Agric. Sci. Technol.* **3**, 017.

Zhou, S., Li, H., Huang, D., Lu, D., Lai, S., Zhou, D., Wang, Z., Fu, C. and Li, K. (2009) Ideal type of rice core germplasm Huanghuazhan and its derivatives with good grain quality. *Chinese J. Rice Sci.* **23**, 153–159.

Zhou, S., Li, H., Huang, D., Lu, D., Li, K., Zhou, D., Lai, S. and Wang, Z. (2010) Breeding and application of Huanghuazhan, a new early, middle or late rice variety with good quality and wide adaptability. *J. Agric. Sci. Technol.* **12**(4), 12–17.

17

Overexpression of a *NF-YC* transcription factor from bermudagrass confers tolerance to drought and salinity in transgenic rice

Miao Chen, Yujuan Zhao, Chunliu Zhuo, Shaoyun Lu* and Zhenfei Guo*

State Key Laboratory for Conservation and Utilization of Subtropical Agro-bioresources, South China Agricultural University, Guangzhou, China

*Correspondence
emails zhfguo@scau.edu.cn;
turflab@scau.edu.cn

Summary

Nuclear factor Y (NF-Y) is a ubiquitous transcription factor formed by three distinct subunits, namely NF-YA, NF-YB and NF-YC. A stress-responsive cDNA of *NF-YC* (*Cdt-NF-YC1*) was isolated from triploid bermudagrass (*Cynodon dactylon* × *Cynodon transvaalensis*), and its role in abiotic stress tolerance was investigated in this study. *Cdt-NF-YC1* transcript was detected in all vegetative tissues with higher levels being observed in roots. Transcription of *Cdt-NF-YC1* in leaves was induced by dehydration, salinity, and treatments with abscisic acid (ABA), hydrogen peroxide (H_2O_2) or nitric oxide (NO), but not altered by cold. The dehydration- or salt-induced transcription of *Cdt-NF-YC1* was blocked by inhibitor of ABA synthesis and scavenger of H_2O_2 or NO, indicating that ABA, H_2O_2 and NO were involved in the dehydration- and salt-induced transcription of *Cdt-NF-YC1*. Overexpression of *Cdt-NF-YC1* resulted in elevated tolerance to drought and salt stress and increased sensitivity to ABA in transgenic rice. Transcript levels of stress/ABA responsive genes (*OsLEA3*, *OsRAB16A*, *OsLIP9* and *OsP5CS1*), ABA synthesis and signalling genes (*OsNCED3* and *OsABI2*), and ABA-independent genes (*OsDREB1A*, *OsDREB1B* and *OsDREB2A*) were substantially higher in transgenic rice than in wild-type plants. The results suggested that that Cdt-NF-YC1 is a good candidate gene to increase drought and salinity tolerance in transgenic rice through modulating gene regulation in both ABA-dependent and ABA-independent pathways.

Keywords: abiotic stress, abscisic acid, bermudagrass, nuclear factor Y, transgenic plants.

Introduction

The nuclear factor Y (NF-Y), also known as HAP, is a CCAAT-specific binding factor composed of three distinct subunits: NF-YA/HAP2, NF-YB/HAP3 and NF-YC/HAP5 (Mantovani, 1999). Each subunit of NF-Y is encoded by a single gene in mammals and yeast, whereas it is encoded by a large gene family in plants. For example, there are ten NF-YA, ten NF-YB and ten NF-YC encoding genes in the Arabidopsis genome (Petroni *et al.*, 2012), and ten NF-YA, 11 NF-YB and seven NF-YC encoding genes in the rice genome (Petroni *et al.*, 2012; Thirumurugan *et al.*, 2008). The multiplicity of the genes implies that NF-Y may act in diverse combinations of each subunit for the transcriptional control (Hackenberg *et al.*, 2012). Regulation of NF-Y in plant growth and development, such as root growth, chloroplast biogenesis, fatty acid biosynthesis, flowering time and embryogenesis, has been well documented (Ballif *et al.*, 2011; Ben-Naim *et al.*, 2006; Cai *et al.*, 2007; Kwong *et al.*, 2003;, Lotan *et al.*, 1998; Miyoshi *et al.*, 2003; Mu *et al.*, 2008, 2013; Petroni *et al.*, 2012; Wei *et al.*, 2010; Wenkel *et al.*, 2006; Yan *et al.*, 2011).

NF-Y is also involved in abiotic stress tolerance. *At-NF-YA5* transcript is strongly induced by drought in an abscisic acid (ABA)-dependent manner; overexpression of *At-NF-YA5* resulted in elevated drought tolerance with induced expression of a number of drought stress-responsive genes in transgenic Arabidopsis (Li *et al.*, 2008). Transcript levels of *AtNF-YA2, 3, 5, 7, 10* are increased between 5- and 50-fold after treatment with low N, 9.4% sucrose or low P, while only *At-NF-YA3, 5* transcripts are increased by twofold after treatment with 100 mM NaCl. Overexpression of *At-NF-YA2, 7* and *10* resulted in enhanced tolerance to several types of abiotic stress (flooding, N starvation, freezing and heat), through participating in modulating gene regulation (Leyva-González *et al.*, 2012). On the contrary, At-NF-YA1 negatively regulates seed germination and postgermination growth under salt stress (Li *et al.*, 2013a). Similarly, Os-NF-YA2 gene is also a negative regulator of abiotic stress tolerance. Disrupting Os-NF-YA2 gene by T-DNA insertion improved drought tolerance in a rice mutant (Najafabadi, 2012). Soybean Gm-NF-YA3 expression is induced by ABA and abiotic stresses (polyethylene glycol, NaCl and cold), and its overexpression resulted in reduced leaf water loss and enhanced drought tolerance in transgenic Arabidopsis and an increased sensitivity to high salinity and exogenous ABA (Ni *et al.*, 2013). An elevated drought tolerance was also observed in transgenic Arabidopsis overexpressing *At-NF-YB1* and in transgenic maize plants overexpressing *Zm-NF-YB2* (Nelson *et al.*, 2007). Poplar Pd-NF-YB7 is induced by osmotic stress and ABA. Ectopic overexpression of *Pd-NF-YB7* in Arabidopsis enhanced drought tolerance and water-use efficiency (Han *et al.*, 2013). Nine *NF-Y* genes were shown to be responsive to drought stress among a total of 37 NF-Y genes in wheat. Only *Ta-NF-YB2* transcript was induced by drought, while the others (*Ta-NF-YA1, Ta-NF-YB3, Ta-NF-YB6, Ta-NF-YB7, Ta-NF-YB8, Ta-NF-YC5, Ta-NF-YC11* and *Ta-NF-YC12*) were down-regulated (Stephenson *et al.*, 2007). In addition, At-NF-YA4, At-NF-YB3 and At-NF-YC2 cooperate with bZIP28 in the endoplasmic reticulum (ER) stress response (Liu and Howell, 2010).

Overexpression of a NF-YC transcription factor from bermudagrass confers tolerance to drought and salinity...

171

Although *At-NF-YC2* shows an induced expression in response to high light, H_2O_2, drought and heat stress, overexpression of this gene in Arabidopsis did not alter oxidative stress tolerance, but led to an early flowering phenotype with an increased *FLOWERING LOCUS T*-transcript levels (Hackenberg et al., 2012). *Picea wilsonii* *Pw-NF-YC* transcripts are induced by salinity, dehydration and ABA; the transgenic Arabidopsis overexpressing *Pw-NF-YC* exhibited an elevated tolerance to salinity and osmotic stress and decreased sensitivity to ABA with higher transcript levels of *COR15a*, *KIN1*, *DREB2A* and *RD29A* genes (Li et al., 2013b). Despite the response of NF-YC to abiotic stress has been reported, the roles of NF-YC in plant tolerance are not consistent (Hackenberg et al., 2012; Li et al., 2013b). Compared to the roles of NF-YA or NF-YB plays in abiotic stress tolerance, function of NF-YC has not been fully understood.

Rice is the most important food crop in Asia. It is planted in the tropics, subtropics, semi-arid tropics and temperate regions, and often exposed to severe abiotic stresses, such as drought, salinity and chilling. Unfortunately, rice plants are sensitive to these stresses. In contrast, bermudagrass (*Cynodon dactylon*) is the most important warm-season turfgrass species with drought and salinity tolerance. ABA treatment significantly increased drought tolerance of triploid bermudagrass (*C. dactylon* × *Cynodon transvaalensis*) (Lu et al., 2009). Considering the important role of NF-Y plays in drought and salinity tolerance, an ABA up-regulated *NF-YC* in bermudagrass was investigated in this study. A cDNA sequence encoding NF-YC, designated as *Cdt-NY-YC1*, with high similarity to *Os-NF-YC4* (Os06g45640) and *At-NF-YC2* (At1g56170) was cloned from triploid bermudagrass. Transgenic rice plants overexpressing *Cdt-NF-YC1* for elevated drought and salinity tolerance were generated and analysed.

Results

Cloning and characterization of *Cdt-NF-YC1*

A 878-bp length cDNA sequence of *Cdt-NF-YC1* (KF939122) was cloned from triploid bermudagrass by RT-PCR using primers designed based on the sequence of *Zm-NF-YC2* cDNA (EU964024.1) 5′ and 3′ untranslated regions. It has an open reading frame of 774 bp and encodes a peptide of 258 amino acid residues with a polypeptide of 28.6 kDa and an isoelectric point (pI) of 5.60. Multiple alignments of Cdt-NF-YC1 and other reported NF-YCs showed that it had the highest identity (86.5%) at AA level with the maize Zm-NF-YC2 (ACG36142). An alignment of Cdt-NF-YC1 with Zm-NF-YC2, Os-NF-YC4/OsHAP5B

(OS06G45640) and At-NF-YC2 (At1g56170) is shown in Figure S1. Cdt-NF-YC1 contains conserved core regions common to NF-YC2 subunits in other organisms (Mantovani, 1999): a P and Q-rich region in the N- and C-terminal regions, NF-YA/NF-YB interaction domains, and the highly conserved AAs in the DNA-binding domain near the N-terminal region. A phylogenetic analysis based on the full-length amino acid sequences showed that Cdt-NF-YC1 has a high identity with Os-NF-YC4 in the Os-NF-YC family, or with At-NF-YC2 compared to other members of At-NF-YC family (Figure S2). Subcellular localization of Cdt-NF-YC1 was determined by transient expression of Cdt-NF-YC1::GFP fusion in transformed onion epidermal cells. Compared to the whole cellular distribution of GFP (Figure 1a), Cdt-NF-YC1::GFP fusion protein fluorescence was localized at nucleus and in the peripheral region of the transformed cells (Figure 1b). To demonstrate that Cdt-NF-YC1 was expressed in cytoplasm, but not in apoplast, the transformed epidermal cells were subjected to plasmolysis, which showed that the green fluorescence was indeed aggregated at nucleus and within cytoplasm (Figure 1c).

Cdt-NF-YC1 transcripts in response to ABA, H_2O_2, NO and abiotic stresses

Under nonstress conditions, higher level of *Cdt-NF-YC1* transcript was detected in rhizomes and roots than in leaves, stems and stolons (Figure 1d). Compared to the unaltered transcripts in water treated leaves (Figure 2a), *Cdt-NF-YC1* transcripts were significantly induced by ABA, H_2O_2 and sodium nitroprusside (SNP, nitric oxide producer) from 30 min to 12 h. The maximum amount of *Cdt-NF-YC1* transcripts was reached at 12 h, the end of the treatment period, with ABA, H_2O_2 or SNP, respectively (Figure 2b–d).

Transcript of *Cdt-NF-YC1* was not altered by chilling treatment (Figure 3a). *Cdt-NF-YC1* transcript was significantly induced after dehydration treatment, with the peak at 1–2 h (Figure 3b). *Cdt-NF-YC1* transcript was also induced by salinity treatment and reached the peak at 0.5 and 1 h, followed by a gradual decrease at 3–12 h (Figure 3c). To better understand whether ABA, H_2O_2 or NO was involved in abiotic stress-induced *Cdt-NF-YC1* transcript, detached leaves were pretreated with naproxen (NAP), inhibitor of ABA biosynthesis, dimethylthiourea (DMTU), scavenger of H_2O_2, or 2-phenyl-4,4,5,5-tetramethylimidazoline-1-oxyl-3-oxide (PTIO), scavenger of NO, followed by dehydration treatment for 2 h, or salinity treatment for 1 h. Similar to the above pattern (Figure 3b,c), dehydration or salinity-induced *Cdt-NF-YC1* transcript as compared with the control. However,

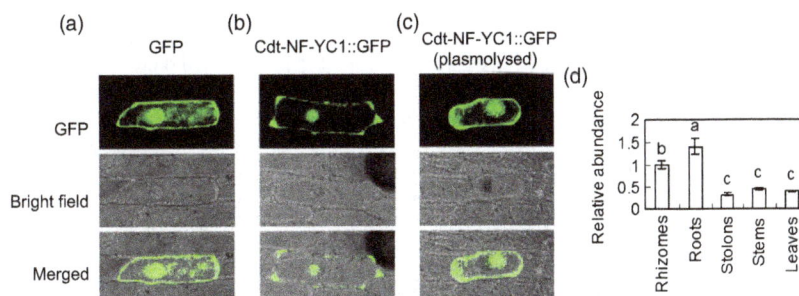

Figure 1 Subcellular localization of Cdt-NF-YC1 in onion epidermal cells and the tissue-specific expression in bermudagrass. Fluorescent microscopic images of GFP protein (control, a) and Cdt-NF-YC1-GFP fusion protein in nonplasmolysed (b) or plasmolysed cells in 30% sucrose solution (c) are presented. Relative expression of CdNY-YC was calculated by defining the normalized transcript level of CdNY-YC in rhizomes as standard (d). Means of three independent samples and standard errors are presented. The same letter above the columns indicates no significant difference at $P < 0.05$.

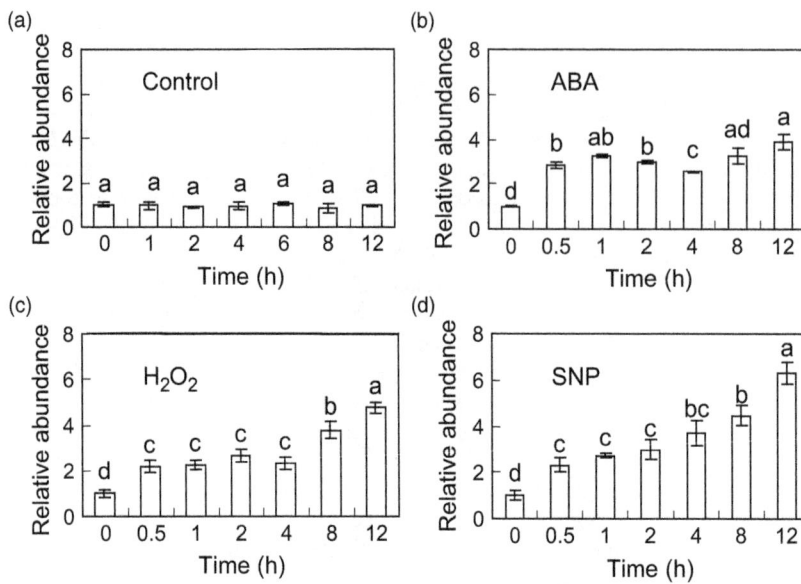

Figure 2 Cdt-NF-YC1 expression as affected by abscisic acid (ABA), H_2O_2 and NO using the methods of real-time quantitative PCR. Leaves were placed in H_2O (as control, a), 100 μM ABA (b), 10 mM H_2O_2 (c) or 100 μM sodium nitroprusside (SNP) (d) solutions as treatments. Actin was detected as reference gene. Relative expression of Cdt-NF-YC1 was calculated by defining the normalized transcript level of Cdt-NF-YC1 before treatments as one. Means of three independent samples and standard errors are presented. The same letter above the columns indicates no significant difference at $P < 0.05$.

induction of Cdt-NF-YC1 expression was suppressed by pretreatment with naproxen, DMTU or PTIO (Figure 3d). The results indicated that ABA, H_2O_2 and NO were essential signals for dehydration- and salt- induced transcription of Cdt-NF-YC1.

Analysis of transgenic rice plants overexpressing Cdt-NF-YC1

To assess the role of Cdt-NF-YC1 in abiotic stress tolerance, transgenic rice plants overexpressing Cdt-NF-YC1 under control of the maize ubiquitin1 promoter were generated. Bermudagrass was not chosen for generation of transgenic plants due to the recalcitrance of its transformation, even though Agrobacterium-mediated transformation of bermudagrass has been reported (Li et al., 2005). DNA hybridization showed that Cdt-NF-YC1 was integrated into the genomes of transgenic plants as one (OE3, 4, 5, 8, and 10) or two transgene copies (OE6), whereas no cross-hybridization was observed in wild type (Figure 4a). Real-time quantitative RT-PCR data showed that Cdt-NF-YC1 transcript

could be detected in the four transgenic lines examined (Figure 4b). Despite ectopic expression of the transgene, transgenic plants had similar morphology to the wild type (Figure S3).

Sensitivity of transgenic rice plants to exogenous ABA

Abscisic acid sensitivity as affected by Cdt-NF-YC1 was measured in transgenic lines OE3 and OE8 in comparison with the wild type. Transgenic plants and the wild type had the similar germination rate under control conditions (Figure 5a). Seed germination was inhibited in all plants in the presence of 1 (Figure 5b) or 3 μM ABA (Figure 5c), while germination of transgenic seeds was inhibited more by ABA than the wild type (Figure 5b,c). Compared to nearly 100% germination rate of the wild-type seeds by day six, for example, only 29% and 31% germination rates were observed in transgenic line OE3 and OE8, respectively, in the presence of 1 μM ABA (Figure 5b). Growth of seedlings was also differentially affected by ABA in transgenic plants and the wild type. All seedlings had similar shoot and root length under control

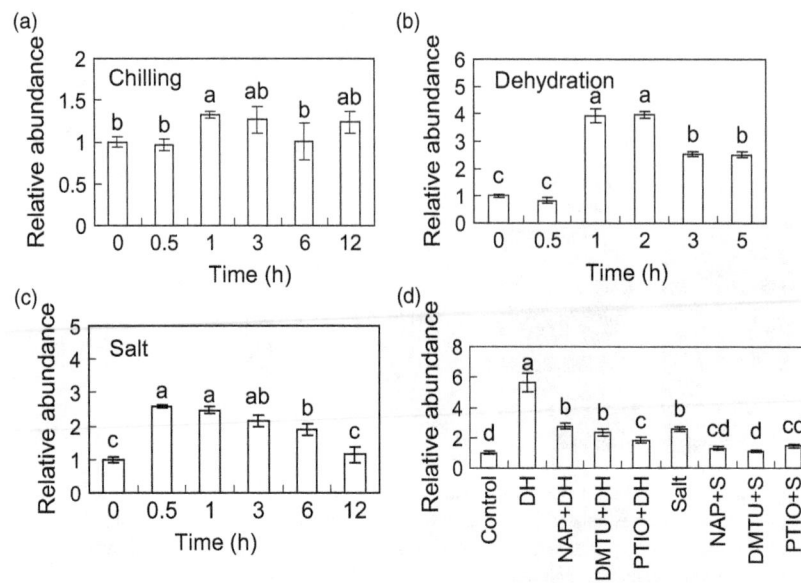

Figure 3 Analysis of Cdt-NF-YC1 transcripts as affected by chilling (a), dehydration (b) and salt stress (c) and the dependence on abscisic acid (ABA), H_2O_2 and NO (d). Pot plants were placed in a growth chamber at 6 °C for chilling treatment (a), or irrigated with 100 mL of 200 mM NaCl solution as salt stress (S) treatment (c). The detached leaves were placed in a hood for gradual dehydration for 5 h as dehydration (DH) treatment (b). Leaflets were placed in H_2O (control), 1 mM naproxen, 5 mM dimethylthiourea (DMTU) or 200 μM 2-phenyl-4,4,5,5-tetramethylimidazoline-1-oxyl-3-oxide (PTIO) for 3 h, followed by transferring into a hood for 2 h for dehydration treatment, or transferring into 200 mM NaCl solution for 2 h for salinity treatment (d). Relative expression was calculated as described in Figure 2.

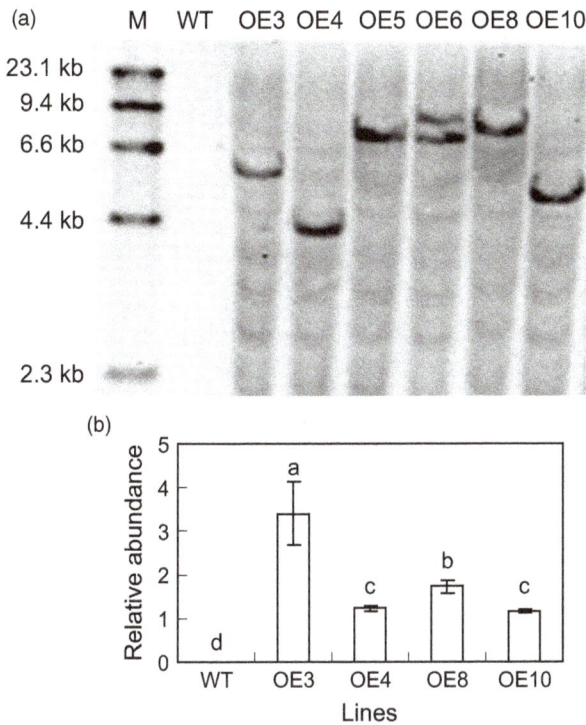

Figure 4 Analysis of the transgenic rice plants overexpressing *Cdt-NF-YC1*. Ten microgram of genomic DNA from each plant line was digested with *Hind*III for DNA blot hybridization (a), and real-time quantitative RT-PCR was used for analysis of *Cdt-NF-YC1* transcription (b). Relative expression was calculated as described in Figure 2. Means of three independent samples and standard errors are presented. The same letter above the columns indicates no significant difference at $P < 0.05$.

conditions, while shorter shoot and roots were observed in transgenic plants than in the wild type in the presence of 1 or 3 μM ABA (Figure 5d,e).

Analysis of plant tolerance to abiotic stresses

The wild-type plants showed extreme wilting 10 days after water withholding, while the transgenic lines still stayed turgid (Figure 6a). While drought stress resulted in increased ion leakage and decreased RWC in all plants, transgenic plants displayed significantly lower levels of ion leakage and higher levels of RWC. For instance, compared to 36% RWC and 68% ion leakage in leaves of the wild-type, transgenic plants maintained 69%–77% RWC and 29%–39% ion leakage 10 days after water withholding (Figure 6b,c). Moreover, detached leaves of transgenic plants lost water more slowly than those of the wild type (Figure 6d).

During the 15 days treatment with 100 mM NaCl, all plants showed a sequential chlorosis from old to young leaves, with the transgenic lines being less damaged by salinity than the wild type (Figure 7a). Chlorophyll and RWC were greatly decreased in the wild-type 10 days after the treatment, while significantly higher levels of chlorophyll and RWC were observed in the transgenic lines (Figure 7b,c). The results indicated that over-expression of *Cdt-NF-YC1* resulted in elevated tolerance to drought and salinity in transgenic plants. However, no significant difference in ion leakage and F_v/F_m was observed between transgenic plants and the wild type after chilling stress (Figure S4), suggesting Cdt-NF-YC1 is not involved in chilling tolerance.

Transcription profiles of stress-related genes regulated by *Cdt-NF-YC1*

To elucidate the role of *Cdt-NF-YC1* in regulation of abiotic stress tolerance, six stress-responsive genes in ABA-dependent pathway (*LEA3*, *RAB16A*, *LIP9*, *P5CS1*, *NCED3* and *ABI2*) and three in ABA-independent pathway (*DREB1A*, *DREB1B* and *DREB2A*) were analysed in response to drought, salinity and ABA, using two transgenic lines in comparison with the wild type. Under control condition, significantly higher levels of *OsLEA3*, *OsRAB16A*, *OsP5CS1*, *OsNCED3*, *OsDREB1A*, *OsDREB1B* and *OsDREB2A* transcripts were observed in transgenic plants than in the wild type (Figure 8a,b,d,e,g–i), while *OsLIP9* and *OsABI2* transcripts showed no significant difference (Figure 8c,f). ABA treatment

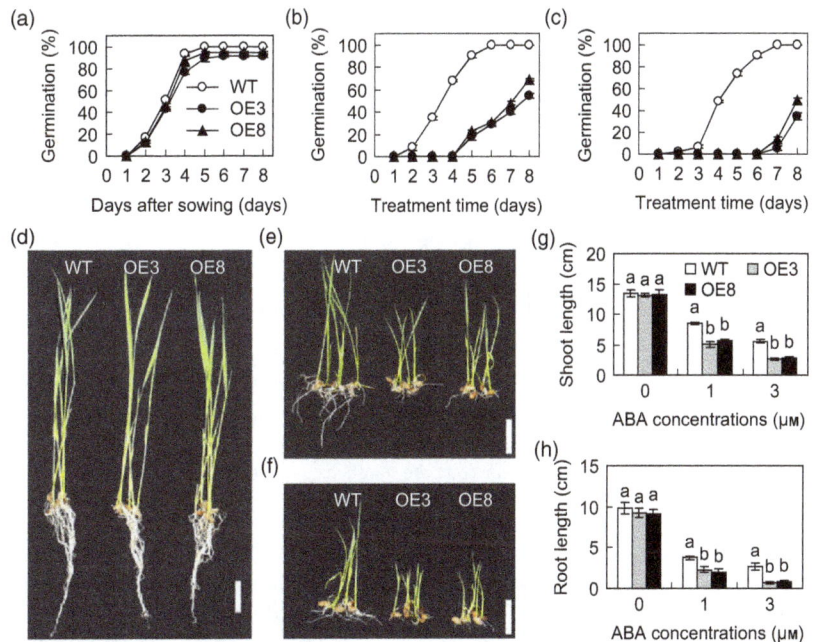

Figure 5 Seed germination and seedling growth of the transgenic rice plants in response to abscisic acid (ABA). Seeds were incubated in Kimura B nutrient solution without ABA (a, d), or containing 1 μM (b, e) or 3 μM ABA (c, f) for germination at 28 °C. Seeds germinated in water were transferred to Kimura B nutrient solution with 0, 1 or 3 μM ABA and grew at 28 °C. The length of shoot (g) and roots (h) was measured 11 days later. Means of three independent samples and standard errors are presented. The same letter above the columns indicates no significant difference at $P < 0.05$.

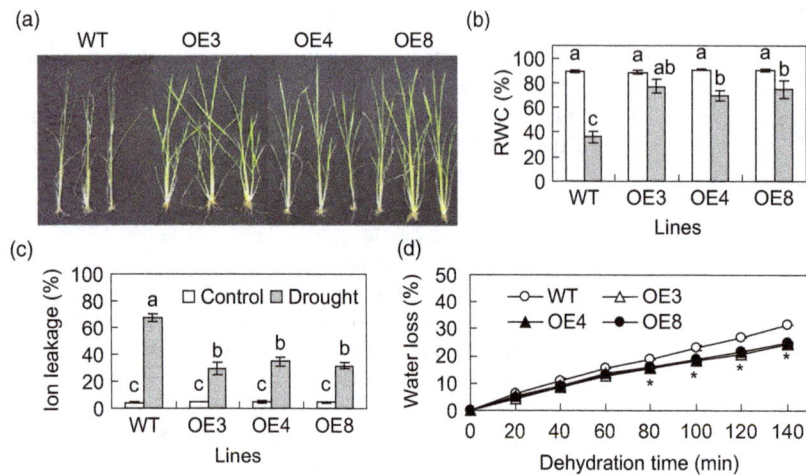

Figure 6 Assessment of drought tolerance in the transgenic rice plants overexpressing *Cdt-NF-YC1*. When the wild type displayed wilting 10 days after withholding irrigation (a), relative water content (RWC, b) and ion leakage (c) were measured. Water loss from detached leaves is expressed as the percentage of initial fresh weight (d). Means of three independent samples and standard errors are presented. The same letter above the columns indicates no significant difference at $P < 0.05$.

induced transcripts of *OsP5CS1*, *OsRAB16A*, *OsLIP9*, *OsLEA3*, and *OsNCED3*, *OsABI2* in the wild-type plants (Figure 8a,b), but did not alter that of *OsDREB1A*, *OsDREB1B* and *OsDREB2A*

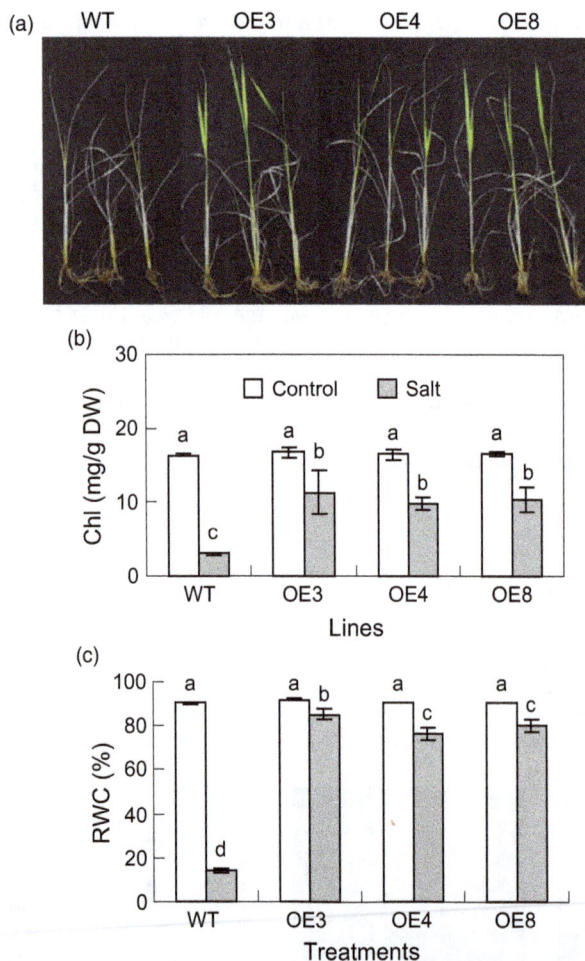

Figure 7 Assessment of salinity tolerance in the transgenic rice plants overexpressing *Cdt-NF-YC1*. When the wild type showed extreme chlorosis 15 days after irrigated with 100 mM NaCl solution (a), chlorophyll (Chl, b) and relative water content (RWC, c) were measured. Means of three independent samples and standard errors are presented. The same letter above the columns indicates no significant difference at $P < 0.05$.

(Figure 8g–i). Moreover, transcripts of all genes were induced by ABA treatment in transgenic plants and maintained at higher levels than those in the wild type (Figure 8). Drought and salinity-induced transcripts of all genes in the wild-type and transgenic plants, and higher levels were maintained in the transgenic lines than in the wild type (Figure 8).

Responses of rice *NF-YCs* to drought and salinity

Transcripts of rice *NF-YCs* in response to dehydration, salinity and ABA were analysed. Transcripts of *Os-NF-YC1*, *4*, *5*, *6* and *7* were induced by treatments with dehydration, salinity and ABA, while those of *Os-NF-YC3* were not affected. *Os-NF-YC2* transcript was decreased by dehydration and ABA, but not affected by salinity treatment (Figure 9).

Discussion

A Cdt-NF-YC1 gene was cloned from triploid bermudagrass. Multiple alignments indicated that Cdt-NF-YC1 showed a high homology in AA sequence with Zm-NF-YC2, Os-NF-YC4 and At-NF-YC2 (Gusmaroli *et al.*, 2002). Cdt-NF-YC1 contains NF-YA/NF-YB interaction domains and a DNA-binding domain, indicating that Cdt-NF-YC1 directly binds to DNA as a subunit of NF-Y heterotrimeric complex to regulate transcription of the down-stream pathway genes. Transient expression of Cdt-NF-YC1::GFP fusion protein in transformed onion epidermal cells indicated that Cdt-NF-YC1 is localized at nucleus and cytoplasm, which is consistent to the proposal that At-NF-YC2 and At-NF-YB3 are translocated from cytosol to nucleus to interact with At-NF-YA4 (Liu and Howell, 2010). Similar to *Pw-NF-YC/PwHAP5* (Li *et al.*, 2013b), *Cdt-NF-YC1* transcript was induced by drought, salinity and ABA. In addition, *Cdt-NF-YC1* transcript was also induced by H_2O_2 and NO, but was not altered by chilling. Higher transcript levels of *Cdt-NF-YC1* were observed in roots and rhizomes than in leaves and stem, supporting the role of *Cdt-NF-YC1* in plant adaptation to drought and salinity. ABA, H_2O_2 and NO are important signals involved in plant adaptation to abiotic stresses. For example, H_2O_2 and NO are essential for cold-induced expression of *myo-inositol phosphate synthase* in *Medicago sativa* ssp. *falcata* (*MfMIPS*), which confers tolerance to cold, drought and salt stresses (Tan *et al.*, 2013). *At-NF-YA5* transcript is strongly induced by drought stress in an ABA-dependent manner (Li *et al.*, 2008). In this study, ABA, H_2O_2 and NO showed to be essential for dehydration- or salinity-induced *Cdt-NF-YC1*,

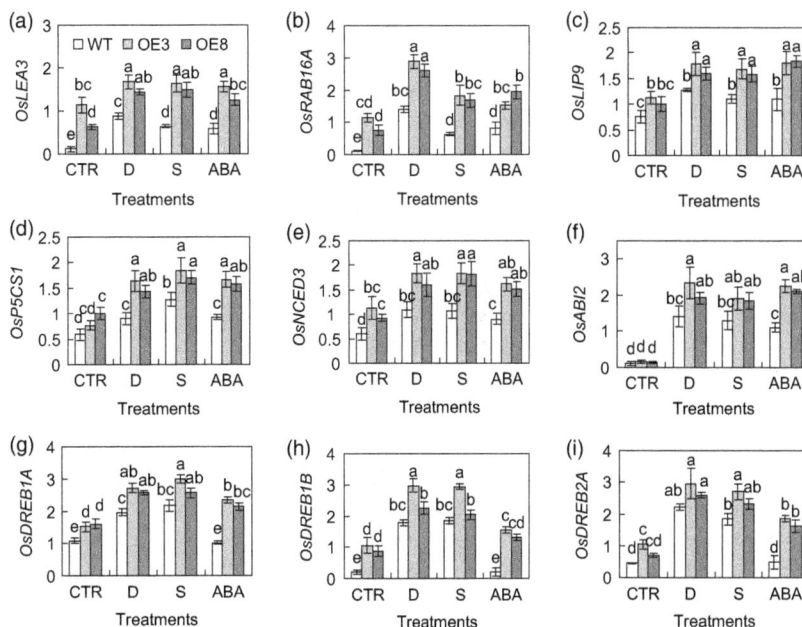

Figure 8 Analysis of transcript levels of *OsLEA3* (a), *OsRAB16A* (b), *OsLIP9* (c), *OsP5CS1* (d), *OsNCED3* (e), *OsABI2* (f), *OsDREB1A* (g), *OsDREB1B* (h) and *OsDREB2A* (i) in the transgenic rice plants overexpressing *Cdt-NF-YC1* in comparison to the wild-type control (WT). Fourteen-day-old seedlings grown in Kimura nutrient solution were placed in 50 μM abscisic acid (ABA) solution for 6 h as ABA treatment, 250 mM NaCl for 5 h as salinity (S) treatment, or were air-dried in a hood for 4 h as drought (D) treatment at 28 °C with those in Kimura nutrient solution being used as a control (CTR). *Actin* was used as a reference gene for real-time quantitative RT-PCR. Means of three independent samples and standard errors are presented. The same letter above the columns indicates no significant difference at $P < 0.05$.

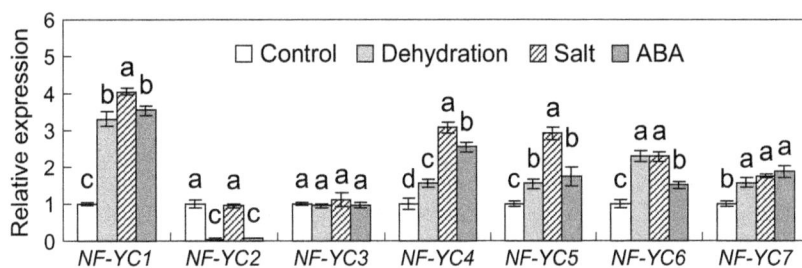

Figure 9 Analysis of transcripts of *Os-NF-YCs* in rice in response to dehydration, salinity and abscisic acid (ABA). The treatments on plants and calculation of relative expression were performed as described in Figure 8. The same letter above the columns indicates no significant difference between treatments for a specific gene at $P < 0.05$.

thus transcripts of *Cdt-NF-YC1* in response to drought and salinity stresses depends on cross-talk of ABA, H_2O_2 and NO signalling.

It has been shown that down-regulation or overexpression of NF-YB and NF-YC genes altered phenotypes and expression of stress-responsive genes associated with ABA signalling (Han *et al.*, 2013; Kumimoto *et al.*, 2008; Li *et al.*, 2008, 2013b; Ni *et al.*, 2013). Similar to transgenic Arabidopsis overexpressing *Gm-NF-YA3* (Ni *et al.*, 2013) or *At-NF-YB2* and *At-NF-YB3* (Kumimoto *et al.*, 2008), transgenic rice plants overexpressing *Cdt-NF-YC1* gene resulted in an increased sensitivity to ABA compared with WT plants based on germination rate. The results suggest that Cdt-NF-YC1 functions downstream of ABA. At-NF-YC2 is involved in ABA-induced expression of a seed storage protein encoding gene *CRUCIFERIN C* (*CRC*) in combination with LEAFY COTYLEDON 1 (LEC1) and LEC1-LIKE (L1L) through the interaction with a seed-specific ABA-response element binding bZIP factor, bZIP67 (Yamamoto *et al.*, 2009).

The role of *Cdt-NF-YC1* in regulation of drought and salinity tolerance was validated in this study using transgenic plants. Transgenic rice plants overexpressing *Cdt-NF-YC1* gene exhibited a decreased water loss and significantly increased tolerance to drought and salinity. The results were consistent with the observation that overexpression of *Pw-NF-YC/PwHAP5* resulted in an elevated tolerance to salinity and osmotic stress in transgenic Arabidopsis (Li *et al.*, 2013b). In addition, most of *OsNF-YCs* were induced by dehydration, salinity stress and ABA, indicating that *OsNF-YC* genes are responsive to drought and

salinity. Our results suggest that, in coordination with NF-YA (Leyva-González *et al.*, 2012; Li *et al.*, 2008; Ni *et al.*, 2013) and NF-YB (Han *et al.*, 2013; Nelson *et al.*, 2007), NF-YC regulates plant adaptation to drought and salinity.

As ABA-induced marker genes (Fukao *et al.*, 2011; Kim *et al.*, 2012; Yamaguchi-Shinozaki and Shinozaki, 2006), transcript levels of rice *LEA3*, *RAB16A*, *LIP9*, *P5CS1*, *NCED3* and *ABI2* genes were induced by ABA, drought and salinity. This case was consistent with the previous observations in other plant species. Up-regulation of these genes led to increased tolerance to drought and/or salinity in transgenic plants (Ohta *et al.*, 2003; RoyChoudhury *et al.*, 2007; Xu *et al.*, 1996). Except for higher *OsABI2* transcript observed in transgenic plants than in the wild type only during ABA treatment, higher transcript levels of *OsLEA3*, *OsRAB16A*, *OsLIP9*, *OsP5CS1* and *OsNCED3* were maintained in transgenic plants than in the wild type under conditions of nonstressed control, drought, salinity and ABA treatment. Although we have not revelled whether the above regulation by Cdt-NF-YC1 is direct or indirect, one to five CCAAT *cis*-acting elements were found in the promoters of the above genes (Table 1) through analysis of the promoter regions using PLACE program (http://www.dna.affrc.go.jp/PLACE/signalscan.html), supporting that they were regulated by NF-Y transcription factors in transgenic rice. The results suggest that Cdt-NF-YC1 up-regulates plant tolerance to drought and salinity through ABA-dependent pathway.

It is well known that *DREB1A*, *DREB2A* and *DREB1B* are involved in ABA-independent regulation of stress-response genes

Table 1 Promoter sequence analysis of genes up-regulated by Cdt-NF-YC1 in transgenic rice

Gene name	Locus name	Motif	Number of CCAAT element
P5CS1	Os05 g0455500	CCAAT	2
RAB16A	Os11 g0454300	CCAAT	1
LIP9	Os02 g0669100	CCAAT	5
LEA3	Os05 g0542500	CCAAT	1
NCED3	Os03 g0645900	CCAAT	1
ABI2	Os01 g0583100	CCAAT	2
DREB1A	Os09 g0522200	CCAAT	4
DREB1B	Os09 g0522000	CCAAT	2
DREB2A	Os01 g0165000	CCAAT	1

(Yamaguchi-Shinozaki and Shinozaki, 2006). In consistence, transcription of *OsDREB1A*, *OsDREB1B* and *OsDREB2A* in the wild type of rice was induced by drought and salinity, but not by ABA treatment as shown in this study. However, their transcript levels could be induced by ABA in transgenic plants, and higher levels were observed in transgenic plants than in the wild type under conditions of nonstressed control, drought, salinity and ABA treatment. The results revealed that Cdt-NF-YC1 is also involved in ABA-independent stress signalling pathways through activating expression of *OsDREB1A*, *OsDREB1B* and *OsDREB2A* directly or indirectly. As an evidence to support this, one or two CCAAT *cis*-acting elements are found in the promoter regions of *OsDREB1A*, *OsDREB1B* and *OsDREB2A* genes (Table 1). It is thus suggested that Cdt-NF-YC1 is a good candidate gene to increase drought and salinity tolerance in transgenic rice through modulating gene regulation in both ABA-dependent and ABA-independent pathways.

Materials and methods

Plant growth and treatments

Bermudagrass plants were grown in plastic pots (10 cm in diameter and 15 cm in depth) containing a mixture of peat and perlite (3 : 1, v/v) in a greenhouse at temperatures of 30/25 °C (day/night) under natural light as described previously (Lu *et al.*, 2009). Plants were grown for 2 weeks after mowing at height of 5 cm by scissors, followed by transferring to a growth chamber for 12 h at 6 °C with a 12 h photoperiod under light of 200 $\mu mol/m^2/s$ as chilling treatment for 12 h, or irrigating with 100 mL of 200 mM NaCl solution as salinity treatment. For dehydration treatment, detached leaves were placed in a Laminar flow hood for gradual dehydration for 5 h. For treatment with chemicals, leaves were excised and placed in distilled water for 1 h to eliminate the potential wound stress influence, and then subjected to the following treatments under light of 200 $\mu mol/m^2/s$: (i) 100 μM ABA, 10 mM H_2O_2 or 100 μM SNP, up to 12 h for detecting their effects on *Cdt-NF-YC1* expression; (ii) H_2O, 1 mM naproxen, 5 mM DMTU or 200 μM 2-phenyl-4,4,5,5-tetramethy-limidazoline-1-oxyl-3-oxide (PTIO) for 3 h, followed by transferring to a hood for 2 h for dehydration treatment, or transferring to 200 mM NaCl solution for 2 h for salinity treatment. The leaves were floated in H_2O under room temperature as nonstressed control.

Transgenic rice lines and the wild-type (*Oryza sativa* L. ssp. *japonica* cv. Zhonghua 11) seedlings were grown in Kimura B solution (compositions: 0.37 mM $(NH_4)_2SO_4$, 0.18 mM KNO_3,

0.21 mM KH_2PO_4, 0.37 mM $Ca(NO_3)_2$, 0.55 mM $MgSO_4$, 0.09 mM EDTA-Fe, 0.0073 mM $MnSO_4$, 9.3 μM HBO_3, 0.015 μM $(NH_4)_6MO_2O_{24}$, 0.15 μM $ZnSO_4$ and 0.16 μM $CuSO_4$) for 14 days, and transplanted to plastic pots (30 cm in diameter and 23 cm in height), filled with rice-field soil. Five plants of each homozygous (T_3) transgenic lines and the wild type were planted in a pot. Plants were grown for 21 days in the greenhouse conditions as described above. For drought stress treatment, water supply was removed from the plants for gradual drying of the soil until plants became wilting. For treatment with salinity, 100 mM NaCl was used to irrigate the plants for 10 days when plants showed obvious injury. For treatment with chilling, plants were placed in a growth chamber with a 12-h photoperiod under light of 800 $\mu mol/m^2/s^1$ at 6 °C for 3 days. For gene transcription analysis under various conditions, 14-days-old rice seedlings were incubated in 50 μM ABA solution for 6 h at 28 °C under light of 200 $\mu mol/m^2/s^1$, or incubated in 250 mM NaCl solutions for 5 h, or placed on paper towels in a hood for 4 h as dehydration treatment. Shoots were sampled for isolation of RNA.

Isolation of *Cdt-NF-YC1* and sequence analysis

Total RNA was isolated using TRIzol reagent (Life Technologies, Carlsbad, CA) according to the manufacturer's protocol, treated with 1 unit/μg of RQ1 DNase (Promega, Madison, WI) to remove genomic DNA, and precipitated with 2 M LiCl. First-strand cDNA was synthesized from 2 μg of total RNA, using M-MLV reverse transcriptase and Oligo (dT) primer as described previously (Zhuo *et al.*, 2013). Polymerase chain reaction (PCR) was conducted to amplify *Cdt-NF-YC1* cDNA using a reaction mixture containing the first-strand cDNA as the template, primers ZG971 (5′-AA GACACAAACAAGAGCTGAGATGGA-3′) and ZG972 (5′-AATTAGC TAACGAGAACCTGAAGTTT-3′) designed based on the sequence of *Zm-NF-YC2* (EU964024.1) untranslated region and *KOD-Plus* DNA polymerase (TOYOBO, Osaka, Japan). The product was cloned to pGEM-T Easy vector (Promega, Madison, WI) for sequencing. The deduced amino acid sequence was used to perform multiple alignment and phylogenetic analysis using DNAMAN software (Lynnon Biosoft, Vaudreuil, QC, Canada). In addition, SignalP4.0 Server (http://www.cbs.dtu.dk/services/SignalP/) was employed to predict signal peptide. Subcellular location of Cdt-NF-YC1 was predicted by using Plant-mPLoc (http://www.csbio.sjtu.edu.cn/bioinf/plant-multi/), WoLF PSORT (http://wolfpsort.org/) and iPSORT prediction (http://ipsort.hgc.jp/).

Subcellular localization of Cdt-NF-YC1:GFP in onion epidermal cells

The coding sequence of *Cdt-NF-YC1* cDNA was fused in frame to the N-terminus of GFP in the vector p35S-GFP. After the fusion gene plasmid (6 μg) was coprecipitated with 3 mg of gold particles, the particles were used for bombarding onion (*Allium cepa*) epidermal cells using the PDS-1000 System (Bio-Rad, Hercules, CA) at 1100 psi helium pressure. The expression of the Cdt-NF-YC1::GFP fusion protein in the onion epidermal cells was observed by confocal laser scanning microscopy (Leica SP, Solms, Germany) after the transformed epidermal cells were incubated on 1/2 MS medium at 22 °C for 30 h. For plasmolysis, the transformed cells were immersed in 30% sucrose for 4 h. The vector p35S-GFP was used as a control.

Real-time quantitative PCR

First-strand cDNA was synthesized from 1 μg of total RNA, using the PrimeScript RT reagent Kit with gDNA Eraser (Takara Bio Inc.,

Otsu, Shiga, Japan). The diluted cDNAs were used as a template in 10-µL PCR reactions containing 15 ng of cDNA, 200 nM each for forward and reverse primers, and 5 µL SYBR Premix *Ex Taq* (Takara). Real-time quantitative PCR was conducted in Mini Option Real-Time PCR System (Bio-Rad) according to the manufacturer's instructions. A negative control without cDNA template was always included. Three technical and two biological replicates were performed in each experiment. Parallel reactions to amplify *actin1* were used to normalize the amount of template. Primers used for *Cdt-NF-YC1* were ZG1637 (5'-AAGATTATGAAGGCT-GATGAG-3') and ZG1638 (5'-TCCACCAAGAAGTCGTAA-3'), while the primers for bermudagrass *actin1* were ZG1603 (5'-CTCTTCCAGCCATCCAT-3') and ZG1604 (5'-CTCATACGGTCAG-CAATG-3'), respectively. All primers were designed using the software tool Beacon Designer (Premier Biosoft International, Palo Alto, CA). The primer specificity was validated by melting profiles, showing a single product specific melting temperature. All PCR efficiencies were around 100%. Bio-Rad CFX Manager (version 1.6) results were exported into Microsoft Excel for further analysis.

Generation of transgenic rice

The coding sequence of *Cdt-NF-YC1* was inserted into an expression vector named pYLox.5 (Zhou *et al.*, 2012) between *KpnI* and *BamHI* restriction sites, under the control of maize Ubi1 promoter. Embryogenic calli were transformed by the method of *Agrobacterium*-mediated transformation, using *Agrobacterium tumefaciens* strain EHA105 harbouring the above construct as previously described (Chen *et al.*, 2013). The positive transgenic plants (T_0) were selected from the regenerated hygromycin-resistance plants using PCR and DNA blot hybridization, and allowed to harvest seeds. By selection with resistance to hygromycin (50 mg/L) at seedlings stage of T_1 and T_2 plants, homozygous transgenic lines were harvested for investigation.

DNA blot hybridization

Genomic DNA was extracted from 1 g of leaves by using the hexadecyltrimethylammonium bromide (CTAB) method. After digested overnight, DNA samples (10 µg) were separated by electrophoresis on 0.8% agarose gel, followed by transferring to Hybond XL nylon membrane (Amersham, GE Healthcare Limited, Buckinghamshire, UK). DNA probes specific to *hpt* (for rice) or *Cdt-NF-YC1* coding sequence were labelled using a PCR digoxigenin (DIG) probe synthesis kit (Roche Diagnostics, Basel, Switzerland) according to the manufacturer's protocol. After hybridization, the DNA filter was washed sequentially with 2× SSC, 0.1% SDS; 1× SSC, 0.1% SDS for 10 min at room temperature; and 0.5× SSC, 0.1% SDS for 15 min at 65 °C. Detection of the signals was carried out using a Lumivision PRO (TAITEC, Saitama, Japan).

Assay of ABA sensitivity during germination and postgermination

After immersed in water overnight, thirty seeds of transgenic lines in comparison with the wild type were placed on paper towel soaked with Kimura solution supplemented with ABA (0, 1 or 3 µM) and incubated in a growth chamber under light of 200 µmol/m²/s with a 12-h photoperiod at 28 °C for germination. Germinated seeds (with radicals >1 mm) were daily counted to calculate germination rate. For testing ABA sensitivity at seedling stage, 30 germinated seeds in water were placed in Kimura solution supplemented with ABA (0, 1 or 3 µM), and

grown in a growth chamber as described above for 11 days. Shoot and root length were measured.

Abiotic stress tolerance assessment

For evaluation of drought tolerance, relative water content (RWC) and ion leakage of the third leaf from the top were determined 10 days after water withholding when plants became wilting as described previously (Guo *et al.*, 2006; Lu *et al.*, 2009). For water loss measurement, the third leaf was detached from plants and weighed immediately, followed by placing in a hood for a series of periods (up to 140 min). The weight was measured every 20 min. Water loss is presented as the percentage of initial fresh weight at each time point. For evaluation of salinity tolerance, chlorophyll and RWC were determined from the third leaf, sampled 15 days after treatment with 100 mM NaCl when plants showed obvious injury, as described previously (Chen *et al.*, 2013). Chlorophyll concentration was calculated as described by Arnon (1949). Chilling tolerance was assessed by measurements of maximal photochemical efficiency (F_v/F_m) and ion leakage of the third leaf as previous described 3 days after chilling treatment at 6 °C as described previously (Guo *et al.*, 2006).

Treatments for analysis of gene expression

Fourteen-day-old homozygous transgenic rice (T_3) and the wild-type seedlings growing in Kimura B complete nutrient solution were transferred to 50 µM ABA solution and incubated for 6 h as ABA treatment, or to 250 mM NaCl for 5 h as salinity treatment, or were air-dried on paper towels in a hood for 4 h as drought stress treatment, at 28 °C under continuous light (200 µmol/m²/s). The second leaves were sampled for isolation of total RNA as described above. The primers used for real-time quantitative RT-PCR are included in Table S1.

Statistical analysis

All the measurements were repeated three times from different individual plants. All data were subjected to analysis of variances according to the model for completely randomized design using an SPSS program (SPSS Inc, Chicago, IL). Differences among means of treatments or plant lines were evaluated by Duncan's test at 0.05 probability level.

Acknowledgements

This work was funded by grants from the Natural Science Foundation of China (31172253, 30972027) and Research Fund for the Doctoral Program of Higher Education of China (20114404110009).

References

Arnon, D.I. (1949) Copper enzymes in isolated chloroplasts. Polyphenoloxidase in *Beta vulgaris*. *Plant Physiol.* **24**, 1–15.

Ballif, J., Endo, S., Kotani, M., MacAdam, J. and Wu, Y. (2011) Over-expression of HAP3b enhances primary root elongation in Arabidopsis. *Plant Physiol. Biochem.* **49**, 579–583.

Ben-Naim, O., Eshed, R., Parnis, A., Teper-Bamnolker, P., Shalit, A., Coupland, G., Samach, A. and Lifschitz, E. (2006) The CCAAT binding factor can mediate interactions between CONSTANS-like proteins and DNA. *Plant J.* **46**, 462–476.

Cai, X., Ballif, J., Endo, S., Davis, E., Liang, M., Chen, D., DeWald, D., Kreps, J., Zhu, T. and Wu, Y. (2007) A putative CCAAT-binding transcription factor is a regulator of flowering timing in Arabidopsis. *Plant Physiol.* **145**, 98–105.

Chen, M., Chen, J.J., Fang, J.Y., Guo, Z.F. and Lu, S.Y. (2013) Down-regulation of *S*-adenosylmethionine decarboxylase genes results in reduced plant length, pollen viability, and abiotic stress tolerance. *Plant Cell Tissue Organ Cult.* **116**, 311–322.

Fukao, T., Yeung, E. and Bailey-Serres, J. (2011) The submergence tolerance regulator SUB1A mediates crosstalk between submergence and drought tolerance in rice. *Plant Cell*, **23**, 412–427.

Guo, Z., Ou, W., Lu, S. and Zhong, Q. (2006) Differential responses of antioxidative system to chilling and drought in four rice cultivars differing in sensitivity. *Plant Physiol. Biochem.* **44**, 828–836.

Gusmaroli, G., Tonelli, C. and Mantovani, R. (2002) Regulation of novel members of the *Arabidopsis thaliana* CCAAT-binding nuclear factor Y subunits. *Gene*, **283**, 41–48.

Hackenberg, D., Keetman, U. and Grimm, B. (2012) Homologous NF-YC2 subunit from Arabidopsis and tobacco is activated by photooxidative stress and induces flowering. *Int. J. Mol. Sci.* **13**, 3458–3477.

Han, X., Tang, S., An, Y., Zheng, D.C., Xia, X.L. and Yin, W.L. (2013) Overexpression of the poplar NF-YB7 transcription factor confers drought tolerance and improves water-use efficiency in Arabidopsis. *J. Exp. Bot.* **64**, 4589–4601.

Kim, H., Hwang, H., Hong, J.W., Lee, Y.N., Ahn, I.P., Yoon, I.S., Yoo, S.D., Lee, S., Lee, S.C. and Kim, B.G. (2012) A rice orthologue of the ABA receptor, OsPYL/RCAR5, is a positive regulator of the ABA signal transduction pathway in seed germination and early seedling growth. *J. Exp. Bot.* **63**, 1013–1024.

Kumimoto, R.W., Adam, L., Hymus, G.J., Repetti, P.P., Reuber, T.L., Marion, C.M., Hempel, F.D. and Ratcliffe, O.J. (2008) The nuclear factor Y subunits NF-YB2 and NF-YB3 play additive roles in the promotion of flowering by inductive long-day photoperiods in Arabidopsis. *Planta*, **228**, 709–723.

Kwong, R.W., Bui, A.Q., Lee, H., Kwong, L.W., Fischer, R.L., Goldberg, R.B. and Harada, J.J. (2003) LEAFY COTYLEDON1-LIKE defines a class of regulators essential for embryo development. *Plant Cell*, **15**, 5–18.

Leyva-González, M.A., Ibarra-Laclette, E., Cruz-Ramírez, A. and Herrera-Estrella, L. (2012) Functional and transcriptome analysis reveals an acclimatization strategy for abiotic stress tolerance mediated by Arabidopsis NF-YA family members. *PLoS ONE*, **7**, e48138.

Li, L., Li, R., Fei, S. and Qu, R. (2005) Agrobacterium-mediated transformation of common bermudagrass (*Cynodon dactylon*). *Plant Cell Tissue Organ Cult.* **83**, 223–229.

Li, W.X., Oono, Y., Zhu, J., He, X.J., Wu, J.M., Iida, K., Lu, X.Y., Cui, X., Jin, H. and Zhu, J.K. (2008) The Arabidopsis NFYA5 transcription factor is regulated transcriptionally and posttranscriptionally to promote drought resistance. *Plant Cell*, **20**, 2238–2251.

Li, Y.J., Fang, Y., Fu, Y.R., Huang, J.G., Wu, C.A. and Zheng, C.C. (2013a) NFYA1 is involved in regulation of postgermination growth arrest under salt stress in Arabidopsis. *PLoS ONE*, **8**, e61289.

Li, L., Yu, Y., Wei, J., Huang, G., Zhang, D., Liu, Y. and Zhang, L. (2013b) Homologous HAP5 subunit from *Picea wilsonii* improved tolerance to salt and decreased sensitivity to ABA in transformed Arabidopsis. *Planta*, **238**, 345–356.

Liu, J.X. and Howell, S.H. (2010) bZIP28 and NF-Y transcription factors are activated by ER stress and assemble into a transcriptional complex to regulate stress response genes in Arabidopsis. *Plant Cell*, **22**, 782–796.

Lotan, T., Ohto, M.A., Yee, K.M., West, M.A., Lo, R., Kwong, R.W., Yamagishi, K., Fischer, R.L., Goldberg, R.B. and Harada, J.J. (1998) Arabidopsis *LEAFY COTYLEDON1* is sufficient to induce embryo development in vegetative cells. *Cell*, **93**, 1195–1205.

Lu, S., Su, W., Li, H. and Guo, Z. (2009) ABA increased drought tolerance in triploid bermudagrass involving H_2O_2 and NO generation. *Plant Physiol. Biochem.* **47**, 132–138.

Mantovani, R. (1999) The molecular biology of the CCAAT-binding factor NF-Y. *Gene*, **239**, 15–27.

Miyoshi, K., Ito, Y., Serizawa, A. and Kurata, N. (2003) *OsHAP3* genes regulate chloroplast biogenesis in rice. *Plant J.* **36**, 532–540.

Mu, J., Tan, H., Zheng, Q., Fu, F., Liang, Y., Zhang, J., Yang, X., Wang, T., Chong, K., Wang, X.J. and Zuo, J. (2008) LEAFY COTYLEDON1 is a key regulator of fatty acid biosynthesis in Arabidopsis. *Plant Physiol.* **148**, 1042–1054.

Mu, J., Tan, H., Hong, S., Liang, Y. and Zuo, J. (2013) Arabidopsis transcription factor genes *NF-YA1, 5, 6, and 9* play redundant roles in male gametogenesis, embryogenesis, and seed development. *Mol. Plant*, **6**, 188–201.

Najafabadi, M.S. (2012) Improving rice (*Oryza sativa* L.) drought tolerance by suppressing a NF-YA transcription factor. *Iran. J. Biotechnol.* **10**, 40–48.

Nelson, D.E., Repetti, P.P., Adams, T.R., Creelman, R.A., Wu, J., Warner, D.C., Anstrom, D.C., Bensen, R.J., Castiglioni, P.P., Donnarummo, M.G., Hinchey, B.S., Kumimoto, R.W., Maszle, D.R., Canales, R.D., Krolikowski, K.A., Dotson, S.B., Gutterson, N., Ratcliffe, O.J. and Heard, J.E. (2007) Plant nuclear factor Y (NF-Y) B subunits confer drought tolerance and lead to improved corn yields on water-limited acres. *Proc. Natl Acad. Sci. USA*, **104**, 16450–16455.

Ni, Z., Hu, Z., Jiang, Q. and Zhang, H. (2013) GmNFYA3, a target gene of miR169, is a positive regulator of plant tolerance to drought stress. *Plant Mol. Biol.* **82**, 113–129.

Ohta, M., Guo, Y., Halfter, U. and Zhu, J.K. (2003) A novel domain in the protein kinase SOS2 mediates interaction with the protein phosphatase 2C ABI2. *Proc. Natl Acad. Sci. USA*, **100**, 11771–11776.

Petroni, K., Kumimoto, R.W., Gnesutta, N., Calvenzani, V., Fornari, M., Tonelli, C., Holt, B.F. III and Mantovani, R. (2012) The promiscuous life of plant NUCLEAR FACTOR Y transcription factors. *Plant Cell*, **24**, 4777–4792.

RoyChoudhury, A., Roy, C. and Sengupta, D.N. (2007) Transgenic tobacco plants overexpressing the heterologous *lea* gene *Rab16A* from rice during high salt and water deficit display enhanced tolerance to salinity stress. *Plant Cell Rep.* **26**, 1839–1859.

Stephenson, T.J., McIntyre, C.L., Collet, C. and Xue, G.P. (2007) Genome-wide identification and expression analysis of the NF-Y family of transcription factors in *Triticum aestivum*. *Plant Mol. Biol.* **65**, 77–92.

Tan, J., Wang, C., Xiang, B., Han, R. and Guo, Z. (2013) Hydrogen peroxide and nitric oxide mediated cold- and dehydration-induced *myo*-inositol phosphate synthase that confers multiple resistances to abiotic stresses. *Plant, Cell Environ.* **36**, 288–299.

Thirumurugan, T., Ito, Y., Kubo, T., Serizawa, A. and Kurata, N. (2008) Identification, characterization and interaction of HAP family genes in rice. *Mol. Genet. Genomics*, **279**, 279–289.

Wei, X., Xu, J., Guo, H., Jiang, L., Chen, S., Yu, C., Zhou, Z., Hu, P., Zhai, H. and Wan, J. (2010) *DTH8* suppresses flowering in rice, influencing plant height and yield potential simultaneously. *Plant Physiol.* **153**, 1747–1758.

Wenkel, S., Turck, F., Singer, K., Gissot, L., Le Gourrierec, J., Samach, A. and Coupland, G. (2006) CONSTANS and the CCAAT box binding complex share a functionally important domain and interact to regulate flowering of Arabidopsis. *Plant Cell*, **18**, 2971–2984.

Xu, D., Duan, X., Wang, B., Hong, B., Ho, T.H.D. and Wu, R. (1996) Expression of a late embryogenesis abundant protein gene, *HVA1*, from barley confers tolerance to water deficit and salt stress in transgenic rice. *Plant Physiol.* **110**, 249–257.

Yamaguchi-Shinozaki, K. and Shinozaki, K. (2006) Transcriptional regulatory networks in cellular responses and tolerance to dehydration and cold stresses. *Annu. Rev. Plant Biol.* **57**, 781–803.

Yamamoto, A., Kagaya, Y., Toyoshima, R., Kagaya, M., Takeda, S. and Hattori, T. (2009) Arabidopsis NF-YB subunits LEC1 and LEC1-LIKE activate transcription by interacting with seed-specific ABRE-binding factors. *Plant J.* **58**, 843–856.

Yan, W.H., Wang, P., Chen, H.X., Zhou, H.J., Li, Q.P., Wang, C.R., Ding, Z.H., Zhang, Y.S., Yu, S.B., Xing, Y.Z. and Zhang, Q.F. (2011) A major QTL, *Ghd8*, plays pleiotropic roles in regulating grain productivity, plant height, and heading date in rice. *Mol. Plant*, **4**, 319–330.

Zhou, H., Liu, Q., Li, J., Jiang, D., Zhou, L., Wu, P., Lu, S., Li, F., Zhu, L., Liu, Z., Chen, L., Liu, Y.-G. and Zhuang, C. (2012) Photoperiod- and thermo-sensitive genic male sterility in rice are caused by a point mutation in a novel noncoding RNA that produces a small RNA. *Cell Res.* **22**, 649–660.

Zhuo, C., Wang, T., Lu, S., Zhao, Y., Li, X. and Guo, Z. (2013) A cold responsive galactinol synthase gene from *Medicago falcate* (*MfGolS1*) is induced by myo-inositol and confers multiple tolerances to abiotic stresses. *Physiol. Plant.* **149**, 67–78.

18

Enhancing blast disease resistance by overexpression of the calcium-dependent protein kinase *OsCPK4* in rice

Mireia Bundó and María Coca*

Centre for Research in Agricultural Genomics (CRAG), CSIC-IRTA-UAB-UB. Edifici CRAG, Bellaterra, Barcelona, Spain

*Correspondence
email
maria.coca@cragenomica.es

Summary

Rice is the most important staple food for more than half of the human population, and blast disease is the most serious disease affecting global rice production. In this work, the isoform OsCPK4 of the rice calcium-dependent protein kinase family is reported as a regulator of rice immunity to blast fungal infection. It shows that overexpression of *OsCPK4* gene in rice plants enhances resistance to blast disease by preventing fungal penetration. The constitutive accumulation of OsCPK4 protein prepares rice plants for a rapid and potentiated defence response, including the production of reactive oxygen species, callose deposition and defence gene expression. *OsCPK4* overexpression leads also to constitutive increased content of the glycosylated salicylic acid hormone in leaves without compromising rice yield. Given that *OsCPK4* overexpression was known to confer also salt and drought tolerance in rice, the results reported in this article demonstrate that OsCPK4 acts as a convergence component that positively modulates both biotic and abiotic signalling pathways. Altogether, our findings indicate that OsCPK4 is a potential molecular target to improve not only abiotic stress tolerance, but also blast disease resistance of rice crops.

Keywords: rice, blast, calcium-dependent protein kinases, defence, resistance, productivity.

Introduction

Rice blast disease, caused by the filamentous ascomycete fungus *Magnaporthe oryzae,* is the most important rice disease due to its severity and wide distribution (approximately 85 countries around the world) (Ou, 1987). *Magnaporthe oryzae* attacks rice plants at all developmental stages, more often during the seedling stage, and it can infect leaves, stems, nodes, collars and panicles (Dean et al., 2012). Rice blast causes severe crop losses varying from 10 to 85% depending on the area and climatology (Skamnioti and Gurr, 2009) (http://www.irri.org/research/better-rice-varieties/disease-and-pest-resistnatn-rice). Resistant cultivars and pesticides have traditionally been used to control this disease. However, the fungus *M. oryzae* overcomes host resistance quickly, and resistant cultivars become ineffective after a few years (Lee et al., 2009). Pesticide use, on the other hand, is costly and environmentally unfriendly. Being rice a paramount source of human food, new strategies providing long-term blast protection should therefore be developed. The study of the plant defence responses offers a vast field of possibilities to improve disease resistance in rice.

In addition to structural barriers and preformed antimicrobial compounds, plants have evolved inducible immune responses to defend themselves against pathogen attack. The defence response starts with the recognition of pathogen-associated molecular patterns (PAMPs) by pattern recognition receptors (PRRs) that activate the PAMP-triggered immunity (PTI) (Boller and He, 2009; Chisholm et al., 2006; Jones and Dangl, 2006). Successful pathogens have evolved to suppress the PTI response by the action of effectors. But plants in turn have evolved a second defence layer, known as effector-triggered immunity (ETI), consisting of resistance proteins that recognize these effectors (Jones and Dangl, 2006). Both PTI and ETI counteract the pathogen attack by inducing immune responses (Tsuda and Katagiri, 2010). The earliest defence reactions include changes in ion fluxes across membranes, an increase in the intracellular calcium concentration, the activation of protein kinases or the synthesis of reactive oxygen species (ROS) (Baxter et al., 2013; Lecourieux et al., 2006; Meng and Zhang, 2013; Seybold et al., 2014; Tena et al., 2011; Torres, 2010). Forward reactions consist of transcriptional reprogramming, alterations in hormone status and cell wall reinforcement through callose depositions and lignifications and in some cases even by cell death at the site of infection (Liu et al., 2014; Luna et al., 2010; Navarro et al., 2004; Tsuda and Katagiri, 2010). Defence responses locally activated in primary pathogen-infected plant tissues are often extended to distal noninfected tissues, conferring systemic acquired resistance (SAR) (Durrant and Dong, 2004; Ryals et al., 1996). This resistance is long-lasting and effective against secondary attack by unrelated pathogens. SAR is associated with the signal molecule salicylic acid (SA) and the accumulation of pathogenesis-related (PR) proteins that are thought to contribute to resistance (Durrant and Dong, 2004).

Calcium influx is one of the earliest events upon pathogen recognition in plant defence response (Ranf et al., 2011). Alterations in calcium concentration are sensed by calcium-binding proteins, including calmodulin, calcium-dependent protein kinases (CDPK or CPKs) and calcineurin B-like proteins, which relay the calcium signal into specific cellular and physiological responses (Dodd et al., 2010; Harper et al., 2004). CPKs represent unique calcium sensors able to translate calcium signals directly into phosphorylation events, because they combine in a single molecule a calcium-binding domain and a serine/threonine kinase domain (Harper et al., 2004). In this sense, genetic and biochemical studies have demonstrated that these plant proteins are important players in numerous signalling pathways and biological processes, including stress signalling cascades and immune signalling responses (Boudsocq and Sheen, 2013; Romeis and Herde, 2014; Schulz et al., 2013).

CPKs are encoded by large gene families, the rice genome containing 31 *CPK* genes (Asano et al., 2005; Ray et al., 2007). In contrast to Arabidopsis CPKs, little is known about the functions of specific rice CPKs. Among the ones functionally characterized are the OsCPK13 (Saijo et al., 2000), OsCPK12 (Asano et al., 2012) and OsCPK9 (Wei et al., 2014) proteins that have been reported as signalling components of abiotic stress responses; the OsCPK10 (Fu et al., 2013) and the OsCPK18 (Xie et al., 2014) were described as positive and negative regulators of M. oryzae resistance, respectively. Only OsCPK12 has been shown to be involved in both abiotic and biotic stress signalling (Asano et al., 2012). Recently, our group reported that OsCPK4 positively regulates salt and drought stress adaptation (Campo et al., 2014). Contrary to OsCPK12 that oppositely modulates the different signalling pathways, the present study reports that OsCPK4 is also a positive regulator of immunity in rice. OsCPK4 overexpression confers an enhanced resistance to blast disease in rice plants by preventing M. oryzae fungal penetration. The enhanced resistance phenotype is associated with the constitutive accumulation of conjugated SA and callose, and a fast and stronger activation of defence responses, including ROS production and defence gene expression, without compromising rice productivity.

Results

OsCPK4 expression is induced by Magnaporthe oryzae infection in rice plants

A search for altered expression genes in a microarray-based global transcriptomic analysis of rice plants in response to M. oryzae elicitors (Campo et al., 2013) identified the OsCPK4 gene as an up-regulated gene in leaves after 2-h treatment (fold change = 1.94; P-value = 0.0002). The OsCPK4 gene (LOC_Os02g03410) encodes a CDPK involved in the adaptation of rice plants to salinity and drought conditions (Campo et al., 2014). To confirm that OsCPK4 gene expression is altered during the defence response of rice plants, it was examined in leaves at different times after inoculation with M. oryzae spores (Figure 1a). OsCPK4 expression was rapid and strongly induced in rice leaves at earlier stages of infection at 6 h postinoculation (hpi), coinciding with the formation of the fungal infective structure, named appressorium (Wilson and Talbot, 2009). OsCPK4 activation increased until 12 hpi (approximately an eightfold increase) and started to decrease at 24 hpi, once fungal penetration had already occurred. These observations show that OsCPK4 is an early-response gene against M. oryzae infection in rice leaves.

OsCPK4 protein accumulation was also examined in blast-infected leaves. In agreement with OsCPK4 transcript levels, Western blot analyses showed an increase in the accumulation of the encoded protein after pathogen inoculation (Figure 1b). These results indicate that OsCPK4 transcriptional activation is translated in the protein accumulation and suggest that the OsCPK4 protein is involved in the defence response of rice plants to M. oryzae infection.

OsCPK4 overexpressor rice plants are more resistant to Magnaporthe oryzae infection

To further investigate the function of OsCPK4 in rice immunity, we used the transgenic OsCPK4-overexpressing rice plants previously described (Campo et al., 2014). These plants were produced in the japonica cultivar Nipponbare and expressed the OsCPK4 full-length cDNA under the control of the strong and constitutive ZmUbi1 promoter. Quantitative RT-PCR analyses

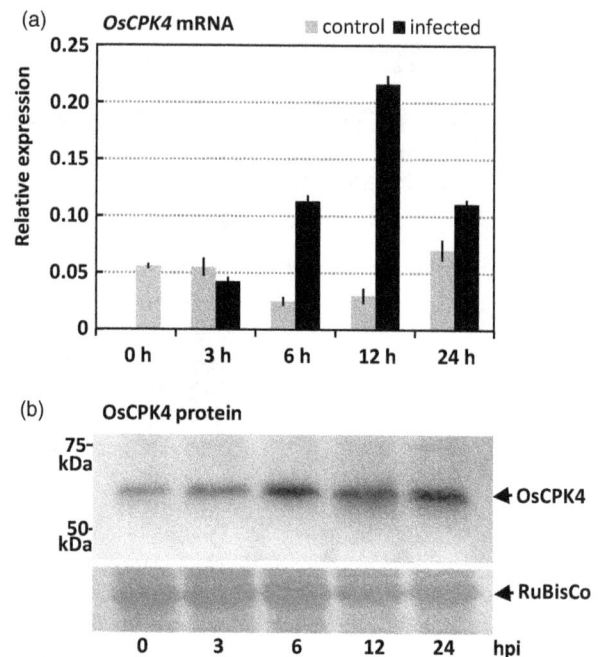

Figure 1 OsCPK4 expression and protein accumulation in response to fungal infection. (a) Transcript levels were determined by qRT-PCR analysis in rice leaves (Oryza sativa cultivar Nipponbare) after inoculation with a Magnaporthe oryzae spore suspension (10^5 spores/mL) at the indicated period of time. Specific primers were used to detect the OsCPK4 mRNA levels that were normalized to the OsUbi5 mRNAs. Error bars indicate SEM of three replicates. (b) OsCPK4 accumulation was determined by Western blot analysis using specific anti-OsCPK4 antibodies at the indicated period of time after inoculation. Lower panel corresponds to Ponceau staining of protein samples (40 µg per lane). Leaves from three different plants grown in soil for 3 weeks were collected in a pool at each different time for total RNA (a) or protein extraction (b). Results are representative of two independent experiments.

confirmed that the expression of OsCPK4 was indeed significantly enhanced in the leaves of OsCPK4-Ox plants in comparison with wild-type or control empty vector plants (Figure S1a), resulting also in an increased accumulation of the corresponding protein (Figure S1b). The activity of the accumulated protein is dependent on the presence of calcium (Figure S1c), suggesting that it remains as a latent protein in the rice leaves prone to be stimulated by calcium changes.

The phenotype of OsCPK4-Ox lines, compared to wild-type or empty vector plants, was then characterized when challenged with the blast fungus using a detached leaf assay (Coca et al., 2004). Following inoculation with the M. oryzae virulent strain FR13, the OsCPK4-Ox leaves developed less severe disease symptoms than control leaves (Figure 2a). At 7 dpi, extensive necrotic lesions with fungal sporulation were macroscopically observed on wild-type and empty vector leaves, whereas only few lesions were developed on the OsCPK4-Ox leaves. The percentage of leaf area affected by blast lesions was determined by image analyses. The results revealed a statistically significant reduction on the lesion area of three independent transgenic lines as compared to control leaves (Figure 2b). In agreement with visual inspection, OsCPK4-Ox leaves contained a significant less fungal biomass than control leaves, as determined by qPCR analysis of M. oryzae DNA (Figure 2c). The enhanced resistance phenotype

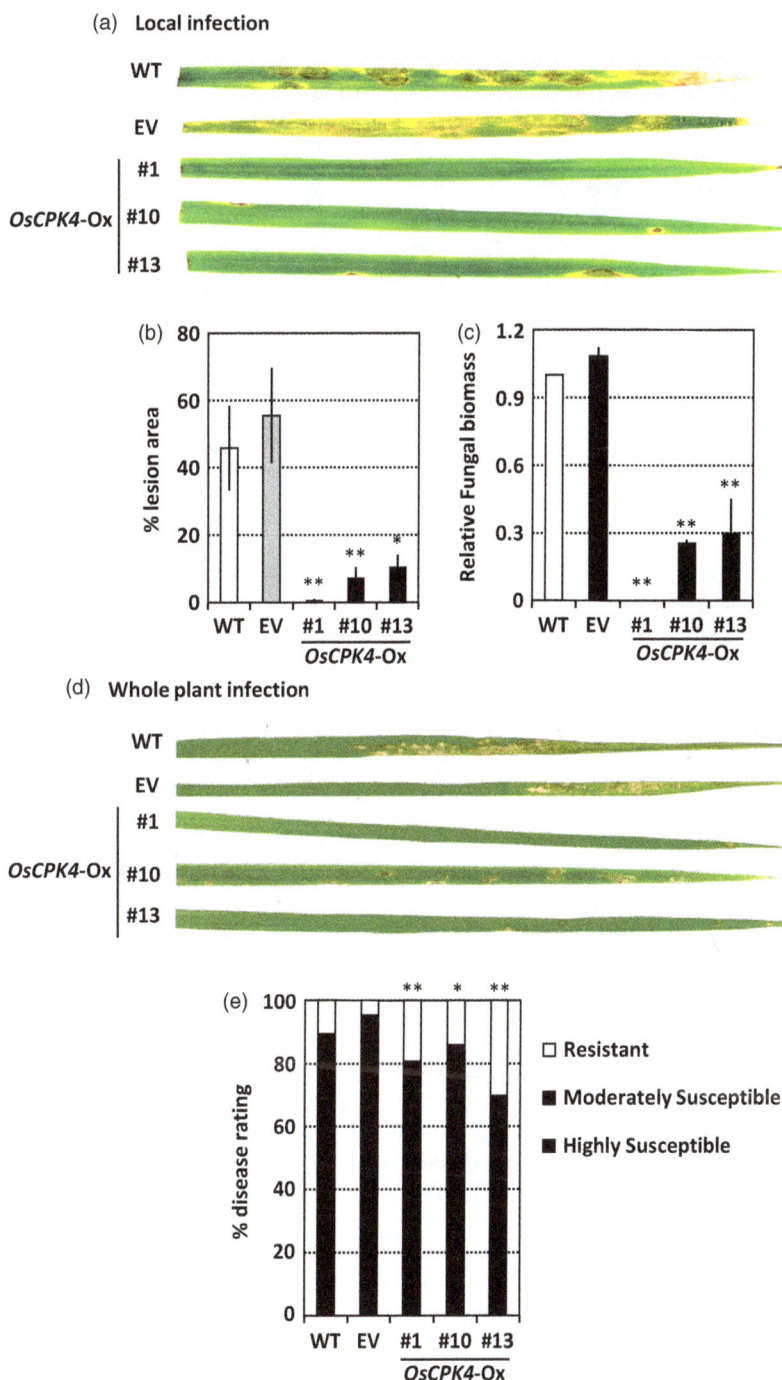

Figure 2 OsCPK4-overexpressing plants are more resistant to Magnaporthe oryzae infection. (a) Rice disease lesions caused by M. oryzae locally inoculated (10^5 spores/mL) on leaves of wild-type (WT), empty vector (EV) and OsCPK4-Ox plants (lines #1, #10 and #13) at 7 dpi. (b) Percentage average of lesion area per leaf of three independent assays with three replicates per line at 7 dpi. (c) Relative fungal amount as determined by qPCR of M. oryzae 26S rDNA gene compared to OsUbi1 gene and referred to WT. Values correspond to the average of three independent assays in which three leaves were used for quantification. (d) Disease lesions on leaves from spray-inoculated whole rice plants with M. oryzae spore suspension (10^5 spores/mL) at 7 dpi. (e) Disease rating for ten plants per line at 7 dpi following the Standard Evaluation System for blast rice disease (IRRI, 2002) based on leaf lesion area percentage. Mean values of 2 independent assays. Asterisks represent significant differences (one-way ANOVA and Tukey's test; $*P \leq 0.05$, $**P \leq 0.01$).

to the blast fungus exhibited by OsCPK4-Ox leaves was then confirmed by whole-plant infection assays. In this case, rice plants were spray-inoculated with a M. oryzae spore suspension, under experimental conditions similar to field conditions. The wild-type and empty vector control plants developed the typical blast disease lesions, whereas the OsCPK4-Ox plants showed clearly less and smaller infection lesions (Figure 2d). Further measure of disease severity showed that a higher percentage of OsCPK4-Ox plants exhibited resistant phenotype (around 22%) than wild-type or empty vector plants (around 5–10%), and a lower percentage exhibited highly susceptible phenotype (around 27%) than control plants (65%) (Figure 2e). Collectively, these results

suggest that OsCPK4 positively mediates an enhanced resistance to blast fungal infection.

To gain more insight into the nature of the enhanced blast resistance observed in the OsCPK4-Ox plants, the infection process and fungal development in rice leaves was investigated by fluorescence microscopy analysis using a GFP-expressing M. oryzae virulent strain (GFP-Guy11). GFP expression is reported not to affect the pathogenicity of M. oryzae fungal strains (Campos-Soriano and San Segundo, 2009; Sesma and Osbourn, 2004). At early infection stages (12 hpi), M. oryzae spores were easily visualized on the leaf surface of the rice plants by fluorescence confocal microscopy (Figure 3a–d). Most of the

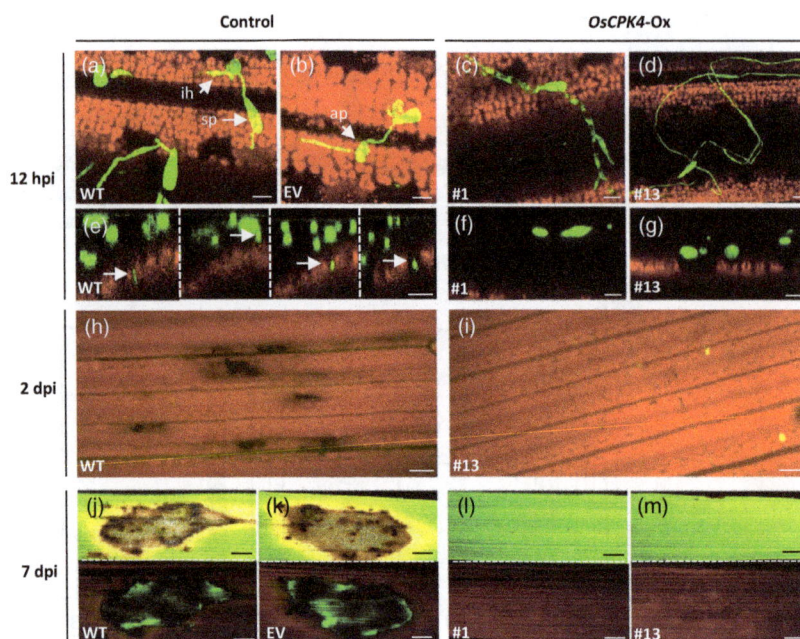

Figure 3 Microscopic analysis of *Magnaporthe oryzae* infection process on rice leaves. Representative images of *OsCPK4* overexpressor (lines 1 and 13), wild-type (WT) and control empty vector (EV) leaves inoculated with the GFP-*M. oryzae* spores (10^5 spores/mL). (a–g) Images of confocal laser microscopy of leaves at 12 hpi, corresponding to projections (a–d) and xz slides (e–g). Epifluorescence images at 2 dpi (h–i) or 7 dpi (j–m, lower panels). (j–m, upper panels) Stereoscopic brightfield images. Bars = 10 μm (a–g), 100 μm (h–i), 1 mm (j–m). Key: sp, spore; ap, appressorium; ih, invasive hypha.

spores on wild-type and empty vector leaves were germinated and produced short germ tubes that developed appressoria and invasive hyphae penetrating into epidermal cells (Figure 3a–b, e). However, *M. oryzae* spores on *OsCPK4*-Ox leaves germinated freely developing abnormal germ tubes—in some cases thick and highly vacuolated (Figure 3c), while, in others, thin and very long (Figure 3d), without visible evidences of penetration events (Figure 3f–g). These observations support that fungal penetration was impaired in *OsCPK4*-Ox leaves. After 2 dpi, infection lesions were visible under fluorescent microscopy in control leaves (Figure 3h), but not in OsCPK4-Ox leaves (Figure 3i). At later stages (7dpi), *M. oryzae* completed its life cycle in wild-type and empty vector leaves showing the typical blast lesions with a bright fluorescent mycelia growing and sporulating (Figure 3j–k). Only small necrotic spots were observed in the OsCPK4-Ox leaves (Figure 3l–m). Our observations indicate that OsCPK4-mediated resistance relies in the interference with fungal penetration rather than colonization.

The resistance of *OsCPK4*-Ox plants to other rice pathogens was also evaluated. Seedlings were assayed against the seed-borne and soil-transmitted fungal pathogen *Fusarium verticillioides*, which has been associated with the bakanae disease in rice (Wulff *et al.*, 2010). Our results indicate that *OsCPK4*-Ox seedlings are as susceptible to *F. verticillioides* infection as control wild-type and empty vector plants (Figure S2). Similarly, *OsCPK4*-Ox seedlings were equally susceptible as control seedlings when challenged with the bacterial pathogen *Dickeya dadantii*, previously known as *Erwinia chrysanthemi*, the causal agent of foot rot in rice (Goto, 1979; Mansfield *et al.*, 2012). These results suggest that the enhanced resistance to *M. oryzae* shown by *OsCPK4*-Ox plants is specific against this fungal pathogen and that it does not affect their defence against other rice pathogens with different pathogenesis mechanisms.

Defence response is early activated in *OsCPK4* overexpressor rice plants

One of the earliest defence reactions is the production of ROS, a hallmark of successful pathogen recognition and activation of

plant defence response (Torres, 2010). Because OsCPK4 interferes with the *M. oryzae* infection process at early stages, the ROS production during defence responses in *OsCPK4*-Ox rice leaves was investigated. ROS formation was monitored *in vivo* using the CM-H$_2$DCFDA probe, a noninvasive fluorescent ROS indicator (Kristiansen *et al.*, 2009). Microscopic analyses showed the induction of fluorescence in rice leaves in response to elicitor treatment, which was faster and stronger in the *OsCPK4* than in wild-type or control empty vector leaves (Figure 4a). Thirty minutes after elicitor treatment, fluorescence was barely visualized in the wild-type or empty vector leaves, but clearly visible in the leaves of two independent *OsCPK4*-Ox lines (Figure 4a, middle panels). At 1-hour treatment, the ROS formation was already detected in the wild-type and empty vector leaves, although a stronger fluorescent labelling was observed in the *OsCPK4* lines (Figure 4a, lower panels). Fluorescence quantification showed significant differences in intensity and timing of ROS formation between *OsCPK4*-Ox and control lines (Figure 4b). Similarly, ROS production was significantly stronger in the *OsCPK4*-Ox leaves compared to control leaves in response to *M. oryzae* spore inoculation (Figure 4c–d). These observations suggest that OsCPK4 accumulation mediates accelerated and potentiated ROS formation in response to *M. oryzae* infection in rice leaves.

Another defence hallmark is the callose deposition to fortify cell walls that avoids pathogen penetration into the plant cell (Luna *et al.*, 2010; Voigt, 2014). Given that *OsCPK4* overexpression prevents fungal penetration, the callose accumulation was analysed in *OsCPK4*-Ox leaves. Callose was clearly visualized after aniline blue staining as intense blue-green fluorescence under UV light in the epidermal cell walls of *OsCPK4*-Ox leaves (Figure 5). Quantification of fluorescent leaf area indicated that callose was more abundantly accumulated in the cell walls of *OsCPK4*-Ox leaves inoculated with *M. oryzae* spores (24 hpi) than in noninoculated leaves (Figure 5b). Under the same experimental conditions, callose fluorescence was not detected in control plant leaves. These observations indicate that *OsCPK4* overexpression mediates the constitutive accumulation of callose, and its

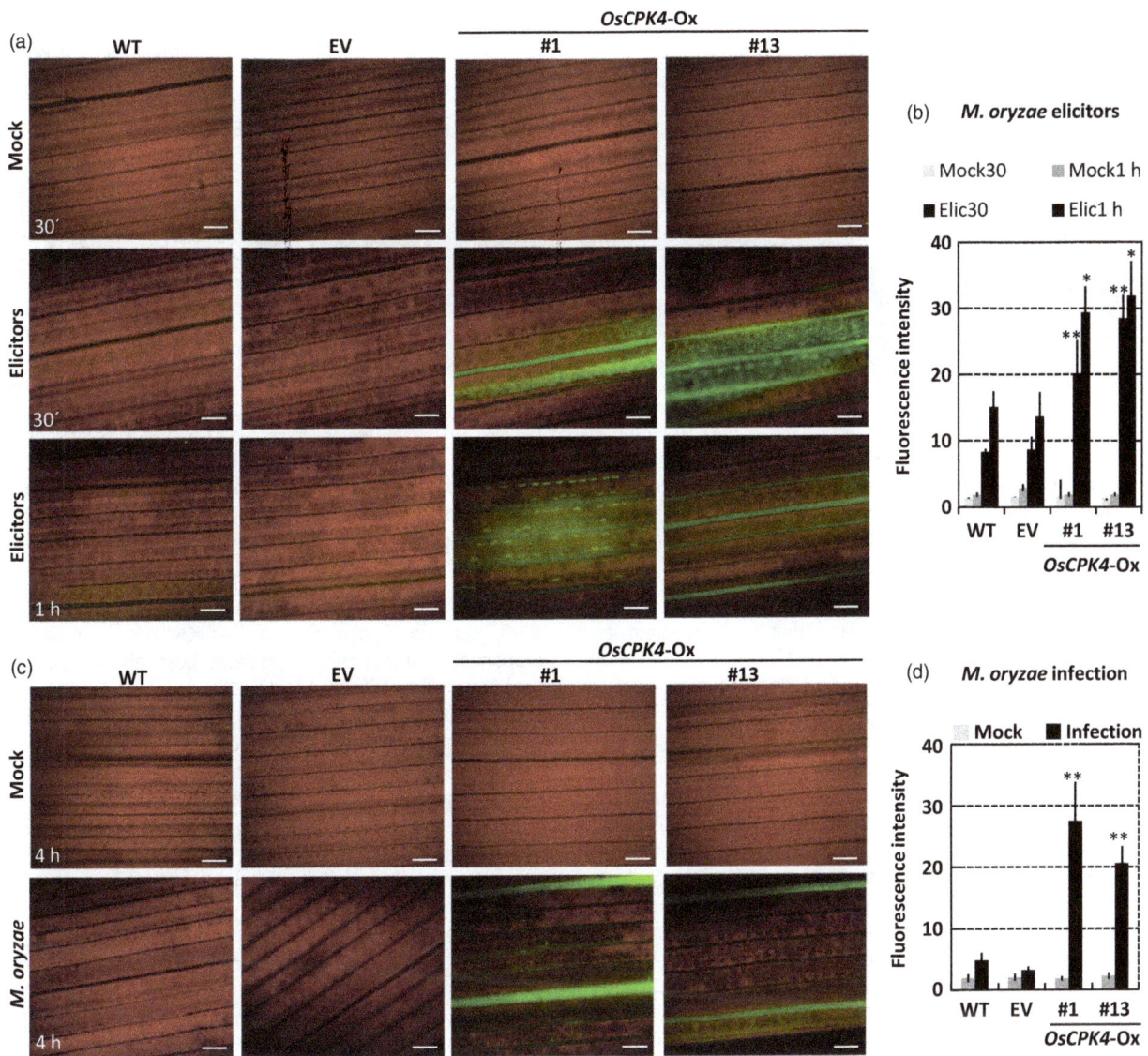

Figure 4 Rapid and strong ROS formation in *OsCPK4*-overexpressing leaves during defence response. Representative epifluorescence microscopy images of wild-type (WT), control empty vector (EV) and *OsCPK4* overexpressor (*OsCPK4*-Ox, lines 1 and 13) leaves after 1-h vacuum infiltration with a 10 μM CM-H$_2$DCFDA solution and treated with (a) *M. oryzae* elicitors (1%) or mock solution; and (c) spore suspension (10^5 spores/mL) or mock solution for the indicated period of time. (b, d) Quantitative comparison of fluorescence intensities in elicitor-treated leaves (b) and fungal-inoculated leaves (d). Values represent the average intensities, and error bars the SD of three independent leaves. Asterisks denote significant differences (one-way ANOVA and Tukey's test, *$P \leq 0.05$, **$P \leq 0.001$). Results are representative of two independent experiments. Scale bar = 200 μm.

stronger deposition in response to pathogen infection in rice leaves.

Defence gene expression is potentiated in *OsCPK4* overexpressor rice plants

To further investigate the mechanism underlying OsCPK4-mediated disease resistance, the expression profile of rice defence genes was analysed in the transgenic plants in response to *M. oryzae* infection. First, the expression of the widely used defence marker *OsPBZ1* and *OsPR5* genes was monitored. These genes encode two SA-regulated pathogenesis-related proteins from the PR10 and PR5 families (Datta *et al.*, 1999; Jwa *et al.*, 2006; Midoh and Iwata, 1996; Rakwal *et al.*, 2001). Stronger induction of these two defence genes was observed in *OsCPK4*-Ox plants when compared to wild-type or empty vector control

plants upon pathogen challenge (Figure 6a–b). These observations suggest that the *OsCPK4*-Ox plants developed a potentiated defence compared to control plants.

Similarly, the analysis of defence signalling components *OsNPR1/OsNH1* and *OsWRKY45* genes showed a stronger induction in the *OsCPK4*-Ox plants than in the control plants (Figure 6c–d). The two genes encode a transcriptional cofactor and transcriptional factor of the SA-mediated defence pathway (Chern *et al.*, 2001; Shimono *et al.*, 2012). Additionally, upstream components, such as the *OsEDS1* gene encoding an activator of SA signalling (Wiermer *et al.*, 2005), or the *OsSID2* gene encoding the isochorismate synthase enzyme responsible for part of SA synthesis in plants (Wildermuth *et al.*, 2001), also showed stronger activation in *OsCPK4*-Ox plants (Figure 6e–f). These results show stronger activation of the SA signalling

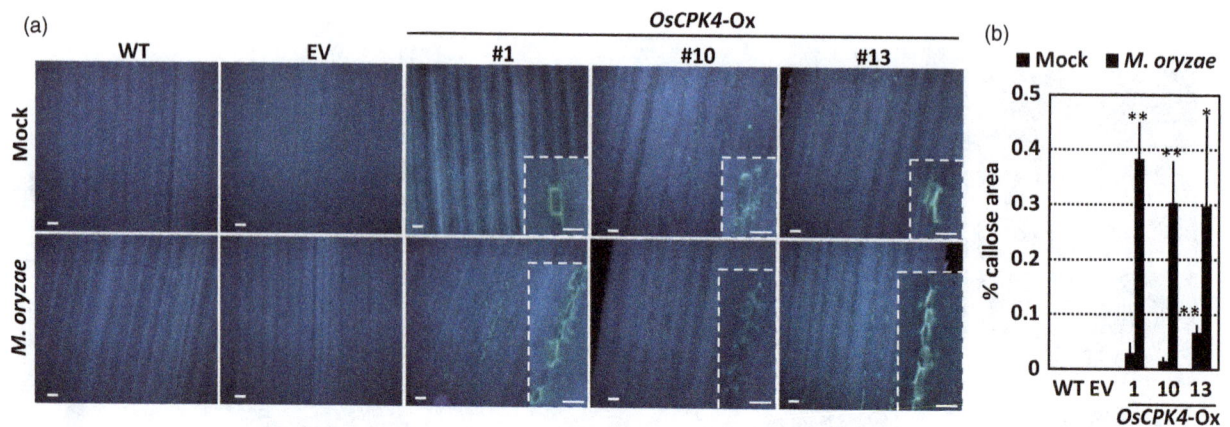

Figure 5 Callose deposition in *OsCPK4*-overexpressing rice leaves. (a) Images of wild-type (WT), empty vector (EV) or *OsCPK4*-overexpressing (*OsCPK4*-Ox) leaves (lines #1, #10 and #13) from 3-week-old plants locally inoculated with *Magnaporthe oryzae* spore suspensions (10^5 spores/mL) or mock solution. Leaves were stained with aniline blue and visualized under UV epifluorescence microscopy at 24 hpi. Magnifications are shown in inset boxes. Bars correspond to 100 μm, and 50 μm in inset boxes. (b) Mean values of the percentage of fluorescent area per leaf of three independent replicas per line in three independent assays (a total of 9 leaves per line). Asterisks denote significant differences (one-way ANOVA and Tukey's test, $*P \leq 0.05$, $**P \leq 0.001$).

defence pathway in *OsCPK4*-Ox plants that might mediate its enhanced resistance to *M. oryzae*.

Overexpression of *OsCPK4* leads to an increased SA content without compromising rice productivity

The observed strong induction of *OsSID2* gene expression, as well as of other genes related to SA defence signalling, prompted us to quantify the SA content in the *OsCPK4*-Ox lines. We determined the levels of free SA and its glucose conjugate (SAG) under control conditions. No significant differences in free SA levels were detected, but *OsCPK4*-Ox leaves accumulated up to twice as much SAG as compared to the control empty vector or wild-type leaves (Figure 7). Our results indicate that the overexpression of *OsCPK4* leads to the accumulation of SAG in rice leaves under control conditions, which in turn results in the strong activation of downstream SA-mediated defence upon pathogen infection, as revealed by our gene expression studies.

The constitutive accumulation of SA is often associated with disease resistance but is also accompanied by fitness costs, that is a penalty in plant growth and productivity (Takatsuji, 2014). To determine the effects of detected high SAG levels in *OsCPK4* rice plants, several fitness parameters of plant growth under controlled conditions were analysed. *OsCPK4*-Ox plants showed a similar appearance than control wild-type and empty vector plants (Figure S3a). They reached the same height at heading time (Figure S3b), flowered at the same period of time after sowing (Figure S3c) and, more importantly, produced a similar grain yield in two different experiments in which plants were grown under random distribution (Figure S3d). Hence, despite the OsCPK4-mediated SAG accumulation, our observations indicate that *OsCPK4* overexpression does not have a negative impact in the growth and productivity of rice plants.

Discussion

The present study reveals that the isoform OsCPK4 from the multigenic family of rice CDPKs has a function in the innate immunity of rice plants. Given that OsCPK4 was also known to participate in the salt and drought stress responses (Campo *et al.*,

2014), our results demonstrate that OsCPK4 is a signalling component that positively modulates both abiotic and biotic stress responses in rice plants. This work shows that the expression of the *OsCPK4* gene was rapidly induced in rice leaves when challenged with the *M. oryzae* pathogen and that *OsCPK4* overexpression conferred enhanced resistance to rice blast disease, together supporting that OsCPK4 mediates the immune response to blast fungus in rice plants. OsCPK4 accumulation is induced at early stages of the infection process, coinciding with pathogen penetration and suggesting that this protein acts at the earliest signalling events initiated upon pathogen recognition. Among the earliest immune reactions, calcium influxes are included (Blume *et al.*, 2000; Jeworutzki *et al.*, 2010; Ranf *et al.*, 2011), which occur through plasma membrane calcium channels activated by the recognition via pathogen recognition receptors (PRRs) of pathogen-associated molecular patterns (PAMPs) (Kurusu *et al.*, 2005). Because OsCPK4 is localized at the plant plasma membrane (Campo *et al.*, 2014), our hypothesized mechanistic model is that OsCPK4 acts as calcium sensor of changes stimulated by pathogen perception that triggers the downstream defence signalling events mediated by phosphorylation cascades (Figure 8). In agreement with the proposed mechanism of action, *OsCPK4*-Ox plants that accumulate constitutively increased levels of the protein exhibited a rapid and potentiated defence response upon pathogen infection. These plants accumulate the full OsCPK4 protein, including the calcium-binding regulatory domain, ready to be stimulated by calcium upon pathogen sensing. Thus, *OsCPK4*-Ox plants showed fast and enhanced ROS production, increased callose deposition and strong defence gene expression when challenged with the *M. oryzae* fungal pathogen. As a result, these plants showed an enhanced disease resistance phenotype against *M. oryzae* as determined by visual inspection, fungal growth quantification and disease lesion measurement. Blast disease resistance was shown not only in detached leaf assays but also in whole-plant infection assays. These results support that OsCPK4 participates in the signal transmission initiated by pathogen perception, and the constitutive increased accumulation of OsCPK4 leads to an accelerated and amplified defence signal.

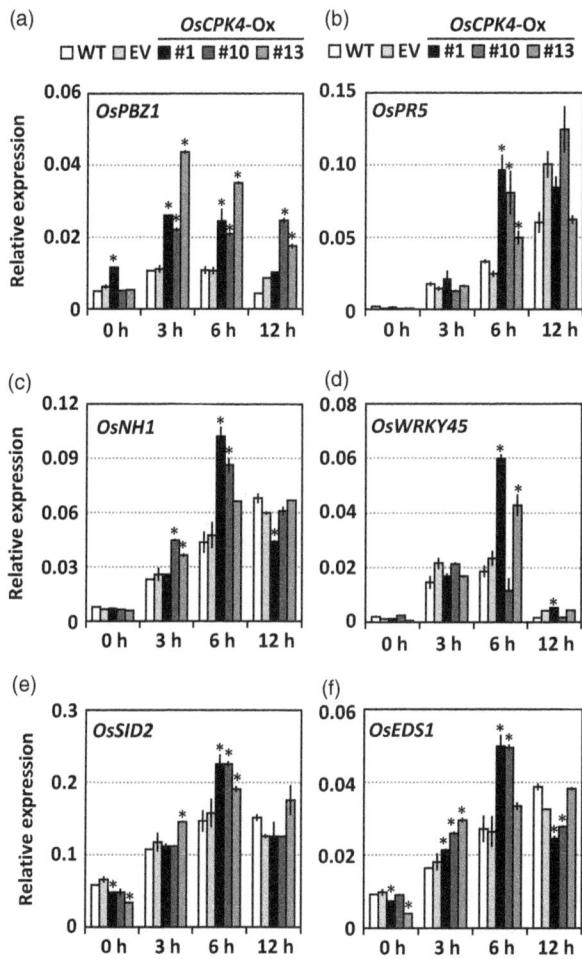

Figure 6 Defence gene expression in *OsCPK4* overexpressor plants in response to *Magnaporthe oryzae* infection. Leaves of wild-type (WT), empty vector (EV) and *OsCPK4*-Ox (lines 1, 10, 13) plants were locally inoculated with a *M. oryzae* spore suspension (10^5 spores/mL) and collected in a pool of 4 leaves at the indicated period of time. Expression levels of indicated defence-related genes were determined by qRT-PCR and normalized to *OsUbi1*. Asterisks denote significant differences (one-way ANOVA and Tukey's test, *$P \leq 0.01$). Results are representative of two independent experiments.

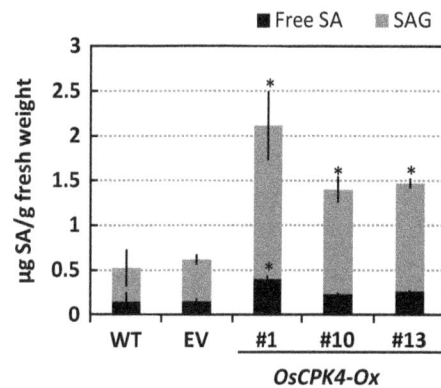

Figure 7 Increased content of total SA, free SA and glucoside conjugate (SAG) in *OsCPK4* overexpressor plants. Data are mean values of two independent quantification in a pool of 3 leaves from 3-week-old wild-type (WT), empty vector (EV) or *OsCPK4* overexpressor (*OsCPK4*-Ox) plants. Asterisk denotes significant differences (one-way ANOVA, *$P < 0.05$).

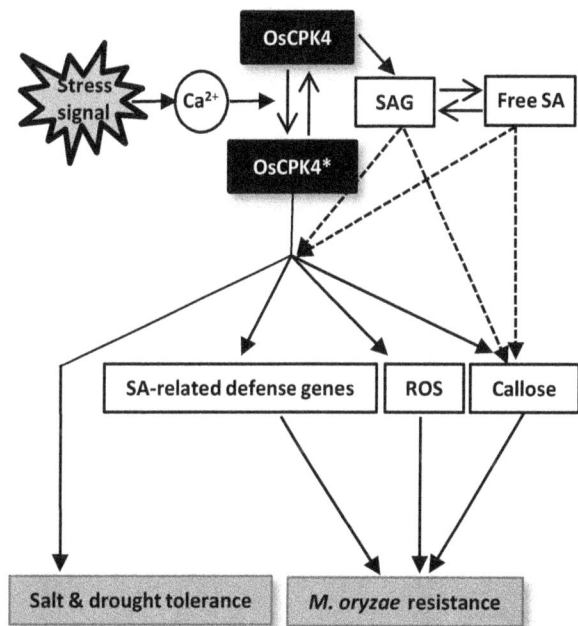

Figure 8 Model for OsCPK4-mediated defence responses. Stress induces Ca^{2+} increase that activates OsCPK4 leading to ROS production, callose deposition and SA-regulated defence gene expression resulting in resistance to *Magnaporthe oryzae* infection. OsCPK4 also mediated the accumulation of SAG.

Our results showed that ROS production was stronger and faster in *OsCPK4*-Ox plants upon elicitor or pathogen perception. ROS levels might reach toxic thresholds for *M. oryzae*, leading to fungal penetration blockage as observed under confocal microscopy. However, the importance of ROS in defence reactions is not only due to their toxicity to pathogens, but also to their role as signalling molecules for local and systemic responses (Mittler *et al.*, 2011). ROS mediate the defensive response through oxidative waves that activate signal transduction through phosphorylation cascades, accompanied by hormonal signalling and the expression of defence-related genes (Baxter *et al.*, 2013; Shetty *et al.*, 2008). Therefore, the increased ROS production might contribute to the enhanced defence responsiveness observed in *OsCPK4*-Ox plants. Be as toxic compound or as signalling molecules, ROS production seems to contribute to the enhanced resistance of *OsCPK4*-Ox plants, and to be activated by OsCPK4 in response to PAMP stimulation. Connections between ROS production and CPKs have been already described in the

literature; these studies report that ectopic expression of constitutively active CPK variants resulted in an increased production of ROS (Dubiella *et al.*, 2013; Kobayashi *et al.*, 2007; Romeis *et al.*, 2001; Xing *et al.*, 2001). Moreover, NADPH oxidases playing a central role in the oxidative burst during immune responses have been reported as CPK targets in potato and Arabidopsis (Dubiella *et al.*, 2013; Kobayashi *et al.*, 2007). Similarly for rice, the plasma membrane NADPH oxidases might be potential targets of the plasma membrane-associated OsCPK4 protein, triggering a fast and strong oxidative burst upon pathogen attack in the plants that constitutively accumulated increased levels of OsCPK4 protein.

Other sources for ROS production also exist in plant cells, such as the peroxidases identified in Arabidopsis as major contributors to ROS production during responses to fungal elicitors (Daudi et al., 2012), and they might be also potential OsCPK4 targets. Future studies will address OsCPK4 target identification.

OsCPK4 overexpressor plants accumulate increased SAG levels, the glycosylated form of SA. SAG is considered a likely storage form of physiologically active free SA, which is accumulated in the vacuole to serve as a source of free SA when required in dicotyledonous plants (Dean et al., 2005; Seo et al., 1995). In rice plants, SAG has been proposed to have per se a role in activating defences for induced resistance (Umemura et al., 2009). This increased accumulation of SAG prepared OsCPK4-overexpressing rice plants for a strong activation of SA-mediated defence signalling upon M. oryzae infection. As a result, intense activation of components of the SA pathway was detected, including the biosynthetic gene OsSID2, the OsNH1 and OsWRKY45 transcriptional activator genes and the end products OsPBZ1 and OsPR5. Another immune response associated with SA is the callose deposition, being promoted by SA (Yi et al., 2014). In agreement with the high SAG content, callose was also accumulated in the OsCPK4-Ox. Callose might represent a physical barrier that prevents fungal penetration leading to the observed resistant phenotype of OsCPK4-Ox plants. Our results reveal that OsCPK4 contributes to the accumulation of SAG and callose in rice plants under noninductive conditions.

Our data suggest that the rice plants overexpressing OsCPK4 are sensitized or preconditioned for a robust and fast immune response by accumulating a signalling component that can be immediately activated upon exposure to stress. Defence responses usually have fitness costs associated with resource allocation for defensive compounds or the toxicity of the defensive products (van Hulten et al., 2006), and the strategies to improve disease resistance in plants based on the constitutive activation of defences are accompanied by negative effects on plant growth and yield (Gust et al., 2010; Takatsuji, 2014). In this sense, we have shown that the overexpression of the OsCPK4 gene in rice plants does not have a negative impact on plant performance, at least under containment conditions. The growth, flowering time and yield fitness parameters of these plants are not significantly different than those of the wild-type plants. This is in agreement with the observation that OsCPK4-overexpressing rice plants did not show the constitutive expression of defence-related genes or ROS accumulation under noninductive conditions, although they do accumulate SAG and callose. This is consistent with the already reported global transcriptomic analyses showing that overexpression of OsCPK4 in rice plants has a low impact in the rice transcriptome (Campo et al., 2014). Altogether, our results support that OsCPK4 might be a good target for blast protection while maintaining rice yield.

OsCPK4-Ox rice plants are also more tolerant to salt and drought stress (Campo et al., 2014). SA, in addition to modulate the immune response in plants, is also known to improve the tolerance to salt and drought stress by preventing membrane damage among other mechanisms (Farooq et al., 2009; Jayakannan et al., 2013). Moreover, SA inhibits lipid peroxidation, thus protecting cell membranes (Dinis et al., 1994; Lapenna et al., 2009). Therefore, the improved tolerance to drought and salinity of OsCPK4-Ox rice plants associated with a reduction in lipid peroxidation could be mediated by the increased content of SAG. This is an interesting result because trade-offs between defence and abiotic stress tolerance have been frequently reported (Sharma et al., 2013). For instance, OsCPK12 oppositely modulates salt stress tolerance and blast disease resistance (Asano et al., 2012). However, crosstalk between biotic and abiotic signalling pathways can result not only in negative but also in positive functional outcomes (Sharma et al., 2013). Our studies demonstrate that OsCPK4 acts as a convergence component that positively modulates both biotic and abiotic signalling pathways, presumably modulating SA levels and suggesting that it is a good molecular target to improve tolerance to different stresses in rice plants.

Experimental procedures

Plant and fungal growth conditions

OsCPK4 overexpressor rice plants were previously generated and described (Campo et al., 2014). They were grown at 28 °C with a 14-h/10-h light/dark photoperiod. Fungal strains of M. oryzae FR13 isolate (provided by D. Tharreau, CIRAD Montpelier, France) and Guy11-GFP (provided by A. Sesma, CBGP Madrid, Spain) were grown in oatmeal agar (72.5 g/L, 30 mg/L chloramphenicol) for 2 weeks at 28 °C using a 16-h/8-h light/dark photoperiod. Their spores were collected in sterile water, filtrated with Miracloth (Calbiochem) and adjusted to the appropriate concentration using a Bürker counting chamber. M. oryzae elicitors were obtained as previously described (Casacuberta et al., 1992). F. verticillioides and D. dadantii strains were grown as previously described (Gómez-Ariza et al., 2007).

RNA isolation and RT-qPCR

Gene expression levels were determined from a pool of four leaves at the same developmental stage of 3-week-old soil-grown plants. Total RNA was extracted using TRIzol reagent (Invitrogen, Basel, Switzerland). DNase-treated RNA (1 μg) was retrotranscribed using the transcriptor first cDNA synthesis kit (Roche, Mannheim, Germany). qRT-PCR analyses were carried out in 96-well optical plates in a LightCycler® 480 System (Roche) according to the following programme: 10 min at 95 °C, 45 cycles of 95 °C for 10 s and 60 °C for 30 s and an additional cycle of dissociation curves to ensure a unique amplification. The reaction mixture contained 5 μL of SYBR Green Master mix reagent (Roche), 2 μL of 1:4 diluted cDNA sample and 300 nm of each gene-specific primer (Table S1) in a final volume of 10 μL. The results for the gene expression were normalized to OsUbi1 (LOC_Os06g46770) and OsUbi5 (LOC_Os01g22490) genes as indicated. Three technical replicates were performed for each sample.

Protein extracts, CPK activity and immunoblot analysis

Protein extracts were obtained from membrane-enriched fractions prepared from leaves in a pool of at least four plants. Samples were ground in liquid nitrogen, thawed in two volumes of extraction buffer (10% sucrose, 50 mM Tris–HCl pH 7.5, 5 mM EDTA, 5 mM EGTA, 5 mM dithiothreitol, 1 mM PMSF) and centrifuged at 15 000 g for 20 min at 4 °C. The pellet was resuspended in 2 volumes of elution buffer (1% Triton X-100, 25 mM Tris–HCl pH 7.5, 1 mM $MgCl_2$, 1 mM PMSF) using a cooled sonication bath. Protein extracts were recovered from the supernatant after centrifugation as before, quantified, separated in SDS-PAGE and transferred to nitrocellulose membranes. Western blot analyses were performed using anti-OsCPK4 antibodies as described (Campo et al., 2014). Two independent experiments with 3 biological repeats on a pool of 3 independent plants for each time point were analysed. Antibodies were raised against the N-terminal variable domain of OsCPK4 (Met1 to Arg58) to specifically recognize this isoform of the conserved OsCPK family protein.

The calcium-dependent kinase activity was analysed as described with minor modifications (Boudsocq et al., 2012). These include that total protein was extracted from rice leaves and immunoprecipitated for 2 h with specific anti-OsCPK4 antibodies bound to Dynabeads® with the antibody coupling kit (Life Technologies, Carlsbad, CA, USA), that the phosphorylation substrates were β-casein peptide (Sigma St. Louis, MO, USA) and myelin basic protein (Invitrogen) and that the unincorporated radioactive nucleotides were discarded using MicroSpin G-25 columns (GE Healthcare Little Chalfont, UK). The concentration of free calcium in each buffer was calculated using MaxChelator (http://maxchelator.stanford.edu/).

Disease resistance assays with rice pathogens

M oryzae infections were performed using a detached leaf infection assay as described (Coca et al., 2004), or a whole-plant infection assay by spraying the fungal spores with an aerograph at 2 atmospheres of pressure. Infection assays were carried out with 3-week-old plants grown in soil, using three pots with 10 plants each per line and 2 mL of spore suspension (10^5 spores/mL) per pot. The plants were maintained for 16 h in a closed plastic bag for high humidity conditions after inoculation. Lesion areas were measured by image analysis software Assess v.2.0 at 7 days postinoculation (dpi). Fungal biomass in rice infected leaves was determined at 7 dpi by qPCR using specific primers for the *26S* ribosomal RNA gene of *M. oryzae* and normalized to *OsUbi1* gene as described (Qi and Yang, 2002). DNA (15 ng per qPCR) was obtained from the rice infected leaves as described (Murray and Thompson 1980), but using MATAB as extraction buffer (0.1 M Tris–HCl, pH 8.0, 1.4 M NaCl, 20 mM EDTA, 2% MATAB, 1% PEG 6000 and 0.5% sodium sulphite). Disease symptoms on whole-plant infection assays were scored at 7 dpi following the Standard Evaluation System for blast rice disease (IRRI, 2002). Three biological replicates were performed for each line and three technical replicates per sample.

Infection assays with *F. verticillioides* were performed as previously described with minor modifications (Bundó et al., 2014), including a seed germination period of 8 h previous inoculation with 10^3 spores/mL suspensions.

Assays with *D. dadantii* were carried out as described with minor modifications (Gómez-Ariza et al., 2007), reducing the seed germination period to 8 h and increasing the inoculation doses to 10^7 CFU.

Fluorescence microscopy

Confocal laser scanning microscopy was performed using an Olympus FV1000 microscope (Tokyo, Japan). GFP was excited with an argon ion laser emitting at 488 nm and fluorescence detected at 500–550 nm. Chlorophyll autofluorescence was visualized at 600–700 nm. Lesions were also observed under a Zoom Stereo Microscope Olympus SZX16 fitted with an Olympus DP72 digital camera.

For ROS detection, leaf segments from at least three different plants were infiltrated with a 10 μM solution of the fluorescent probe CM-H$_2$DCFDA (Molecular Probes) in 100 mM phosphate buffer pH 7.2 for 2 h. The leaves were then treated with a 1% *M. oryzae* elicitor solution in sterile water or inoculated with a 10^5 spores/mL suspension. ROS was monitored over the time

using an Axiophot Zeiss epifluorescent microscope, and fluorescent signals were quantified by image analysis using the ImageJ software.

Callose accumulation was visualized by fluorescence under epifluorescence microscopy after aniline blue staining of leaf segments from at least three different plants as previously described (Luna et al., 2010). The fluorescent area per leaf segment was quantified also using the ImageJ software.

Salicylic acid quantification

Free SA and SA β-glucoside (SAG) content in rice leaves was determined as previously described with some minor modifications (Coca and San Segundo, 2010). Total SA was obtained from 1 g of fresh grinded leaves by two consecutive methanol and ethanol extractions (3 mL each). After alcohol evaporation, the extracts were resuspended in water and separated into two parts, one to determine free SA and the other for SAG. SAG samples were digested with 10 U/mL of β-glucosidase from almonds (Sigma) at 37 °C during 16 h. After digestion, the samples were filled up to 1 mL with milli-Q water, and HCl 37% (50 μL) was added. They were subjected to two consecutive extractions with ethyl acetate/cyclopentane/isopropanol (2 mL, 50:50:1). Organic phases were evaporated and resuspended in methanol (25 μL) for the HPLC analysis using a Zorbax Eclipse XDB-C18 column (Agilent Technologies, Santa Clara, CA, USA). Two biological replicates were performed for each independent line.

Acknowledgements

We thank B. San Segundo for scientific advice; M. Alborno, S. Campo and L. Campos-Soriano for critical reading of the manuscript; and M. Amenós for technical assistance with confocal microscopy. This work was supported by grants BIO2009-08719 and BIO2012-32838 from 'Ministerio de Economía y Competitividad' (MINECO, Spain) with European Regional Development Funds (FEDER).

References

Asano, T., Tanaka, N., Yang, G., Hayashi, N. and Komatsu, S. (2005) Genome-wide identification of the rice calcium-dependent protein kinase and its closely related kinase gene families: comprehensive analysis of the CDPKs gene family in rice. *Plant Cell Physiol.* **46**, 356–366.

Asano, T., Hayashi, N., Kobayashi, M., Aoki, N., Miyao, A., Mitsuhara, I., Ichikawa, H. et al. (2012) A rice calcium-dependent protein kinase OsCPK12 oppositely modulates salt-stress tolerance and blast disease resistance. *Plant J.* **69**, 26–36.

Baxter, A., Mittler, R. and Suzuki, N. (2013) ROS as key players in plant stress signalling. *J. Exp. Bot.* **65**, 129–1240.

Blume, B., Nürnberger, T., Nass, N. and Scheel, D. (2000) Receptor-mediated increase in cytoplasmic free calcium required for activation of pathogen defense in parsley. *Plant Cell*, **12**, 1425–1440.

Boller, T. and He, S.Y. (2009) Innate immunity in plants: an arms race between pattern recognition receptors in plants and effectors in microbial pathogens. *Science*, **324**, 742–744.

Boudsocq, M. and Sheen, J. (2013) CDPKs in immune and stress signaling. *Trends Plant Sci.* **18**, 30–40.

Boudsocq, M., Droillard, M.J., Regad, L. and Laurière, C. (2012) Characterization of Arabidopsis calcium-dependent protein kinases: activated or not by calcium? *Biochem. J.* **447**, 291–299.

Bundó, M., Montesinos, L., Izquierdo, E., Campo, S., Mieulet, D., Guiderdoni, E., Rossignol, M. et al. (2014) Production of cecropin A antimicrobial peptide in rice seed endosperm. *BMC Plant Biol.* **14**, 102.

Campo, S., Peris-Peris, C., Siré, C., Moreno, A.B., Donaire, L., Zytnicki, M., Notredame, C. et al. (2013) Identification of a novel microRNA (miRNA) from rice that targets an alternatively spliced transcript of the Nramp6 (Natural resistance-associated macrophage protein 6) gene involved in pathogen resistance. New Phytol. **199**, 212–227.

Campo, S., Baldrich, P., Messeguer, J., Lalanne, E., Coca, M. and San Segundo, B. (2014) Overexpression of a calcium-dependent protein kinase confers salt and drought tolerance in rice by preventing membrane lipid peroxidation. Plant Physiol. **165**, 688–704.

Campos-Soriano, L. and San Segundo, B. (2009) Assessment of blast disease resistance in transgenic PRms rice using a gfp-expressing Magnaporthe oryzae strain. Plant. Pathol. **58**, 677–689.

Casacuberta, J.M., Raventos, D., Puigdomenech, P. and San Segundo, B. (1992) Expression of the gene encoding the PR-like protein PRms in germinating maize embryos. Mol. Gen. Genet. **234**, 97–104.

Chern, M.S., Fitzgerald, H.A., Yadav, R.C., Canlas, P.E., Dong, X. and Ronald, P.C. (2001) Evidence for a disease-resistance pathway in rice similar to the NPR1-mediated signaling pathway in Arabidopsis. Plant J. **27**, 101–113.

Chisholm, S.T., Coaker, G., Day, B. and Staskawicz, B.J. (2006) Host-microbe interactions: shaping the evolution of the plant immune response. Cell, **124**, 803–814.

Coca, M. and San Segundo, B. (2010) AtCPK1 calcium-dependent protein kinase mediates pathogen resistance in Arabidopsis. Plant J. **63**, 526–540.

Coca, M., Bortolotti, C., Rufat, M., Penas, G., Eritja, R., Tharreau, D., del Pozo, A.M. et al. (2004) Transgenic rice plants expressing the antifungal AFP protein from Aspergillus giganteus show enhanced resistance to the rice blast fungus Magnaporthe grisea. Plant Mol. Biol. **54**, 245–259.

Datta, K., Velazhahan, R., Oliva, N., Ona, I., Mew, T., Khush, G.S., Muthukrishnan, S. et al. (1999) Over-expression of the cloned rice thaumatin-like protein (PR-5) gene in transgenic rice plants enhances environmental friendly resistance to Rhizoctonia solani causing sheath blight disease. Theor. Appl. Genet. **98**, 1138–1145.

Daudi, A., Cheng, Z., O'Brien, J.A., Mammarella, N., Khan, S., Ausubel, F.M. and Bolwell, G.P. (2012) The apoplastic oxidative burst peroxidase in Arabidopsis is a major component of pattern-triggered immunity. Plant Cell, **24**, 275–287.

Dean, J., Mohammed, L. and Fitzpatrick, T. (2005) The formation, vacuolar localization, and tonoplast transport of salicylic acid glucose conjugates in tobacco cell suspension cultures. Planta, **221**, 287–296.

Dean, R., Van Kan, J.A., Pretorius, Z.A., Hammond-Kosack, K.E., Di Pietro, A., Spanu, P.D., Rudd, J.J. et al. (2012) The Top 10 fungal pathogens in molecular plant pathology. Mol. Plant Pathol. **13**, 414–430.

Dinis, T.C.P., Madeira, V.M.C. and Almeida, L.M. (1994) Action of phenolic derivatives (acetaminophen, salicylate, and 5-aminosalicylate) as inhibitors of membrane lipid peroxidation and as peroxyl radical scavengers. Arch. Biochem. Biophys. **315**, 161–169.

Dodd, A.N., Kudla, J. and Sanders, D. (2010) The language of calcium signaling. Ann. Rev. Plant Biol. **61**, 593–620.

Dubiella, U., Seybold, H., Durian, G., Komander, E., Lassig, R., Witte, C.P., Schulze, W.X. et al. (2013) Calcium-dependent protein kinase/NADPH oxidase activation circuit is required for rapid defense signal propagation. Proc. Natl Acad. Sci. USA **110**, 8744–8749.

Durrant, W.E. and Dong, X. (2004) Systemic acquired resistance. Ann. Rev. Phytopathol. **42**, 185–209.

Farooq, M., Basra, S.M.A., Wahid, A., Ahmad, N. and Saleem, B.A. (2009) Improving the drought tolerance in rice (Oryza sativa L.) by exogenous application of salicylic acid. J. Agr. Crop Sci. **195**, 237–246.

Fu, L., Yu, X. and An, C. (2013) Overexpression of constitutively active OsCPK10 increases Arabidopsis resistance against Pseudomonas syringae pv. tomato and rice resistance against Magnaporthe grisea. Plant Physiol. Biochem. **73**, 201–210.

Gómez-Ariza, J., Campo, S., Rufat, M., Estopa, M., Messeguer, J., San Segundo, B. and Coca, M. (2007) Sucrose-mediated priming of plant defense responses and broad-spectrum disease resistance by overexpression of the maize pathogenesis-related PRms protein in rice plants. Mol. Plant Microbe Interact. **20**, 832–842.

Goto, M. (1979) Bacterial foot rot of rice caused by a strain of Erwinia chrysanthemi. Phytopathology, **69**, 213–216.

Gust, A.A., Brunner, F. and Nürnberger, T. (2010) Biotechnological concepts for improving plant innate immunity. Curr. Opin. Biotechnol. **21**, 204–210.

Harper, J.F., Breton, G. and Harmon, A. (2004) Decoding Ca2+ signals through plant protein kinases. Ann. Rev. Plant Biol. **55**, 263–288.

van Hulten, M., Pelser, M., van Loon, L.C., Pieterse, C.M.J. and Ton, J. (2006) Costs and benefits of priming for defense in Arabidopsis. Proc. Natl Acad. Sci. USA **103**, 5602–5607.

IRRI, International Rice Research Institute. (2002) Standard Evaluation System for Rice. Los Baños, Philippines: IRRI, International Rice Research Institute.

Jayakannan, M., Bose, J., Babourina, O., Rengel, Z. and Shabala, S. (2013) Salicylic acid improves salinity tolerance in Arabidopsis by restoring membrane potential and preventing salt-induced K(+) loss via a GORK channel. J. Exp. Bot. **64**, 2255–2268.

Jeworutzki, E., Roelfsema, M.R., Anschütz, U., Krol, E., Elzenga, J.T., Felix, G., Boller, T. et al. (2010) Early signaling through the Arabidopsis pattern recognition receptors FLS2 and EFR involves Ca2 + -associated opening of plasma membrane anion channels. Plant J. **62**, 367–378.

Jones, J.D.G. and Dangl, J.L. (2006) The plant immune system. Nature, **444**, 323–329.

Jwa, N.S., Agrawal, G.K., Tamogami, S., Yonekura, M., Han, O., Iwahashi, H. and Rakwal, R. (2006) Role of a defense/stress-related marker genes, proteins and a secondary metabolites in a defining rice self-defense mechanisms. Plant Physiol. Biochem. **44**, 261–273.

Kobayashi, M., Ohura, I., Kawakita, K., Yokota, N., Fujiwara, M., Shimamoto, K., Doke, N. et al. (2007) Calcium-dependent protein kinases regulate the production of reactive oxygen species by potato NADPH oxidase. Plant Cell, **19**, 1065–1080.

Kristiansen, K.A., Jensen, P.E., MØller, I.M. and Schulz, A. (2009) Monitoring reactive oxygen species formation and localisation in living cells by use of the fluorescent probe CM-H$_2$DCFDA and confocal laser microscopy. Physiol. Plant. **136**, 369–383.

Kurusu, T., Yagala, T., Miyao, A., Hirochika, H. and Kuchitsu, K. (2005) Identification of a putative voltage-gated Ca2 + channel as a key regulator of elicitor-induced hypersensitive cell death and mitogen-activated protein kinase activation in rice. Plant J. **42**, 798–809.

Lapenna, D., Ciofani, G., Pierdomenico, S.D., Neri, M., Cuccurullo, C., Giamberardino, M.A. and Cuccurullo, F. (2009) Inhibitory activity of salicylic acid on lipoxygenase-dependent lipid peroxidation. Biochim. Biophys. Acta **1790**, 25–30.

Lecourieux, D., Ranjeva, R. and Pugin, A. (2006) Calcium in plant defence-signalling pathways. New Phytol. **171**, 249–269.

Lee, F., Cartwright, R.D., Jia, Y. and Correll, J.C. (2009) Field resistance expressed when the Pi-ta gene is compromised by Magnaporthe oryzae. In Advances in Genetics, Genomics and Control of Rice Blast Disease (Wang, G.L. and Valent, B., eds), pp. 281–289. London, UK: Springer Netherlands.

Liu, W., Liu, J., Triplett, L., Leach, J.E. and Wang, G.L. (2014) Novel insights into rice innate immunity against bacterial and fungal pathogens. Annu. Rev. Phytopathol. **52**, 213–214.

Luna, E., Pastor, V., Robert, J., Flors, V., Mauch-Mani, B. and Ton, J. (2010) Callose deposition: a multifaceted plant defense response. Mol. Plant Microbe Interact. **24**, 183–193.

Mansfield, J., Genin, S., Magori, S., Citovsky, V., Sriariyanum, M., Ronald, P., Dow, M. et al. (2012) Top 10 plant pathogenic bacteria in molecular plant pathology. Mol. Plant Pathol. **13**, 614–629.

Meng, X. and Zhang, S. (2013) MAPK cascades in plant disease resistance signaling. Ann. Rev. Phytopathol. **51**, 245–266.

Midoh, N. and Iwata, M. (1996) Cloning and characterization of a probenazole-inducible gene for an intracellular pathogenesis-related protein in rice. Plant Cell Physiol. **37**, 9–18.

Mittler, R., Vanderauwera, S., Suzuki, N., Miller, G., Tognetti, V.B., Vandepoele, K. and Van Breusegem, F. (2011) ROS signaling: The new wave? Trends in Plant Science **16**(6), 300–309. doi:10.1016/j.tplants.2011.03.007.

Murray, M.G. and Thompson, W.F. (1980) Rapid isolation of high molecular weight plant DNA. Nucleic Acids Research **8**(19), 4321–4325.

Navarro, L., Zipfel, C., Rowland, O., Keller, I., Robatzek, S., Boller, T. and Jones, J.D.G. (2004) The transcriptional innate immune response to flg22. Interplay and overlap with Avr gene-dependent defense responses and bacterial pathogenesis. *Plant Physiol.* **135**, 1113–1128.

Ou, S.H. (1987) *Rice Diseases.* Surrey, UK: Commonwealth Mycological Institute.

Qi, M. and Yang, Y. (2002) Quantification of *Magnaporthe grisea* during infection of rice plants using real-time polymerase chain reaction and Northern blot/phosphoimaging analyses. *Phytopathology*, **92**, 870–876.

Rakwal, R., Agrawal, G.K. and Agrawal, V.P. (2001) Jasmonate, salicylate, protein phosphatase 2A inhibitors and kinetin up-regulate OsPR5 expression in cut-responsive rice (*Oryza sativa*). *J. Plant Physiol.* **158**, 1357–1362.

Ranf, S., Eschen-Lippold, L., Pecher, P., Lee, J. and Scheel, D. (2011) Interplay between calcium signalling and early signalling elements during defence responses to microbe- or damage-associated molecular patterns. *Plant J.* **68**, 100–113.

Ray, S., Agarwal, P., Arora, R., Kapoor, S. and Tyagi, A. (2007) Expression analysis of calcium-dependent protein kinase gene family during reproductive development and abiotic stress conditions in rice (*Oryza sativa* L. ssp. *indica*). *Mol. Genet. Genomics.* **278**, 493–505.

Romeis, T. and Herde, M. (2014) From local to global: CDPKs in systemic defense signaling upon microbial and herbivore attack. *Curr. Opin. Plant Biol.* **20**, 1–10.

Romeis, T., Ludwig, A.A., Martin, R. and Jones, J.D. (2001) Calcium-dependent protein kinases play an essential role in a plant defence response. *EMBO J.* **20**, 5556–5567.

Ryals, J.A., Neuenschwander, U.H., Willits, M.G., Molina, A., Steiner, H.Y. and Hunt, M.D. (1996) Systemic acquired resistance. *Plant Cell*, **8**, 1809–1819.

Saijo, Y., Hata, S., Kyozuka, J., Shimamoto, K. and Izui, K. (2000) Over-expression of a single Ca^{2+}-dependent protein kinase confers both cold and salt/drought tolerance on rice plants. *Plant J.* **23**, 319–327.

Schulz, P., Herde, M. and Romeis, T. (2013) Calcium-dependent protein kinases: hubs in plant stress signaling and development. *Plant Physiol.* **163**, 523–530.

Seo, S., Ishizuka, K. and Ohashi, Y. (1995) Induction of salicylic acid β-Glucosidase in tobacco leaves by exogenous salicylic acid. *Plant Cell Physiol.* **36**, 447–453.

Sesma, A. and Osbourn, A.E. (2004) The rice leaf blast pathogen undergoes developmental processes typical of root-infecting fungi. *Nature*, **431**, 582–586.

Seybold, H., Trempel, F., Ranf, S., Scheel, D., Romeis, T. and Lee, J. (2014) Ca^{2+} signalling in plant immune response: from pattern recognition receptors to Ca^{2+} decoding mechanisms. *New Phytol.* **204**, 782–790.

Sharma, R., De Vleesschauwer, D., Sharma, M.K. and Ronald, P.C. (2013) Recent advances in dissecting stress-regulatory crosstalk in rice. *Mol. Plant* **6**, 250–260.

Shetty, N., JØrgensen, H., Jensen, J., Collinge, D. and Shetty, H.S. (2008) Roles of reactive oxygen species in interactions between plants and pathogens. *Eur. J. Plant Pathol.* **121**, 267–280.

Shimono, M., Koga, H., Akagi, A., Hayasi, N., Goto, S., Sawada, M., Kurihara, T. *et al.* (2012) Rice WRKY45 plays important roles in fungal and bacterial disease resistance. *Mol. Plant Pathol.* **13**, 83–94.

Skamnioti, P. and Gurr, S.J. (2009) Against the grain: safeguarding rice from rice blast disease. *Trends Biotechnol.* **27**, 141–150.

Takatsuji, H. (2014) Development of disease-resistant rice using regulatory components of induced disease resistance. *Front Plant Sci.* **5**, 630.

Tena, G., Boudsocq, M. and Sheen, J. (2011) Protein kinase signaling networks in plant innate immunity. *Curr. Opin. Plant Biol.* **14**, 519–529.

Torres, M.A. (2010) ROS in biotic interactions. *Physiol. Plant.* **138**, 414–429.

Tsuda, K. and Katagiri, F. (2010) Comparing signaling mechanisms engaged in pattern-triggered and effector-triggered immunity. *Curr. Opin. Plant Biol.* **13**, 459–465.

Umemura, K., Satou, J., Iwata, M., Uozumi, N., Koga, J., Kawano, T., Koshiba, T. *et al.* (2009) Contribution of salicylic acid glucosyltransferase, OsSGT1, to chemically induced disease resistance in rice plants. *Plant J.* **57**, 463–472.

Voigt, C.A. (2014) Callose-mediated resistance to pathogenic intruders in plant defense-related papillae. *Front. Plant Sci.* **5**, 168.

Wei, S., Hu, W., Deng, X., Zhang, Y., Liu, X., Zhao, X., Luo, Q. *et al.* (2014) A rice calcium-dependent protein kinase OsCPK9 positively regulates drought stress tolerance and spikelet fertility. *BMC Plant Biol.* **14**, 133.

Wiermer, M., Feys, B.J. and Parker, J.E. (2005) Plant immunity: the EDS1 regulatory node. *Curr. Opin. Plant Biol.* **8**, 383–389.

Wildermuth, M.C., Dewdney, J., Wu, G. and Ausubel, F.M. (2001) Isochorismate synthase is required to synthesize salicylic acid for plant defence. *Nature*, **414**, 562–565.

Wilson, R.A. and Talbot, N.J. (2009) Under pressure: investigating the biology of plant infection by *Magnaporthe oryzae*. *Nat. Rev. Microbiol.* **7**, 185–195.

Wulff, E.G., Sorensen, J.L., Lubeck, M., Nielsen, K.F., Thrane, U. and Torp, J. (2010) *Fusarium* spp. Associated with rice Bakanae: ecology, genetic diversity, pathogenicity and toxigenicity. *Environ. Microbiol.* **12**, 649–657.

Xie, K., Chen, J., Wang, Q. and Yang, Y. (2014) Direct phosphorylation and activation of a mitogen-activated protein kinase by a calcium-dependent protein kinase in rice. *Plant Cell*, **26**, 3077–3089.

Xing, T., Wang, X.J., Malik, K. and Miki, B.L. (2001) Ectopic expression of an Arabidopsis calmodulin-like domain protein kinase-enhanced NADPH oxidase activity and oxidative burst in tomato protoplasts. *Mol. Plant Microbe Interact.* **14**, 1261–1264.

Yi, S.Y., Shirasu, K., Moon, J.S., Lee, S.G. and Kwon, S.Y. (2014) The Activated SA and JA signaling pathways have an influence on flg22-triggered oxidative burst and callose deposition. *PLoS ONE*, **9**, e88951.

19

Targeted promoter editing for rice resistance to *Xanthomonas oryzae* pv. *oryzae* reveals differential activities for *SWEET14*-inducing TAL effectors

Servane Blanvillain-Baufumé[1,†], Maik Reschke[2,‡], Montserrat Solé[2,§], Florence Auguy[1], Hinda Doucoure[1], Boris Szurek[1], Donaldo Meynard[3], Murielle Portefaix[3], Sébastien Cunnac[1,*], Emmanuel Guiderdoni[3], Jens Boch[2,‡] and Ralf Koebnik[1,*]

[1]*UMR Interactions Plantes Microorganismes Environnement (IPME), IRD-CIRAD-Université, Montpellier, France*
[2]*Institut für Biologie, Institutsbereich Genetik, Martin-Luther-Universität Halle-Wittenberg, Halle (Saale), Germany*
[3]*CIRAD, UMR AGAP (Amélioration génétique et Adaptation des Plantes), Montpellier, France*

*Correspondence
email
sebastien.cunnac@ird.fr

email koebnik@gmx.de
[†]Present address: LabEx CeMEB, Université de Montpellier, Montpellier, France.
[‡]Present address: Institut für Pflanzengenetik, Leibniz Universität Hannover, Hannover, Germany
[§]Present address: Sustainable Agro Solutions S.A., Almacelles (Lleida), Spain

Keywords: bacterial leaf blight, susceptibility gene, genome engineering, TALEN, transgene-free plants.

Summary

As a key virulence strategy to cause bacterial leaf blight, *Xanthomonas oryzae* pv. *oryzae* (*Xoo*) injects into the plant cell DNA-binding proteins called transcription activator-like effectors (TALEs) that bind to effector-binding elements (EBEs) in a sequence-specific manner, resulting in host gene induction. TALEs AvrXa7, PthXo3, TalC and Tal5, found in geographically distant *Xoo* strains, all target *OsSWEET14*, thus considered as a pivotal TALE target acting as major susceptibility factor during rice–*Xoo* interactions. Here, we report the generation of an allele library of the *OsSWEET14* promoter through stable expression of TALE-nuclease (TALEN) constructs in rice. The susceptibility level of lines carrying mutations in AvrXa7, Tal5 or TalC EBEs was assessed. Plants edited in AvrXa7 or Tal5 EBEs were resistant to bacterial strains relying on the corresponding TALE. Surprisingly, although indels within TalC EBE prevented *OsSWEET14* induction in response to BAI3 wild-type bacteria relying on TalC, loss of TalC responsiveness failed to confer resistance to this strain. The TalC EBE mutant line was, however, resistant to a strain expressing an artificial *SWEET14*-inducing TALE whose EBE was also edited in this line. This work offers the first set of alleles edited in TalC EBE and uncovers a distinct, broader range of activities for TalC compared to AvrXa7 or Tal5. We propose the existence of additional targets for TalC beyond *SWEET14*, suggesting that TALE-mediated plant susceptibility may result from induction of several, genetically redundant, host susceptibility genes by a single effector.

Introduction

Preventing colonization by pathogenic microorganisms is one of the major challenges for plants during development. Classically, plant resistance traits are governed by the so-called dominant *R* genes that typically encode nucleotide-binding leucine-rich repeat (NB-LRRs) proteins, which detect the molecular activity of pathogen effector proteins in the plant cell (Cui *et al.*, 2015). Alternatively, recessive immunity to adapted pathogenic microbes can emerge from the mutation, or the loss, of a susceptibility (*S*) gene that acts as a basic host–pathogen compatibility factor to promote disease. In breeding for resistance, altering *S* genes to counteract the infection strategy represents an interesting and potentially more durable alternative to the introduction of dominant *R* genes (van Schie and Takken, 2014).

Bacterial leaf blight (BLB) is a widespread vascular rice disease caused by *Xanthomonas oryzae* pv. *oryzae* (*Xoo*), which severely reduces grain yield and represents a major threat for global food security. In Asia, BLB control strategies rely essentially on genetic resistance. African BLB pathogens were found to be genetically distinct from Asian isolates, and effective rice resistances against African isolates have not yet been deployed (Gonzalez *et al.*, 2007; Poulin *et al.*, 2015; Verdier *et al.*, 2012). *Xoo* pathogenicity

depends on a specific class of virulence factors, called TALEs (transcription activator-like effectors), which resemble eukaryotic transcriptional activators (for review, see Hutin *et al.*, 2015a). Upon translocation into the plant cell and import in the nucleus, TALEs bind to specific promoter elements (effector-binding elements, EBEs) following a DNA recognition code where the repeat-variable diresidues (RVDs) of each repeat forming the TALE DNA-binding domain interact with a specific nucleotide (Boch *et al.*, 2009; Moscou and Bogdanove, 2009). This recognition initiates transcription of the targeted gene, whose function often determines the outcome of the interaction.

Abundant genetic data suggest that rice resistance mechanisms to *Xoo* exhibit atypical features (Zhang and Wang, 2013). The rice genome encodes over 400 NB-LRRs proteins but only one of them (*Xa1*) has been shown to confer resistance to a few *Xoo* strains (Yoshimura *et al.*, 1998). Instead, rice resistance to *Xoo* often relies on executor (*E*) genes distinct from classical *R* genes, whose transcriptional activation by TALEs triggers immunity, leading to dominant resistance (for review, see Zhang *et al.*, 2015). Alternatively, resistance can be conferred by recessive alleles corresponding to mutated forms of susceptibility genes (for review, see Iyer-Pascuzzi and McCouch, 2007; Kottapalli *et al.*, 2007) and results in this case from the loss of induction of a gene essential to disease (Hutin *et al.*, 2015a). This type of resistance alleles includes promoter-mutated forms of the nodulin *MtN3/SWEET* gene family, occurring in the rice natural diversity, which function as TALE-unresponsive resistance alleles against Asian *Xoo* strains due to DNA polymorphism in the EBEs recognized by the cognate TALEs. For example, it was shown that the recessive *xa13* resistance allele was derived from a mutation in the promoter region of *Os8N3/SWEET11* recognized by the TALE PthXo1 from the Philippine strain PXO99[A] (Chu *et al.*, 2006). Similarly, whereas PthXo2 from *Xoo* strain JXO1[A] (Japan) drives *Os12N3/SWEET13/Xa25* expression in the susceptible *indica* rice variety IR24 through direct binding to a 22-bp EBE, *japonica* varieties (including Nipponbare) that are resistant to *Xoo* bacteria relying on PthXo2 display a single-nucleotide polymorphism (SNP) at the 4th position of the PthXo2 EBE within the *SWEET13* promoter, thus preventing its induction upon infection (Richter *et al.*, 2014; Zhou *et al.*, 2015). As they govern situations of recessive resistance or susceptibility (Hutin *et al.*, 2015a; Zhang *et al.*, 2015), polymorphic promoter sequences of *SWEET* genes can be of special interest for resistance engineering strategies.

SWEET11, *SWEET13* and *SWEET14*, belonging to *SWEET* family clade III, have been shown to be targeted by several TALEs (Antony *et al.*, 2010; Yang *et al.*, 2006; Zhou *et al.*, 2015), and systematic analysis of rice *SWEET* paralogs further revealed that all, and only, clade-III members can act as susceptibility genes (Streubel *et al.*, 2013). Because they encode sugar transporters mediating glucose and sucrose export, *SWEET* gene induction by TALEs is thought to trigger sugar release to the apoplast, providing a nutrient source to the pathogen (Chen, 2014; Chen *et al.*, 2015; Cohn *et al.*, 2014).

Os11N3/SWEET14 stands out as an interesting example of convergent evolution because it is targeted by unrelated TALEs from multiple, phylogenetically distinct *Xoo* strains: AvrXa7 from strain PXO86 (Philippines), PthXo3 from strain PXO61 (Philippines), Tal5 from strain MAI1 (Mali) and TalC from strain BAI3 (Burkina Faso) (Antony *et al.*, 2010; Chu *et al.*, 2006; Streubel *et al.*, 2013; Yu *et al.*, 2011; Zhou *et al.*, 2015). Interestingly, EBEs recognized by these four TALEs were found to overlap or to be in a close vicinity (Hutin *et al.*, 2015a). In particular, TalC directly activates *SWEET14* through recognition of a DNA box located upstream from the AvrXa7, PthXo3 and Tal5 EBEs (Yu *et al.*, 2011). Engineering mutations within AvrXa7 EBE in the *Os11N3/SWEET14* promoter resulted in disease resistance against an Asian *Xoo* strain carrying the AvrXa7 effector (Li *et al.*, 2012). In addition, a naturally occurring deletion encompassing AvrXa7 and Tal5 EBEs in the *Oryza barthii* wild rice species was recently shown to confer broad-spectrum resistance to bacterial blight (Hutin *et al.*, 2015b). These data support the current view that major virulence TALEs target a single major susceptibility gene.

The TalC effector from African *Xoo* strain BAI3 has been identified in a mutant screen for loss of virulence on susceptible rice varieties (Yu *et al.*, 2011). As a *talC* mutant is severely affected in virulence and *talC trans*-complementation restores virulence, TalC is considered as BAI3's major virulence TALE. However, because no mutation in TalC EBE has been engineered nor identified in the rice natural diversity so far, data are still lacking to formally attest that TalC virulence activity solely consists in *SWEET14* induction. To address this question and to generate sources of resistance to *Xoo* African strains relying on TalC, we have deployed a genome editing approach based on TALE-nucleases (TALENs). TALENs are fusions between designer TALE modules with customized recognition specificity and the nuclease

domain of the type IIS restriction enzyme *Fok*I (Chen and Gao, 2013; Li *et al.*, 2011; Sun and Zhao, 2013). Target site recognition and TALEN dimerization triggers a double-strand break (DSB), which in turn induces non-homologous end joining (NHEJ)-mediated DNA repair pathways and generates small random insertions or deletions at the cleavage site, resulting in an 'edited' sequence.

To compare the relative contributions of multiple EBEs within the *SWEET14* promoter, we targeted the AvrXa7, Tal5 and TalC EBEs for mutagenesis. Expression and pathogenicity assays revealed that disruption of AvrXa7- and Tal5-mediated *SWEET14* induction rendered edited rice plants resistant to *Xoo* infection. Surprisingly, modifications of the TalC EBE failed to confer resistance to bacteria relying on TalC for infection, thus suggesting that this major virulence TALE can mediate plant disease through induction of more than a single susceptibility gene.

Results

Generation of TALEN-expressing transgenic rice and selection of lines edited in AvrXa7, Tal5 and TalC EBEs

To modify three EBEs within the *SWEET14* promoter (the two overlapping AvrXa7 and Tal5 EBEs and the more upstream TalC EBE), we conducted two sets of experiments, targeting AvrXa7 and TalC EBEs, respectively. To do so, we assembled two TALEN pairs designed to recognize sequences on both sides of each EBE (Figure 1, black dashed lines). For each pair, the left (L-) and right (R-) TALEN-encoding genes were cloned into distinct binary vectors and subsequently introduced into *Agrobacterium tumefaciens*. The two resulting *A. tumefaciens* strains were mixed prior to transformation of the Kitaake rice cultivar. PCR-based analysis of T0-regenerated seedlings showed that 84% and 90% of the plants studied (for AvrXa7 and TalC EBE mutagenesis, respectively) had integrated both transgenes. To characterize edition events, 342 base-pair (bp) PCR fragments amplified from double transgenic individuals and encompassing the three EBEs were Sanger sequenced and the resulting chromatograms were manually deconvoluted to resolve the sequence of the edited alleles. We found that 51% and 30% of the plants carrying both T-DNAs (for AvrXa7 and TalC EBEs, respectively) were edited at one or both gene copies. Mutations were either deletions of up to 51 bp (68% and 73% of all mutations for AvrXa7 and TalC EBEs, respectively), insertions of up to 22 bp (7% in both cases) or combined deletion plus insertion events (26% and 19%). Altogether, we obtained 41 distinct mutations within AvrXa7 EBE, 13 of which also affected the Tal5 EBE, and 26 distinct mutations within TalC EBE (Table 1). We selected several T0 lines, corresponding to overall 19 AvrXa7-Tal5 EBE alleles (Figure 1a) and 16 TalC EBE alleles (Figure 1b), to study mutation transmission and segregation at the T1 generation (Table 1, see also Supplementary Tables S1 and S2). *SWEET14* promoter sequencing of the T1 plants revealed that all mutations were transmitted to the next generation and that their segregation pattern was consistent with classical Mendelian inheritance, with a few exceptions. Three lines carrying biallelic mutations in T0 produced T1 progenies with one to two additional mutations, suggesting that TALENs were still active on the edited sequences. Transgenic T1 plants carrying each mutation at the homozygous state were selected and propagated. At that stage, we also selected individuals deprived of both T-DNAs using a PCR approach, which was successful for 10 of 11 AvrXa7 EBE-edited T1 lines, and five of 10 TalC EBE-edited T1 lines (Table 1, S1 and S2). All

(a)

	TalC EBE		AvrXa7 EBE	Tal5 EBE	

SWEET14 CACACACCATAAGGGCATGCATGTCAGCAGCTGGTCATGTGTGCCTTTTCATTCCCTTCTTCCTTCCTAGCACTATATAAACCCCCTCCAACCAGGTGCTAAGCTCATCAAGCCTTCAAGCAA WT

sweet14-1 CACACACCATAAGGGCATGCATGTCAGCAGCTGGTCATGTGTGCCTTTTCATTCCCTTCTTCCTTCCTAGCACTATATAAACCCCCTCCAA----GTGCTAAGCTCATCAAGCCTTCAAGCAA −4 bp
sweet14-2 CACACACCATAAGGGCATGCATGTCAGCAGCTGGTCATGTGTGCCTTTTCATTCCCTTCTTCCTTCCTAGCACTATATAAACCCCCTCCAAC----TGCTAAGCTCATCAAGCCTTCAAGCAA −4 bp
sweet14-3 CACACACCATAAGGGCATGCATGTCAGCAGCTGGTCATGTGTGCCTTTTCATTCCCTTCTTCCTTCCTAGCACTATATAAACCCCCTCCAA-----TGCTAAGCTCATCAAGCCTTCAAGCAA −5 bp
sweet14-4 CACACACCATAAGGGCATGCATGTCAGCAGCTGGTCATGTGTGCCTTTTCATTCCCTTCTTCCTTCCTAGCACTATATAAACCCCCTCCA-----GTGCTAAGCTCATCAAGCCTTCAAGCAA −5 bp
sweet14-5 CACACACCATAAGGGCATGCATGTCAGCAGCTGGTCATGTGTGCCTTTTCATTCCCTTCTTCCTTCCTAGCACTATATAAACCCCCTCCA------TGCTAAGCTCATCAAGCCTTCAAGCAA −6 bp
sweet14-6 CACACACCATAAGGGCATGCATGTCAGCAGCTGGTCATGTGTGCCTTTTCATTCCCTTCTTCCTTCCTAGCACTATATAAACCCC-------CAGGTGCTAAGCTCATCAAGCCTTCAAGCAA −7 bp
sweet14-7 CACACACCATAAGGGCATGCATGTCAGCAGCTGGTCATGTGTGCCTTTTCATTCCCTTCTTCCTTCCTAGCACTATATAAACCCCCT------------AAGCTCATCAAGCCTTCAAGCAA −13 bp
sweet14-8 CACACACCATAAGGGCATGCATGTCAGCAGCTGGTCATGTGTGCCTTTTCATTCCCTTCTTCCTTCCTAGCACTATATAAACCCCCTCCAA--------------------GCCTTCAAGCAA −20 bp
sweet14-9 CACACACCATAAGGGCATGCATGTCAGCAGCTGGTCATGTGTGCCTTTTCATTCCCTTCTTCCTTCCTAGCACTATATAAA--------------------GCTCATCAAGCCTTCAAGCAA −21 bp
sweet14-10 CACACACCATAAGGGCATGCATGTCAGCAGCTGGTCATGTGTGCCTTTTCATTCCCTTCTTCCTTCCTAGCACTATATAAACCCCCTCCAA----------------------GCAA −28 bp
sweet14-11 CACACACCATAAGGGCATGCATGTCAGCAGCTGGTCATGTGTGCCTTTTCATTCCCTTCTTCCTCCT---------------------------------AGCTCATCAAGCCTTCAAGCAA −33 bp
sweet14-12 CACACACCATAAGGGCATGCATGTCAGCAGCTGGTCATGTGTGCCTTTTCATTCCCTTCTTCTTT------------------------------------CTCATCAAGCCTTCAAGCAA −38 bp
sweet14-13 CACACACCATAAGGGCATGCATGTCAGCAGCTGGTCATGTGTGCCTTTTCATTCCCTTCTTCCTTCCTAGCA------------------------------AGCCTTCAAGCAA −38 bp
sweet14-14 CACACACCATAAGGGCATGCATGTCAGCAGCTGGTCATGTGTGCCTTTTCATTCCCTTCTT--CTCATCAAGCCTTCAAGCAA −42 bp
sweet14-15 CACACACCATAAGGGCATGCATGTCAGCAGCTGGTCATGTGTGCCTTTTCATTCCCTTCTT---CCTTCAAGCAA −51 bp

sweet14-16 CACACACCATAAGGGCATGCATGTCAGCAGCTGGTCATGTGTGCCTTTTCATTCCCTTCTTCCTTCCTAGCACTATATAAACCCCCTCCAACCAGGTGCTAAGCTCATCAAGCCTTCAAGCAA +1 bp
sweet14-17 CACACACCATAAGGGCATGCATGTCAGCAGCTGGTCATGTGTGCCTTTTCATTCCCTTCTTCCTTCCTAGCACTATATAAACCCCCTCCAAGCTAAGT-----CTCATCAAGCCTTCAAGCAA −12+7 bp
sweet14-18 CACACACCATAAGGGCATGCATGTCAGCAGCTGGTCATGTGTGCCTTTTCATTCCCTTCTTCCTTCCTTCATGT--------------------GTGCTAAGCTCATCAAGCCTTCAAGCAA −27+6 bp
sweet14-19 CACACACCATAAGGGCATGCATGTCAGCAGCTGGTCATGTGTGCCTTTTCATTCCCTTCTTCCTTCCTAGCACTATATAAACCCCCTTCAAGCTTCAAGCAAAGCAAAC------TCAAGCAA −28+22 bp

(b)

	TalC EBE		AvrXa7 EBE	Tal5 EBE	

SWEET14 CACACACCATAAGGGCATGCATGTCAGCAGCTGGTCATGTGTGCCTTTTCATTCCCTTCTTCCTTCCTAGCACTATATAAACCCCCTCCAACCAGGTGCTAAGCTCATCAAGCCTTCAAGCAA WT

sweet14-20 CACACACCATAAGGGCATGCATGT---CAGCTGGTCATGTGTGCCTTTTCATTCCCTTCTTCCTTCCTAGCACTATATAAACCCCCTCCAACCAGGTGCTAAGCTCATCAAGCCTTCAAGCAA −3 bp
sweet14-21 CACACACCATAAGGGCATGCATGT----AGCTGGTCATGTGTGCCTTTTCATTCCCTTCTTCCTTCCTAGCACTATATAAACCCCCTCCAACCAGGTGCTAAGCTCATCAAGCCTTCAAGCAA −4 bp
sweet14-22 CACACACCATAAGGGCATGCAT----GCAGCTGGTCATGTGTGCCTTTTCATTCCCTTCTTCCTTCCTAGCACTATATAAACCCCCTCCAACCAGGTGCTAAGCTCATCAAGCCTTCAAGCAA −4 bp
sweet14-23 CACACACCATAAGGGCATGCAT------AGCTGGTCATGTGTGCCTTTTCATTCCCTTCTTCCTTCCTAGCACTATATAAACCCCCTCCAACCAGGTGCTAAGCTCATCAAGCCTTCAAGCAA −6 bp
sweet14-24 CACACACCATAAGGGCATGCA--------GCTGGTCATGTGTGCCTTTTCATTCCCTTCTTCCTTCCTAGCACTATATAAACCCCCTCCAACCAGGTGCTAAGCTCATCAAGCCTTCAAGCAA −8 bp
sweet14-25 CACACACCATAAGGGCATGCATG----------GTCATGTGTGCCTTTTCATTCCCTTCTTCCTTCCTAGCACTATATAAACCCCCTCCAACCAGGTGCTAAGCTCATCAAGCCTTCAAGCAA −10 bp
sweet14-26 CACACACCATAAGGGC----------GCAGCTGGTCATGTGTGCCTTTTCATTCCCTTCTTCCTTCCTAGCACTATATAAACCCCCTCCAACCAGGTGCTAAGCTCATCAAGCCTTCAAGCAA −10 bp
sweet14-27 CACACACCATAAGGGG----------AGCAGCTGGTCATGTGTGCCTTTTCATTCCCTTCTTCCTTCCTAGCACTATATAAACCCCCTCCAACCAGGTGCTAAGCTCATCAAGCCTTCAAGCAA −10 bp
sweet14-28 CACACACCATAAGGGCATGC-----------TGGTCATGTGTGCCTTTTCATTCCCTTCTTCCTTCCTAGCACTATATAAACCCCCTCCAACCAGGTGCTAAGCTCATCAAGCCTTCAAGCAA −11 bp
sweet14-29 CACACACCATAAGGGCATGCAT-------------CATGTGTGCCTTTTCATTCCCTTCTTCCTTCCTAGCACTATATAAACCCCCTCCAACCAGGTGCTAAGCTCATCAAGCCTTCAAGCAA −13 bp
sweet14-30 CACACACCATAAGGGCATGCATGT---------------GTGTGCCTTTTCATTCCCTTCTTCCTTCCTAGCACTATATAAACCCCCTCCAACCAGGTGCTAAGCTCATCAAGCCTTCAAGCAA −16 bp
sweet14-31 CACACACCATAAGGGCATG-------------------TGTGCCTTTTCATTCCCTTCTTCCTTCCTAGCACTATATAAACCCCCTCCAACCAGGTGCTAAGCTCATCAAGCCTTCAAGCAA −20 bp
sweet14-32 CACACACCATAAGGGCATGCAT-----------------------TTTCATTCCCTTCTTCCTTCCTAGCACTATATAAACCCCCTCCAACCAGGTGCTAAGCTCATCAAGCCTTCAAGCAA −24 bp

sweet14-33 CACACACCATAAGGGCATGCATGTCAGCAGCTGGTCATGTGTGCCTTTTCATTCCCTTCTTCCTTCCTAGCACTATATAAACCCCCTCCAACCAGGTGCTAAGCTCATCAAGCCTTCAAGCAA +1 bp
sweet14-34 CACACACCATAAGGGCATGCATGTCAGCAGCTGGTCATGTGTGCCTTTTCATTCCCTTCTTCCTTCCTAGCACTATATAAACCCCCTCCAACCAGGTGCTAAGCTCATCAAGCCTTCAAGCAA +3 bp
sweet14-35 CACACACCATAAGGCAATAT-------AGCTGGTCATGTGTGCCTTTTCATTCCCTTCTTCCTTCCTAGCACTATATAAACCCCCTCCAACCAGGTGCTAAGCTCATCAAGCCTTCAAGCAA −14+6 bp

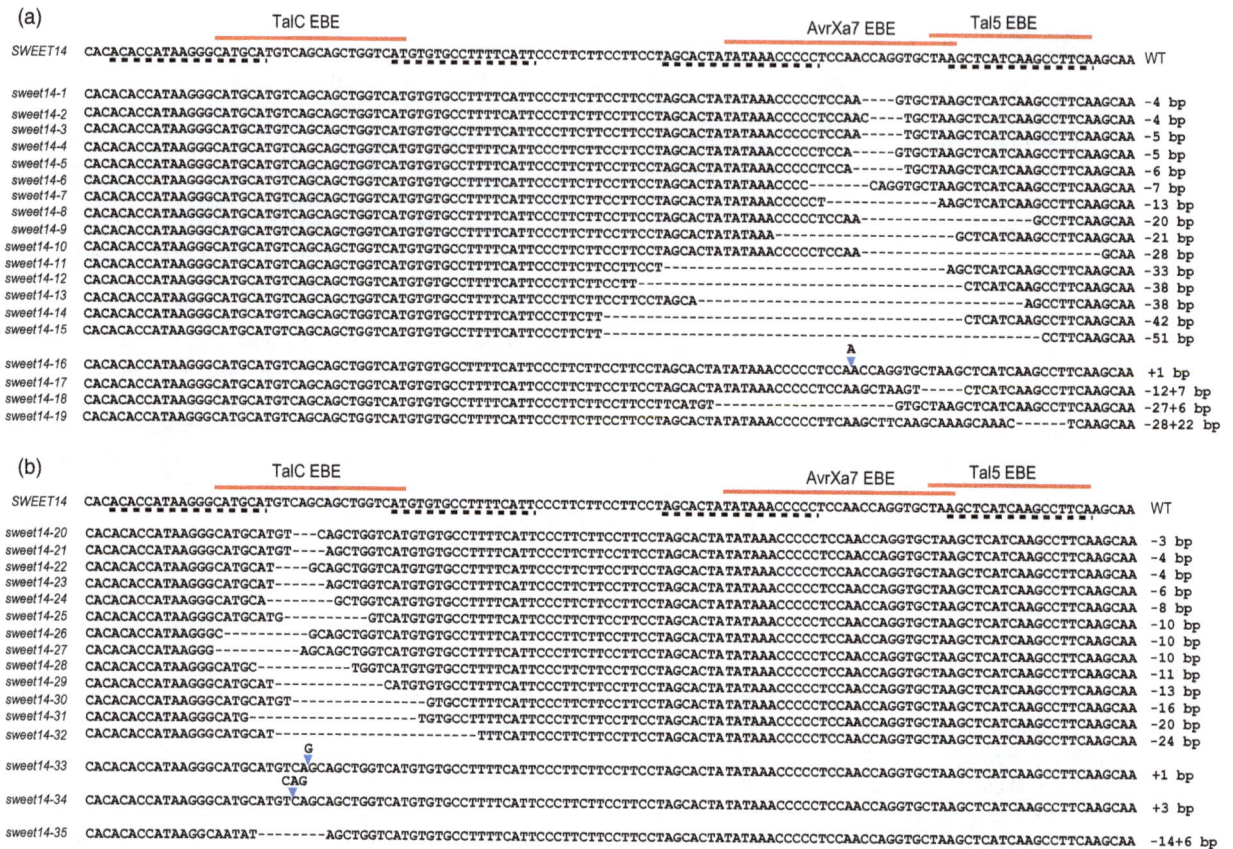

Figure 1 Alignment of a *SWEET14* promoter fragment in selected rice (cv. Kitaake) lines edited in AvrXa7 (a) or TalC (b) EBEs. On top, the 118 bp-*SWEET14* sequence corresponds to the −328 to −211 promoter fragment relative to the ATG start codon. These mutant alleles were brought to the homozygous stage. Allele names are provided on the left side of each sequence and the nature of the associated mutation on the right. Dashed lines below the *SWEET14* wild-type sequence represent the binding sites of the TALEN pairs used in this study. Deletions and insertions are represented by black dashes and blue letters, respectively.

phenotypic analyses described below were performed on T2 or T3 plants that carried homozygous mutations, as confirmed by promoter sequencing.

Mutations within AvrXa7 and Tal5 EBEs lead to resistance against bacterial strains relying on the corresponding effectors

It was previously shown that TALEN-mediated mutations within AvrXa7 EBE are able to defeat an *Xoo pthXo1* mutant strain expressing *avrXa7* in *trans* (Li *et al.*, 2012). Furthermore, our group recently demonstrated that in some accessions of the African wild rice *O. barthii* and its domesticated progeny (*Oryza glaberrima*), *xa41*, a naturally occurring *SWEET14* allele consisting in a deletion encompassing AvrXa7 and Tal5 EBEs, confers resistance to various *Xoo* strains including an *avrXa7* expressing strain (Hutin *et al.*, 2015b). To check whether (i) engineered mutations within AvrXa7 EBE could also confer resistance to the original PXO86 wild-type strain from which *avrXa7* was isolated, and (ii) mutations resembling *xa41* could also confer resistance to *avrXa7* or *tal5* expressing strains in an *Oryza sativa* background, we focused on two Kitaake edited lines, carrying the *sweet14-10* (a 28-bp deletion equivalent to the 18-bp *xa41* deletion) and *sweet14-11* (a 33-bp deletion removing the entire AvrXa7 EBE and the first two base pairs of Tal5 EBE) alleles, respectively (Figure 2a).

SWEET14 expression was induced more than 1500-fold upon infiltration with PXO86 or a BAI3 *talC⁻* insertion mutant carrying a vector encoding *talC*, *avrXa7* or *tal5*. Whereas this induction was not affected in either of the two edited lines in response to BAI3, it was abolished in both of them after infection with PXO86, BAI3 *talC⁻*/*avrXa7* and BAI3 *talC⁻*/*tal5* (Figure 2b). This suggests that both *sweet14-10* and *sweet14-11* mutations prevent EBE recognition by the AvrXa7 and Tal5 TALEs, but retain TalC responsiveness. Interestingly, these data further show that a mutation of the first two base pairs of Tal5 EBE is sufficient to prevent recognition by this effector.

Disease caused by *Xoo* bacteria is typically assessed using two methods: the first one consists of syringe infiltration of bacteria into leaves and following the development of water-soaked lesions around the infiltration area. The second approach, called leaf clipping, consists of cutting the tip of the leaves using scissors previously dipped in the bacterial suspension, and measuring the length of the chlorotic to necrotic lesions developing along the leaf blade. Using these methods, both edited lines were unable to develop water-soaking or vascular lesions in response to PXO86 bacteria (Figure 2c, d and e). When testing the BAI3 *talC⁻* complemented strains, only the strain expressing TalC was virulent on the edited lines (Figure 2c, f and g), suggesting that both *sweet14-10* and *sweet14-11* alleles are able to defeat strains relying on AvrXa7 and Tal5.

Table 1 Summary of the results obtained for editing of AvrXa7 and TalC EBEs in *O. sativa* cv. Kitaake using TALENs

	AvrXa7 EBE	TalC EBE
Number of T0 plants studied	135	171
Number of T0 plants carrying both transgenes	113	154
Number of T0 edited plants retrieved	58	46
Edition efficiency (edited *vs.* carrying both transgenes)	51%	30%
Number of independent T0 lines retrieved	30	23
Number of distinct edited alleles and classification (deletions/insertions/combined events)	41 (27/3/11)	26 (19/2/5)
Range of the deletions	−1 to −51 bp	−3 to −24 bp
Range of the insertions	+1 to +23 bp	+1 to +10 bp
Number of independent T0 lines genotyped at the T1 generation	11	10
Number of independent edited alleles available at the homozygous stage (deletions/insertions/combined events)	19 (15/1/3)[a]	16 (13/2/1)[b]
Number of transgene-free T1 lines carrying a unique edited allele at the homozygous stage	10	5

[a]Alleles represented in Figure 1a (see also detailed information in Table S1).
[b]Alleles represented in Figure 1b (see also detailed information in Table S2).

As previously observed on the rice accession IR24, expression of neither *talC*, *avrXa7* nor *tal5* from a plasmid is able to fully complement a BAI3 *talC⁻* mutant (Streubel *et al.*, 2013). At 21 dpi, a time point where disease was more pronounced, infection with the strain delivering TalC triggered similar symptoms in Kitaake wild-type and edited plants. However, lesions caused by *avrXa7* or *tal5*-complemented strains were >twofold shorter in the edited lines *vs.* wild-type plants (Figure 2f). At 14 dpi, quantification of bacterial populations in a 5-cm leaf section located 5 cm away from the clipping site showed that both edited lines supported ca. 1000-fold less bacterial colonization than wild-type plants when using *avrXa7* or *tal5*-complemented strains (Figure 2g). These data, relying on a distinct set of bacterial strains, distinct plant genetic backgrounds and novel edited alleles confirm and broaden previous studies showing that loss of *SWEET14* induction in AvrXa7 or Tal5 EBE-edited lines correlates with loss of susceptibility in response to the TALEs AvrXa7 and Tal5 (Hutin *et al.*, 2015b; Li *et al.*, 2012).

SWEET alleles mutated in the TalC EBE lose TalC responsiveness but do not confer resistance

We next studied four lines carrying independent mutations within TalC EBE (Figure 3a). *SWEET14* expression in the four edited lines in response to BAI3 was reduced about 1000-fold relative to Kitaake and resembled that of wild-type plants inoculated with the BAI3 *talC⁻* strain (Figure 3b), showing that TALEN-mediated EBE disruption successfully rendered this gene non-responsive to TalC. Surprisingly, symptoms caused by wild-type BAI3 bacteria on the TalC EBE-edited lines were hardly reduced compared to those observed on Kitaake plants, irrespective of the inoculation method used. Vascular lesions measured 14 days after leaf clipping were marginally shorter in two independent replicates

of the experiment shown in Figure 3c and d or basically identical in one replicate of the same experiment (not shown). Furthermore, water-soaking symptoms at 6 dpi were either unaltered or only slightly reduced (Supplementary Figure 1). These results show that loss of *SWEET14* induction upon TalC delivery in Kitaake plants fails to confer disease resistance.

As the genome sequence of cultivar Kitaake is not available, we cannot exclude that a gene duplication of *SWEET14* had happened in this genetic background. We therefore examined whether the 'retained susceptibility' phenotype was restricted to edited plants of this cultivar or shared by other accessions, thus representing an intrinsic feature of TalC-triggered susceptibility. To this end, we used the same TALEN pair as before to mutagenize TalC EBE in the reference *O. sativa* Nipponbare background, and studied a line carrying an homozygous −332 + 17 indel affecting the 5′ end of TalC EBE (Figure 4a), named *sweet14-36*. In these analyses, we used the Kitaake/*sweet14-32* line for comparison. Gene expression studies performed in the same conditions as before confirmed non-induction of *SWEET14* in response to BAI3 in the edited Nipponbare line, *sweet14-36*, compared to wild-type plants (Figure 4b). In leaf clipping assays, even if lesions obtained on Nipponbare were shorter than those typically measured on Kitaake in response to BAI3, TalC EBE disruption in Nipponbare *sweet14-36* only allowed a very slight gain of resistance in terms of symptom development (Figure 4d and e). Bacterial growth assays performed with all lines at 14 dpi did not reveal any significant difference in bacterial population sizes in both edited lines *vs.* wild-type plants in the 10- to 15-cm leaf section beyond the inoculation site (Figure 4f). These data demonstrate that Nipponbare behaves similar to Kitaake and that a hypothesized *SWEET14* gene duplication cannot account for the uncoupling between disease resistance and loss of TalC responsiveness at the *SWEET14* locus. We therefore conclude that loss of *SWEET14* induction alone is not sufficient to confer strong resistance to *talC*-expressing bacteria.

The susceptibility of TalC EBE-edited lines does not require *SWEET14* induction but specifically depends on *talC*

The unexpected susceptibility of TalC EBE-edited lines could be an unintended consequence of the editing process due to 'off-target' editing or tissue culture somaclonal variation. For example, the immune system of an edited rice line may have been accidentally modified, leading to strong susceptibility to *Xoo* infection. Alternatively, the regulation of genetically redundant susceptibility gene(s) whose activity can compensate for the loss of *SWEET14* induction may have been affected. If susceptibility to BAI3 resulted from the misregulation of such susceptibility gene(s), we would expect complementation of the loss of TalC activity in a BAI3 *talC⁻* mutant, rescuing virulence on the *sweet14-32* line.

To address this question, we first used leaf clipping to inoculate the *sweet14-32* line, the AvrXa7/Tal5 EBEs-edited line *sweet14-15* (Figure 1a) and wild-type Kitaake. In addition to the BAI3 *talC⁻* mutant, these plants were also challenged with BAI3 and PXO86 as positive controls and water as a negative control. As shown in Figure 5a, wild-type BAI3 caused similarly severe BLB symptoms with lesion lengths reaching more that 12 cm on all three host genotypes while leaves inoculated with the BAI3 *talC⁻* mutant remained essentially symptomless, comparable to the water controls. Lesions produced by PXO86 on *sweet14-32* and Kitaake leaves were similar to those caused by BAI3. As observed

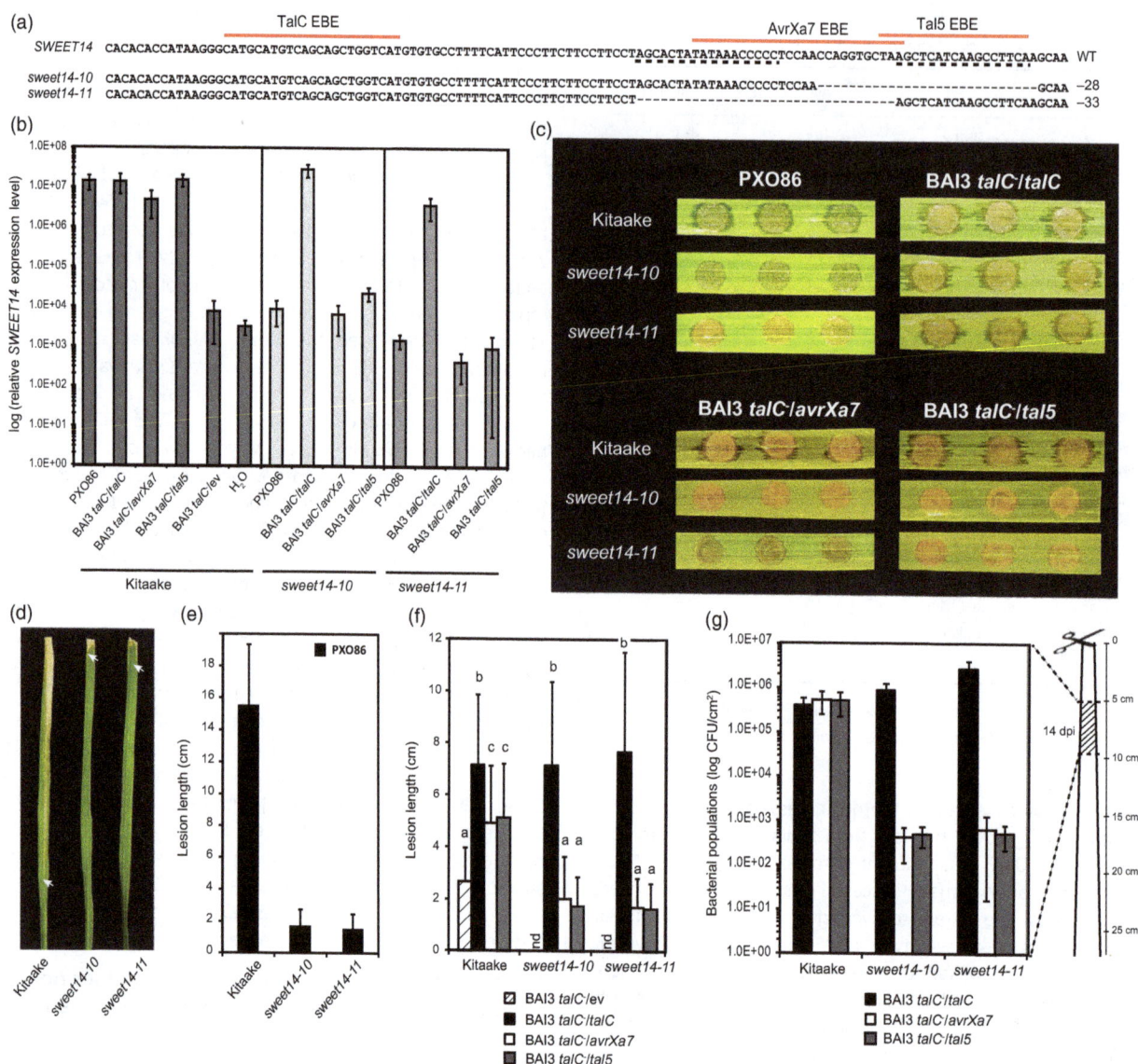

Figure 2 Functional analysis of two Kitaake edited lines carrying deletions in AvrXa7 and/or Tal5 EBEs. (a) Location of the studied mutations on the *SWEET14* promoter (same fragment as in Figure 1). (b) *SWEET14* expression pattern obtained by RT-qPCR 2 days post-leaf infiltration with the indicated strains. The graph uses a base-10 logarithmic scale. Bars represent the average expression obtained from four independent RNA samples, with standard deviation. (c) Water-soaking symptoms obtained after leaf infiltration with the indicated strain. Pictures were taken 5 days post-inoculation (dpi). (d and e) Qualitative (d) and quantitative (e) evaluation of the disease symptoms obtained 14 days post-leaf clipping inoculation with the PXO86 wild-type strain. For d, lesions were photographed at 14 dpi and arrow heads indicate the end of the lesion. On e, bars represent the average and standard deviation obtained from N > 30 symptomatic leaves. (f) Lesion length measured 21 days post-leaf clipping inoculation with the indicated BAI3 derivative strains. Bars represent the average and standard deviation obtained from N > 20 symptomatic leaves. The letters above the bars represent the result of a Tukey's HSD statistical test. Identical letters indicate means that are not significantly different from each other (α = 0.05). 'nd', not determined. (g) Bacterial populations extracted at 14 days post-leaf clipping inoculation from the 5-cm leaf segment as depicted on the cartoon. The graph uses a base-10 logarithmic scale. Bars represent the average and standard error obtained from four independent leaf samples.

before on AvrXa7/Tal5 EBE-edited lines (Figure 2), PXO86 did not cause lesions on *sweet14-15* plants. These results demonstrate that susceptibility of the edited lines to BAI3 depends on TalC and that the TalC EBE-edited line *sweet14-32* is not generally impaired in immunity to *Xoo* infection.

To further examine whether these lines can mount resistance to an *Xoo* infection and to uncouple *SWEET14* induction from other TalC activities, we used a *talC⁻* mutant strain expressing an artificial *SWEET14*-targeting TALE, *artTAL14-2* (Streubel *et al.*, 2013). Because of the position of its binding site, which

is essentially unrelated to and located downstream of the TalC EBE (see the dotted line in Figure 4a), *SWEET14* promoter recognition by ArtTAL14-2 in the Kitaake/*sweet14-32* line should also be abolished and prevent gene induction in response to this strain. RT-qPCR experiments confirmed that *SWEET14* expression in the edited Kitaake line was not induced upon infection with BAI3 or BAI3 *talC⁻/artTAL14-2* (Figure 4b). By contrast, strong induction was obtained in the edited Nipponbare line *sweet14-36* carrying an intact ArtTAL14-2 EBE in the promoter of *SWEET14* (Figure 4b). In parallel, we assessed the

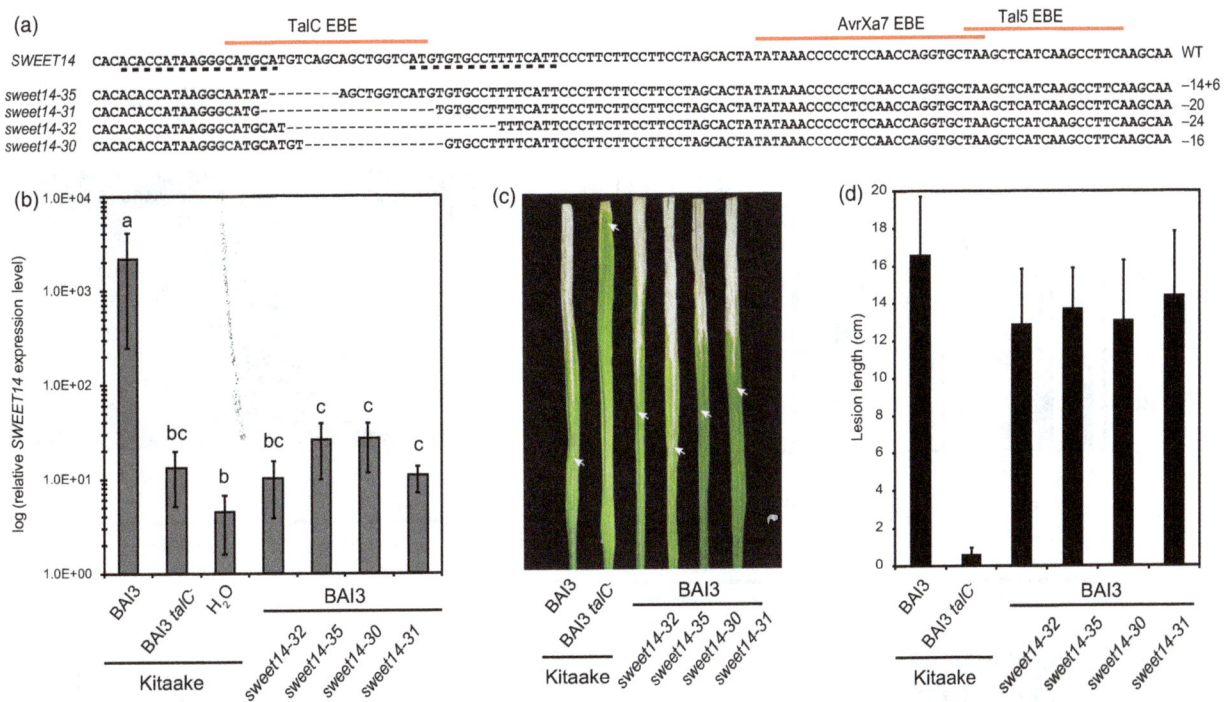

Figure 3 Functional analysis of four Kitaake edited lines carrying mutations in TalC EBE. (a) Location of the studied mutations on the *SWEET14* promoter (same fragment as in Figure 1). (b) *SWEET14* expression pattern obtained by RT-qPCR 2 days post-infiltration with the indicated strains. Bars represent the average expression obtained from four independent biological experiments, each including four independent RNA samples (16 in total), with standard deviation. The letters above the bars represent the result of a Tukey's HSD statistical test. Identical letters indicate means that are not significantly different from each other (α = 0.05). (c and d) Lesion length photographed (c) or measured (d) 14 days post-leaf clipping inoculation with the BAI3 wild-type or *talC* mutant strain. On (c), arrow heads indicate the end of the lesion. On (d), bars represent the average and standard deviation obtained from N > 30 symptomatic leaves.

resistance levels of the TalC EBE-edited lines in response to BAI3 *talC⁻/artTAL14-2*. While water-soaking symptoms elicited by this strain after infiltration of the Nipponbare/*sweet14-36* line were similar to the corresponding wild-type plants, they were dramatically reduced on Kitaake/*sweet14-32* plants (Figure 5b). Following leaf clipping, lesion length was reduced by half (Figure 5c) and bacterial population sizes were reduced more than 1000-fold (Figure 5d). Hence, although the *artTAL14-2* construct only partially complemented the BAI3 *talC⁻* strain in leaf clipping assays as previously reported (Streubel *et al.*, 2013), Kitaake/*sweet14-32* plants were significantly and reproducibly more resistant to BAI3 *talC⁻/artTAL14-2* compared to wild-type plants.

Altogether, these data show that non-induction of *SWEET14* upon delivery of ArtTAL14-2 in the *sweet14-32* line correlates with elevated disease resistance levels. This finding contrasts with results obtained with *talC* where *SWEET14* induction and susceptibility are clearly uncoupled (Figure 4). In conclusion, the *sweet14-32* line is not generally impaired in immunity to *Xoo* infection and TalC-dependent disease occurs independent of *SWEET14* gene induction.

The susceptibility of TalC EBE-edited lines does not appear to be due to induction of a clade-III *SWEET* susceptibility gene

The *talC*-dependent susceptibility of TalC EBE-edited lines prompted us to consider the possibility that the TalC regulon extends to other, genetically redundant gene(s) beyond

SWEET14. SWEET genes form a multigenic family in rice. Notably, clade-III *SWEETs* have been shown to be functionally equivalent susceptibility genes in the rice–*Xoo* interaction (Streubel *et al.*, 2013). Clade-III *SWEETs* other than *SWEET14* could thus potentially act as redundant, TalC-targeted susceptibility genes.

To test this, we first examined available transcriptomic data obtained after infection with BAI3 or a BAI3 *talC⁻* mutant (Yu *et al.*, 2011). We found that among the five clade-III *SWEETs* (*SWEET11* to *SWEET15*), only *SWEET14* is significantly and strongly induced in a *talC*-dependent manner upon BAI3 infection (Figure 6a). This observation indicates that TalC does not target any additional clade-III *SWEET* gene beyond *SWEET14* in the susceptible Nipponbare cultivar. Even if unlikely, we cannot *a priori* exclude the possibility that TalC binds to an alternate, low-affinity DNA box at another genomic locus when its primary target EBE is destroyed, thus directly or indirectly inducing another clade-III *SWEET* gene and compensating the loss of *SWEET14* induction. In order to examine this possibility, we performed RT-qPCR to compare clade-III *SWEET* gene expression following leaf infiltration of the TalC EBE-edited line *sweet14-32* versus wild-type Kitaake. First, to ensure that the qPCR primer pairs designed to monitor *SWEET* gene expression (Streubel *et al.*, 2013) also perform well in our genetic backgrounds, we confirmed the induction of individual clade-III *SWEET* genes upon infiltration with a *Xoo* strain expressing a TALE (natural or artificial) targeting the corresponding *SWEET* gene (Supplementary Figure 2). Next, we infiltrated the *sweet14-32* line and Kitaake with either BAI3 or the BAI3 *talC⁻* mutant and monitored

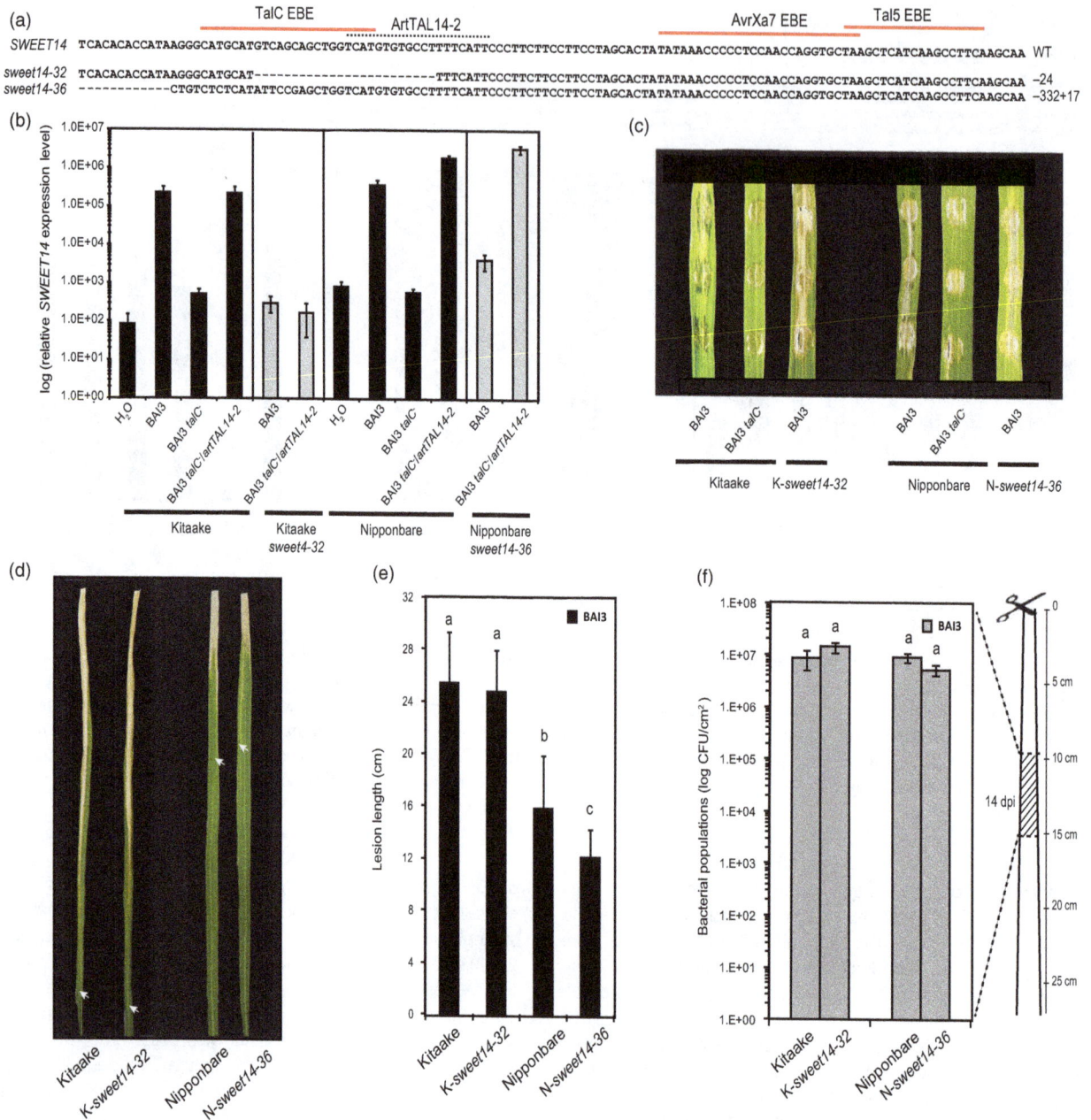

Figure 4 Functional analysis of edited rice lines carrying mutations in TalC EBE in the Kitaake (*sweet14-32*) or Nipponbare (*sweet14-32*) backgrounds. (a) Location of the studied mutations on the *SWEET14* promoter (same fragment as in Figure 1). (b) *SWEET14* expression pattern obtained by RT-qPCR 2 days post-infiltration with the indicated strains. The graph uses a base-10 logarithmic scale. Bars represent the average expression obtained from four independent RNA samples, with standard deviation. (c) Symptoms obtained 5 days post-leaf infiltration of the indicated strain. The letter in front of each allele indicates the cultivar, K- for Kitaake and N- for Nipponbare. (d and e) Lesion length photographed (d) or measured (e) 14 days post-leaf clipping inoculation of the BAI3 wild-type strain. On (d), arrow heads indicate the end of the lesion. On (e) bars represent the average and standard deviation obtained from N > 30 symptomatic leaves, and letters above the bars represent the result of a Tukey's HSD statistical test. Identical letters indicate means that are not significantly different ($\alpha = 0.05$). (f) Bacterial populations extracted at 14 days post-leaf clipping inoculation from a 5-cm leaf segment as depicted on the cartoon. The graph uses a base-10 logarithmic scale. Each bar represents the average obtained from four independent samples, with standard error. A Tukey's HSD test performed on these data indicated no statistically significant differences ($\alpha = 0.05$).

expression of *SWEET11* to *SWEET15* (Figure 6b). As observed previously (Figure 4), *SWEET14* was highly induced in Kitaake 48 h post-BAI3 inoculation as compared to the BAI3 *talC⁻* strain, and this induction was abrogated in the *sweet14-32* edited background. For the remaining four clade-III *SWEET* genes, testing for mean differences across bacterial strain–rice genotype

combinations did not reveal any statistically significant differences in transcript abundance (Figure 6b) in any of the three biological experiments that were independently performed. Thus, retained susceptibility of edited lines at the TalC EBE of *SWEET14* is unlikely to result from TalC-mediated induction of a clade-III *SWEET* gene.

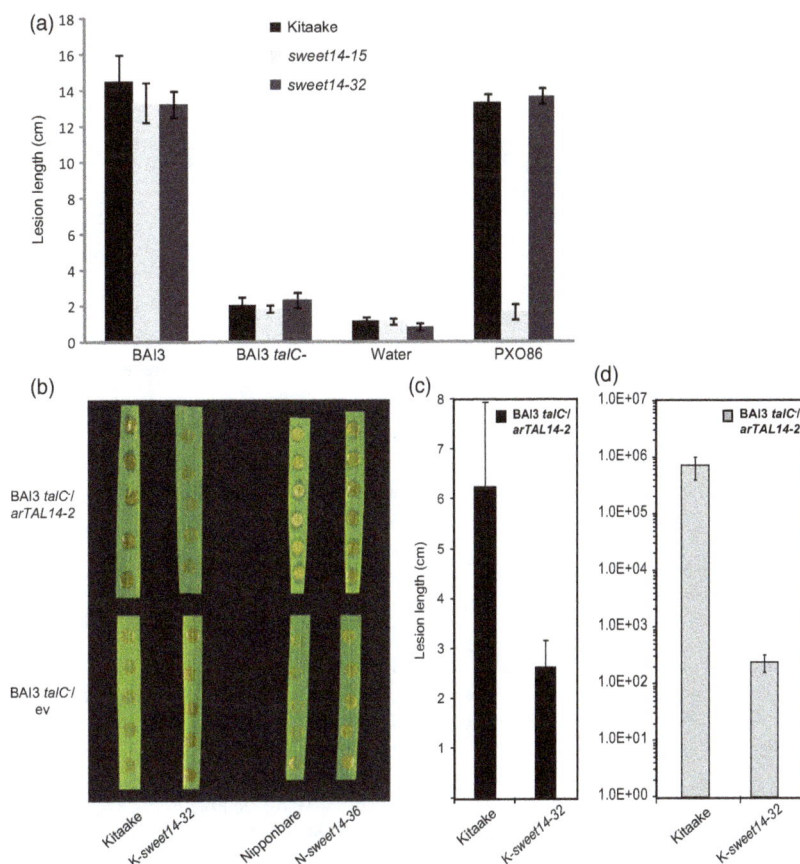

Figure 5 TalC EBE-edited plants are resistant to the BAI3 *talC* mutant and to bacteria expressing ArtTAL14-2. (a) Lesion length measured 14 days post-leaf clipping inoculation of the edited lines *sweet14-15* and *sweet14-32* together with the control background genotype Kitaake. Bacterial or mock treatments are indicated under the *x*-axis. Bars represent the average and standard deviation obtained from N > 3 symptomatic leaves. This experiment was repeated three times with similar results. (b) Water-soaking symptoms obtained after infiltration of the BAI3 *talC* mutant expressing in *trans* the artificial TAL gene *artTAL14-2* (binding site represented in Figure 4a) or the corresponding empty vector (ev). Pictures were taken 5 dpi. (c) Lesion length measured 21 days post-leaf clipping inoculation of the Kitaake edited line with BAI3 *talC⁻/artTAL14-2* strain. Bars represent the average and standard deviation obtained from N > 20 symptomatic leaves. (d) Bacterial populations extracted at 14 days post-leaf clipping inoculation from a 5-cm leaf segment collected as depicted on the cartoon in Figure 2g. The graph uses a base-10 logarithmic scale. Each bar represents the average obtained from four independent samples, with standard error.

Discussion

Given enormous potential, genome editing has attracted biologists of a broad range of disciplines and underwent an explosive development in the past years. In plants, while genome editing strategies including zinc-finger nucleases, TALENS or CRISPR-Cas systems have been deployed in a number of studies, most of them have essentially addressed method implementation and analyses of the nature and heritability of the generated mutations. Yet, very few studies have provided significant added value to functional genomics and/or breeding.

Using TALENs, we have efficiently edited the promoter of a central BLB susceptibility gene at two distinct target sequences. In a simple experimental set-up based on two TALEN-expressing transgenes and two *A. tumefaciens* strains, we have obtained mutation frequencies (up to 51%) that had not been reached in studies based on expression of both TALENs from a single T-DNA: 19-36% (Zhang *et al.*, 2013), 21%–25% (Zhang *et al.*, 2016) or 30% (Shan *et al.*, 2015).

The nature of the mutations obtained in our study, as well as the proportion of deletions *vs.* insertions or indels, followed a distribution resembling previous reports (Zhang *et al.*, 2016), although it is interesting to note that we did not obtain any event of the 'substitution' type. Instead, striking differences are observed regarding mutation zygosity. In previous studies, the presence, or even prevalence, of chimeras (at least three different sequences including the wild-type sequence) in T0-regenerated plantlets, suggests that TALEN-induced mutations likely occur rather late during differentiation of the embryogenic cells into plants (Shan *et al.*, 2015; Zhang *et al.*, 2016). In contrast, all our T0 edited plants carried monoallelic or biallelic mutations,

suggesting that the TALEN pairs used have worked efficiently at an early stage of the transformation process. Driving TALENs transcription with the strong maize ubiquitin promoter may have contributed to the absence of chimera and high level of biallelic edition in the regenerated plants in our experiments.

Transmission to subsequent generations through Mendelian inheritance was achieved in all published studies including ours. Like previous reports (Shan *et al.*, 2015), additional mutations were in a few cases detected in the offspring of heterozygous T0 lines carrying biallelic mutations, suggesting that in some cases, in particular when TALEN-binding sites were not disrupted in the first round of mutations, further TALEN DBS activity has occurred in gamete progenitor cells leading to T1 seeds. Furthermore, our simple but highly efficient 'two-strain' strategy does not preclude the possibility to use T-DNA segregation at the T1 generation to identify transgene-free edited plants. This was achieved for a large proportion of our edited lines (Supplementary Tables 1 and 2).

TALEN-induced mutations engineered at AvrXa7 EBE in the Kitaake background were shown to render rice plants resistant to a *pthXo1* mutant expressing *avrXa7* or *pthXo3* (Li *et al.*, 2012). In addition, a 18-bp deletion encompassing AvrXa7 and Tal5 EBEs in *O. glaberrima* and some accessions of *O. barthii* confers resistance to *avrXa7*- or *tal5*-expressing bacteria (Hutin *et al.*, 2015b). Using systematic EBE editing on *O. sativa* cv. Kitaake plants, we show that indels affecting AvrXa7 and Tal5 EBEs can defeat the corresponding strains by making the major susceptibility gene *SWEET14* unresponsive to the TALEs. Collectively, three studies therefore provide evidence that AvrXa7 and Tal5 are major virulence TALEs, with *SWEET14* being their unique susceptibility target. In addition, our work offers a large repertoire of alleles

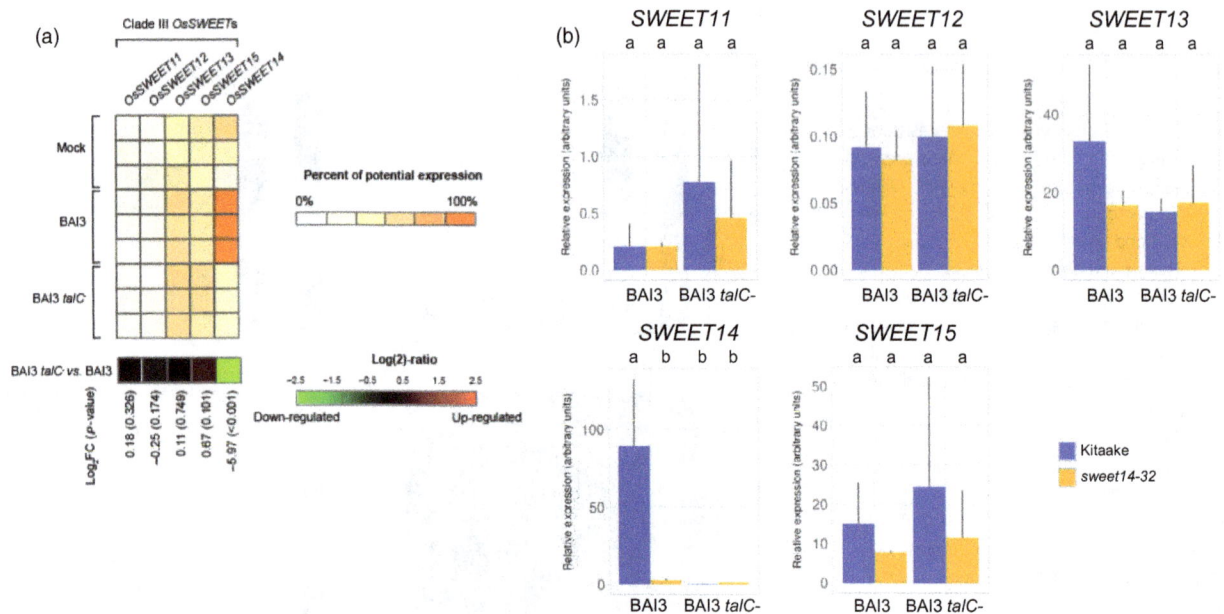

Figure 6 Expression of the five clade-III *OsSWEET* genes in response to BAl3 or a *talC* mutant. (a) Gene expression derives from previously published microarray data and analysed using Genevestigator. The heat-map displays the percent of potential expression for *SWEET11* (*LOC_Os08 g42350*), *SWEET12* (*LOC_Os03 g22590*), *SWEET13* (*LOC_Os12 g29220*), *SWEET15* (*LOC_Os02 g30910*) and *SWEET14* (*LOC_Os11 g31190*). Each row refers to an independent replicate of the microarray experiment. The last row of the heat-map represents the $\log_2$ of the expression ratio between conditions 'BAl3 *talC⁻* mutant' *vs.* 'BAl3 WT', with associated *P*-value (into brackets). (b) Clade-III *SWEET* genes expression was measured by RT-qPCR 2 days post-infiltration of the *sweet14-32* line or the Kitaake background with the wild-type BAl3 or the BAl3 *talC⁻* mutant strains. Bars represent average expression obtained from three independent RNA samples, with standard deviation. Letters above the bars represent the result of a Tukey's HSD test. Identical letters correspond to means that are not significantly different from each other ($\alpha = 0.05$). This experiment was repeated three times with similar results.

that can be used in breeding for resistance to AvrXa7- or Tal5-expressing strains. Because the function of dominant *S* genes is critical for bacterial virulence, mutated *S* gene alleles have been proposed as a potentially more durable resistance strategy compared to dominant resistance executor (*E*) genes (Gawehns *et al.*, 2013; Gust *et al.*, 2010; Iyer-Pascuzzi and McCouch, 2007; Leach *et al.*, 2001). Nonetheless, TALE-dependent loss of susceptibility can be overridden through multiple evolutionary routes (Hutin *et al.*, 2015a). Predicting the durability of natural or engineered recessive resistance and long-term field testing of resistant lines remains a great challenge for EBE-based resistance breeding.

Contrary to AvrXa7- or Tal5 EBEs, the biological significance of TalC-triggered *SWEET14* induction has never been addressed so far. Here, we provide a set of *SWEET14* alleles disrupted in TalC EBE and their functional characterization. Unexpectedly, editing of this EBE had only a little impact, if any, on resistance, as opposed to modification of AvrXa7 or Tal5 EBEs, which results in a strong gain in resistance. While *SWEET14* induction by TalC or artificial TALEs is sufficient to trigger disease (Streubel *et al.*, 2013; Yu *et al.*, 2011), preventing TalC-mediated *SWEET14* activation does not compromise plant susceptibility. Similar results were obtained in Kitaake and in the reference rice cultivar Nipponbare which carries a single *SWEET14* gene, thereby ruling out the possibility that a gene duplication event in Kitaake has masked the effect of the EBE disruption.

The unaffected susceptibility of a TalC EBE-edited line in response to BAl3 could originate from unspecific modifications of the genome or epigenome introduced either by the nucleases (off-targets) or by the genetic transformation process itself (somaclonal variation), that could translate into impaired antibacterial immunity or compensation for the absence of *SWEET14* induction (through enhanced apoplastic relase of carbohydrates for example). However, the TalC EBE-edited line *sweet14-32* was as resistant as Kitaake against a BAl3 *talC⁻* mutant strain (Figure 5a) and even more resistant than the wild type when this strain carried *artTAL14-2* (Figure 5b-d). These findings clearly argue against unintended modifications of the genome elsewhere than at the *SWEET14* promoter being causal to the retained susceptibility phenotype of TalC EBE-edited lines.

Importantly, both TalC and ArtTAL14-32 fail to mediate *SWEET14* induction in the *sweet14-32* line because their cognate EBEs are largely deleted in this allele. Yet, while susceptible to wild-type BAl3 bacteria expressing TalC (Figure 3c-d), this edited line was resistant to BAl3 *talC⁻* bacteria expressing the artificial TALE (Figure 5b-d), demonstrating that ArtTAL14-2 and TalC have separable virulence activities. Because both proteins possess distinct RVD arrays, and are, therefore, unlikely to share the same targets beyond *SWEET14*, we favour the hypothesis that TalC targets additional, genetically redundant, susceptibility gene(s). Genetic interplay between TalC susceptibility targets may involve some degree of additive effect rather than strict redundancy because the BAl3 *talC⁻/talC* strain reproducibly induced stronger symptoms and its population reached higher levels *in planta* compared to BAl3 *talC⁻* strains expressing *avrXa7* or *tal5* (as exemplified in Figure 2f on Kitaake). Members of the clade-III *SWEET* family were obvious candidates to be tested as susceptibility genes in the absence of *SWEET14* induction. However, neither analysis of public microarray data obtained in a wild-type Nipponbare background (Figure 6a) nor our own transcript profiling in the Kitaake TalC EBE-edited background (Figure 6b) detected significant *talC*-dependent upregulation of a clade-III

SWEET besides SWEET14. Yet, even if very unlikely, the possibility that a clade-III SWEET gene is a bona fide TalC target cannot be completely excluded on the sole basis of this data. Clearly, more work is required to unambiguously identify the redundant TalC target(s) and to validate its/their function as susceptibility gene(s).

Although major virulence TALEs are known to have direct targets with no or unexplored biological activity in addition to their primary S gene target (Boch et al., 2009; Cernadas et al., 2014; Li et al., 2012), no TALEs have been reported so far to possess more than a single biologically significant target. It will be critical to determine whether this feature is shared by other TALEs from Xanthomonas strains infecting crops. If so, resistance strategies based on engineering single TALE-unresponsive S gene alleles may not be as straightforward as originally anticipated (Hutin et al., 2015a; Li et al., 2012) and will require a much finer understanding of virulence TALE targets.

Experimental procedures

Design and assembly of TALENs targeting AvrXa7 and TalC EBEs

We designed TALENs pairs with DNA-binding domains composed of 16–18 repeats. Our main criteria for selecting both binding sequences were the following: (i) a T at position zero, (ii) a 15-bp spacer region to allow FokI dimerization and (iii) a minimum of three strong RVDs (HD or NN) in the repeat array (corresponding in the target sequence to C and G, respectively; Streubel et al., 2012). Modular assembly in a compatible ENTRY vector was performed using the GoldenTal method (Geißler et al., 2011). The N-terminal domain of the TALENs contained a portion of TALE Hax3 (amino acids 153–288), a SV40 nuclear localization signal and an epitope tag (left TALEN: c-myc; right TALEN: HA). The C-terminal domain contained Hax3 amino acids 1–63 and an heterodimeric 'sharkey' FokI nuclease domain [left TALEN: DS variant; right TALEN: RR variant (Guo et al., 2010)]. The left and right TALENs were inserted between the maize ubiquitin promoter and the NOS terminator into the pCAMBIA2300 (geneticin resistance) and pCAMBIA5300 (hygromycin resistance) binary plasmids (http://www.cambia.org/), respectively, using GATE-WAY® cloning (INVITROGEN). The resulting constructs were mobilized into Agrobacterium tumefaciens strain EHA105 by electroporation.

Rice stable transformation with TALEN constructs

TALEN expression in rice was accomplished through A. tumefaciens-mediated stable transformation, as previously described (Sallaud et al., 2003). Two A. tumefaciens strains, each harbouring one of the two TALENs, were mixed prior to transformation of Oryza sativa L. ssp. japonica (cvs. Kitaake or Nipponbare) via co-culture of bacteria with seed-embryo calli. The calli were transferred to plates containing 50 mg/L hygromycin for 2–3 weeks, until hygromycin-resistant cell lines develop, then for three additional weeks on plates supplemented with both antibiotics (50 mg/L hygromycin and 100 mg/L geneticin), in order to select transformed cell lines carrying both T-DNAs.

Molecular analysis of T0 plants and chromatogram-based detection of mutations

Leaf samples were collected from T0 plants and subjected to DNA extraction using a standard MATAB-based protocol (Romero et al., 2014). A portion of the OsSWEET14 promoter (342 bp in wild-type plants) was amplified by PCR using primers 5'-TCCAGGGTCACACACCATAAG and 5'-TGCAGCAAGATCTT-GATTAACTA. For analysis of the Nipponbare allele sweet14-36, the reverse primer used was 5'-TTGCGGCTCATCAGTTTCTC). After DNA sequencing of the PCR products, chromatograms that harboured off-set traces due to sequence heterozygosity were resolved manually and sequences of the edited alleles were deduced by alignment to the wild-type promoter sequence. The presence of the T-DNA originating from pCAMBIA5300 was monitored by PCR using the 5'-CTGAACTCACCGCGACGTCTG and 5'-GGCGTCGGTTTCCACTATCG primers specific for the Hpt hygromycin resistance marker gene, and presence of the T-DNA originating from pCAMBIA2300 using the 5'-GCGATAGAAGGC GATGCG and 5'-CCGGCTACCTGCCCATTCGA primers specific for the NptII geneticin resistance marker gene.

Bacterial strains, plant inoculations and growth of bacteria in planta

All Xoo strains used in this study are published: wild-type BAI3 (Gonzalez et al., 2007), wild-type PXO86 (Vera Cruz, 1989), PXO99A (Hopkins et al., 1992), BAI3 talC⁻ mutant (Yu et al., 2011) and BAI3 talC⁻ complemented with plasmids containing talC, avrXa7, tal5, artTAL12-2, artTAL13-2, artTAL14-2 or artTAL15-1 coding sequences (Streubel et al., 2013). Rice cultivation and disease assays were performed as previously described (Hutin et al., 2015b; Yu et al., 2011). Bacteria were inoculated at an optical density (OD_{600}) of 0.5 (infiltrations) or 0.4 (leaf clipping) in water. For quantification of bacterial populations, 5-cm leaf sections were collected at 5 dpi. Each processed sample contained leaf sections from three independent inoculations, and four to six independent samples were processed as per condition. Samples were frozen in liquid nitrogen, ground using metal beads, diluted in water and spotted on PSA medium containing the appropriate antibiotics.

Gene expression analyses

A 4-cm leaf section was entirely infiltrated, and collected at 48 h post-inoculation (hpi) for RNA extraction. Each individual sample contained three independent infiltration areas, and three to four independent replicate samples were processed. After sample grinding, total RNA was extracted from plant leaves using TRI-reagent (EUROMEDEX), and further purified using the RNA Clean-Up & Concentration kit (ZYMO RESEARCH). After TURBO DNase treatment (AMBION), 3 µg RNA were reverse transcribed into cDNA using SuperScriptIII (INVITROGEN). All gene expression studies were performed by quantitiative real-time PCR in a LightCycler (ROCHE), using SYBR-based Mesa Blue qPCR Master-mix (EUROGENTEC). Average transcript levels were calculated using the ΔCt method from three to four independent cDNA samples using the OsEF-1α gene for normalization. OsSWEET14 transcript was studied using primers 5'-ACTTGCAAGCAAGAA CAGTAGT and 5'-ATGTTGCCTAGGAGACCAAAGG and OsEF-1α transcript using primers 5'-GAAGTCTCATCCTACCTGAAGAAG and 5'-GTCAAGAGCCTCAAGCAAGG.

Acknowledgements

This work was funded by a Phase I 'Grand Challenge Exploration Grant' of the Bill and Melinda Gates Foundation (OPP1060078), an IRD postdoctoral fellowship attributed to SB-B, a 'Chercheur d'Avenir' grant from the Region Languedoc-Roussillon attributed to SC and a grant from the European Regional Development Fund of the European Commission to JB. This work benefited from interactions promoted by COST Action FA 1208 (https://www.cost-sustain.org).

References

Antony, G., Zhou, J., Huang, S., Li, T., Liu, B., White, F. and Yang, B. (2010) Rice *xa13* recessive resistance to bacterial blight is defeated by induction of the disease susceptibility gene *Os-11N3*. *Plant Cell*, **22**, 3864–3876.

Boch, J., Scholze, H., Schornack, S., Landgraf, A., Hahn, S., Kay, S., Lahaye, T. *et al.* (2009) Breaking the code of DNA binding specificity of TAL-type III effectors. *Science*, **326**, 1509–1512.

Cernadas, R.A., Doyle, E.L., Niño-Liu, D.O., Wilkins, K.E., Bancroft, T., Wang, L., Schmidt, C.L. *et al.* (2014) Code-assisted discovery of TAL effector targets in bacterial leaf streak of rice reveals contrast with bacterial blight and a novel susceptibility gene. *PLoS Pathog.* **10**, e1003972.

Chen, L.-Q. (2014) SWEET sugar transporters for phloem transport and pathogen nutrition. *New Phytol.* **201**, 1150–1155.

Chen, K. and Gao, C. (2013) TALENs: customizable molecular DNA scissors for genome engineering of plants. *J. Genet. Genomics*, **40**, 271–279.

Chen, H.-Y., Huh, J.-H., Yu, Y.-C., Ho, L.-H., Chen, L.-Q., Tholl, D., Frommer, W.B. *et al.* (2015) The Arabidopsis vacuolar sugar transporter SWEET2 limits carbon sequestration from roots and restricts *Pythium* infection. *Plant J.* **83**, 1046–1058.

Chu, Z., Yuan, M., Yao, J., Ge, X., Yuan, B., Xu, C., Li, X. *et al.* (2006) Promoter mutations of an essential gene for pollen development result in disease resistance in rice. *Genes Dev.* **20**, 1250–1255.

Cohn, M., Bart, R.S., Shybut, M., Dahlbeck, D., Gomez, M., Morbitzer, R., Hou, B.-H. *et al.* (2014) *Xanthomonas axonopodis* virulence is promoted by a transcription activator-like effector-mediated induction of a SWEET sugar transporter in cassava. *Mol. Plant Microbe Interact.* **27**, 1186–1198.

Cui, H., Tsuda, K. and Parker, J.E. (2015) Effector-triggered immunity: from pathogen perception to robust defense. *Annu. Rev. Plant Biol.* **66**, 487–511.

Gawehns, F., Cornelissen, B.J.C. and Takken, F.L.W. (2013) The potential of effector-target genes in breeding for plant innate immunity. *Microb. Biotechnol.* **6**, 223–229.

Geißler, R., Scholze, H., Hahn, S., Streubel, J., Bonas, U., Behrens, S.-E. and Boch, J. (2011) Transcriptional activators of human genes with programmable DNA-specificity. *PLoS ONE* **6**, e19509.

Gonzalez, C., Szurek, B., Manceau, C., Mathieu, T., Séré, Y. and Verdier, V. (2007) Molecular and pathotypic characterization of new *Xanthomonas oryzae* strains from West Africa. *Mol. Plant Microbe Interact.* **20**, 534–546.

Guo, J., Gaj, T. and Barbas, C.F. (2010) Directed evolution of an enhanced and highly efficient *FokI* cleavage domain for zinc finger nucleases. *J. Mol. Biol.* **400**, 96–107.

Gust, A.A., Brunner, F. and Nürnberger, T. (2010) Biotechnological concepts for improving plant innate immunity. *Curr. Opin. Biotechnol.* **21**, 204–210.

Hopkins, C.M., White, F.F., Choi, S.H., Guo, A. and Leach, J.E. (1992) Identification of a family of avirulence genes from *Xanthomonas oryzae* pv. *oryzae*. *Mol. Plant Microbe Interact.* **5**, 451–459.

Hutin, M., Pérez-Quintero, A.L., Lopez, C. and Szurek, B. (2015a) MorTAL Kombat: the story of defense against TAL effectors through loss-of-susceptibility. *Front Plant Sci.* **6**, 535.

Hutin, M., Sabot, F., Ghesquière, A., Koebnik, R. and Szurek, B. (2015b) A knowledge-based molecular screen uncovers a broad spectrum *OsSWEET14* resistance allele to bacterial blight from wild rice. *Plant J.* **84**, 694–703.

Iyer-Pascuzzi, A.S. and McCouch, S.R. (2007) Recessive resistance genes and the *Oryza sativa-Xanthomonas oryzae* pv. *oryzae* pathosystem. *Mol. Plant Microbe Interact.* **20**, 731–739.

Kottapalli, K.R., Kottapalli, P., Agrawal, G.K., Kikuchi, S. and Rakwal, R. (2007) Recessive bacterial leaf blight resistance in rice: complexity, challenges and strategy. *Biochem. Biophys. Res. Commun.* **355**, 295–301.

Leach, J.E., Cruz, C.M.V., Bai, J.F. and Leung, H. (2001) Pathogen fitness penalty as a predictor of durability of disease resistance genes. *Annu. Rev. Phytopathol.* **39**, 187–224.

Li, T., Huang, S., Jiang, W.Z., Wright, D., Spalding, M.H., Weeks, D.P. and Yang, B. (2011) TAL nucleases (TALNs): hybrid proteins composed of TAL effectors and *FokI* DNA-cleavage domain. *Nucleic Acids Res.* **39**, 359–372.

Li, T., Liu, B., Spalding, M.H., Weeks, D.P. and Yang, B. (2012) High-efficiency TALEN-based gene editing produces disease-resistant rice. *Nat. Biotechnol.* **30**, 390–392.

Moscou, M.J. and Bogdanove, A.J. (2009) A simple cipher governs DNA recognition by TAL effectors. *Science*, **326**, 1501.

Poulin, L., Grygiel, P., Magne, M., Gagnevin, L., Rodriguez-R, L.M., Forero Serna, N., Zhao, S. *et al.* (2015) New multilocus variable-number tandem-repeat analysis tool for surveillance and local epidemiology of bacterial leaf blight and bacterial leaf streak of rice caused by *Xanthomonas oryzae*. *Appl. Environ. Microbiol.* **81**, 688–698.

Richter, A., Streubel, J., Blücher, C., Szurek, B., Reschke, M., Grau, J. and Boch, J. (2014) A TAL effector repeat architecture for frameshift binding. *Nat. Commun.* **5**, 3447.

Romero, L.E., Lozano, I., Garavito, A., Carabali, S.J., Triana, M., Villareal, N., Reyes, L. *et al.* (2014) Major QTLs control resistance to rice hoja blanca virus and its vector *Tagosodes orizicolus*. *G3 Genes Genomes Genet.* **4**, 133–142.

Sallaud, C., Meynard, D., van Boxtel, J., Gay, C., Bes, M., Brizard, J.P., Larmande, P. *et al.* (2003) Highly efficient production and characterization of T-DNA plants for rice (*Oryza sativa* L.) functional genomics. *Theor. Appl. Genet.* **106**, 1396–1408.

van Schie, C.C.N. and Takken, F.L.W. (2014) Susceptibility genes 101: how to be a good host. *Annu. Rev. Phytopathol.* **52**, 551–581.

Shan, Q., Zhang, Y., Chen, K., Zhang, K. and Gao, C. (2015) Creation of fragrant rice by targeted knockout of the *OsBADH2* gene using TALEN technology. *Plant Biotechnol. J.* **13**, 791–800.

Streubel, J., Blücher, C., Landgraf, A. and Boch, J. (2012) TAL effector RVD specificities and efficiencies. *Nat. Biotechnol.* **30**, 593–595.

Streubel, J., Pesce, C., Hutin, M., Koebnik, R., Boch, J. and Szurek, B. (2013) Five phylogenetically close rice *SWEET* genes confer TAL effector-mediated susceptibility to *Xanthomonas oryzae* pv. *oryzae*. *New Phytol.* **200**, 808–819.

Sun, N. and Zhao, H. (2013) Transcription activator-like effector nucleases (TALENs): a highly efficient and versatile tool for genome editing. *Biotechnol. Bioeng.* **110**, 1811–1821.

Vera Cruz, C. (1989) How variable is *Xanthomonas campestris* pv. oryzae? Bact. Blight Rice, Proceedings Int. Work. Int. Rice Res. Inst. Manila (Philippines), pp 153–166

Verdier, V., Vera Cruz, C. and Leach, J.E. (2012) Controlling rice bacterial blight in Africa: needs and prospects. *J. Biotechnol.* **159**, 320–328.

Yang, B., Sugio, A. and White, F.F. (2006) *Os8N3* is a host disease-susceptibility gene for bacterial blight of rice. *Proc. Natl Acad. Sci. USA*, **103**, 10503–10508.

Yoshimura, S., Yamanouchi, U., Katayose, Y., Toki, S., Wang, Z.X., Kono, I., Kurata, N. *et al.* (1998) Expression of *Xa1*, a bacterial blight-resistance gene in rice, is induced by bacterial inoculation. *Proc. Natl Acad. Sci. USA*, **95**, 1663–1668.

Yu, Y., Streubel, J., Balzergue, S., Champion, A., Boch, J., Koebnik, R., Feng, J. *et al.* (2011) Colonization of rice leaf blades by an African strain of *Xanthomonas oryzae* pv. *oryzae* depends on a new TAL effector that induces the rice nodulin-3 *Os11N3* gene. *Mol. Plant Microbe Interact.* **24**, 1102–1113.

Zhang, H. and Wang, S. (2013) Rice versus *Xanthomonas oryzae* pv. *oryzae*: a unique pathosystem. *Curr. Opin. Plant Biol.* **16**, 188–195.

Zhang, Y., Shan, Q., Wang, Y., Chen, K., Liang, Z., Li, J., Zhang, Y. *et al.* (2013) Rapid and efficient gene modification in rice and *Brachypodium* using TALENs. *Mol. Plant* **6**, 1365–1368.

Zhang, J., Yin, Z. and White, F. (2015) TAL effectors and the executor *R* genes. *Front. Plant Sci.* **6**, 641.

Zhang, H., Gou, F., Zhang, J., Liu, W., Li, Q., Mao, Y., Botella, J.R. *et al.* (2016) TALEN-mediated targeted mutagenesis produces a large variety of heritable mutations in rice. *Plant Biotechnol. J.* **14**, 186–194.

Zhou, J., Peng, Z., Long, J., Sosso, D., Liu, B., Eom, J.-S., Huang, S. *et al.* (2015) Gene targeting by the TAL effector PthXo2 reveals cryptic resistance gene for bacterial blight of rice. *Plant J.* **82**, 632–643.

PERMISSIONS

LIST OF CONTRIBUTORS

Niranjan Baisakh, Mangu V. RamanaRao and Prasanta Subudhi
School of Plant, Environmental and Soil Sciences, Louisiana State University Agricultural Center, Baton Rouge, LA, USA

Kanniah Rajasekaran
Southern Regional Research Center, Agricultural Research Service, United States Department of Agriculture, New Orleans, LA, USA

Jaroslav Janda and David Galbraith
Department of Plant Sciences, University of Arizona, Tucson, AZ, USA

Cheryl Vanier
University of Nevada, Las Vegas, NV, USA

Andy Pereira
Department of Crop, Soil and Environmental Sciences, University of Arkansas, Fayetteville, AK, USA

Peng-Kai Sun
Institute of Molecular Biology, Academia Sinica, Taipei, Taiwan

Yi-Shih Chen
Institute of Molecular Biology, Academia Sinica, Taipei, Taiwan
Department of Life Sciences, National Central University, Jhongli City, Taiwan

Shuen-Fang Lo
Institute of Molecular Biology, Academia Sinica, Taipei, Taiwan
Agricultural Biotechnology Center, National Chung Hsing University, Taichung, Taiwan

Su-May Yu
Institute of Molecular Biology, Academia Sinica, Taipei, Taiwan
Agricultural Biotechnology Center, National Chung Hsing University, Taichung, Taiwan
Department of Life Sciences, National Chung Hsing University, Taichung, Taiwan

Chung-An Lu
Department of Life Sciences, National Central University, Jhongli City, Taiwan

Tuan-Hua D. Ho
Agricultural Biotechnology Center, National Chung Hsing University, Taichung, Taiwan
Institute of Plant and Microbial Biology, Academia Sinica, Taipei, Taiwan
Department of Life Sciences, National Chung Hsing University, Taichung, Taiwan

Longlong Zhu
State Key Laboratory of Crop Genetics and Germplasm Enhancement, Ministry of Agriculture, Nanjing Agricultural University, Nanjing, China

Jingguang Chen, Yong Zhang, Yawen Tan, Guohua Xu and Xiaorong Fan
State Key Laboratory of Crop Genetics and Germplasm Enhancement, Ministry of Agriculture, Nanjing Agricultural University, Nanjing, China
Key Laboratory of Plant Nutrition and Fertilization in Low-Middle Reaches of the Yangtze River, Ministry of Agriculture, Nanjing Agricultural University, Nanjing, China

Min Zhang
Key Laboratory of Plant Nutrition and Fertilization in Low-Middle Reaches of the Yangtze River, Ministry of Agriculture, Nanjing Agricultural University, Nanjing, China

Karabi Datta and Moumita Ganguly
Department of Botany, University of Calcutta, Kolkata, India

Sellapan Krishnan
Department of Botany, Goa University, Goa, India

Kazuko Yamaguchi Shinozaki
Graduate School of Agricultural and Life Sciences, University of Tokyo, Yayoi, Bunkyo-Ku, Tokyo, Japan

Swapan K. Datta
Department of Botany, University of Calcutta, Kolkata, India
Indian Council of Agricultural Research, Krishi Bhawan, New Delhi, India

Xianxin Dong, Xiaoyan Wang, Zhengting Yang, Xiaoyun Xin, Jianxiang Liu, Jinshui Yang and Xiaojin Luo
State Key Laboratory of Genetic Engineering, Institute of Genetics, Institute of Plant Biology, School of Life Sciences, Fudan University, Shanghai, China

Liangsheng Zhang
State Key Laboratory of Genetic Engineering, Institute of Genetics, Institute of Plant Biology, School of Life Sciences, Fudan University, Shanghai, China
Department of Biology, The Pennsylvania State University, University Park, PA, USA

Shuang Wu and Chuanqing Sun
The Department of Plant Genetics and Breeding, China Agricultural University, Beijing, China

Qiang He and Kyu-Won Kim
Department of Plant Resources, College of Industrial Science, Kongju National University, Yesan, 32439, Korea

Yong-Jin Park
Department of Plant Resources, College of Industrial Science, Kongju National University, Yesan, 32439, Korea
Center for crop genetic resource and breeding (CCGRB), Kongju National University, Cheonan, 31080, Republic of Korea

Emily E. Helliwell, Qin Wang and Yinong Yang
Department of Plant Pathology and Huck Institutes of Life Sciences, Pennsylvania State University, University Park, PA, USA

Fengcheng Li, Mingliang Zhang, Zhen Hu, Ran Zhang, Yongqing Feng, Xiaoyan Yi, Weihua Zou, Lingqiang Wang, Guosheng Xie and Liangcai Peng
National Key Laboratory of Crop Genetic Improvement and National Centre of Plant Gene Research, Huazhong Agricultural University, Wuhan, China
Biomass and Bioenergy Research Centre, Huazhong Agricultural University, Wuhan, China
College of Plant Science and Technology, Huazhong Agricultural University, Wuhan, China Beijing, China

Kai Guo
National Key Laboratory of Crop Genetic Improvement and National Centre of Plant Gene Research, Huazhong Agricultural University, Wuhan, China
Biomass and Bioenergy Research Centre, Huazhong Agricultural University, Wuhan, China
College of Life Science and Technology, Huazhong Agricultural University, Wuhan, China

Changyin Wu
National Key Laboratory of Crop Genetic Improvement and National Centre of Plant Gene Research, Huazhong Agricultural University, Wuhan, China
College of Life Science and Technology, Huazhong Agricultural University, Wuhan, China

Jinshan Tian
Yichang Academy of Agricultural Science, Yichang, China

Tiegang Lu
Biotechnology Research Institute, Chinese Academy of Agricultural Sciences/National Key Facility for Gene Resources and Genetic Improvement,

Fabien Lombardo and Takashi Akiyama
Division of Applied Genetics, Institute of Agrobiological Sciences, National Agriculture and Food Research Organization (NARO), Ibaraki, Japan

Hitoshi Yoshida
Division of Applied Genetics, Institute of Agrobiological Sciences, National Agriculture and Food Research Organization (NARO), Ibaraki, Japan
Division of Crop Development, Central Region Agricultural Research Center, NARO, Niigata, Japan

Hiroyuki Shimizu and Tomohito Ikegaya
Division of Crop Breeding Research, Hokkaido Agricultural Research Center, NARO, Hokkaido, Japan

Makoto Kuroki
Division of Crop Breeding Research, Hokkaido Agricultural Research Center, NARO, Hokkaido, Japan
Division of Rice Research, Institute of Crop Science, NARO, Ibaraki, Japan

Takami Hayashi
Division of Crop Breeding Research, Hokkaido Agricultural Research Center, NARO, Hokkaido, Japan
Division of Agro-Production Technologies and Management Research, Tohoku Agricultural Research Center, NARO, Iwate, Japan

Shan-Guo Yao, Mayumi Kimizu, Shinnosuke Ohmori and Osamu Yatou
Division of Crop Development, Central Region Agricultural Research Center, NARO, Niigata, Japan

Tomoya Yamaguchi and Setsuo Koike
Division of Agro-Production Technologies and Management Research, Tohoku Agricultural Research Center, NARO, Iwate, Japan

Qiwei Shan, Yi Zhang, Kunling Chen and Caixia Gao
State Key Laboratory of Plant Cell and Chromosome Engineering, Institute of Genetics and Developmental Biology, Chinese Academy of Sciences, Beijing, China

Kang Zhang
State Key Laboratory of Plant Cell and Chromosome Engineering, Institute of Genetics and Developmental Biology, Chinese Academy of Sciences, Beijing, China

Beijing Genovo Biotechnology Co. Ltd, Beijing, China

Takeshi Shiraya, Kazusato Oikawa and Kentaro Kaneko
Department of Applied Biological Chemistry, Niigata University, Niigata, Japan

Toshiaki Mitsui
Department of Applied Biological Chemistry, Niigata University, Niigata, Japan
Graduate School of Science and Technology, Niigata University, Niigata, Japan

Taiki Mori, Tatsuya Maruyama, Maiko Sasaki, Takeshi Takamatsu and Kimiko Itoh
Graduate School of Science and Technology, Niigata University, Niigata, Japan

Hiroaki Ichikawa
Division of Plant Sciences, National Institute of Agrobiological Sciences, Tsukuba, Japan

Lei Wang, Jie Zheng, Yanzhong Luo, Tao Xu, Qiuxue Zhang, Lan Zhang, Miaoyun Xu, Chunyi Zhang and Yunliu Fan
Biotechnology Research Institute, The National Key Facility for Crop Gene Resources and Genetic Improvement, Chinese Academy of Agricultural Sciences, Beijing, China

Jianmin Wan
Crop Sciences Institute, The National Key Facility for Crop Gene Resources and Genetic Improvement, Chinese Academy of Agricultural Sciences, Beijing, China

Ming-Bo Wang
CSIRO Plant Industry, Canberra, NSW, Australia

Huihui Yu, Jing Li and Fasong Zhou
Life Science and Technology Center, China National Seed Group Co., Ltd, Wuhan, China

Weibo Xie and Qifa Zhang
National Key Laboratory of Crop Genetic Improvement, National Center of Crop Molecular Breeding, Huazhong Agricultural University, Wuhan, China

Hui Zhang, Botao Zhang, Yanfei Mao, Lan Yang, Heng Zhang and Nanfei Xu
Shanghai Center for Plant Stress Biology, Chinese Academy of Sciences, Shanghai, China

Jinshan Zhang, Pengliang Wei and Feng Gou
Shanghai Center for Plant Stress Biology, Chinese Academy of Sciences, Shanghai, China

University of Chinese Academy of Sciences, Shanghai, China

Zhengyan Feng
Shanghai Center for Plant Stress Biology, Chinese Academy of Sciences, Shanghai, China
University of Chinese Academy of Sciences, Shanghai, China
Institute of Plant Physiology and Ecology, Shanghai Institutes for Biological Sciences, Chinese Academy of Sciences, Shanghai, China

Jian-Kang Zhu
Shanghai Center for Plant Stress Biology, Chinese Academy of Sciences, Shanghai, China
Department of Horticulture and Landscape Architecture, Purdue University, West Lafayette, IN, USA

Kang Zhang, Qian Song, Qiang Wei, Chunchao Wang, Liwei Zhang, Wenying Xu and Zhen Su
State Key Laboratory of Plant Physiology and Biochemistry, College of Biological Sciences, China Agricultural University, Beijing, China

Degui Zhou, Chongrong Wang, Hong Li and Kanghuo Li
Guangdong Provincial Key Laboratory of New Technology in Rice Breeding, Rice Research Institute, Guangdong Academy of Agricultural Sciences, Guangzhou, China

Shaochuan Zhou
Guangdong Provincial Key Laboratory of New Technology in Rice Breeding, Rice Research Institute, Guangdong Academy of Agricultural Sciences, Guangzhou, China
Agricultural College, Hunan Agricultural University, Changsha, China

Zechuan Lin
State Key Laboratory of Protein and Plant Gene Research, Peking–Tsinghua Center for Life Sciences, School of Advanced Agricultural Sciences and College of Life Sciences, Peking University, Beijing, China

Wei Chen
State Key Laboratory of Protein and Plant Gene Research, Peking–Tsinghua Center for Life Sciences, School of Advanced Agricultural Sciences and College of Life Sciences, Peking University, Beijing, China
School of Information and Engineering, Wenzhou Medical University, Wenzhou, China

Haodong Chen Renbo Yu and Gang Zhen
State Key Laboratory of Protein and Plant Gene Research, Peking–Tsinghua Center for Life Sciences, School of Advanced Agricultural Sciences and College of Life Sciences, Peking University, Beijing, China
Shenzhen Institute of Crop Molecular Design, Shenzhen, China

Hang He and Xing Wang Deng
State Key Laboratory of Protein and Plant Gene Research, Peking–Tsinghua Center for Life Sciences, School of Advanced Agricultural Sciences and College of Life Sciences, Peking University, Beijing, China
Shenzhen Institute of Crop Molecular Design, Shenzhen, China
Frontier Laboratories of Systems Crop Design Co., Ltd., Beijing, China

Xiaoyan Tang
Shenzhen Institute of Crop Molecular Design, Shenzhen, China
Frontier Laboratories of Systems Crop Design Co., Ltd., Beijing, China

Fengyun Zhang and Junliang Yi
Agricultural College, Hunan Agricultural University, Changsha, China

Yaoguang Liu
College of Life Sciences, South China Agricultural University, Guangzhou, China

William Terzaghi
Department of Biology, Wilkes University, Wilkes-Barre, PA, USA

Miao Chen, Yujuan Zhao, Chunliu Zhuo, Shaoyun Lu and Zhenfei Guo
State Key Laboratory for Conservation and Utilization of Subtropical Agro-bioresources, South China Agricultural University, Guangzhou, China

Mireia Bundó and María Coca
Centre for Research in Agricultural Genomics (CRAG), CSIC-IRTA-UAB-UB. Edifici CRAG, Bellaterra, Barcelona, Spain

Servane Blanvillain-Baufumé, Florence Auguy, Hinda Doucoure, Boris Szurek, Sebastien Cunnac and Ralf Koebnik
UMR Interactions Plantes Microorganismes Environnement (IPME), IRD-CIRAD-Université, Montpellier, France

Maik Reschke, Montserrat Solé and Jens Boch
Institut fur Biologie, Institutsbereich Genetik, Martin-Luther-Universit at Halle-Wittenberg, Halle (Saale), Germany

Donaldo Meynard, Murielle Portefaix and Emmanuel Guiderdoni
CIRAD, UMR AGAP (Amélioration génétique et Adaptation des Plantes), Montpellier, France

Index